Fire in Mediterranean Ecosystems

Ecology, Evolution and Management

Exploring the role of fire in each of the five mediterranean-type climate ecosystems, this book offers a unique view of the evolution of fire-adapted traits and the role of fire in shaping Earth's ecosystems. Analyzing these geographically separate but ecologically convergent ecosystems provides key tools for understanding fire regime diversity and its role in the assembly and evolutionary convergence of ecosystems.

Topics covered include regional patterns; the ecological role of wildfires; the evolution of species within those systems, and the ways in which societies have adapted to living in fire-prone environments. Outlining complex processes clearly and methodically, the discussion challenges the belief that climate and soils alone can explain the global distribution and assembly of plant communities.

An ideal research tool for graduates, researchers and fire managers, this study provides valuable insights into the requirements for regionally tailored approaches to fire management across the globe.

Jon E. Keeley is a Research Scientist with the U.S. Geological Survey, Western Ecological Research Center, Sequoia-Kings Canyon Field Station in Three Rivers, California, and an Adjunct Professor in the Department of Ecology and Evolutionary Biology, University of California, Los Angeles, USA.

William J. Bond is a Professor in the Department of Botany, University of Cape Town, South Africa.

Ross A. Bradstock is Director of the Centre for Environmental Risk Management of Bushfires, University of Wollongong, New South Wales, Australia.

Juli G. Pausas is a Scientist at the Centro de Investigación sobre Desertificación of the Spanish National Research Council (CIDE-CSIC) in Valencia, Spain.

Philip W. Rundel is a Distinguished Professor in the Department of Ecology and Evolutionary Biology, University of California, Los Angeles, USA.

Fire in Mediterranean Ecosystems

Ecology, Evolution and Management

JON E. KEELEY
United States Geological Survey, California, USA

WILLIAM J. BOND
University of Cape Town, South Africa

ROSS A. BRADSTOCK
University of Wollongong, New South Wales, Australia

JULI G. PAUSAS
Spanish National Research Council (CIDE-CSIC), Valencia, Spain

PHILIP W. RUNDEL
University of California, Los Angeles, USA

CAMBRIDGE UNIVERSITY PRESS
Cambridge, New York, Melbourne, Madrid, Cape Town,
Singapore, São Paulo, Delhi, Tokyo, Mexico City

Cambridge University Press
The Edinburgh Building, Cambridge CB2 8RU, UK

Published in the United States of America
by Cambridge University Press, New York

www.cambridge.org
Information on this title: www.cambridge.org/9780521824910

© J. E. Keeley, W. J. Bond, R. A. Bradstock, J. G. Pausas and P. W. Rundel 2012

This publication is in copyright. Subject to statutory exception
and to the provisions of relevant collective licencing agreements,
no reproduction of any part may take place without
the written permission of Cambridge University Press.

First published 2012

Printed in the United States of America by Edwards Brothers Incorporated

A catalog record for this publication is available from the British Library

Library of Congress Cataloging-in-Publication Data

Fire in Mediterranean ecosystems : ecology, evolution and management / Jon E. Keeley... [et al.].
    p. cm.
  ISBN 978-0-521-82491-0 (Hardback)
  1. Fire ecology. 2. Mediterranean-type ecosystems. 3. Plants–Effect of fires on.
I. Keeley, Jon E.
  QH545.F5F5747 2011
  577.3'824–dc23
                                                           2011013006

ISBN 978-0-521-82491-0 Hardback

Cambridge University Press has no responsibility for the persistence or
accuracy of URLs for external or third-party internet websites referred to
in this publication, and does not guarantee that any content on such
websites is, or will remain, accurate or appropriate.

Contents

Section I Introduction		1
1	Mediterranean-type Climate Ecosystems and Fire	3
2	Fire and the Fire Regime Framework	30
3	Fire-related Plant Traits	58
Section II Regional Patterns		81
4	Fire in the Mediterranean Basin	83
5	Fire in California	113
6	Fire in Chile	150
7	Fire in the Cape Region of South Africa	168
8	Fire in Southern Australia	201
Section III Comparative Ecology, Evolution and Management		231
9	Fire-adaptive Trait Evolution	233
10	Fire and the Origins of Mediterranean-type Vegetation	275
11	Plant Diversity and Fire	310
12	Alien Species and Fire	330

13	**Fire Management of Mediterranean Landscapes**	349
14	**Climate, Fire and Geology in the Convergence of Mediterranean-type Climate Ecosystems**	388
	References	398
	Index	498

Section I

Introduction

A significant portion of Earth is subject to periodic fires. There is a paradox in that these fire-prone landscapes have historically been interpreted in terms of just climate and geology, with limited consideration of the evolutionary and ecological role that fire has had in shaping functional types and community assembly. The theme throughout this book is that plant traits and plant communities over much of the fire-prone portions of the globe cannot be understood without consideration of the climate–fire–geology filter that controls the assembly of these systems. One rationale for focusing on mediterranean-type climate (MTC) regions is that vast portions of these landscapes are annually subjected to high fire risk. High fire danger is a consequence of climate and plant structure and many features of both are shared between the five MTC regions. These ecosystems in widely disjunct parts of the globe are tied together by a long history of convergent evolution/ecology studies that began with nineteenth-century geographic comparisons of plant morphologies. These early geographers were "fire-blind" in failing to recognize some of the most critical factors responsible for convergence. Indeed, even very keen observers such as Charles Darwin visited the highly fire-prone Eucalyptus *woodland of Australia and failed to appreciate the extraordinary story of fire adaptation on this landscape. Substantial scientific focus on these ecosystems over the past several decades has provided a wealth of background information necessary for interpreting the role of fire in driving the degree of convergence in plant traits and community assembly, as well as insights into reasons for examples of non-convergence between these regions.*

In this first section we provide the necessary foundation for understanding the ecological, evolutionary and management issues involving fire and fire-adaptive traits. Section II examines the relevant fire issues in each of the five MTC regions and this is built on in Section III with a synthesis focused on revealing emergent patterns that come from a global comparison across these widely disjunct regions.

1 Mediterranean-type Climate Ecosystems and Fire

This book is about fire and the ecosystem role it plays in plant communities with distributions centered in one of the five mediterranean-type climate (MTC) regions of the world (Fig. 1.1). These landscapes are related by their marked climatic seasonality, with precipitation in the winter under mild temperatures and drought in the summer coupled with high temperatures (Box 1.1). MTCs are regions where precipitation exceeds potential evapotranspiration during the rainy season (Rundel 2010), resulting in sufficient plant growth that becomes highly flammable during the summer dry season, a unifying factor that has played out in common ecological responses to fire. Collectively these regions comprise only about 2% of the land area of the world but they house more than 15% of the total vascular plant flora (Rundel 2004). All are dominated by fire-prone ecosystems often juxtaposed with major metropolitan centers (Fig. 1.2) and are dominated by fire-adapted vegetation resulting from a long evolutionary association with fire (Pausas & Keeley 2009).

Although our focus is on the highly fire-prone landscapes with MTC, fire is a global ecosystem process and one whose role in shaping the distribution of fauna and flora is widely underappreciated. Over half of the land surface of Earth is considered to be fire-prone (Fig. 1.3), with perhaps a third of the land mass experiencing frequent intensive burning (Chuvieco *et al.* 2008). The emerging discipline of *pyrogeography* emphasizes the necessity for considering fire in understanding local ecological interactions as well as global earth system processes (Bowman *et al.* 2009; Moritz *et al.* 2010a). Fire is not a new ecosystem process but rather one that has been part of land plant evolution since the Paleozoic (Pausas & Keeley 2009; Bond & Scott 2010). Historically the disciplines of ecology, biogeography and paleoecology have considered climate and geology to be the key factors determining ecosystem assembly and distribution. But on many of these landscapes fire is a factor of equal or greater importance and interactions between all three factors determine the potential pool of available functional types in both flora and fauna (Fig. 1.4).

Through feedback processes fire, climate and geology are connected by different functional types and attempts to understand community assembly without considering the interrelationships between these three factors may lead to misleading conclusions about their origins and distribution. For example, small leaves are

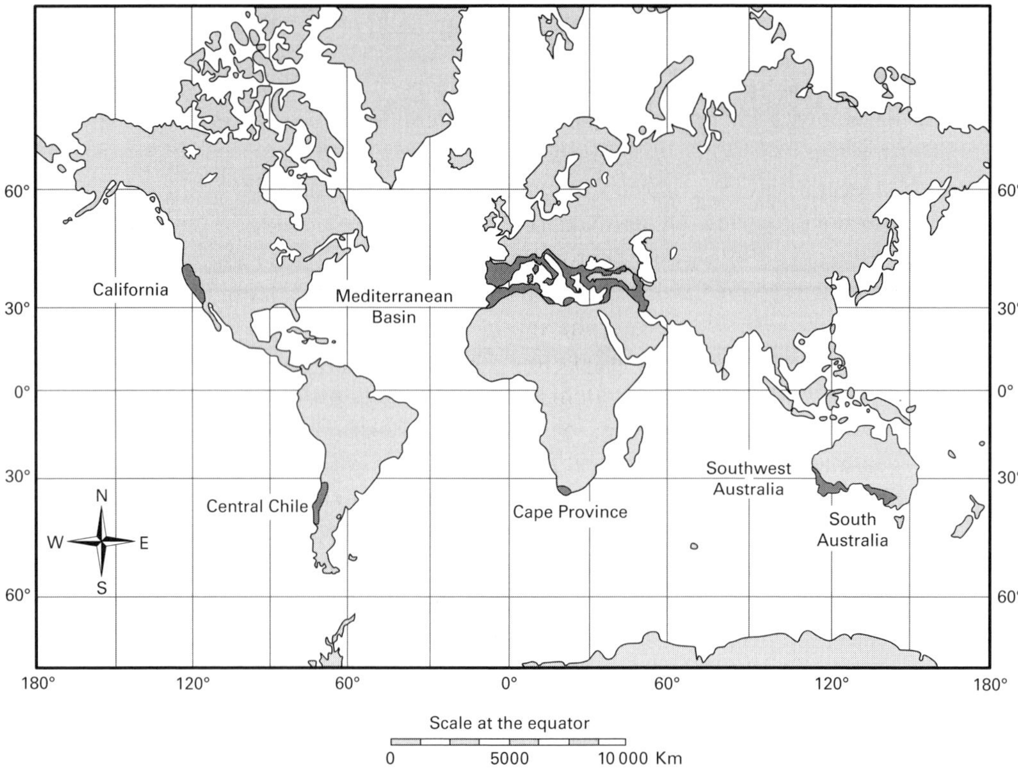

Fig. 1.1 *The five mediterranean-type climate (MTC) regions are distributed on the western sides of continents between 30° and 40° latitude.*

Fig. 1.2 *Massive wildfire near the major metropolitan area of Los Angeles, California, illustrating the juxtaposition of wildland fires and major population centers in most MTC regions of the world. Unlike the majority of large fires in the region, this 65 000-ha Station Fire in August 2009 was not driven by Santa Ana winds and hence the vertical smoke plume. (Photo by Bob Ginn.)*

Box 1.1 The Mediterranean-type Climate

The mediterranean-type climate (MTC) is characterized by winter rains and summer drought. Both winter growing conditions and summer drought contribute to making these landscapes some of the most fire-prone in the world. Coastal influence moderates winter temperatures so that rains coincide with suitable growing temperatures (Fig. B1.1.1). Due to this extended winter–spring growing season, primary productivity is moderately high for semi-arid regions and vegetation forms dense, sometimes impenetrable thickets that, when dry, contribute to fire spread. The annual summer drought reduces fuel moisture to levels conducive to rapid ignition. This MTC is the result of global circulation patterns that generate a summer high pressure cell of dry sinking air that blocks incoming summer storms on the western sides of continents concentrated between 32° and 38° N or S latitude. Globally there are five subtropical high pressure cells that lie within these latitudes during the summer and migrate toward the poles during the winter, creating a MTC in: the Mediterranean Basin, California, central Chile, the southwestern portion of the Western Cape Province of South Africa, and the southwestern portion of Western Australia and South Australia plus adjoining portions of the province of Victoria.

The seasonal distribution of precipitation and temperature illustrated for Los Angeles (Fig. B1.1.1) is mirrored in the other MTC regions. However, despite the similarities, there are climatic differences (Table B1.1.1) that may be expected to affect fire regimes. For example, much of California and parts of Chile and the Mediterranean Basin have the most severe summer droughts, often with several months of no rainfall, although there is much intraregional variation. For example, interior parts of the western Mediterranean Basin may have significant summer precipitation; summer rain as a percentage of annual total is 21% for Valencia, Spain, 12% Marseille, France, or 10% Rome, Italy (Müller 1982). In Chile, the MTC spans a large latitudinal range although the proportion of summer precipitation increases poleward. In the south, where the Andean range is lower, the MTC may extend eastward into Argentina, as is evident at Bariloche, Argentina (41° S latitude) with about 700 mm rainfall and less than 10% in summer (Müller 1982). In both South Africa and southern Australia summer precipitation increases eastward, although seasonal patterns are less predictable and summer droughts occur in some years. In South Africa the MTC region is more or less restricted to the southwestern Cape region; just 4 degrees of longitude east of Cape Town, the amount of summer rain more than triples at Oudtshoorn (Table B1.1.1).

Winter precipitation likewise varies between MTC regions comprising over two thirds of the annual total in the eastern Mediterranean Basin, California and central Chile. In Seville, Spain, the autumn precipitation equals the winter total and in Adelaide, South Australia, spring and autumn precipitation are equal and

Continued

Box 1.1 (cont.)

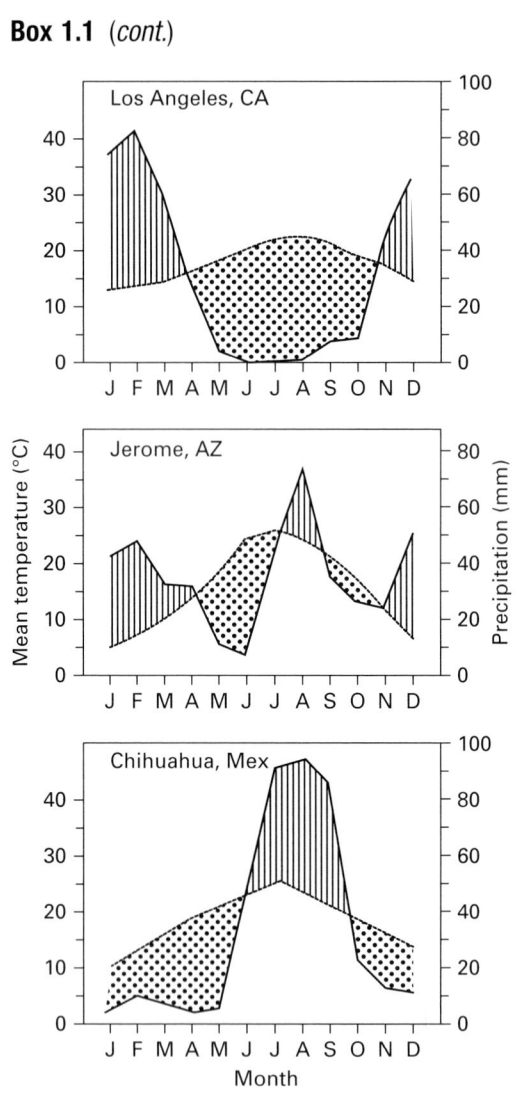

Fig. B1.1.1 *Climate diagrams contrasting the mediterranean-type climate of Los Angeles, California, USA, with two other non-MTC regions in western North America all with MTV (solid line, monthly precipitation; dashed line, monthly mean daytime temperature; area filled with hatching, periods when precipitation exceeds evaporation; dotted area, period when precipitation < evaporation): Los Angeles, California (100 m, 34°5′ N, 118°5′ W), Jerome, Arizona (1600 m, 34°5′ N, 112°7′ W), and Chihuahua, Mexico (1350 m, 28°2′ N, 105°7′ W). In addition to differences in seasonal distribution of precipitation, the annual variance in precipitation is much greater in California than interior regions. For example, 40 years of data for Los Angeles showed 5 months with a coefficient of variation between 220% and 350%, whereas the Arizona site had no month with more than 120%. (From Keeley 2000.)*

Continued

Table B1.1.1 Climate characteristics for representative stations in the five MTC regions: northern hemisphere summer (Jun, Jul, Aug) and winter (Dec, Jan, Feb); southern hemisphere summer (Dec, Jan, Feb) and winter (Jun, Jul, Aug).

MTC region	Sevilla, Spain	Haifa, Israel	San Francisco, California, USA	Los Angeles, California, USA	Santiago, Chile	Concepción, Chile	Cape Town, South Africa	Oudtshoorn, South Africa	Perth, Western Australia	Adelaide, South Australia
Latitude (°)	37N	33N	38N	34N	33S	37S	34S	34S	38S	35S
Longitude (°)	6W	35E	122W	118W	70W	73W	18E	22E	115E	145E
Elevation (m)	30	10	15	100	520	15	20	335	60	45
Ppt Total (mm)	535	670	530	375	360	1295	505	255	890	525
Summer ppt (%)	3	1	2	1	2	22	6	21	4	14
Winter ppt (%)	39	70	60	79	67	52	47	22	57	35
Summer T mean (°C)	26.8	27.1	15.1	22.5	19.4	17.4	20.9	23.1	22.9	21.4
Winter T mean (°C)	11.4	14.8	11.7	13.9	8.5	9.3	13.1	12.2	13.4	11.8

Source: From Müller (1982).

Box 1.1 (*cont.*)

about midway between summer and winter. In addition, between MTC regions there is much variation in mean winter and summer temperatures (Table B1.1.1).

Rainfall reliability appears to vary across MTC regions and it has been hypothesized to account for differences in fire responses (Cowling *et al.* 2005), although most fire-related traits likely predate current precipitation regimes (see Chapter 10). The impact of regional differences in rainfall patterns needs to be considered in the context of differences in substrate (see Fig. 1.5) and topography, all of which affect soil water-holding capacity and may select for differences in drought tolerance between plant taxa in different MTC. The nexus between climate and fire is complicated and needs to be considered in a multivariate context.

The boundaries around MTC regions are not uniformly agreed upon by all sources. Some limit MTC to areas that not only exhibit the characteristic winter rain/summer drought pattern but include total precipitation limits and temperature limits. A more liberal interpretation than used in this book would include much greater portions of southwest Asia in the Mediterranean Basin, a substantially greater part of the Pacific Northwest and Great Basin in western North America, much of the eastern foothills of Argentina in South America, and portions of New South Wales and Victoria in Australia (Le Houerou 2004).

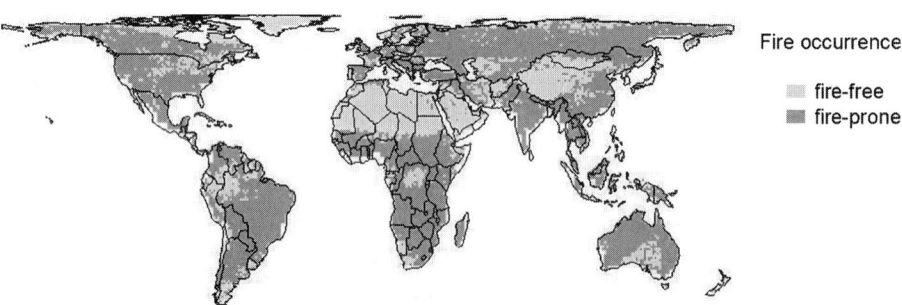

Fig. 1.3 *Global distribution of fire-prone and fire-free landscapes. (From Krawchuk et al. 2009.)*

common in MTC regions and are interpreted as responses to low rainfall and low nutrient availability (Yates *et al.* 2010). However, small leaves also increase flammability and thus, as selection for small sclerophyll leaves increases, landscape flammability increases, affecting ignitions and fire spread, and the predictability of fire in the environment, which in turn affects trait selection (e.g. Ojeda *et al.* 2010).

The thesis of this book is that explaining the origin of plant traits and assembly of communities based on climate and/or geology alone is inadequate, particularly

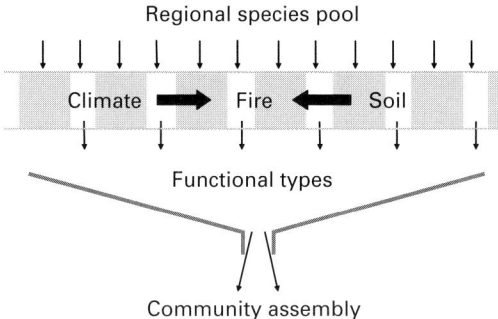

Fig. 1.4 *The environmental template in fire-prone landscapes can be thought of as a process in which the regional species pool is filtered to a subset of functional types. The combination of climate and geology places bounds on the range of plant traits that control fire regimes and feedback from fire further affects the pool of available functional types. This climate, fire, geology interaction acts as a control on community assembly, with the important emergent property of fuel types and ignition probabilities determining fire regime.*

in fire-prone regions. Reconstructions of plant evolution in MTC floras (e.g. Raven & Axelrod 1978; Axelrod 1980, 1989; Ackerly 2009; Hopper 2009) have either ignored fire or treated it as an incidental process and not adequately considered the immense impact of feedback processes between fire and climate or fire and geology. By excluding fire, studies have derived incomplete, and in some cases perhaps spurious, conclusions. For example, the recent contention that resprouting of top-killed Hawaiian trees reflects an adaptation to drought because it increases along a gradient of increasing aridity (e.g. Busby *et al.* 2010) is confounded by the fact that fires in the Hawaiian Islands likewise increase along this same gradient during El Niño/Southern Oscillation (ENSO) events (Chu *et al.* 2002; Weise *et al.* 2010). Even on these tropical islands, sources of natural ignitions from lightning (Pessi *et al.* 2004) have created a fire regime with the potential for driving the selection of resprouting, particularly since fires need only occur once in the life span of an individual to select for this trait.

In this book we eschew the notion that any single factor such as drought or oligotrophic soils provides a sufficient explanation for ecological patterns on fire-prone landscapes. Instead we show how syndromes of fire, soils and climate can act synergistically to shape vegetation, particularly the sclerophyllous woody plants that dominate MTC, and sometimes far beyond those climatic boundaries. The interplay of fire, climate and geology is well illustrated by the convergent patterns of plant traits and fire response in MTC ecosystems. A primary factor tying these regions together is the intense summer drought, which potentially contributes to greater predictability of fires. Winter rainfall under mild temperatures plays a major role in determining levels of primary productivity and thus fuel structure, which is an additional factor tying together MTC ecosystems. Subtle differences in climates (Box 1.1) and not so subtle differences in soils (Fig. 1.5) as well as different phylogeographic histories (see Chapter 10) have

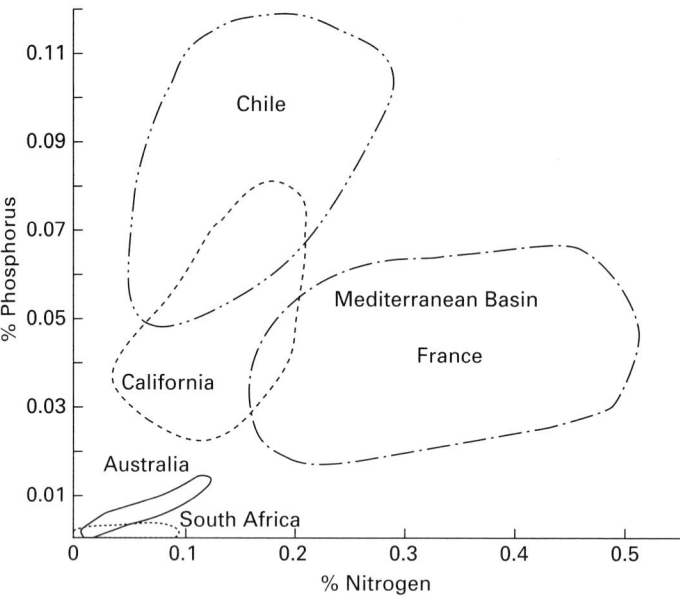

Fig. 1.5 *Levels of phosphorus and nitrogen in the soils of the five MTC regions. (Redrawn from di Castri 1981.)*

contributed to differences in fire regimes and trait evolution, which are also part of this story.

Mediterranean-type Climates and Ecosystem Convergence

The five MTC regions are: the Mediterranean Basin, California, central Chile, the western half of the Western Cape Province of South Africa, and much of southern Australia, including the southwestern portion of Western Australia, South Australia and adjacent regions in Victoria. The Mediterranean Basin contains by far the largest expanse of MTC landscapes followed by Australia, while South Africa represents the smallest (Table 1.1). In all MTC regions shrublands are an important component and these are typically evergreens with broad or small, stiff and sclerophyllous leaves on woody stems 1–3 m (sometimes 5 m) tall (Fig. 1.6). Some of the regions have a significant component of smaller stature, fine-leaved sclerophylls often referred to as heathlands, a vegetation that is widespread outside MTC regions, including subarctic and tropical alpine environments (Specht 1979). Broadleaf evergreen sclerophyll woodlands are abundant in the Mediterranean Basin, California and Australia, of lesser importance in central Chile and rather depauperate in the Cape region (Table 1.1). Taller stature conifer forests are abundant in California (Fig. 1.7a) and to a lesser extent in the Mediterranean Basin. Broadleaf forests (Fig. 1.7b) are present in Chile and Western Australia (Table 1.1) and of increasing dominance in the Australian states of South Australia and Victoria.

Table 1.1 Vegetation comparisons between the five MTC regions of the world

	Mediterranean[a]	California[b]	Chile	South Africa	Australia[c]
Total vegetated area (km^2)	1 286 950	98 575	111 725	68 175	430 875
Grassland (%)	15	9	0	0	4
Shrubland (%)[d]	45	34	79	97	63
Woodland (%)	25	35	8	2	27
Forest (%)	15	22	13	< 1	6

[a] Due the extensive east–west range of the Mediterranean Basin there are some important differences in vegetation; for example grasslands are a minor component of the western basin but comprise 18% of the vegetation in the eastern basin, and shrublands follow an opposite pattern.

[b] Includes Baja California but not the northwest or Central Valley or higher elevations of the Sierra Nevada.

[c] In this book the MTC in Australia is considered to include more forested landscape in South Australia and Victoria than represented by these data. MTV further east in NSW is dominated by forests.

[d] Includes scrub and succulent associations.

Source: From Underwood *et al.* (2009), modified for the Mediterranean Basin by MODIS and CORINE remote imaging analysis (accessed Jan 2010).

Sclerophyllous vegetation dominant in MTC regions sometimes extends beyond these climatic boundaries at similar latitudes, and it is preferable to refer to these as mediterranean-type vegetation (MTV). For example, in most MTC regions the evergreen sclerophyll shrublands extend eastward into regions with summer rains (Box 1.1). In the past these have been referred to as mediterranean-type ecosystems (MTEs), but this term should be avoided because generally the designation is based on the presence of MTV, and seldom do we have sufficient information to confirm ecosystem similarities. When MTV outside of MTC regions has been studied in detail, such vegetation often comprises ecosystems that are rather different from those in MTC regions (e.g. Fotheringham 2009). For example, the same species of trees and shrubs may dominate MTV in both summer drought and summer rain climatic regimes, but frequently the latter will have an understory rich in C_4 grasses and the former will not. Thus, the focus in this book is on ecosystems in MTC regions with consideration of similar sclerophyllous MTV outside that climatic region when it can add to our understanding of fire. Biogeographical patterns are very different from one region to the next. In North America evergreen sclerophyll shrublands exist under the MTC of California and many of the same MTV species occur disjunct 500 km eastward in Arizona. Other pockets of MTV shrublands are found on well-drained substrates in the southeastern part of the continent under an aseasonal rainfall regime. Similar patterns are evident in the distribution of sclerophyllous-leaved shrublands in southern Australia, dominating in the MTC region of Western

Fig. 1.6 Shrublands dominating the five MTC regions of (a) matorral in Portugal representative of the western Mediterranean Basin, (b) maquis and oak woodland in Israel typical of the eastern Mediterranean Basin, (c) southern Californian chaparral, (d) central Chilean matorral with Chilean wine palm in the monotypic genus Jubaea, (e) South African Cape fynbos, and (f) Western Australian heathland. (Photos by Jon Keeley.)

Australia and South Australia with an increasingly patchy distribution in the aseasonal rainfall regime of the state of New South Wales.

An alternative to the climatic approach is the mediterranean zone concept that considers all landscape regardless of climate as included if it comprises critical MTV components (Le Houerou 2004). Although this approach may work well for regions such as Eurasia that share many vegetation features, it becomes more difficult to apply globally.

Fig. 1.7 *Forests are widespread in some MTC regions; in particular: (a)* Pinus-*dominated conifer forest in California, and (b)* Eucalyptus-*dominated broadleaf forest in southern Australia. These forest types are widespread far outside the MTC on both continents. Similar conifer-dominated forests occur in the Mediterranean Basin and broadleaf and conifer forests in southern Chile, but such forests are largely lacking in the South African Cape region. (Photos by Jon Keeley.)*

A central concept that ties together the five MTC regions is the hypothesis that these ecosystems exhibit a remarkable degree of evolutionary convergence in the structure and function of many taxa. The idea originated with the nineteenth-century geographers and botanists Grisebach (1872) and Schimper (1903). They recognized that disparate regions of the world possessed similar sclerophyllous woodlands, and that the greatest centers of evergreen sclerophyllous leaf development were in regions of MTC. Similarities between these regions were evident at the community scale as all comprised vegetation with similar growth forms, including a high preponderance of multistemmed shrubby phanerophytes (see Raunkiaer growth form entry in Box 1.2) and geophytes. Despite these remarkable similarities, MTC regions were floristically very different and such observations formed the basis for one of the classical examples of evolutionary convergence. This concept provided an important opportunity for testing ideas of adaptation in intensive comparative studies of MTC ecosystems beginning in the early 1970s by Mooney and colleagues (di Castri & Mooney 1973; Mooney 1977a). This International Biological

Box 1.2 Terms Used Frequently in This Book

aseasonal climate: long-term average rainfall similar in all seasons and thus lacking a predictable annual drought, although, at somewhat longer timescales anomalous rainfall years generate drought in different seasons; such environments experience fire-prone conditions but not annually

BP: abbreviation for years before present, or **kBP**, thousands of years before present

ENSO: acronym for El Niño–Southern Oscillation; El Niño is the anomalous warming of surface water of the eastern tropical Pacific Ocean and cooling of the surface water in the western Pacific, resulting in changes in winds that affect distant weather events. This occurs periodically every 4–6 yrs and alternates with La Niña cooling in the central Pacific; this periodic altering between El Niño and La Niña is known as the Southern Oscillation

fire-dependent recruitment: plant functional type that delays seedling recruitment to a single pulse in the first growing season after fire

fire-independent recruitment: plant functional type that does not delay seedling recruitment until the immediate postfire conditions and recruits more or less continuously between fires

fire intensity: energy release from a fire

fire return interval: time between fires at a defined place, also expressed as inter-fire interval

fire rotation interval: time to burn an area equivalent to the size of a particular unit in question, such as a county, province or state

fire severity (or **burn severity**): impact of fire intensity on plant biomass loss both aboveground and belowground

kyr (or **ka**): abbreviation for a thousand years

lignotuber: a basal woody tuber produced in seedlings and saplings as a normal developmental stage, which contains adventitious buds that generate resprouts after fire

Ma: abbreviation for million years ago

malacophyllous leaves: short-lived semi-fleshy leaves often in arid-adapted summer-deciduous subshrubs

Mg ha^{-1}: megagrams per hectare and equivalent to tonnes per hectare

MTC: acronym for mediterranean-type climate of winter rains with mild temperatures alternating with late spring and summer drought and high temperatures

MTE: acronym for mediterranean-type ecosystem. Often applied to ecosystems dominated by evergreen sclerophyllous-leaved plants typical of mediterranean-type climates (MTC). This term is best restricted to MTC regions because such mediterranean-type vegetation (MTV) when it occurs outside MTC regions is often assembled into very different plant communities with rather different ecosystem properties (e.g. Fotheringham 2009)

Continued

> **Box 1.2** (*cont.*)
>
> **MTV**: acronym for mediterranean-type vegetation often forming closed-canopy shrublands, woodlands and forests dominated by sclerophyllous-leaved evergreen plants and dominant in MTCs but can be important in non-MTC regions
>
> **obligate resprouter**: a plant that regenerates after fire solely by vegetative resprouts and recruits seedlings during the inter-fire interval (see fire-independent recruitment)
>
> **postfire seeder**: a plant that recruits seedlings in a postfire pulse of recruitment. Taxa may be either "obligate seeder" that does not resprout, or "facultative seeder" that recruits seedlings in a postfire pulse of recruitment and resprouts after fire
>
> **Raunkiaer life forms**: a French system for classifying plant growth forms and largely replaced in this book as follows:
> phanerophytes with buds > 25 cm above ground – trees and shrubs
> chamaephytes with buds near the ground – suffrutescents of low subshrubs
> hemicryptophytes with buds at the soil surface – herbaceous perennials
> cryptophytes with buds below ground – geophytes
> therophytes have no perenniating buds – annuals
>
> **sclerophyllous leaves**: evergreen leaves with hard surface layers of cells often with embedded hard sclerid cells
>
> **WUI**: acronym for wildland–urban interface, the boundary between wildlands and urban, suburban or peri-urban (transition zone between urban and rural) development

Program (IBP) project, funded by the U.S. National Science Foundation and led by Dr. Harold Mooney, comprised one of the most intensive studies of the idea that phylogenetically unrelated organisms occupying similar ecological niches in similar environments become more similar than their ancestors (Blondel 1991). This IBP project extended the concept by testing the hypothesis that entire communities of plants and animals would converge under similar environmental conditions. Certainly as important as the outcome of those studies was the fact that this research initiated a trajectory of research collaborations that resulted in an extraordinary transfer of ideas across these five MTC regions of the world (Appendix 1.1).

The ecosystem convergence concept has been lauded as having led to important comparative findings on the structure and function of different ecosystems developing under different climates (Mooney & Dunn 1970; Orians & Solbrig 1977; Cody & Mooney 1978). The concept is viewed as independent evidence that characteristics of organisms are predictable from features of their environment, and thus rightly viewed as adaptive traits. This perspective, however, has not been universal, and the concept of MTC ecosystem convergence has been criticized (Barbour & Minnich 1990).

To be sure, one of the main limitations to the study of convergence in MTC regions is the inordinate attention to comparing landscapes with broad climatic similarities, but often with marked differences in other environmental factors: for example, soil texture and nutrients (Fig. 1.5), topography, fire, or subtle variations in rainfall distribution, all of which contribute to different selective pressures in the five MTC regions. These other factors lower the expectation of convergence in community structure and function (Blondel 1991). Thus, it is important to stress that ecosystem convergence is to be expected for organisms in similar environments, and this includes a multitude of environmental factors beyond just climate.

Ecosystem or community convergence may be the result of evolutionary or ecological processes. If phylogenetically unrelated organisms that occupy similar ecological niches in similar environments become more similar than their ancestors, this is termed evolutionary convergence. Evolutionary convergence is one form of *homoplasy*, a term for the generation of similar structures or functions in organisms that does not imply the mechanism of origin. Homoplasy may arise through multiple evolutionary mechanisms, including selection (both convergent and parallel evolution), genetic drift, and migration (Wake 1991; Leroi et al. 1994). Homoplasy may also arise through purely competitive interactions in ecological communities, a process known as ecological sorting (Wilson 1999). In the former case convergence is viewed as an indicator of the efficacy of natural selection, whereas in the latter case it is a measure of competitive displacement, evident in the overdispersion of plant traits in a community (Pausas & Verdú 2010). Regardless of the origin, convergence phenomena may provide useful tests of the predictability of functional types (Reich et al. 1997), community assembly rules (Schluter & Ricklefs 1993; Wilson 1999) and development of the ecological niche concept (Harmon et al. 2005).

Demonstrating evolutionary convergence requires that one have some information on the ancestral condition and evidence that it has changed in response to a particular environment. This is generally not possible at the community scale, but with a combination of phylogenetics and fossils, demonstrating changes in individual species' traits is possible. However, results of such comparisons must recognize that limitations to similarity may be imposed by different genetic or developmental constraints within lineages (Wake 1991) and historical events not replicated in the different environments (Peet 1978). The role of phylogeny and other historical factors has been addressed by testing the null hypothesis that communities in similar environments do not differ from control communities in very dissimilar environments (Crowder 1980). With this approach it is possible to partition the variance among communities into phylogenetic effects and habitat effects in order to compare assumed convergent communities (Schluter 1986).

Ackerly (2004a) investigated the relative importance of phylogeny in determining specific leaf area (leaf area per unit mass) in California chaparral shrubs by asking the question "which came first: the trait or the environment?" His phylogenetic analysis concluded that in most cases low specific leaf area predated the origins of the MTC; consequently it was presumed that this trait was not the result of adaptation to the current environment but rather due to ecological sorting processes. The early origin for such sclerophyllous leaves is consistent with the fossil record (Axelrod 1973), but the conclusion that this leaf type is not an adaptation to the current environment needs some qualification. As discussed in Chapter 10, the timing of the origin of the MTC was quite possibly much earlier than is often assumed. Also, demonstrating that a particular leaf type predates the current MTC is not equivalent to demonstrating it predates the present environment. Such a conclusion assumes that the MTC summer drought is the primary driver of the evolution of specific leaf mass, when in fact it may be some other contemporary environmental factor, for example drought at any time of the year or shallow substrates or even the winter rainfall pattern, all of which predate the MTC and argue against categorically excluding an adaptive origin for small sclerophyllous leaves. Similar caveats apply to other MTC studies that have dismissed adaptive origins for contemporary traits (Herrera 1992).

We view convergence as most usefully measured at the level of communities and recognized when similar environments generate communities comprising species with similar structures and functions. Quantifying this convergence requires testing with null models against other communities under different environments. Separating evolutionary convergence from ecological convergence requires some information on the phylogenetic changes observed in lineages of the component species of the communities, although such comparative methods have been challenged (Leroi *et al.* 1994; Westoby *et al.* 1995). Most important is the need to closely evaluate the critical environmental factors contributing to producing similar environments. One example of matching similar environments was the search for convergence in MTC sites in southwestern Australia and South Africa by Cowling & Witkowski (1994). They matched substrate characteristics between the two regions, ensuring a more precise comparison of similar environments. Additional factors that need to be considered in comparing MTC regions include variation in the timing of rainfall and severity of summer drought (Box 1.1).

With respect to convergence in MTC ecosystems we see that there are varying levels of similarity in environments and that similarity varies with the organism and scale of focus and consequently differing expectations of convergence. For example, with respect to soil animals, humus development appears to be a critical factor determining "similar" environments, perhaps more so than the climate (di Castri 1973). Also, convergence should not be expected at all scales. For example, postfire species diversity at the point scale of 1 m^2 is strikingly similar between MTC regions, but at the community scale of 0.1 ha, it is remarkably different (see Chapter 11).

Many studies have focused on convergent aspects of vegetation response to fire in MTC ecosystems. As discussed in more detail in Chapters 3 and 9 these have largely focused on modes of postfire regeneration. In this book we will examine ecosystem responses to fire in the five MTC regions and related vegetation (MTV) that often extends outside the MTC (Box 1.2).

Five MTC Regions

The largest MTC region is the Mediterranean Basin, which covers an expanse of more than 3500 km^2, with increases in summer aridity from north to south and west to east (Quézel 1981; Grove & Rackham 2001). It includes portions of Portugal, Spain, France, Italy, Greece and Turkey on the northern side of the Mediterranean Sea, Morocco, Algeria and Tunisa on the southern side, and at the eastern end of the basin Israel and adjacent parts of the Middle East. This vast distribution across multiple ethnic regions has led to multiple names applied to vegetation types. Evergreen broadleaf *matorral*, *maquis*, *macchia* and *garrigue* shrublands (Fig. 1.6a) transition on arid sites to a lower growing drought-deciduous spiny formation known as *tomillares*, *phrygana* or *batha*. Geophytes are common throughout the region and on disturbed sites there is a rich flora of annuals. On arid sites low levels of disturbance are sufficient to displace shrublands with annual grasslands.

Shrublands are dominated by evergreen sclerophyllous-leaved shrubs commonly differentiated based on their distribution on calcareous or non-calcareous substrates. All of these taxa resprout after fire and some have well developed basal lignotubers (see Chapter 3) and relatively few are postfire seeders (Box 1.2). Evergreen woodlands of low broadleaf trees such as *Quercus ilex*, *Q. pyrenaica* and *Q. suber* (Fig. 1.6b), once widespread, still persist in more mesic sites. Conifer forests of the moderate-sized serotinous-cone (see Chapter 3) *Pinus halapensis* are widespread throughout the northern side of the basin and sometimes replaced at the eastern end of the basin by the closely related *P. brutia*. In addition, in the western portion of the basin are populations of non-serotinous *P. nigra*, *P. pinea* and *P. sylvestris*.

Some forests such those dominated by *P. halapensis* typically burn in high-intensity crown fires often associated with maquis fires, whereas others such as *P. nigra* or *Quercus pyrenacica* burn in understory surface fires. Due to the long history of human civilization in this region, woodland distribution has greatly contracted and it is often difficult to distinguish between natural pine stands and plantations, some over 1000 years old (Grove & Rackham 2001). This MTV extends eastward into the summer-rain climate of eastern Turkey. Most wildland fires are started by people, although lightning-ignited fires occur in most mountainous regions of the basin (Vázquez & Moreno 1998). Fires tend to be small due to the extreme habitat fragmentation, although in southern France the frequent foehn winds known locally as the mistral winds, or the meltemia winds of Greece (Box 1.3), are capable of rapidly spreading fires over extensive portions of the landscape.

Box 1.3 Fire Weather Winds in MTC Regions

Regardless of where you are in the world, large uncontrollable fires usually are driven by high winds. Most MTC regions have extreme wind events that typically last a few days and may occur many times in a year, although are usually concentrated in particular seasons (Table B1.3.1). When these winds coincide with droughts and ignitions they are associated with extreme fire events, characterized by high rates of spread, long flame lengths, long-distance spotting and presence of fire whirls and other unpredictable wind patterns.

Katabatic or downslope winds are pushed over the leeward sides of mountains and heat at the dry adiabatic lapse rate as they descend. They are typically associated with a reversal from onshore to offshore airflow. Sometimes these are localized events such as the Sundowner Winds near the town of Santa Barbara in California (Blier 1998).

Other winds develop from synoptic weather conditions known as foehn wind events where interior high pressure cells are juxtaposed with low pressure troughs on the coastal side of mountains. Examples include the southern Europe mistral winds and southern California Santa Ana winds among others noted in Table B1.3.1. Topography plays a key role in the development of these foehn winds and their ultimate trajectory to the coast (Fosberg *et al.* 1966; Schroeder & Buck 1970). The mistral winds are funneled down the Rhone River Valley and the Santa Ana winds follow the Santa Ana River Valley, although these winds follow other drainages as well and their ultimate manifestation is a result of local terrain (Whiteman 2000; Moritz *et al.* 2010b).

When fires are ignited during these extreme winds they produce severe burning conditions sometimes referred to as *firestorms* (e.g. Fig. B1.3.1). Foehn winds are associated with extreme fire weather in many parts of the world. They contribute to most of the area burned in regions such as southern California (Keeley 2006a; Moritz *et al.* 2010b), in part because the winds are an annual event each autumn (Lessard 1988; Raphael 2003). Mistral winds and the similar sharav winds in the eastern Mediterranean Basin are also annual events contributing to extreme fire danger every year (Kutiel & Kutiel 1991) and they too account for the bulk of area burned (Levin & Saaroni 1999). The bergwinds of South Africa are associated with fires frequent enough to shape landscape patterns of forests and shrublands (Geldenhuys 1994). Southern-ocean cold fronts in spring and summer produce strong, hot dry brickfielder winds across southeastern Australia and promote the rapid development of large, intense fires, which are invariably associated with major losses of life and property (Hasson *et al.* 2009). However, in this non-MTC region, these are not annual events (Sharples *et al.* 2010).

Other weather anomalies appear to be equally important as foehn winds in fire activity on the Iberian Peninsula (Millán *et al.* 1998). In southwest Australia roughly once a decade strong hot winds associated with tropical cyclones bring gale force winds onshore during summer and create extremely hazardous fire conditions resulting in firestorms (McCaw & Hanstrum 2003).

Continued

Box 1.3 (cont.)

Table B1.3.1 Extreme gale-force winds with high temperatures and low humidity in MTC regions that are often associated with massive fires

MTC region	Subregion	Wind	Season	Characteristics
Mediterranean	Eastern Spain	Poniente	Summer	Katabatic foehn winds with westerly offshore flow mostly at night from low pressure in the Mediterranean Sea
Mediterranean	Southern France	Mistral	Late autumn – spring	Katabatic foehn winds with southerly offshore flow from high pressure system over the Alps and low pressure in the Mediterranean Basin
Mediterranean	Greece	Meltemia	Summer – autumn	Annual winds known as Etesian winds developing from high pressure ridge and blowing in a northwesterly direction toward a low pressure system arising from localized heating
Mediterranean	North Africa and southern Spain	Sirocco or Leveche	Spring and autumn	Katabatic winds with northerly flow from the Sahara Desert
Mediterranean	Israel, Lebanon	Sharav, Aka Khamsin	Late spring and autumn	Katabatic foehn winds with westerly flow from the Arabian Desert
California	Ventura Co. to Ensenada, MX	Santa Ana	Autumn – spring	Katabatic foehn winds with easterly or northeasterly offshore flow arising from high pressure system in the Great Basin and strongest in early morning
California	San Francisco Bay area and northern Sierra Nevada Mtns	Diablo or Mono	Autumn – spring	Katabatic foehn winds with easterly or northeasterly offshore flow arising from high pressure system in the Great Basin and strongest in early morning
California	Central coast	Sundowner	All seasons	Katabatic localized wind with offshore flow resulting from mesoscale pressure gradients and strongest in late afternoon
Chile	South central	Puelche	Autumn	Katabatic foehn winds with easterly offshore flow arising from high pressure system to the east of the Andes and western shore lows
South Africa	Cape Region	Bergwind	Autumn and winter	Katabatic foehn winds with northeasterly offshore flow arising from a Kalahari high

Continued

Box 1.3 (cont.)
Table B1.3.1 (cont.)

MTC region	Subregion	Wind	Season	Characteristics
Australia	Victoria and NSW	Brickfielder	Summer	pressure and coastal low pressure systems Katabatic foehn winds with northwesterly offshore flow from high pressure interior and coastal low pressure systems

Katabatic winds are downslope winds; foehn winds are due to synoptic conditions from interior high pressure cells and coastal low pressure troughs.

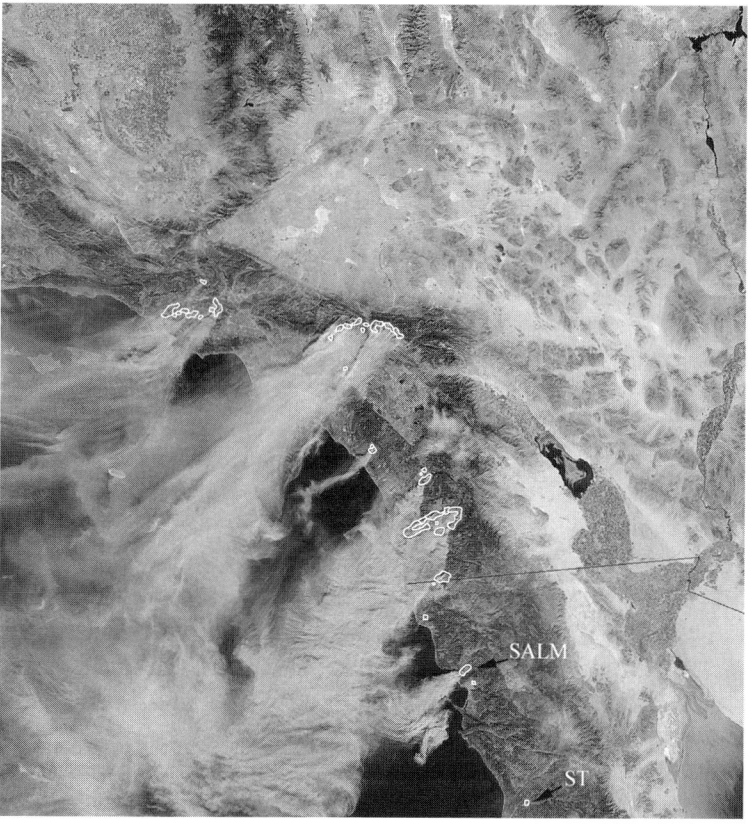

Fig. B1.3.1 *Santa Ana wind-driven fires and smoke in 2003 from Ventura County, USA, to San Antonio de Las Minas near Ensenada, Mexico (SALM arrow). Note the apparent lack of Santa Ana winds on the fire further south near Santo Tomás (ST arrow at bottom of panel) due to effects of the Gulf of California and San Pedro Mártir (see Keeley & Fotheringham 2001a, 2001b). (Image captured by the Moderate Resolution Imaging Spectroradiometer (MODIS) on the Terra satellite on October 26, 2003; http://earthobservatory.nasa.gov/NaturalHazards.)*

Continued

Box 1.3 (cont.)

Fig. B1.3.2 *Tree damage resulting from a mistral wind event following wildfire in a* Pinus halapensis *forest in southern France. (Photo by J.E. Keeley.)*

Santa Ana winds are the best studied of the foehn winds and they are similar in some but not all respects to foehn wind events in other MTC regions. These winds peak in frequency in the autumn and spring. It is the former wind events that produce extremely dangerous fire weather because they follow on the heels of a 6-month or more annual drought. When anomalous winter drought conditions are followed by spring Santa Ana wind events this may lead to significant out-of-season fires (Keeley *et al.* 2009b). Regardless of season these winds typically have less than 10% relative humidity and produce gusts that exceed 100 km per hour (Fosberg *et al.* 1966; Ryan 1969). Although referred to as "desert winds," the high temperatures and low humidity are the result of compression as air descends to form the basin air mass (Mitchell 1969) and, on a local scale, as it descends through coastal passes (Krick 1933).

Santa Ana and other foehn wind events typically last for several days. Multiple such events occur annually and there are usually 40–50 days per year with these wind events, although the frequency of southern European mistral winds (Fig. B1.3.2) may be double that number (Weber & Kaufmann 1998). Although these wind events are described as lasting for a period of days, they actually wax and wane during this period. Santa Ana winds develop overnight and peak in the early morning (Edinger *et al.* 1964), a pattern also evident with the European mistral winds (Weber & Kaufmann 1998). With daytime convectional heating these winds are often held aloft and onshore flow may predominate during the afternoon, which can change fire spread direction and greatly complicate firefighting activities (Keeley & Zedler 2009). Santa Ana wind events include

Continued

Box 1.3 (cont.)

patterns of wind speed, direction, relative humidity and temperature and these are not always tightly coupled. For example, Santa Ana wind flow pushes dry hot air masses offshore for several days, but when the offshore flow ceases, the onshore flow carries dry hot air back on land and thus high temperatures and low humidity may persist for many days after the Santa Ana winds have subsided (Fig. B1.3.3).

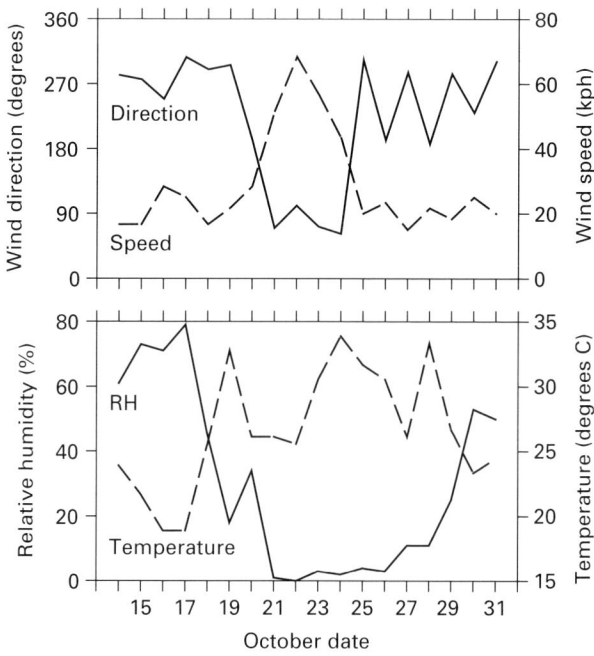

Fig. B1.3.3 *Mid-day (11 am–12 pm) wind direction, wind speed, relative humidity (RH) and temperature before, during and after the Santa Ana wind event that initiated the 2007 Witch Fire in San Diego, California, at 33° 13′ 34″ N/116° 59′ 32″ W, elevation 418 m. (From Keeley et al. 2009b.)*

In California the MTC extends along the Pacific Ocean from about 30° to 43° N. At the southern end of its distribution in Baja California, Mexico, the common evergreen shrublands and forests of California are restricted to upper elevations in mountain ranges and at the northern end to drier interior parts of southern Oregon (Rundel 2004). In California the MTC distribution is synonymous with the California Floristic Province (Raven & Axelrod 1978), which includes plant assemblages from the coast to the crest of the interior ranges of the Sierra Nevada. Along this elevational and moisture gradient are plant communities that exhibit many similarities to the Mediterranean Basin, and include some of the same

Table 1.2 Floristic comparisons of the shrubland plant families dominant in the five MTC regions of the world

Mediterranean	California	Chile	South Africa	Australia
Anacardiaceae	Anacardiaceae	Anacardiaceae	Asteraceae	Casurinaceae
Cistaceae	Ericaceae	Asteraceae	Ericaceae	Ericaceae
Ericaceae	Fabaceae	Fabaceae	Fabaceae	Fabaceae
Fabaceae	Fagaceae	Fagaceae	Proteaceae	Myrtaceae
Fagaceae	Rhamnaceae	Lauraceae	Rhamnaceae	Proteaceae
Rhamnaceae	Rosaceae	Rhamnaceae	Rutaceae	Xanthorrhoeaceae

genera and most of the same families (Table 1.2). At the most arid end of the gradient are low stature semi-deciduous *sage scrub* shrublands and these are displaced by larger stature evergreen sclerophyllous-leaved *chaparral* shrublands (Fig. 1.6c) on north-facing slopes near the coast, and on most slope faces between roughly 600 m and 1500 m (Keeley 2000). Many shrub species resprout after fire and some have lignotubers, but a substantial number of species do not resprout and regenerate after fire strictly by seeds. Shrublands form complex mosaics with evergreen *oak woodlands* and alien-dominated annual *grasslands*. Above about 1500 m shrublands give way to forests with a mixture of deciduous hardwoods (*Quercus kelloggi*) and pines (*Pinus ponderosa*), and higher still *mixed conifer forests* (Fig. 1.7a) that include other pines, firs (*Abies*) and several other conifer genera (Holland & Keil 1995).

East of the California Floristic Province on the rain shadow side of the interior mountain ranges are desert scrub types that are generally not considered part of the MTC due to very low rainfall, which is exceeded by the rainy season potential evapotranspiration (Rundel 2010). On these landscapes the MTC summer drought dominates the climate and this drought pattern continues further north into eastern Washington and Idaho. In the southwestern USA chaparral shrublands occur in disjunct pockets in central and southern Arizona, which has bimodal annual rainfall (Fotheringham 2009), and in the eastern Sierra Oriental of mainland Mexico with winter drought and summer rain (Keeley 2000). Similar fire-prone MTV shrublands continue eastward controlled largely by edaphic conditions (Menges & Kohfeldt 1995; Carrington & Keeley 1999). Evergreen sclerophyllous forests dominated by the genus *Pinus* are widely distributed in non-MTC regions of western North America and further east along the southern part of the USA, forming similar fire-prone forests to those in the MTC (e.g. Platt 1999). This extension of MTV far outside strictly MTC regions illustrated by *Pinus* resembles a similar pattern with evergreen sclerophyllous *Eucalyptus* in Australia.

In California along the elevational gradient as shrublands merge into woodlands and then into forests, fuel structure changes markedly, as do fire regimes (see Chapter 2). The lower elevation chaparral burns in high-intensity crown fires whereas the forests are more prone to low-intensity surface fires. The proportion of human-ignited fires vs. lightning-ignited fires varies from mostly human ignitions in the lower foothill shrublands to a substantial proportion of lightning-ignited fires in the higher montane sites. Southern California is noted for its autumn Santa Ana winds (Box 1.3) and when they coincide with ignitions the result is firestorms that cover vast areas in a brief period of time.

The MTC region of central Chile occurs from roughly 30° to 40° S and closely matches the California landscape with a north–south trending central valley between lower coastal ranges and the much higher interior Andes (Mooney 1977a; Arroyo *et al.* 1995; Dallman 1998). As in California, lower-stature matorral shrublands with many drought-deciduous species are found near the coast, with evergreen sclerophylls dominating sites further inland (Fig. 1.6d). Much of this landscape has experienced widespread human disturbance and forms an open mosaic of shrubland and alien-dominated annual grassland. There is a much greater overlap with plant families in the northern hemisphere than with other MTC regions in the southern hemisphere (Table 1.2), perhaps reflecting the early breakup of Gondwana continents (see Chapter 10). As in the previously discussed regions, geophytes and annuals comprise an important part of these ecosystems. Broadleaf evergreen woodlands of *Cryptocarya alba* in the laurel family occupy moister sites up to 1400 m, and in valley bottoms, and extending to higher elevations are forests dominated by winter-deciduous *Nothofagus obliqua*. Unlike those in California, evergreen conifer forests do not dominate at elevations above these evergreen and deciduous woodlands, however; further south and still within the MTC are patches of the very distinctive fire-adapted conifer *Araucaria araucana* (Aagesen 2004).

Most fires are started by humans and are generally small due to the mosaic of shrublands with grasslands that are intensively grazed and exploited for firewood, reducing fuel continuity (Zunino & Riveros 1990). Lightning-ignited fires are rare because there are few summer thunderstorms due to the massive Andes escarpment that blocks monsoonal air masses from penetrating into Chile (Aschmann & Bahre 1977). Despite the apparent lack of a predictable natural source of fires, sclerophyll shrublands are highly resilient to fire and most all-woody taxa resprout after fire, some from basal lignotubers, but few recruit seedlings after fire.

In South Africa the MTC is concentrated in the southwestern portion of the country known as the Cape region or botanically as the Cape Floristic Region. Vegetation is largely one of two evergreen shrublands, *fynbos* (Fig. 1.6e) and the lower stature *renosterveld* (Cowling *et al.* 1997a). The former dominates on quartzites and other substrates producing very coarse-textured oligotrophic or low-nutrient soils (Fig. 1.5), whereas renosterveld is restricted to more fine-grained and more fertile soils derived from shales. Floristically these vegetation types are markedly different, with fynbos dominated by sclerophyllous-leaved southern hemisphere families such as the Proteaceae and Restionaceae, and renosterveld by more cosmopolitan plant families (Table 1.2). Small patches of broadleaf evergreen woodlands and forests persist in isolated riparian areas or steep cliffs known as kloofs. These sites are considered refugia where these fire-sensitive forests have persisted (Moll *et al.* 1980; Manders 1991; Geldenhuys

1994) and from which they expand outwards in the rare absence of fires (Luger & Moll 1993).

Most fires are started by people, although lightning-ignited fires occur with some frequency in the mountains (Horne 1981; Manry & Knight 1986) and occasionally such fires can exceed 100 000 ha (Versfeld et al. 1992). Strong winds, known locally as berg winds, play a critical role in determining fire size, and the highest winds are during the summer drought (Box 1.3). As in other MTC regions, many fynbos species resprout from lignotubers and have dormant seedbanks that are stimulated by fire, plus a very rich geophyte flora that resprouts and flowers after fire. Unlike California, annuals represent a less impressive part of the postfire flora. Components of fynbos vegetation persist on low-nutrient soils to the east of the Western Cape Province, and outside the MTC, and this MTV extends in patches into the summer-rain climate of eastern South Africa.

Western Australia and South Australia have a well-developed MTC and landscapes are dominated by evergreen sclerophyllous heathlands (Fig. 1.6f), mallee shrublands and *Eucalyptus* woodlands and forests with an abundance of geophytes and other herbaceous perennials. The highly weathered coarse-grained oligotrophic substrates are similar to pockets of such soil in South Africa (Fig. 1.5) and this is one factor accounting for both regions sharing many plant families (Table 1.2). The very low nutrient soils in the southwestern portion of Western Australia support a sclerophyllous heathland vegetation known as *kwongan* that has many floristic and structural relationships with South Africa's fynbos (Pate & Beard 1984). As in South Africa these shrublands are resilient to very high fire frequencies. On drier or sometimes more fertile sites various species of *Eucalyptus* form mallee vegetation characterized by extensive coppicing from large lignotubers after fires (Parsons 1981). A significant portion of the Southwestern Botanical Province includes closed-canopy *Eucalyptus* forests that tower above the associated shrubland communities (Rundel 2004). The MTC extends eastward into South Australia and western Victoria, with landscapes dominated by complexes of heathlands, mallee and woodland communities. Eucalypt forest with a heath understory is widely distributed across southern Australia into the non-MTC southeast with an aseasonal climate (Box 1.2).

Conclusions

A significant portion of terrestrial vegetation is fire-prone. Climate and geology interact with fire to create the environmental template that determines plant traits, community assembly and plant and animal distributions. Mediterranean-type climates dominate five widely separate parts of the globe in temperate latitudes on the western ocean-facing sides of continents in both the northern and southern hemispheres. Early observations of the dominance of evergreen sclerophyllous-leaved shrublands and woodlands in these climatic zones illustrate a remarkable level of convergence. Coupling of seasonal drought with high summer temperatures has created expansive landscapes dominated by sclerophylls and some of the most fire-prone regions of the world. This book uses these ecosystems as a focal point for discussion of the integral role that fire plays in the ecology and evolution of plant traits, community assembly and contemporary management responses.

Appendix 1.1 Ecosystem Convergence and MEDECOS

It is curious that research scientists in other ecosystems, such as tropical rainforests, coniferous forests and summer monsoon grasslands, have not made equal use of the comparative biology of their systems as is the case with the five mediterranean-type climate ecosystems. Part of the reason may be that these five MTC regions are major population centers with academic institutions and both academic and government scientists intensely interested in the lessons to be learned from comparison with other MTC ecosystems. This book in part results from collaborations developed through MEDECOS conferences.

Mediterranean Ecosystem (MEDECOS) conferences have been held every 3–5 yrs since 1971, rotating venues through all five MTC regions. The first meeting in Valdivia in 1971 was at the beginning of the U.S. National Science Foundation funded integrated research program comparing Chile and California within the framework of the International Biological Program (IBP). This research took advantage of these two widely separate regions, with very similar physical environments, to address the question of whether similar environments comprising phylogenetically different biotas will produce structurally and functionally similar ecosystems.

MEDECOS conferences typically include scientists and resource managers from 15 or more countries who assemble for several days of formal presentations and informal discussions. The term "MEDECOS" was first used for the fourth meeting when it became apparent that there was widespread support for continuing these conferences at periodic intervals. Themes and immediate products from MEDECOS conferences are as follows:

1971 [MEDECOS I]. Valdivia, Chile. Theme: Convergent Evolution
di Castri, F. & Mooney, H.A. (eds) (1973) *Mediterranean type ecosystems: origin and structure*, 405 pp. Springer, New York.

1977 [MEDECOS II]. Stanford, California. Theme: Ecosystem Role of Fire
Mooney, H.A. & Conrad, C.E. (eds) (1977) *Proceedings of the symposium on the environmental consequences of fire and fuel management in mediterranean ecosystems*, GTR-WO-3, 498 pp. USDA Forest Service, Washington, DC.

1980 [MEDECOS III]. Stellenbosch, South Africa. Theme: Convergence and Role of Nutrients

Day, J.A. (ed.) (1983) *Mineral nutrients in mediterranean ecosystems*. South African National Scientific Programmes Report 71.

Kruger, F.J., Mitchell, D.T. & Jarvis, J.U.M. (eds) (1983) *Mediterranean-type ecosystems: the role of nutrients*, 552 pp. Springer, New York.

1984 MEDECOS IV. Perth, Australia. Theme: Ecosystem Resilience

Dell, B. (ed.) (1984) *MEDECOS IV: Proceedings of the 4th international conference on mediterranean ecosystems held at Perth, Western Australia, August 13–17, 1984*. University of Western Australia, Nedlands, Australia.

Dell, B., Hopkins, A.J. M. & Lamont, B.B. (eds) (1986) *Resilience in mediterranean-type ecosystems*, Tasks for Vegetation Science 16, 168 pp. Dr. W. Junk, Dordrecht, The Netherlands.

1986 MEDECOS V. Montpellier, France. Theme: Time Scales and Water Stress

di Castri, F., Floret, C., Rambal, S. & Roy, J. (eds) (1988) *Time scales and water stress: Proceedings of 5th international conference on mediterranean ecosystems*, 678 pp. International Union of Biological Sciences, Paris.

Roy, J., Aronson, J. & di Castri, F. (eds) (1995) *Time scales of biological responses to water constraints: the case of Mediterranean biota*, 243 pp. SPB Academic Publishing, Amsterdam.

1991 MEDECOS VI. Crete, Greece. Theme: Animal–Plant Interactions

Thanos, C.A. (ed.) (1992) *MEDECOS VI: Proceedings of the 6th international conference on mediterranean climate ecosystems "Plant–animal interactions in mediterranean type ecosystems" held at Maleme, Crete (Greece) September 23–27, 1991*, 389 pp. University of Athens, Greece.

1994 MEDECOS VII. Viña del Mar, Chile. Theme: Landscape Degradation

Rundel, P.W., Montenegro, G. & Jaksic, F.M. (eds) (1998) *Landscape disturbance and biodiversity in mediterranean-type ecosystems*, Ecological Studies 136, 447 pp. Springer, New York.

1997 MEDECOS VIII. San Diego, California Theme: Global Change

[No publication]

2000 MEDECOS IX. Stellenbosch, South Africa. Theme: Mediterranean-type Ecosystems: Past, Present and Future

[Collection of papers published in *Journal of Mediterranean Ecology*, **2** (2001) edited by D.M. Richardson, K.J. Esler & R.M. Cowling].

2004 MEDECOS X. Rhodes, Greece Theme: Conservation and Management

Arianoutsou, M. & Papanastasis, V.P. (eds) (2004) *Ecology, Conservation and management of mediterranean climate ecosystems: Proceedings of the 10th international conference on mediterranean climate ecosystems, April 25–May 1, 2004, Rhodes, Greece*. Millpress, Rotterdam, The Netherlands.

2007 MEDECOS XI. Perth, Western Australia [no theme]

Rokich, D., Wardell-Johnson, G., Yates, C., *et al.* (eds) (2007) *Proceedings of the international mediterranean ecosystems conference: Medecos XI*. Kings Park and Botanic Garden, Perth, Australia.

2011 MEDECOS XII. Los Angeles, California. Theme: Linking Science With Resource Management

MEDECOS conferences have always been informally arranged by consensus opinion of participants at the previous conference. At MEDECOS X a vote was taken to formally establish an organization to be named the International Society of Mediterranean Ecologists (acronym ISOMED).

Other organizations that have taken an active interest in the study and management of MTC ecosystems include the International Union for the Conservation of Nature (IUCN) whose 2004 "Malibu Declaration" is a call-to-arms in response to the "rampant urbanization" that threatens natural areas in these MTC regions (www.interenvironment.org/med-5/malibudeclaration.htm).

2 Fire and the Fire Regime Framework

A global view of potential vs. actual vegetation distributions points to fire as a major driver of biome distribution and determinant of community structure (Bond et al. 2005). In ecological terms, fire acts much like an herbivore, consuming biomass and competing with biotic consumers for resources, and in this sense is an important part of trophic ecology (Bond & Keeley 2005). As in other competitive interactions, not only can fire competitively exclude herbivores by temporarily eliminating resources, but intensive grazing is known to exclude fire by consuming herbaceous ground fuels (Savage & Swetnam 1990). Coexistence is often enhanced by temporal separation of trophic niches, with herbivores grazing early in the spring on green herbaceous material that is unavailable for burning, whereas later in the season the remaining dry thatch is readily consumed by fire. In many respects fire is a more potent competitor because it is not limited by either toxins or protein deficiency and readily consumes dead woody biomass, but by contrast it is often limited by ignition sources and continuity of fuels.

Fire scientists have long symbolized the critical elements of fire in a triangle of fuel, oxygen and heat (Pyne et al. 1996). These are indeed necessary for fire ignition and propagation but are insufficient for predicting the global distribution of fire-prone ecosystems. The conditions both necessary and sufficient to explain the ecological distribution of fire activity can be summarized by four parameters: biomass, seasonality, ignitions and fuel structure (Fig. 2.1). In addition to biomass fuels to spread a fire there must be a dry season that converts potential fuels to available fuels. In mediterranean-type climate (MTC) ecosystems summer drought results in high fire hazard on an annual basis, in contrast to many temperate forests that are only periodically vulnerable to fire in response to decadal or longer oscillations in climate. Vegetation only burns when ignitions are present to initiate the combustion process and landscapes vary markedly in the potential for natural ignitions from lightning, and in the extent of anthropogenic ignition sources. However, understanding the ecosystem distribution of fire requires consideration of a fourth parameter, fuel structure, which is fundamental to recognizing how different fire regimes develop.

Many landscapes are dominated by ecosystems where fire is a natural and necessary process for long-term sustainability of those systems. Despite the obvious resilience of many communities to periodic fire, it is misleading to think of species as being fire adapted; rather, they are adapted to a particular temporal and

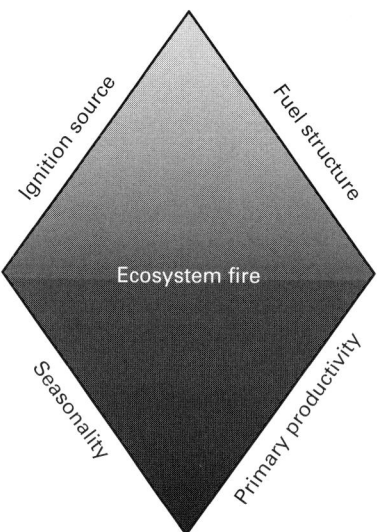

Fig. 2.1 *Fire diamond schematic of factors necessary and sufficient for predicting the distribution of fire as an ecosystem process. A certain level of primary productivity is needed in order to spread fire, and climatic seasonality, with annual to decadal cycles of drying, is required to convert these potential fuels to available fuels that will combust. Fire regimes are controlled strongly by ignition frequency and fuel structure with important feedback loops between all four factors.*

spatial pattern of burning. This is captured in the concept of a fire regime, which includes the fuel types consumed, frequency and timing of burning, intensity of the fire and the spatial distribution of individual fire events (Keeley et al. 2009a). Fires are often referred to as disturbances, but in many plant communities fire is an integral ecosystem process and disturbances are perturbations to the fire regime that lie outside the historic realm. Such disturbances include increased fire frequency, as well as suppression and exclusion of fire.

Fire Regimes

Generalizations about fire effects on plant trait distribution based on fire frequency or intensity are widespread in the literature. However, these seldom are broadly applicable beyond local settings because they ignore the fact that fire regimes change across landscapes and these changes often represent substantial shifts in both fire behavior and plant responses. Thus, the fire regime concept is fundamental to understanding fire and fire effects, and switches in fire regimes such as from a surface fire to a crown fire regime need to be factored into any global generalizations.

On the basis of research in Australian sclerophyll forests and shrublands, Gill (1973) introduced the concept of a fire regime, which comprised patterns of fire frequency, intensity and seasonality. It has since been recognized as a critical attribute for describing fire-prone ecosystems (e.g. Heinselman 1981), although it is an ecosystem attribute with both temporal and spatial variation (Morgan *et al.* 2001). Building on Gill's original concept, a more complete picture of a fire regime includes five parameters (Keeley *et al.* 2009a): (1) fuel consumption and fire spread patterns, (2) fire intensity and severity, (3) frequency, (4) burn patch size and distribution, and (5) fire seasonality. The key concept is that these factors act in concert to produce the fire regime and it is the entire fire regime that constrains functional types, community assembly patterns and biome distribution. Dissecting out one factor and evaluating responses to that factor alone will lead to erroneous conclusions unless one is cognizant of how other attributes of the fire regime covary. For example, as discussed in Chapter 9, attempts to define global patterns of resprouting response to fire frequency or fire severity have proven elusive and that is because changes in other parameters such as types of fuels consumed produce thresholds of response that may reverse the relationship.

Patterns of Fuel Consumption and Fire Spread

Ecosystems differ greatly in both the horizontal and vertical pattern of biomass distribution, and thus in fuel structure (Fig. 2.1), and this has a profound impact on fire spread characteristics. On the basis of patterns of fuel consumption, fires are often categorized into classes of fire regimes, which include *surface fires*, *crown fires* and *ground fires*. Within these classes, regimes may vary markedly; for example, both South African fynbos and California chaparral are crown fire regimes but differ greatly in fire frequency and intensity and seasonal distribution.

Surface fires, which are sometimes referred to as *understory fires*, spread by fuels on the ground and apply to many forest types (Fig. 2.2a) where there is a discontinuity between surface fuels and tree canopy fuels (Fig. 2.2b). Closed-canopy forests typically have dead leaf and stem surface fuels whereas more open-canopy forests have standing herbaceous surface fuels. The type of surface fuels can result in significant differences in fire regime. For example, herbaceous fuels respond to high rainfall years by increasing fire activity in subsequent years, but the same is not observed in systems where surface fuel structure is dominated by downed leaves and branches (see discussion of antecedent climate effects below).

Crown fires burn in the canopies of the dominant growth forms (Fig. 2.2d), and the term is most usefully applied to shrub and tree dominated vegetation (Fig. 2.2c). This is the predominant fire behavior in MTC regions. In closed-canopy shrublands the general lack of surface fuels, coupled with the shrub fuel structure, results in fires spreading as *independent crown fires* or *running crown fires*. This type of fire behavior is dependent on either strong winds, steep terrain or a high proportion of dead to live canopy biomass. Other vegetation types such as North American lodgepole pine, Australian *Eucalyptus* or Mediterranean oak

Fig. 2.2 *Examples of a surface fire regime and a crown fire regime. North American (a) mixed conifer forest historically burned by (b) low-intensity surface fires and (c) chaparral shrubland historically burned by (d) high-intensity crown fires. (Photos by Jon Keeley, a,c; by U.S. National Park Service, b; and U.S. Forest Service, d.)*

woodlands and forests share many of the same fuel characteristics with closed-canopy shrublands (i.e. high density of trees and accumulation of dead fuels in the canopy), and thus are highly susceptible to crown fires (Bradstock *et al.* 2002; Keeley *et al.* 2009a). In some forests, fires may spread by a combination of surface and canopy fuels and are termed *active crown fires*. In more open forests where fires are spread largely by surface fuels, localized accumulation of understory "ladder" fuels may carry fires into the canopy of individual trees or small groups of trees and these are termed *passive or dependent crown fires* (van Wagner 1977). Often times these crown fires are referred to as *stand-replacing fires*, although this is largely a northern hemisphere perspective that describes crown fires in conifer forests where entire stands burned in high-intensity fires are killed. Australian *Eucalyptus* forests and woodlands are an exception in that even when burned in high-intensity crown fires, close to 100% of the trees resprout epicormically (Gill 1997), rapidly returning the forest to its original state (see Fig. 3.3d). Northern hemisphere oak woodlands have a similar capacity for epicormic resprouting and resilience to high-intensity crown fires.

Ground fires spread by smoldering combustion through duff and can be sustained at relatively high fuel high moisture conditions (Miyanishi 2001). In mires with deep peat layers such fires may smolder for months until extinguished by rainfall. In forests, such fires play a potential role in collecting ignitions at one point in time, and later irrupting into surface or crown fires when the weather changes.

Some ecosystems are characterized by either surface fires or crown fires, but in many systems mixtures of both fire type occur. The proportion of landscape burning in one or the other fire type is a function of the time since last burning, rate of fuel accumulation, antecedent drought and severity of fire weather. Sometimes a single fire will comprise a mixture of surface and crown fires and has been termed *stand-thinning* fires (Keeley & Zedler 1998). Some ecosystems experience a temporal mix of surface fires alternating over time with high-intensity crown fires (Zimmerman & Omi 1998). These mixed fire regimes may perform important ecosystem functions by creating landscape mosaics critical for plant regeneration and animal habitats as discussed in Chapter 3.

Fire Intensity and Severity

Fire intensity describes the physical combustion process of energy release from organic matter. In physical terms it is the energy per unit volume multiplied by the velocity at which the energy is moving, measured as watt m^{-2}. This represents the reaction intensity in Rothermel's (1972) fire spread models, which forms the basis for most fire behavior models. However, in fire science the term fire intensity often takes on other meanings.

One example is Byram's (1959) fireline intensity, which is the rate of heat transfer per unit length of the fireline:

$$I = HWR,$$

where H equals heat of combustion (kJ kg^{-1} of fuel), W is consumed fuel (kg m^{-2}), and R is the rate of fire spread (m s^{-1}), giving a fireline intensity (I) in kW m^{-1}. This is the radiant energy release in the flaming front and is a good measure of fire propagation, which is critical for fire suppression activities and has been incorporated into fire danger rating calculations (Hirsch & Martell 1996; Weber 2001). It can vary from 1000 kW m^{-1} in forests with understory burning of surface fuels to 20 000 kW m^{-1} in shrublands with active crown fires (Agee 1993; van Wilgen et al. 1985).

Direct measurement of fireline intensity requires that one distinguish fuels consumed by the flaming front from the total fuel consumption. However, practical measures of fuel consumption are based on the difference between prefire and postfire fuel inventories, and this inflates estimates of fireline intensity (Alexander 1982; Scott & Reinhardt 2001). As a result flame length is used as a surrogate for fireline intensity since there is a significant relationship between these parameters in forest and shrubland ecosystems (van Wilgen 1986; Burrows 1995; Fernandes et al. 2000). However, in vegetation with a mixture of fine fuels and woody fuels or on steep slopes this relationship is not always reliable (Weise & Biging 1996; Keeley 2009). Since fireline intensity is determined by both rate of spread and fuels consumed, it is possible for two very different fires to have similar fireline intensities: for example, low heat output with a high rate of spread and high heat output with a low rate of spread (Pyne et al. 1996).

Fireline intensity is an established metric in forested ecosystems as there is a well-documented relationship between flame length and scorching height of tree crowns. However, fireline intensity often cannot explain mortality patterns since mortality may be more a function of total heat output reflected in flame residence time or a function of smoldering combustion in the duff after the flame front passes (Wade 1993). Many other fire effects are not well predicted by fireline intensity. Soil duff consumption, for example, is more related to temperatures at the soil surface and the duration of heating (Ryan & Frandsen 1991; Miyanishi 2001). Also, survival of seedbanks or rhizomes may be more closely tied to duration of heating as well as maximum soil temperatures than to fireline intensity (Beadle 1940; Flinn & Wein 1977; Auld & O'Connell 1991; Bradstock & Auld 1995). This should come as no surprise since often very little radiant or convective heat from combustion of aerial fuels is transferred to the soil, and generally soil temperatures are more dependent on consumption of fine fuels on the surface (Bradstock & Auld 1995). Although fireline intensity provides information for fire fighters concerned with fire containment, resource managers may be more concerned with temperature and duration of heating (residence time) as these may be critical to retention of sensitive ecosystem components. In the future, fire managers will likely depend heavily on remote imaging technologies for fire intensity and these do not always scale with fireline intensity (Smith et al. 2005).

Due to the difficulties of measuring fire intensity, particularly for unplanned wildfires, the terms fire severity or burn severity have been used as a postfire indicator of intensity. Some definitions of fire severity have been rather general statements about broad impacts of fires, such as the degree of environmental change caused by fire (e.g. White & Pickett 1985; Simard 1991; Jain et al. 2004), and consequently have not lent themselves to operationally useful metrics. Most empirical studies that have measured fire severity have had a common basis that centers on the loss or destruction of aboveground or belowground organic matter (Keeley 2009). In forests, height of bole scorch or crown volume scorch are two common measures and these correlate with fire intensity (Cheney 1981; McCaw et al. 1997; Catchpole 2000). In shrublands, fire severity is commonly assessed using the twig diameter on standing skeletons, which correlates with a crude measure of fire intensity (Moreno & Oechel 1989; Pérez and Moreno 1998) and similar measures have been used in heathlands (Whight & Bradstock 1999).

On landscapes prone to large wildfires, assessing fire severity has benefited from the use of Landsat satellite imagery, which generates indices correlated with ground measures of fire severity (Conard et al. 2002; Miller & Yool 2002; Chafer et al. 2004; Hammill & Bradstock 2006; Keeley et al. 2008; Veraverbeke et al. 2010). Indices based on the Normalized Differenced Vegetation Index (NDVI) derived from the Landsat thematic mapper (TM) sensor seem to hold the most promise (but see Gitas *et al.* 2009 for a review of the ability of other sensors to detect fire severity). One of the more widely utilized Landsat-derived indices is the differenced Normalized Burn Ratio (dNBR), based on the difference between near infrared (NIR) and middle-wave infrared (MIR), which is used to calculate the Normalized Burn Ratio (NBR) before and after fire and is defined as

$$\text{NBR} = \frac{(\text{NIR} - \text{MIR})}{(\text{NIR} + \text{MIR})}.$$

Indices of fire severity reflect the difference between prefire NBR and postfire NBR and are termed the differenced Normalized Burn Ratio or dNBR, defined as:

$$\text{dNBR} = \text{prefire NBR} - \text{postfire NBR}.$$

This technique generates a spatially explicit estimate of fire severity based on the detection of change in vegetation cover and biomass, the exposure of soil and alteration of soil color. The images resulting from this analysis reveal that the burned area within the perimeter of a fire comprises a mosaic of different fire severities (Fig. 2.3). There is some evidence that the calibration of the remote signals to fire severity may differ between vegetation types (Hammill & Bradstock 2006; Roy *et al.* 2006). However, even across broad landscapes, satellite-based indices of fire severity are strongly correlated with parameters assumed to be related to fire severity. For example, Bradstock *et al.* (2010) found such measures of fire severity to increase under severe fire weather and decrease with time since last fire in Australian *Eucalyptus*-dominated landscapes. In southern California chaparral there is a highly significant decline in this measure of severity with number of times a site burned (Keeley *et al.* 2008).

Miller & Thode (2007) have proposed that there are advantages to use of the relative dNBR, where the dNBR ratio is expressed relative to the prefire NBR (i.e. a measure of prefire biomass cover),

$$\text{RdNBR} = \text{dNBR}/\text{prefire NBR}.$$

It appears that the absolute dNBR gives an estimate of fire intensity as it expresses a quantity related to the loss of total biomass, whereas the RdNBR gives a measure more closely associated with the percentage mortality of aboveground vegetation. As a result, RdNBR is widely used in forested areas subject to a combination of surface and crown fires as it estimates tree mortality; however, in non-forested crown fire regimes where complete top-kill is common, RdNBR may not be very informative (Keeley *et al.* 2008).

Controls on fire intensity and fire severity include both endogenous and exogenous ecosystem factors. Fuel loads and fuel moisture are key endogenous controls and antecedent climate and fire weather as well as topographic location are critically important exogenous factors (Odion *et al.* 2004; Keeley *et al.* 2008; Bradstock *et al.* 2010). Past fire history as well leaves a legacy of standing and downed fuels that affects severity of subsequent fires (Thompson & Spies 2010).

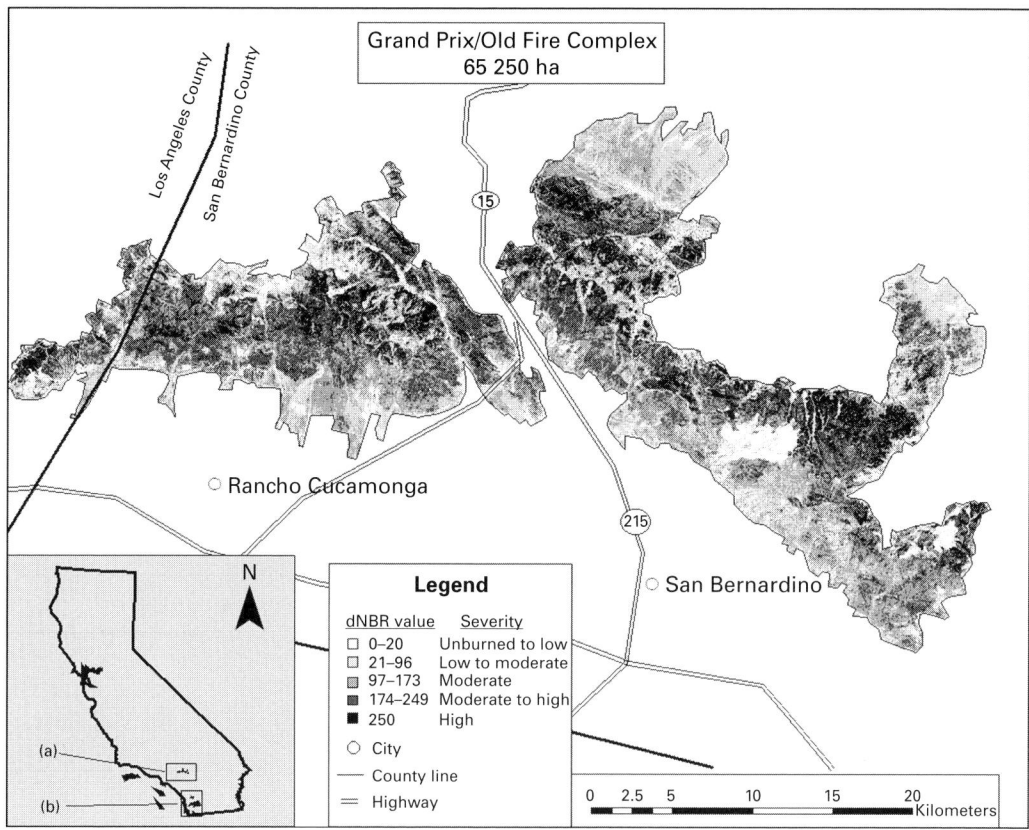

Fig. 2.3 *Landsat TM-based differenced Normalized Burn Ratio (dNBR) indices for two large wildfires illustrating mosaic of different fire severities. Inset shows these fires (a) and additional fires burning at the same time (b), all of which are included in the study of fire severities in Keeley* et al. *2008. (From Keeley* et al. *2008.)*

Fire intensity and fire severity are operationally tractable measures, but they are largely of value only so far as they can predict ecosystem responses such as soil erosion or natural regeneration (Fig. 2.4). Such predictions are complicated by the fact that many biotic and abiotic factors enter into the relationship between fire intensity, fire severity and ecosystem response (e.g. Peterson & Ryan 1986; Neary *et al.* 1999; Lentile *et al.* 2006). For example, alien plant invasion may increase with increased fire severity in forests (Turner *et al.* 1999) but show the opposite relationship in shrublands (Keeley *et al.* 2008). Degree of fire severity can affect the extent of postfire resprouting in herbs and shrubs (Flinn & Wein 1977; Keeley 2006c) and seedling recruitment (Whelan 1995; Bond & van Wilgen 1996; Ryan 2002; Pausas *et al.* 2003; Keeley *et al.* 2005a; Johnstone & Chapin 2006) but there are significant species-specific differences; high fire severity may be inhibitory in some but stimulatory in others. In some forests fire severity may be correlated with long-lasting impacts on regeneration (Lecomte *et al.* 2006) but in other

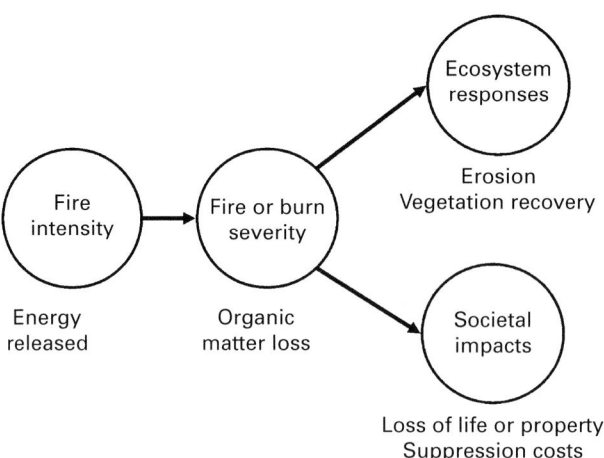

Fig. 2.4 *Schematic representation relating the energy output from a fire (fire intensity), the impact as measured by organic matter loss (fire severity, also known as burn severity), and ecosystem responses and societal impacts that may vary in response to different patterns of fire severity.*

ecosystems severity is not closely tied to short-term or long-term changes in recovery (Box 2.1). In North America some agencies utilize a ground measurement of fire severity known as the CBI or Composite Burn Index (Key & Benson 2006) but correlation between this metric and dNBR ranges from weak to good depending on the ecosystem (Gitas *et al.* 2009). Weak correlations are likely tied to the fact that in some ecosystems CBI mixes both fire severity and ecosystem response variables (Keeley 2009).

A major management reason for postfire assessments of fire or burn severity is because it is thought to be an important indicator of watershed stability (Robichaud *et al.* 2000; Wilson *et al.* 2001; Ruiz-Gallardo *et al.* 2004). Loss of aboveground biomass exposes more soil surface and increases the kinetic force of precipitation on the soil surface, which can increase overland water flow (Moody & Martin 2001). Also, loss of soil organic matter alters the binding capacity of soil and results in other structural changes that can affect erosional processes. Postfire increase in soil water repellency due to hydrophobic soil layers is tied, albeit sometimes weakly, to fire severity, although in some ecosystems soil hydrophobicity is unrelated to fire severity (Doerr *et al.* 2006). In summary, fire *per se* does affect ecosystem hydrology, but the degree of fire severity may not play a major role relative to the multitude of other factors that determine postfire hydrological functioning (Robichaud *et al.* 2000; Doerr *et al.* 2006; González-Pelayo *et al.* 2006; Pausas *et al.* 2008).

Fire Frequency

Fire frequency is the number of occurrences of fire within a circumscribed area and time period of interest. It is commonly expressed as the *fire rotation interval*, which

Box 2.1 Use of Landsat dNBR to Determine Fire Severity and Ecosystem Response in Crown Fire Shrublands (from Keeley 2009)

In late October 2003 five large wildfires burned more than 200 000 ha in southern California. A total of 250 tenth-hectare plots were sampled in these burned areas to assess fire severity and vegetation recovery (Keeley et al. 2008). Fire severity was assessed using the twig diameter method commonly used in crown fire ecosystems (Moreno & Oechel 1989) on multiple samples of the same shrub (*Adenostoma fasciculatum*). Vegetation recovery was based on plant cover in the first spring following fires. The Landsat TM assessment known as the differenced Normalized Burn Ratio (dNBR) was provided by the U.S. Geological Survey EROS data center (Sioux Falls, SD). This remote sensing index was strongly correlated with field measurement of fire severity (Fig. B2.1.1a), explaining over a third of the variation between these sites. However, when dNBR was used to predict ecosystem response variables related to recovery there was little to no relationship. Total vegetative recovery (Fig. B2.1.1b) was very weakly related to dNBR and explained only about 1% of the variation, and there was no significant relationship with woody cover ($P = 0.94$, not shown), or percentage of resprouting by the dominant shrub (Fig. B2.1.1c).

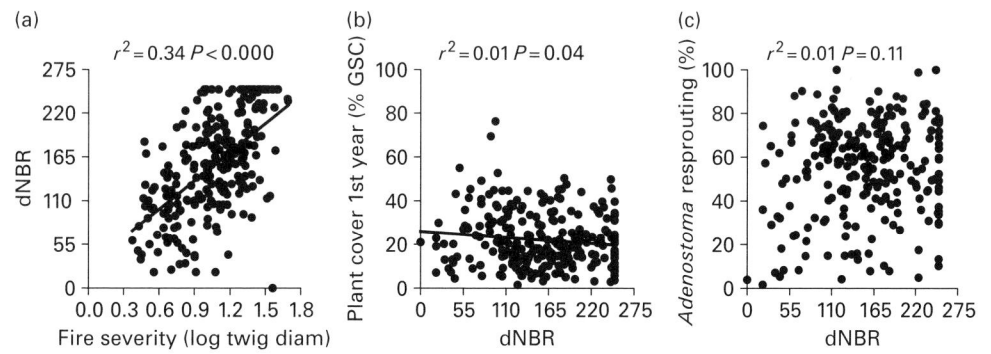

Fig. B2.1.1 *Relationship between (a) field measurement of fire severity and early assessment dNBR, (b) dNBR and first-year plant cover and (c) dNBR and postfire resprouting by the shrub* Adenostoma fasciculatum.

is the time required to burn the equivalent of a specified area, or *fire return interval*, which is the spatially explicit time between fires in a specified area, sometimes expressed as the inter-fire interval. For example, wildlands in southern California have an average fire rotation interval of 36 yrs, but the actual fire return interval can vary from fires every few years at some sites to fires every 100 yrs at other sites (Keeley et al. 1999a). The variance in fire return interval is critical to ecosystem resilience and species differ markedly in tolerance to high and low departures from the mean fire return interval (see Chapter 3).

One of the difficulties with precise usage of the term fire frequency is that many landscapes have a mixed fire regime that comprises both surface and

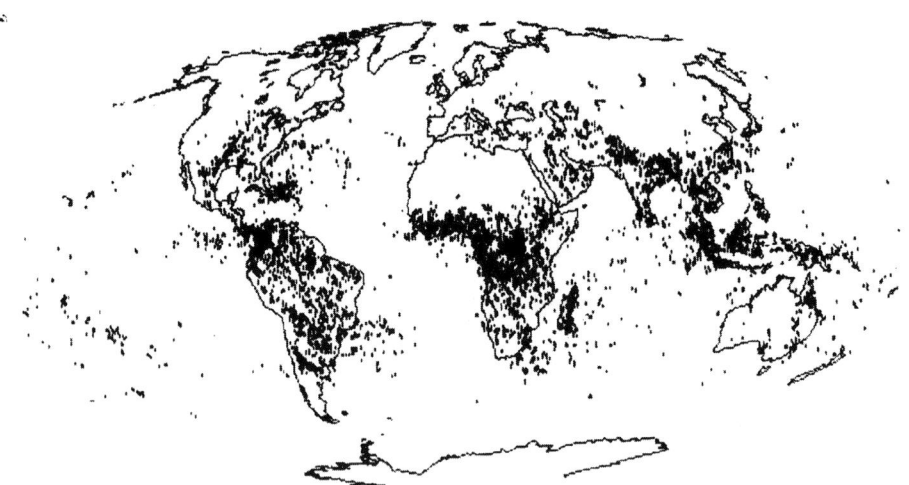

Fig. 2.5 *Global distribution of lightning during the period March to May 1987. (From Goodman & Christian 1993.)*

crown fires as well as unburned patches. Stand averages of fire frequency obscure some of the important ecosystem processes inherent in heterogeneous burning patterns at fine scales. A further complication is that fire frequency may change over time due to changes in climate and anthropogenic increases in ignitions.

All four parameters of the fire diamond (Fig. 2.1) potentially constrain fire frequency. However, given sufficient available fuel, ignitions will be a major determinant. Lightning is the most abundant natural source of fires and its frequency exhibits striking global differences (Fig. 2.5). With the noticeable exception of central Chile (see Figure 10.4), all MTC regions experience frequent lightning strikes, although on a global basis none of the MTC regions would be considered lightning hot spots and of those that do occur in MTC regions, a significant proportion occur during winter storms and seldom contribute to fires.

The relative impact of human ignitions is largely a function of the background lightning ignition frequency and the impact is geographically quite variable. In many mountainous environments the natural lightning strike density is high and human population density is low, and thus humans have had minimal impact on fire frequency. MTC regions have been particularly vulnerable to human ignitions due to the high population centers in coastal portions of all five regions and relatively moderate lightning ignition frequency in these lowland areas (Fig. 2.6).

In terms of understanding human impacts on fire regimes, the length of human occupation, and thus period of time over which humans have altered fire frequency may be an important factor. For example, the Mediterranean Basin has potentially had fire regimes altered by human ignitions for more than 50 000 yrs

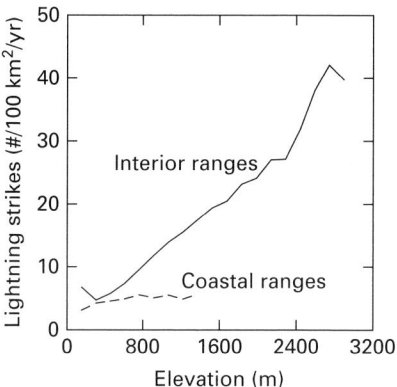

Fig. 2.6 *Elevational distribution of lightning strike density for the interior and coastal ranges of southern California. (From Keeley 2006a.) Similar elevational patterns of lightning-ignited fires are reported for other MTC regions (e.g. Horne 1981; Vázquez & Moreno 1998; Mazarakis et al. 2008).*

(Ne'eman *et al.* 2004), whereas that impact has been about one tenth as long in California. Prior to human occupation in California, ignitions in coastal regions were potentially less frequent and fires were dependent on foehn or other offshore winds (see Box 1.3) spreading lightning-ignited fires from the interior to the coast. On many landscapes humans have altered fuel structure in ways that also might impact lightning-ignited fires, contributing to new patterns of distribution (e.g. Arienti *et al.* 2009). It has been suggested that the local probability of ignition by lightning is reflected in the evolutionary development of vegetation responses across broad landscapes (Manry & Knight 1986).

Estimating past frequency on a landscape is an important means of assessing ecosystem tolerances to fire return intervals. Techniques for measuring fire frequency vary in their strengths and weaknesses. The method applicable to most landscapes is estimating fire frequency from observational records of past fire events. Although such records may extend back to the nineteenth century or slightly earlier (Keeley & Zedler 2009), systematic records of comparative value generally began sometime in the twentieth century, with timing varying in different MTC regions. In California the U.S. Forest Service (USFS) has one of the longest periods of documented records, with most forests established sometime between 1906 and 1911 (Cermak 2005). Although it is a widely held view by many reputable scientists that the record keeping on these forests in the early years was unreliable, there is no documented evidence for this opinion. Early twentieth-century records likely lacked a full accounting of wilderness fires, most of which were lightning-ignited fires, because these events typically occurred in isolated areas that were not readily accessed and often burned out before detection. However, a study of early USFS records suggests there was a very careful reporting of fires on those landscapes where rangers had access by foot or horseback.

In other MTC regions annual fire records generally do not extend back more than a few decades; however, other historical records may be useful. For example, a 130-yr fire history was compiled for the Valencia region of Spain based on an extensive review of old forest administration dossiers and newspapers (Pausas & Fernández-Muñoz 2011). Given the high population density of the area during the period 1873–2006, and the importance of fires in the media, the compiled fire history is considered a reliable estimate of area burned, but perhaps less reliable in terms of the number of fires.

A widely applied technique for estimating fire histories in North American coniferous forests is fire scar dendrochronology (Dieterich & Swetnam 1984). It is primarily applicable in surface fire regimes where trees are scarred but survive low intensity understory fires. Like written records it can provide annual and even seasonal resolution of past fires but has the added advantage of recording fire histories over a span of hundreds or sometimes thousands of years (Swetnam 1993). The primary limitation of this methodology is that it provides a point record of some but not all past fires and requires some untested assumptions in drawing conclusions about spatial patterns of past fires (Baker & Ehle 2001; Falk & Swetnam 2003; Veblen 2003; Hessl et al. 2004; van Horne & Fúle 2006). It also is unable to capture the past history on landscapes previously subjected to logging, which makes highly managed forests, such as those in the Mediterranean Basin or portions of the western USA, unsuitable for characterizing fire regime. The primary limitation to this technique in the study of MTC regions is that it requires forests with a history of surface fire regimes and thus it has not received wide application in most MTC regions.

Fire scar dendrochronology has been used in some Western Australia forests, although sometimes utilizing different assumptions in the interpretation of fire scars (e.g. Burrows et al. 1995). These investigators interpreted an increase in fire scars following European colonization to be the result of less frequent fires. They reasoned that prior to colonization high rates of burning by Aboriginal people kept fuel loads so low that they rarely scarred trees. However, tests of this interpretation of fire scar data failed to support it (Richards 2000) and other studies have found that fire scars are a consistent indicator of fire frequency in Western Australian *Callitris* trees (O'Donnell et al. 2010). More widely used on these landscapes are fire scars left in the leaf base patterns in grasstrees (*Xanthorrhoea* species). Although lacking the precision and reliability of tree ring fire scar records, they are apparently useful in drawing broad generalizations about past fires (Miller et al. 2007).

Charcoal and pollen deposits can provide fire frequency estimates covering the past 10 000 yrs or longer, but typically at temporal resolutions of decades to centuries (Clark & Robinson 1993; Millspaugh et al. 2004). Fire frequency estimates based on charcoal deposition are affected by wind patterns that affect dispersion of particles, which in turn are affected by particle size, which in turn is a function of fuel type, as well as sediment movement. Charcoal abundances in sediment cores may be functions of both fire frequency and severity, with concentrated charcoal layers (or charcoal peaks) in the time series reflecting either individual high severity events, frequent fire periods, concentrated erosion periods, or all of these processes in combination. These studies have shown vegetation changes in concert with changes in climate and fire (Whitlock *et al.* 2004; Power *et al.* 2008; Marlon *et al.* 2009). Although these studies have sampled over a broad regional and global scale, they are biased toward reflecting fire regimes on landscapes with Holocene lake deposits. On some landscapes these are primarily high-elevation or high-latitude areas that are characteristic of conifer forests with high-frequency surface fire regimes. One of the conclusions drawn from these studies is that Holocene climate was the only important driver of fire regimes and humans had little impact. In lightning-saturated environments this was probably true. However, many of the landscapes not represented in the charcoal record are non-forested environments where lightning ignitions are limiting and Holocene population densities were high, as were human sources of ignition (Keeley 2002b). Recent charcoal studies in the Mediterranean Basin have detected clear signs of both climate and humans affecting Holocene fire frequency (Gil-Romera *et al.* 2010).

Fire Patch Size and Distribution

Fire size varies over many orders of magnitude, from a lightning-ignited fire that remains localized around the tree it strikes, to massive crown fires that burn hundreds of thousands of hectares. On some landscapes a small percentage (5% or less) of fires account for 95% of the area burned (Strauss *et al.* 1989). This means that it is primarily the very large fires in the tail of the size distribution that determine the age distribution and spatial age mosaic of the landscape.

Distributions of fire size vary regionally and between surface fire and crown fire regimes (Fig. 2.2). Within fire perimeters the size of different fire severity patches may vary greatly, creating a mosaic of patches (Fig. 2.3). Many forests exhibit complicated patterns of fuel consumption, comprising a mixture of surface fire, crown fire and unburned patches. This heterogeneity is important to ecosystem processes such as tree recruitment, which often requires gaps in the forest canopy coupled with patches of surviving parent seed trees (see Chapter 3). Fire-induced gaps in the forest canopy provide high light environments for successful seedling establishment, but equally important they also accumulate surface fuels at a slower rate, and thus contribute to patchiness of burning in subsequent fires, providing safe sites for saplings until they reach sufficient size to survive fires (Keeley & Stephenson 2000).

MTC shrublands commonly experience large crown fires that cover vast areas and often in a rather coarse-grained pattern of uniform high severity. This poses little threat to most plant species in these systems because regeneration mostly depends on dormant seedbanks and resprouting from basal lignotubers (Box 2.1).

This generalization, however, may not apply to fauna that must recolonize burned sites and whose long-term persistence is controlled by metapopulation dynamics. On highly contemporary fragmented landscapes, metapopulation dynamics may be compromised and pose severe threats to postfire recovery of some animal populations (Walter 1977; Main 1981; see also Chapter 8).

MODIS (Moderate Resolution Imaging Spectroradiometer) is a key instrument aboard the NASA Terra (EOS AM) and Aqua (EOS PM) satellites that has found wide use globally in the mapping of wildland fires (Giglio et al. 2006; Roy et al. 2008), among many other applications. Terra's orbit around Earth is timed so that it passes from north to south across the equator in the morning, while Aqua passes south to north over the equator in the afternoon. MODIS image data are acquired in 36 co-registered spectral bands at moderate spatial resolutions (250, 500, and 1000 meters). Thermal information at 1000-m spatial resolution is collected twice daily by each sensor (one daytime and one nighttime observation) in the mid to high latitudes.

The U.S. Forest Service and the international fire monitoring community use MODIS imagery to provide a near real-time geospatial overview of current wildland fire occurrence at regional, national and international scales. These fire data are integrated with various sources of contextual spatial data and information in a suite of geospatial data and mapping products and utilized by fire managers to assess the current fire situation (http://activefiremaps.fs.fed.us/). These data additionally serve as a decision support tool in strategic decisions relevant to resource allocation in relation to fire-suppression activities.

Fire Seasonality

As illustrated in Fig. 2.1, seasonality is a necessity for converting potential fuels into available fuels. Rarely are fire-prone ecosystems vulnerable to burning year round, although some grassland ecosystems may qualify. Most ecosystems have a particular season when fires are most likely and it is highly variable. For example, the season of greatest area burned in Mediterranean Basin ecosystems is in the summer, whereas in southern California it is in the autumn. Fires in non-MTC ecosystems may peak in other seasons, for example in the winter or spring in strongly monsoonal climates. Fire seasonality is a function of the coincidence of ignitions with condition of the fuels. Generally fire seasons center around the driest time of the year and it is noteworthy that the peak numbers of ignitions do not always coincide with peak area burned. In the southwestern USA generally the largest fires are during the driest time of the year, which is late spring and early summer, whereas the peak numbers of ignitions (mostly from lightning) are in summer during the monsoon season (late June to early July). Even in regions where humans dominate the fire ignitions such as in southern California, the highest density of fires occur in June whereas the greatest area burned usually occurs in October or November (Keeley & Fotheringham 2003a).

Seasonality in some fire-prone landscapes is not annual but rather arises from decadal-scale droughts that create high fire hazard in some years but not in others. These aseasonal climates have on average significant rainfall each season, but in anomalous years there may be an extended drought conducive to widespread fires (Vines 1974). Although lacking a predictable annual drought as is present in MTC regions, these aseasonal climates have a predictable drought associated with El Niño/Southern Oscillation (ENSO) events that occur once or twice a decade and create a major fire risk (Verdon et al. 2004).

One of the potentially important impacts of humans is to provide ignition sources during seasons when lightning ignitions are unlikely. The impact of out-of-season fires is only beginning to be studied but is potentially very important. In western coniferous forests it affects the types of fuels consumed (Knapp et al. 2005). Out-of-season winter burning has potential safety advantages for controlling fires in these highly hazardous fuels (Keeley 2002a). However, there are noteworthy examples of unexpectedly poor regeneration following such out-of-season burns (Keeley 2006b). Seasonality effects on postfire recovery are evident in most MTC ecosystems (Papanastasis 1980; Bond 1984; Hobbs & Atkins 1990). In MTC regions one impact of out-of-season burning is that it cuts short the postfire growing season; for example, natural summer or autumn fires are followed by an approximately six month growing season for resprout growth and seedling recruitment, whereas winter burns may cut this growing season in half. Out-of-season spring fires also may have negative consequences on the fauna, due to disruptions of nesting season, an impact not experienced during normal summer and autumn fires.

Emergent Properties of Fire Regimes

Fire regimes are not just the sum total of the five factors described above, but an emergent property of these and other biotic and abiotic ecosystem attributes (Gill & Bradstock 2003). As a result, each site has its own unique fire regime; however, broadly speaking there are three categories worth recognizing for the purposes of contrasting ecosystems: *surface* fire, *crown* fire and *mixed* (surface and crown) fire regimes. Although there are multiple parameters that define a fire regime, two factors that play key roles are site productivity and disturbance frequency (Keeley & Zedler 1998; Pausas & Bradstock 2007). These two factors play key roles in circumscribing fire regimes (Fig. 2.7) and lead to distinct patterns of fuel consumption and fire spread. For example surface fire regimes are usually constrained to generating low-intensity fires whereas crown fire regimes, due to the higher fuel loads being consumed, generate much higher intensity fires. These in turn both affect and are affected by fire frequency, which in turn may be constrained by seasonality. Generally surface fires are smaller and patchier than

Fire and the Fire Regime Framework

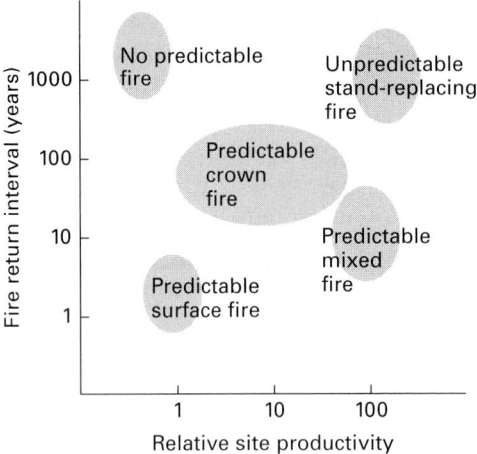

Fig. 2.7 *Fire regimes generated by patterns of productivity and disturbance frequency. (From Keeley & Zedler 1998.)*

crown fires, in part because fire spread in crown fires is often facilitated by, and sometimes dependent upon, high winds capable of overcoming potential barriers to fire spread such as patches of reduced fuel loads, or physical barriers such as streams, rock outcrops and fuel breaks.

Fire regimes may be altered over time as climates change; however, human impacts are presently one of the major drivers of fire regime change. MTC regions in particular may be characterized as having experienced substantial shifts in fire regime due to human occupation (Keeley & Zedler 2009; Pausas & Fernández-Muñoz 2011). Anthropogenic fire regimes are often thought of as disruptions in natural fire regimes and exhibit two extremes. On the one hand humans are responsible for greatly increasing fire frequency on landscapes where lightning-ignited fires are infrequent, and on the other hand humans suppress fires and greatly diminish the historical or *natural* fire frequency (Keeley *et al.* 2009a). The relative importance of these two impacts varies markedly with the fire regime. For example, in conifer forests with low-intensity surface fires, fire suppression has been highly effective and consequently has accomplished near total fire exclusion for a century or more. However, in crown fire shrublands, high-intensity fires driven by high winds have proven very difficult to suppress and thus fire exclusion has happened only on a very limited scale. Indeed, on these landscapes, since they are often juxtaposed with urban environments, they have experienced an increase, not decrease, in fire frequency (see Fig. 5.10).

Physical scientists interested in finding common underlying patterns in complex systems have turned their attention to biological systems including ecosystem fire activity. Across a wide range of fire regimes in North America and Australia, wildfires exhibit a power law distribution of frequency vs. area (Malamud *et al.* 1998). Such patterns suggest to some that there is a common underlying mechanism that explains fire activity. Malamud *et al.* (1998; Malamud & Turcotte 1999) showed that a forest model driven solely by rate of tree growth, which was considered equivalent to fuel accumulation, likewise fit a power model for fire size and frequency. Since their model was driven solely by internal ecosystem processes, a phenomenon known as self-organized behavior to illustrate the lack of exogenous drivers, they concluded that fuel accumulation alone was a widespread driver of fire regimes, and referred to this as self-organized criticality.

However, a great many biological data sets with a large variance will fit a power law relationship. We contend that very different fire regimes show a similar power law relationship between number of fires and size of fires, and the mechanisms are likely to be very different. For example, in forests with surface fire regimes, where fuels are often limiting to fire spread, fire frequency may alter fuels in ways that impact potential fire size. But in shrubland crown fire regimes where external factors such as extreme weather are a more important determinant of fire spread, fire frequency may affect fire size through the probability of ignitions during severe fire weather.

The self-organized behavior of fire regimes was questioned by Boer *et al.* (2009) who found that fire behavior could be accounted for by external factors such as the length and intensity of weather events, which also show a power law distribution. However, Carlson & Doyle (2002) proposed that complex systems can be best modeled when viewed as driven by both endogenous and exogenous factors. They developed the highly optimized tolerance (HOT) model, aimed at optimizing the trade-offs between system yield and tolerance to risks. Moritz *et al.* (2005) used this model with southern California fire history data and found a far better fit than with the self-organized criticality model.

Fire Behavior

Fuels, weather and topography are the major determinants of fire behavior. Under severe fire weather conditions of high wind and low humidity fuels become less of a determinant of fire behavior (Cary *et al.* 2006; Bradstock *et al.* 2010) and steep rugged terrain may have similar effects. However, under other conditions, fuels control fire behavior by a combination of inherent structural characteristics, moisture content, live or dead status, and their quantity and arrangement at a local and landscape scale.

Fuel Attributes

Characterizing fuels in MTC regions in meaningful ways that reflect different fire regimes is complicated. Although primary productivity (Fig. 2.1) is a key determinant of fuel loads, not all biomass represents available fuel. In savannas most of the aboveground herbaceous biomass may be available fuel whereas in some closed-canopy forests fuels include only the sloughed-off leaves and stems that

Table 2.1 Fuel characteristics of selected vegetation types
In crown fire ecosystems not all of the aboveground biomass is consumed in the fire, rather a subset of the total comprises the available fuel and is typically the small diameter stems and dead biomass, and the proportion varies with fire weather. In these crown fire regimes surface fuels are not heavily involved in fire spread. In contrast in closed-canopy forests with surface fire regimes the primary driver of fire spread is the dead surface fuels, and unless fire behavior switches to a crown fire, much of the aboveground biomass is not burned, although a proportion of understory herbs, shrubs and saplings may also be available fuels.

Vegetation	Total aboveground biomass (Mg ha^{-1})	Dead surface fuels (Mg ha^{-1})	Source[a]
Mediterranean Basin			
Phrygana	6–10		1
Quercus coccifera garrigue			
3 yr postfire	7		2
unburned >30 yrs	28–35		1,3
Mixed maquis	10–79		1,4
Ulex parviflorus shrublands			
3 yr postfire	13		5
9 yr postfire	36		5
17 yr postfire	59		5
Ulex shrublands	28–45		6
Erica shrublands	13–110		7,8
Perennial grassland			
unburned 6 yr	18		9
Quercus ilex woodlands	100–165		10
California			
Sage scrub	15–19	2–8	11
Baccharis scrub	25–40		12
Adenostoma chaparral			
2–6 yr	6–8		
> 30 yr	14–22		13
Mixed chaparral	28–49		14
Arctostaphylos chaparral	72		14
Ceanothus megacarpus chaparral			
5 yr	21		15
21–22 yr	49–76		15,16
Ceanothus oliganthus chaparral			
6 yr	8–24		17
21 yr	43–115		17
Annual grassland			
Sandstone	3–6		18
Pinus ponderosa forest			
10 yr		160	19
unburned > 100 yr		191	19
Abies concolor forest			
1 yr		20–60	20
10 yr		160	19
unburned > 100 yr		163–250	19
unburned > 100 yr		180–214	20

Table 2.1 (cont.)

Vegetation	Total aboveground biomass (Mg ha^{-1})	Dead surface fuels (Mg ha^{-1})	Source[a]
Chile			
Matorral	9–17		21
South Africa			
Fynbos			
4 yr	6–7		22
unburned >35 yr	43–76		22
unburned	10–34		23
unburned	6–35		24
Renosterveld	14		24
Australia			
Heathland			
5 yr	9–10		25
15 yr	18–19		25
Banksia heath	14	4	26
Banksia woodland			
4 yr	5	2	27
12–22 yr	7	2	27
Jarrah (*Eucalyptus*) forest			
9–18 yr	22	28	
Various *Eucalyptus* forest types		10–70	29
Eucalyptus forest	228	37 (+ bark)	30
Tropical grasslands	9–11		31

[a] (1) Dimitrakopoulos (2002); (2) Malanson & Trabaud (1988); (3) Trabaud (1991); (4) Sağlam *et al.* (2008); (5) Baeza *et al.* (2006); (6) de Luís *et al.* (2004); (7) Viegas *et al.* (2002); (8) Moreno *et al.* (2004); (9) Caturla *et al.* (2000); (10) Ibáñez *et al.* (1999); (11) Westman *et al.* (1981); (12) Russell & Tompkins (2005); (13) Rundel & Parsons (1979); (14) Riggan & Dunn (1982); (15) Schlesinger *et al.* (1982); (16) Gray (1982); (17) Riggan *et al.* (1988); (18) McNaughton (1968); (19) Keifer *et al.* (2006); (20) Knapp *et al.* (2005); (21) Mooney *et al.* (1977); (22) van Wilgen (1982); (23) van Wilgen *et al.* (1985); (24) Stock & Allsopp (1992); (25) Bell *et al.* (1984); (26) Low & Lamont (1990); (27) Burrows & McCaw (1990); (28) Smith *et al.* (2004); (29) Good (1996); (30) Walker (1981); (31) Long *et al.* (1989).

accumulate on the soil surface. In crown fire shrublands, dead biomass retained in the canopy is a readily available fuel but the availability of living foliage varies seasonally and with stand age and ratio of dead to live fuels in the canopy.

In MTC regions, fuels (Table 2.1) are lowest in grasslands and other herbaceous-dominated vegetation, with the lowest fuel loads typically around 5 Mg ha^{-1} (= metric tons ha^{-1}). Lower-stature shrublands such as Mediterranean phrygana, California sage scrub, South African fynbos and Australian heathlands, are several times higher than these grasslands, and dense shrublands such as garrigue, chaparral and matorral weigh in between

20–100 Mg ha^{-1}. On more oligotrophic soils typical of Gondwana landscapes of the Western Cape of South Africa or Western Australia, shrubland fuel loads are markedly lower. South African fynbos typically burns with return intervals around 15 yrs and at this frequency may have fuel loads around 10–20 Mg ha^{-1}, but in rare instances when left unburned for much longer periods may approach fuel loads more typical of California chaparral (van Wilgen *et al.* 2010).

These biomass levels don't tell the whole story as not all aboveground stems are equally vulnerable to igniting and to spreading fire. Typically, the available fuels are considered to be dead stems and live stems < 6 mm in diameter plus foliage. However, severe fire weather can increase the range of stem sizes that are available fuels. Since a greater proportion of early seral stages of shrub succession comprises smaller diameter stems, flammability may be relatively high (see Fig. 5.3). This contradicts the dogma that fire hazard increases with stand age, which seems like an overly simplistic view (e.g. Paysen & Cohen 1990; M.J. Baeza *et al.* 2011).

Many factors affect the extent to which fuels contribute to fire behavior. Live fuels and dead fuels have very different characteristics. Climate plays a key role by affecting fuel moisture, and this is particularly important in determining flammability of live fuels (see next subsection). Dead fuels are key because they ignite most readily and in shrublands the ratio of dead to live fuels affects the extent to which live fuels will ignite and this ratio varies by species and age. The size of dead fuels is also very critical because smaller diameter stems will ignite faster than larger diameter stems, due in part to the fact that they dry faster and heat penetrates faster. Thus, vegetation types such as heathland on poor soils that generate very fine fuels may be much more fire-prone than coarser vegetation of more fertile soils such as chaparral (Ojeda *et al.* 2010). Even within a plant community there are marked structural differences that make some species much more fire-prone than others, for example retention of dead branches as opposed to shedding them (see Chapter 9). These generalizations apply to crown fire regimes and very different patterns are to be expected in vegetation types where fires are driven by surface fuels (e.g. Scarff and Westoby 2006).

Total aboveground biomass in most forests is substantially greater than in shrublands but for many forests with a history of surface fire regimes, it is the dead surface fuels that are considered available fuels, although large logs may not be included. In such forests the surface fuels can range from 10 Mg ha^{-1} to more than 200 Mg ha^{-1} (Table 2.1), dependent on site productivity and time since fire. In some forests the fuel load also includes live and dead standing understory shrubs and tree saplings that can more than double the fuel load (Keifer 1998; Sackett & Haase 1998), but are not always included in fuel inventories.

Structural characteristics also may affect surface fuel arrangement as in the greater packing density of small fir needles, compared with the larger fascicles of pine needles that do not compact as densely on the forest floor. This packing density affects airflow and heat propagation in ways that can have diverse effects on fire behavior. Alternatively, widely spaced canopy branches may inhibit fire spread relative to more densely packed canopies; for example, the dense canopies of many *Pinus* make them more susceptible to high-intensity burning compared with the open spreading canopies of many *Eucalyptus* species.

One of the key factors affecting both ignition and combustion is canopy porosity, measured as the canopy volume/leaf-and-stem volume (Rundel *et al.* 1980). High canopy porosity allows fine fuels to react more rapidly to changes in relative humidity and also increases oxygen flow around the fuel. Although high canopy porosity increases flammability, it leads to lower bulk density (mass/volume) and fuel loading (mass/area), reducing the total energy available for combustion. For example, the high canopy porosity of the needle-leaved chaparral shrub *Adenostoma fasciculatum* increases flammability under a wide range of conditions, whereas the lower canopy porosity, but higher bulk density, of the associated shrub *Quercus berberidifolia* limits the range of conditions suitable for burning, although under the severest conditions *Quercus* fuels should be expected to generate higher fire intensities. Chemical composition of fuels may likewise affect flammability by volatilizing at lower temperatures, and combustion of these gases contributes to heating of structural fuels (Pyne *et al.* 1996). However, the chemical make-up of foliage, and the potential heat energy content, exhibits far less variation between species than structural characteristics (e.g. Dimitrakopoulos & Panov 2001), and the latter may contribute more to different patterns of fire behavior (Dimitrakopoulos 2001).

As elaborated in Chapter 9, some species possess structural characteristics that make them more flammable than others and this may have adaptive significance (Bond & Midgely 1995). For example, retention of dead branches contributes to spreading fire into canopies of some trees and shrubs. When this trait is coupled with fire-dependent regeneration it may be viewed as an adaptive trait, as in serotinous pines (Keeley & Zedler 1998), and when supported by phylogenetic analysis it may be interpreted as an adaptation selected for by fire (Schwilk & Ackerly 2001); see Chapter 9.

Antecedent Climate

Climate in the months or even years prior to a fire affects fire behavior by (1) altering fuel moisture and (2) altering fuel loads. Fuel moisture is particularly critical because it has a major effect on combustion and is often a major determinant of fire spread and fuel consumption (e.g. Bilgili & Saglam 2003; Anderson & Anderson 2010). Factors affecting fuel moisture vary between dead and live fuels. Dead fuel moisture is largely affected by weather and fuel thickness, with smaller fuels requiring less time under low humidity and high temperatures to dry sufficiently for combustion. Live fuel moisture is affected not just by climate but by access to soil moisture, which is determined by rooting depth and water use characteristics. Deeply rooted plants are likely to have greater access to soil moisture far longer into the dry season than shallow-rooted plants and this can affect flammability (Green 1981). Fuel moisture can also vary with patterns of stomatal control and water use (e.g. Jacobsen *et al.* 2008). Even on the same sites there may be substantial variation in fuel moisture of particular species (Weise *et al.* 1998). Since live stems can withstand lower moisture levels than leaves, overall plant fuel moisture will be affected by structural characteristics such as the leaf to stem ratio.

In MTC regions such as the eastern Mediterranean Basin or California where the summer droughts are particularly severe (see Box 1.1), the live fuel moisture of many shrubs in most years is at the lowest level of tolerance ($\sim 60\%$) during late summer and autumn. However, in certain years late spring rains can reduce the length of the fire season through effects on live fuel moisture in summer and autumn (Dennison *et al.* 2008). In the MTC Cape region of South Africa and in southwestern Australia, the summer drought on average is not as dry as in California and thus live fuel moisture levels are generally higher (van Wilgen *et al.* 1990b).

In addition to affecting fuel moisture, antecedent climate plays a key role in producing short-term changes in fuel loads. For example, in open forests with surface fire regimes driven by herbaceous understory, high-rainfall years will increase these fuels and lead to increased burning in subsequent years (Westerling *et al.* 2002; Pausas 2004). The potential for short-term changes in fuels is largely a function of fuel structure. For example, in California conifer forests the lower-elevation ponderosa pine community typically has fires driven by herbaceous fuels whereas the higher-elevation and more closed-canopy mixed conifer forest has surface fires that feed on dead leaves and branches and other forest floor litter. In the ponderosa forest historical fire scar records show significantly elevated fire activity 1 to 2 yrs following high-precipitation years (Fig. 2.8b) (Swetnam & Betancourt 1998). However, the mixed conifer forests lack herbaceous surface fuels and do not exhibit such a lag effect (Fig. 2.8a). Similar patterns have been observed in Australian semi-arid landscapes (Bradstock & Cohn 2002a).

In shrubland crown fire regimes, drought can alter dead fuel loads and this plays a key role because the dead fuels dry rapidly and combust readily, whereas live fuels absorb heat and tend to suppress fire. Thus, fire behavior is markedly affected by the ratio of dead/live fuel. Although it is widely accepted that this ratio increases with stand age (Baeza *et al.* 2002; Regelbrugge & Conard 2002), less well appreciated is the extent of short-term changes in dead fuels due to drought. In assessing the contribution of dead fuels on fire behavior this is generally considered to be captured by stand age, which is how it is commonly factored into fire danger indices. However, this has limitations in shrublands because the rate of dead fuel accumulation is often not a linear response to age (Keeley & Fotheringham 2003a; Baeza *et al.* 2011) and can experience short-term changes due to drought-caused dieback (Keeley & Zedler 2009). Altering fire danger indices by including fuel inventories (Woodall *et al.* 2005) are an improvement, but to truly capture periodic drought-induced changes in dead fuels we ultimately will need to develop remote imagery that can capture the dead to live signal in shrublands. There are

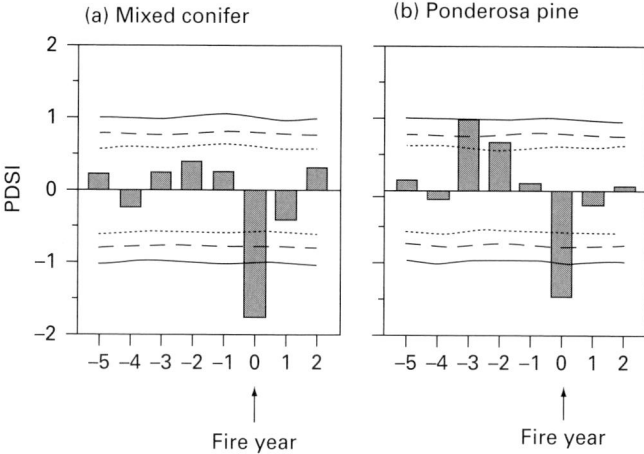

Fig. 2.8 *Presentation known as the superposed epoch analysis of fire activity and Palmer Drought Severity Index (PDSI); bars above zero indicate a wet year and below zero a drought year, and the horizontal solid, dashed and dotted lines indicate 95%, 99% and 99.9% confidence levels, for (a) mixed conifer forests and (b) ponderosa pine forests in the MTC of California. In both forest types years of low precipitation are years of high fire activity. However, only in ponderosa pine forests is fire activity correlated with high precipitation in prior years, which is due to the increased herbaceous fuels produced by high rainfall. This effect of antecedent rainfall is not evident in mixed conifer forests because they largely lack herbaceous surface fuels. (From Swetnam & Betancourt 1998.)*

some promising leads on imagery for capturing short-term changes in hazardous fuels and related drought effects (Ustin *et al.* 2009; García *et al.* 2010).

Drought is the most commonly observed climatic effect on fire activity. Severe droughts cause fuels to dry earlier in the year and result in much lower fuel moisture for both live and dead fuels and are routinely associated with years of high fire activity (Fig. 2.8), regardless of the type of fuels. However, drought can also stress plants and increase dead fuel volumes on landscapes. For example, for most of the first decade of the twenty-first century in southern California there was an anomalously severe drought, which resulted in substantial dieback of vegetation in some years. It has been hypothesized that this increased dead fuel load was the major factor contributing to five of the nine historical megafires (>50 000 ha) occurring during this period (Keeley & Zedler 2009). Most of these megafires occurred during the autumn when live fuel moisture is normally at its lowest level of tolerance; thus, live fuel moisture cannot explain why these fires were larger than those in an average year. However, increased dead fuel loads not only directly increase the rate of fire spread but also can contribute to increased incidence of spread through long-distance spot fires that increase when dead fuels increase. Since dead fuels decompose slowly in these arid environments they leave a legacy that may affect fire behavior even when followed by normal rainfall years. A similar pattern of intense droughts and large fires was also observed in southern Australia during this same period.

Fire Weather

Most large fires owe their origin to severe fire weather that includes high temperatures, low humidity and wind. These conditions contribute to drying of live fuels, thus reducing the heating required to drive off moisture and effect combustion. Wind accelerates fire spread rate by both increasing oxygen supply and carrying heated air to adjacent fuels on the downwind side, thus raising the fuel temperature and driving off water vapor. Most MTC regions have extraordinarily severe wind events during either summer or autumn (see Box 1.3), which greatly exacerbates the fire danger.

Wind is particularly critical because it is capable of carrying firebrands, which often occur when gusts are greater than 16 km hr^{-1} (Green 1981). Such firebrands may be carried more than a kilometer from the fire front and ignite new fires. Firebrands are lifted by the convective buoyancy of the flaming front and thus fast-moving crown fires are particularly likely to create spot fires far ahead of the front (Pyne *et al.* 1996). The extent to which firebrands effectively spread fire is a function of the fuel type and relative humidity. Fine fuels such as grass do not produce firebrands as effectively as woody fuels. Spot fires are a function of not just the proper fuels for creating flying embers but the state of fuels where they land. Landing in dead fuels is generally required for ignition of spot fires and such ignitions are greatly affected by relative humidity.

In the absence of substantial winds, flames are nearly vertical and fuels ahead of the fire front are not preheated by hot gases, making such plume-driven fires far more dependent on fuel structure (Morvan & Dupuy 2004). Other factors that affect plume-driven fires are the fact that such plumes may rise to more than 10 000 m (see Fig. 1.2) and contribute to long-distance firebrand transport and spot fires. In addition, on some landscapes they can create hazardous internal winds. This commonly occurs when substantial fuel loads contribute to a massive plume that rises faster than the ambient winds and increases airflow into the fire. Also, this plume may collapse as the fire moves into lower-volume fuels, causing a downward rushing of air in every direction and spreading fire rapidly outward (Mutch 2003). A similar phenomenon may be driven by rapid cooling as convectional air masses form ice at higher altitudes, generating microbursts of wind blowing outward at ground level (Pyne *et al.* 1996).

Topography

Steep terrain may act to spread fire much like winds do because the incline increases the extent to which the flaming front parallels the ground and heats fuels ahead of the front. Indeed, fire spread roughly doubles for each 13-degree rise in slope (Green 1981). In addition, topographic features may cause unstable changes in velocity and direction as winds adapt to the topography. On coastal-facing slopes onshore winds may be channeled up-canyon and produce eddies at ridgelines that then become turbulent and erratic. These erratic winds converge at ridgetops and can contribute to long-distance spotting (Pyne *et al.* 1996).

Barriers of reduced fuel loading, which could include rocks, rivers, alluvial fans and other terrain features, may inhibit fire spread. This topographic effect varies through the season; as drought progresses, different portions of the landscape are added as available fuels, contributing to a temporal component of landscape fuel heterogeneity. Steep terrain also enhances long-distance spread of firebrands.

Fire Behavior and Landscape Models

Models have played a fundamental role in fire science and fire management. Of widest use are fire behavior models that predict fire activity based on equations that relate fire spread to surface fuel characteristics and environmental factors (Rothermel 1972). Many models have been developed but two of the more widely used are BEHAVE (Andrews 1986) and FARSITE (Finney 1998). The former predicts fire characteristics associated with different categories of fuel and environmental data and the latter gives spatially explicit predictions of fire spread. One of the major limitations to modeling fire behavior is the lack of detailed fuel bed data for most landscapes. As a consequence there has been extensive study of fuel characteristics for a wide range of vegetation types and these have been formulated into fuel models that can be used in fire spread models (Scott & Burgan 2005). In recent years there has been a proliferation of fire behavior models designed to overcome certain limitations in previous models (see review in Peterson *et al.* 2007).

One of the primary limitations of fire behavior models is that they often are based on physical models of fire spread in dry surface fuels and are not easily adapted to dealing with crown fires in live and dead shrublands. Alternative approaches to this "first principles" approach are probabilistic models of fire behavior based on empirically determined relationships between shrubland fuels, environmental parameters and observed fire characteristics (e.g. Bilgili & Saglam 2003).

Fire spread models have found a wide range of uses. Managers often use these to predict fire spread during fire events in order to more effectively deploy fire suppression crews. Both managers and researchers use these models to evaluate fuel management impacts on subsequent fire behavior (e.g. van Wagtendonk 1996; Graham *et al.* 1999).

Percolation models have proved useful in understanding fire behavior and impacts on vegetation in crown fire ecosystems by testing the influence of fuels and weather on fire spread between adjacent cells (Turner & Romme 1994). Such models have been applied to questions of prescription burning impacts (Bradstock *et al.* 1998a) or to test assumptions about the role of fuel age mosaics in controlling the size of wildfires (Zedler & Seiger 2000; Keeley & Zedler 2009).

Models such as FIRESCAPE, LANDSUM or FATELAND have been developed to investigate landscape patterns of burning under different fire regimes

(Keane *et al.* 2003; Pausas 2006; Pausas & Lloret 2007). These models simulate ecological succession scenarios based on different fire regime characteristics and allow scientists to examine the relative role of different factors. For example, Cary *et al.* (2006) found that for five very different forest types, and with different landscape models, predictions of fire activity were generally more sensitive to variations in climate and weather than to complexity of the terrain or fuels. Such landscape models are sometimes used to estimate fire sensitivity to predicted climate change scenarios (Cary & Banks 1999).

LANDIS is a forest succession model that predicts the impact of fire and other disturbances on successional changes (Mladenoff 2004). It has been modified for crown fire shrublands and used to predict the impact of different fire frequencies on plant life histories (Syphard *et al.* 2006). FATELAND (Pausas 2006) is a landscape model that also includes a fire regime and a vegetation dynamics simulator, and supports many types of vegetation. The vegetation dynamics of this model are based on key plant functional attributes related to disturbance (Noble & Slatyer 1980; Moore & Noble 1990), and thus it is very appropriate for predicting vegetation changes in MTV under different disturbance regimes (Pausas *et al.* 2006c; Pausas & Lloret 2007).

Mediterranean-type Climate Ecosystem Fire Regimes

Mediterranean-type climate shrublands and woodlands are largely dominated by crown fire regimes that have historically burned in high-intensity fires and these ecosystems are remarkably resilient to such conditions. MTC savannas and some forests are more commonly burned by low-intensity surface fires.

MTC vegetation is one of the most fire-prone ecosystems, perhaps only outdone by tropical C_4 grasslands with their capacity for high annual production that rapidly dries to become available fuels. In terms of woody-dominated vegetation, MTC regions differ from other ecosystems in that the climate imposes an annual fire risk, unlike other regions that may be dependent on infrequent decadal-scale climate anomalies to create high fire danger.

MTC ecosystems illustrate the critical importance of seasonality in creating highly fire-prone ecosystems. Sufficient primary productivity to fuel fires results from the relatively moderate winter and spring growing conditions. At these latitudes and with varying levels of marine influence, portions of these landscapes typically initiate the growing season with the onset of rains and growth is sustained in most years for 4–6 months. Thus, for part of the year these ecosystems sustain productivity long enough to produce rather dense stands of contiguous biomass of potential fuels. The predictable summer drought couples greatly diminished rainfall with high temperatures, which together convert a substantial fraction of these potential fuels into available fuels. The widespread importance of high winds in many MTC regions further exacerbates the fire potential (see Box 1.3).

MTC ecosystems all share the feature that anthropogenic ignitions are currently abundant and responsible for the vast majority of fires. Many factors, which will be discussed in later chapters, contribute to contemporary patterns of ignition, not the least of which is the high population density of most MTC regions, which arises from a climate ideally suited for human comfort. Lightning is the only natural source of ignition that is predictable enough in time to influence evolutionary patterns and its presence is spatially variable. Presently its importance varies between regions as well as within regions and will be discussed in subsequent chapters.

Conclusions

The fire regime concept is fundamental to understanding fire behavior and plant responses to fire. Depending on fuel structure, fires in MTV may be either crown fires that top kill most of the vegetation or surface fires that burn in the understory of forests. Overlying this distinction are countless variations in seasonal timing, intensity and frequency that are profoundly important in understanding fire responses. Failure to consider these distinctions has contributed to limitations in drawing global generalizations and unifying theories of wildland fire.

3 Fire-related Plant Traits

As illustrated in Fig. 2.1 there are four environmental parameters that are necessary to determine the distribution of fire-prone ecosystems. However, they are insufficient to predict ecosystem responses to fire without a detailed understanding of the fire regime (see Fig. 2.7). Different fire regimes have very different potentials for recovery and place very different premiums on specific plant traits. For example, those traits contributing to the persistence of species in crown fire regimes will often be very different from those in surface fire regimes. In short, organisms are not adapted to fire *per se*, but rather to a particular fire regime. Plant traits that are adaptive in fire-prone environments are discussed here. The evolution of such traits and the extent to which they represent adaptations to fire are considered in Chapter 9.

Plant populations exhibit four modes of recovery following fire:

(1) endogenous regeneration from resprouts or fire-triggered seedling recruitment,
(2) delayed seedling recruitment from postfire resprout seed production,
(3) delayed seedling recruitment from *in situ* surviving parent plants, or
(4) colonization from unburned metapopulations.

These regeneration modes apply to mediterranean-type climate (MTC) ecosystems as well as non-MTC ecosystems and generally sort into vegetation types with different fire regimes, and there is some degree of convergence in regeneration modes between plant communities with the same fire regime. Plants adapted to crown fire regimes typically recover from fire by a combination of endogenous regeneration from seedlings and resprouts as well as delayed seedling recruitment from resprout seed production. In contrast, in surface fire regimes the dominant tree species depend on survival of parent seed trees for seedling recruitment during early succession, and the understory species recover by combination of colonization and endogenous regeneration.

Endogenous Postfire Regeneration

Resprouting

Resprouting refers to the initiation of new shoots, usually from existing meristems in woody plants, following fire or other disturbances. This term is preferable to

Table 3.1 Widespread postfire obligate resprouting shrubs in mediterranean-type climate regions

These taxa are resilient to fire by vegetative resprouting but have not modified their reproductive cycle to delay reproduction to postfire conditions, rather seedlings recruit sporadically during the fire-free interval.

Mediterranean	California	Chile	South Africa	Australia
Arbutus	*Cercocarpus*	*Colliguaya*	*Diospyros*	
Daphne	*Comarostaphylos*	*Kageneckia*	*Heeria*	
Lonicera	*Heteromeles*	*Lithraea*	*Maytenus*	
Myrtus	*Prunus*	*Quillaja*	*Myrsine*	
Phillyrea	*Rhamnus*	*Schinus*	*Rhus*	
Pistacia	*Rhus*			
Quercus	*Quercus*			
Rhamnus	*Styrax*			
Ruscus	*Sambucus*			
Styrax	*Xylococcus*			
Viburnum				

sprouting, which refers to initiation of new shoots throughout the life cycle of a plant. Resprouting from vegetative structures that survive fire may occur on aboveground stems known as *epicormic resprouts*, from *basal resprouts* on basal burls or lignotubers (see below), bulbs, corms, rhizomes or roots. In some rhizomatous species, resprouting may be coupled with vegetative spread as in most species of *Rosa* and *Rubus* or in some species of *Eucalyptus* (Lacey 1974). Basal resprouting is a nearly universal trait in perennial dicotyledonous (dicot) plants (Wells 1969), although in broadleaf trees it is often restricted to the sapling stage (Del Tredici 2001; Bond & Midgley 2003).

Resprouting is perhaps a near universal recovery mechanism following top-kill from fire. However, it is also of adaptive value following other disturbances such as freezing, wind, drought and browsing and has led to the conclusion that even on fire-prone landscapes it is not an adaptation to fire *per se* (Wells 1969; Axelrod 1973; Mooney 1977b; Keeley 1981; Lloret *et al.* 1999a). However, in light of the evidence for fire as a potential selective agent throughout the evolutionary history of land plants, this conclusion is likely not true for all lineages with resprouting taxa (see Chapter 9).

On fire-prone landscapes many species regenerate after fire from resprouting as well as recruiting seedlings in a single postfire pulse of seed germination, and these are termed *facultative seeders*. *Obligate seeders* are woody species that lack resprouting capacity and are dependent entirely on postfire seedling recruitment. *Obligate resprouters* are species that are present in the first growing season after fire solely as vegetative resprouts with no seedling recruitment at that time. Such obligate resprouters are rare in some MTC ecosystems such as kwongan in Western Australia but very common in the Mediterranean Basin, Californian and Chilean shrublands (Table 3.1). Obligate resprouting is a highly conserved trait at the generic level.

There are many lineages where one could make the argument that postfire resprouting evolved as a response to top-kill from fire. However, an obvious case where resprouting is not likely an adaptation selected for by fire is in herbaceous perennials (hemicryptophytes; see Box 1.2). This growth form is most abundant in highly seasonal environments such as MTC regions, where the common cycle is summer dieback followed by winter or spring sprouting, and thus sprouting is a natural phenological stage that may occur annually regardless of fire. However, although most herbaceous perennials may resprout in the winter and spring without fire, they typically flower much more than usual following fire. In all MTC regions geophytes (cryptophytes) are the most common herbaceous perennial and typically comprise a conspicuous part of the postfire flora from resprouts (Le Maitre & Brown 1992; Rundel 1996; Parsons & Hopper 2003; Proches & Cowling 2004). In the absence of fire in these closed-canopy shrublands, geophytes typically become dormant for extended periods of time, and many of these respond to fire by coupling resprouting with profuse flowering that is often synchronized across the population in the first growing season after fire (Le Maitre & Brown 1992; Tyler & Borchert 2002; Borchert & Tyler 2009), suggesting some fire-adapted modification of the annual sprouting cycle. This growth form seldom maintains dormant seedbanks and thus it is normally an obligate resprouter (Keeley & Bond 1997; Keeley *et al.* 2006b). In some cases it appears that soil microclimate changes produced by removal of the shrub cover are the cue for triggering postfire flowering (Stone 1951), which is possibly a general seasonal trigger for flowering in grasslands and other open sites in the absence of fire.

In woody plants a seemingly specialized resprouting mode is evident in a diverse array of woody species that resprout from swollen lignified structures at the base of the stems, known as basal burls or lignotubers (Fig. 3.1). Depending on the species and number of fire cycles these structures may vary from a few centimeters to more than a meter in diameter (Fig. 3.2c) or even larger in some *Eucalyptus* species (Lacey 1983). These burls possess storage carbohydrates, inorganic nutrients and adventitious buds that are considered critical to postfire resprouting (Carr *et al.* 1984; James 1984). Basal burls are common in MTC regions (Keeley 1981) but occur in other ecosystems, often induced as a wound response to repeated coppicing or other disturbance and can be found in tropical forests (Johnston & Lacey 1983), tropical grasslands (Davy 1922; Lawson *et al.* 1968; White 1976), and temperate forests (Cant 1937; Ekanayake 1962; Mallik 1993; Stone & Cornwall 1968). Such structures also need not be basal as evidenced by tropical "living fences" that resprout and form burls from repeated coppicing at fence height nearly 2 m above ground (Sauer 1979).

Somewhat unique among taxa in MTC regions is that many woody resprouters (Table 3.2) produce these burls as a normal ontogenetic stage in early development (Fig. 3.2b). They commonly are initiated from axillary buds of the cotyledons and expand until they coalesce into a swelling that encircles the stem (Kerr 1925), but there is significant variation in this ontogeny (Lacey & Jahnke 1984; Mibus & Sedgley 2000). Canadell & Zedler (1995) suggest that the term *lignotuber* be reserved for these structures that are genetically determined and the term (basal)

Fig. 3.1 *Resprouting lignotuber of* Erica australis *from Portugal in the western Mediterranean Basin. (Photo by P. Maia with permission from Prensa Científica S.A.)*

burl be used to describe all woody swellings, genetic or induced. In most cases lignotubers are initiated in the first year of development (Kerr 1925; Wieslander & Schreiber 1939; Montenegro *et al.* 1983; Dodd *et al.* 1984; Molinas & Verdaguer 1993; Del Tredici 2001), although in *Eucalyptus* species there is apparently some degree of flexibility in the timing of lignotuber initiation (Carr *et al.* 1982). Lignotubers enhance the ability of resprouting individuals to persist indefinitely on a site and preempt resources for considerable periods of time. The ability to resprout over countless fire cycles (e.g. Fig. 3.2c) may allow individuals to persist for hundreds if not thousands of years (Canadell & Zedler 1995; Nicolle 2006).

Epicormic resprouting occurs from dormant buds under the bark of scorched trunks and canopy stems. It is very widespread in arboreal forms of the northern hemisphere's *Quercus* (Fig. 3.3a) and many other hardwood trees in both hemispheres. It is not common in coniferous trees, but does occur in *Pinus canariensis* from the Canary Islands and in *Pseudotsuga macrocarpa*, associated with California chaparral (Fig. 3.3b). In the Western Cape of South Africa there is one fynbos arborescent shrub, *Protea nitida*, that resprouts epicormically, as do most *Eucalyptus* in Australia (Fig. 3.3c). Unlike basal resprouting that is tolerant of high-intensity crown fires, epicormic sprouting in *Quercus* species is common when canopies are scorched but may fail following high-intensity fires. However, some Australian *Eucalyptus*

Fig. 3.2 *Examples of lignotubers at different stages of development in the California chaparral resprouter* Arctostaphylos glandulosa; *(a) typical rounded basal burl, (b) burls produced in the first couple of years of development, and (c) platform-like structure resulting from resprouting after repeated fire cycles. (Photos by Jon Keeley.)*

Table 3.2 Lignotuber taxa where the structure appears to be a normal developmental stage and not the result of coppicing

Mediterranean	California	Chile	South Africa	Australia	non-MTCs
Shrubs:					
Erica spp.	*Adenostoma* spp.	*Colliguaya odorifera*	*Audouinia capitata*	*Adenanthos* spp.	*Kalmia latifolia*
Phillyrea spp.	*Arctostaphylos* spp.	*Nothofagus antarctica*	*Berzelia* spp.	*Allocasuarina*	*Prosopis* spp.
	Ceanothus spp.	*Trevoa trinervia*	*Brunia laevis*	*Banksia* spp.	*Rhododendron* spp.
	Garrya spp.		*Erica* spp.	*Calothamnus* spp.	
	Xylococcus bicolor		*Leucadendron* spp.	*Conospermum* spp.	
			Leucospermum spp.	*Conostephium*	
			Protea spp.	*Darwinia*	
			Staavia spp.	*Dryandra* spp.	
				Eremaea	
				Eucalyptus spp.	
				Grevillea spp.	
				Hakea spp.	
				Isopogon spp.	
				Lambertia spp.	
				Melaleuca spp.	
				Persoonia spp.	
				Petrophile spp.	
				Podocarpus drouynians	
				Verticordia spp	
Trees:					
Arbutus unedo	*Aesculus californica*	*Cyptocarya alba*		*Eucalyptus* spp.	*Betula populifolia*
Juniperus oxycedrus	*Sequoia sempervirens*				*Ginkgo biloba*
Olea europea	*Umbellularia californica*				*Juniperus deppeana*
Quercus suber					
Tetraclinis articulata					

are extraordinarily tolerant of high-intensity fires and entire stands often recover rapidly from epicormic resprouts after high-intensity crown fires (Fig. 3.3d). This capacity appears to involve unique anatomical features that enhance this sprouting capacity (Burrows 2002).

Both basal and epicormic resprouting is limited by fire intensity, which affects tissue necrosis, and by season, which affects carbohydrate reserves and possibly cell hydration. Long fire-free intervals potentially inhibit resprouting by generating higher fire intensities from the greater fuel accumulation, and there is a strong interaction between resprouting and plant size and age (Noble 1984; Moreno &

Fig. 3.3 *Epicormic sprouting in (a) Mediterranean Basin oak* (Quercus suber), *characteristic of most arborescent oaks, (b) California chaparral conifer,* Pseudotsuga macrocarpa, *an uncommon trait in gymnosperms, (c)* Eucalyptus *from Australia, and (d) a forest of epicormically resprouting* Eucalyptus *from Victoria, Australia, illustrating stand level recovery 11 months after a high-intensity fire in February 2009. (Photos by Felipe Catry, a, and Jon Keeley, b–d.)*

Oechel 1994). Climate plays a role since intense droughts induce vegetation dieback, thus increasing dead fuels, as well as reducing live fuel moisture, both of which will increase fire intensity and reduce resprouting. Shorter fire intervals, and fires at certain seasons, potentially deplete carbohydrate reserves and this is tied to reduced resprouting (Rundel *et al.* 1987; Haidinger & Keeley 1993; Canadell & López-Soria 1998; Cruz *et al.* 2003a), although dependence on stored

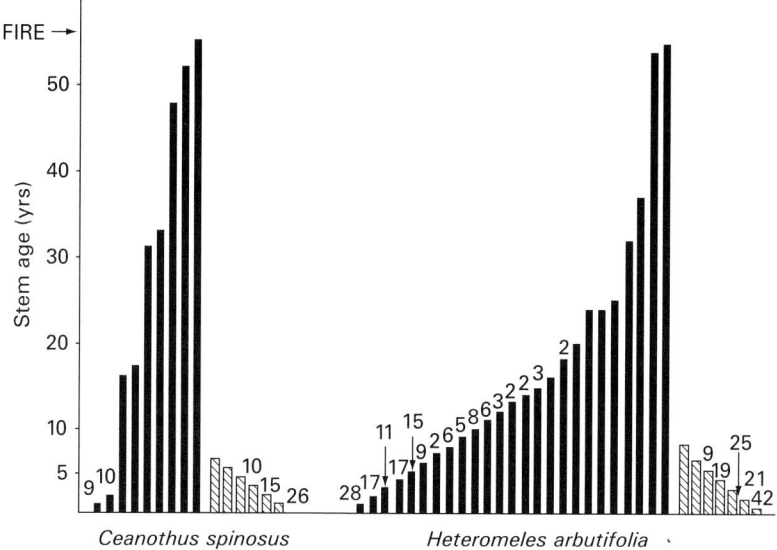

Fig. 3.4 *Distribution of stem ages on two individual resprouting shrubs: the facultative seeder* Ceanothus spinosus *and the obligate resprouter* Heteromeles arbutifolia, *both from California chaparral. Bar indicates stem age and each bar represents one stem, unless topped with a number; hatched bars indicate dead stems. (From Keeley 1992c.)*

carbohydrates is variable (Bond & Midgley 2003; Cruz et al. 2003b). Other seasonal effects include tissue hydration since hydrated cells are less tolerant of high temperatures (Whelan 1995).

There are age effects on resprouting capacity. For example, some arboreal oaks (*Quercus* spp.) resprout from the base as young saplings but at maturity lack this resprouting capacity (Longhurst 1956). This is consistent with the wider distribution of basal resprouting, which is common in crown fire regimes but less common in surface fire regimes. In essence oak saplings experience crown fires and thus resprouting capacity has substantial selective value, but as they mature they are exposed to surface fires, and this has not selected for basal resprouting at this stage. Thus, in the rare event fires are carried into the canopy these mature trees generally do not resprout from the base. This may be the selective basis for the widespread distribution of this age effect on resprouting in North American trees (Del Tredici 2001) and may also explain a very similar age-related change in resprouting in tropical savanna trees (Bond & Midgley 2003).

Age effects are also evident in other growth forms. California subshrubs resprout vigorously at a young age but later lose the resprouting capacity (Hobbs & Mooney 1985; Keeley 2006c). It has been hypothesized that this is due to phylogenetic constraints because the genera that exhibit this pattern are from herbaceous perennial lineages (e.g. *Artemisia*, *Eriogonum*) and these subshrubs have lost the resprouting capacity due to secondarily evolved lignification that buries adventitious buds in older stems (Keeley 2006c).

One complication in understanding the origins of postfire resprouting is that sprouting is a feature of canopy rejuvenation (Fig. 3.4) that often occurs in the absence of disturbance, both in highly fire-prone shrublands (Mesléard & Lepart 1989; Keeley 1992a) and mesic rainforests (Marrinan *et al.* 2005). Thus, many shrubs that resprout after fire continue to produce new basal sprouts throughout their life span, although depending on growing conditions many of these never recruit into the canopy.

Generally, resprouting is a species-specific trait; however, there are notable examples where different populations of the same species have diverged in resprouting capacity, with both resprouting and non-resprouting populations (Keeley 2000; Ojeda *et al.* 2005). The relevant selective factors might be expected to vary with different lineages since resprouting is associated with a different set of traits in different floras (Pausas *et al.* 2004b).

Seedling Recruitment

Many fire-prone ecosystems in the world are resilient to fires but most species have not capitalized on fire as an opportunity for seedling recruitment. In contrast, many species in MTC ecosystems and related mediterranean-type vegetation (MTV) have fire-dependent reproduction. This is most prominent in closed-canopy shrublands where a predictable fire cycle produces postfire conditions that represent resource-rich sites for seedling recruitment. This includes creation of canopy gaps, reduced competition for soil moisture resources, accelerated mineralization, reduced herbivore populations and synchronized recruitment that satiates predators.

However, there are potential costs to postfire recruitment. One is delayed reproduction, often for many decades; rather than spreading recruitment risks out over climatically different years, species gamble on a single pulse of recruitment in the first postfire year. The potential costs of delayed reproduction are of such a magnitude that they are not likely to evolve except under extraordinary conditions (Gadgil & Bossert 1970), and these are discussed in Chapter 9. In MTC regions other costs of recruiting on recently burned sites are associated with greater exposure to more intense summer drought stress, which requires specialized physiological and anatomical traits (S.D. Davis *et al.* 1998; Pratt *et al.* 2008). Specialization to postfire recruitment is linked to traits that diminish recruitment success in the shrub understory and thus many species have specialized on postfire recruitment whereas other species have a very different character syndrome specialized on fire-independent recruitment in the understory (Keeley 1998).

Postfire seedling recruitment requires seedbanks that are cued to germinate by fire. Some species exhibit fire-dependent reproduction because essentially all seedling recruitment is restricted to the first postfire year. This pulse of recruitment results in stands of single-aged cohorts as with many woody plants (Bond & van Wilgen 1996; Keeley *et al.* 2006b). Other species are more opportunistic and couple postfire recruitment with seedling recruitment between fires as with many

Table 3.3 Example of recruitment patterns for different life histories taken from a five-year study of postfire seedling recruitment patterns for representative woody species in California chaparral
Precipitation was below average in years 1, 3, and 4 and substantially above average in years 2 and 5.

Species	Growth form[a]	No. of sites	Total seedling recruitment (ha^{-1}) X ± S.E.	Percentage by year 1	2	3	4	5
Obligate seeders								
Ceanothus crassifolius	s	10	62 100 ± 17 800	99	1	0	0	0
C. greggii	s	8	18 500 ± 18 500	100	0	0	0	0
C. oliganthus	s	6	103 900 ± 95 500	100	0	0	0	0
Helianthemum scoparium	ss	12	33 500 ± 17 500	73	13	7	0	7
Lotus scoparius	ss	39	46 100 ± 8400	72	9	2	1	16
Facultative seeders								
Adenostoma fasciculatum	s	31	104 500 ± 22 800	94	3	0	0	3
Ceanothus spinosus	s	9	36 600 ± 25 100	92	3	0	0	5
Artemisia californica	ss	12	11 600 ± 4800	41	42	13	1	4
Eriogonum fasciculatum	ss	24	8600 ± 2900	71	21	5	0	3
Salvia mellifera	ss	27	41 900 ± 16 100	73	12	3	3	9
Obligate resprouters								
Cercocarpus betuloides	s	8	200 + 100	0	33	0	33	34
Rhamnus crocea	s	21	400 ± 100	4	0	15	4	77
Encelia californica	ss	2	497 300 ± 487 800	0	67	15	0	18
Hazardia squarrosa	ss	13	37 700 ± 17 300	0	54	34	6	6

[a] s, shrub, large stature, typically > 2 m, hard wood and long lived; ss, subshrub or suffrutescent growth forms that are smaller stature with light wood and short lived.
Source: From Keeley *et al.* (2006b).

suffrutescents or subshrubs that grow rapidly and set seed before the canopy closes in (Table 3.3).

In more open shrub communities where canopies do not close in, seedling recruitment may occur both after fire and continuously between fires in gaps between shrubs (e.g. Lloret 1998; Carrington & Keeley 1999; DeSimone & Zedler 1999; Holmes & Newton 2004). Some of these species have long-lived seedbanks and others more transient seedbanks that require continuous replenishing in order to maintain a seedbank for postfire recruitment.

Although seeds of some species may colonize burned sites (see below), most recruitment derives from *in situ* dormant seedbanks that accumulate between fires. This conclusion is based on seedbank studies that show that most species recruiting after fire are present in the soil prior to fire (Zammit & Zedler 1988; Parker & Kelly 1989; Auld 1994; Enright *et al.* 2004) and on inferences based on the timing of dispersal, fire and recruitment. Most taxa disperse seeds in spring or summer and most fires occur after dispersal is completed, thus making it unlikely that species present in the first growing season after fire are the result of colonization. This potentially changes after the first year, but dispersal into burned areas is affected by the perimeter to burned area ratio, and sources of colonizers along the perimeter. In a detailed study in California chaparral it was found that five years after fire about half of the species were not present on the site in year 1 (Keeley *et al.* 2005a). These *apparent colonizers*, however, did not represent colonization of the burned area because almost all of these species were present in year 1 on other sites within the burned perimeter. Thus, these were postfire species that were relatively rare across the burned area in year 1 and through mass action effects ended up in new sites by year 5.

Species that recruit seedlings in the first growing season after fire from dormant seedbanks are widespread in MTC regions. Some of these woody species regenerate after fire both by seedlings and resprouting (facultative seeders); however, a significant number only regenerate with seedlings and are termed obligate seeding species. These latter species lack all capacity for resprouting and are a major exception to Vesk & Westoby's (2004) supposition that all species resprout and only differ in level of resprouting. Obligate seeding species were recognized by Wells (1969) as being unique because resprouting capacity is nearly universal in woody dicots and he attributed the obligate seeding habit in shrubs as a trait selected for by fire. Obligate seeders comprise a small subset of woody plants in crown fire regimes (Table 3.4). One of the key attributes of obligate seeders is their sensitivity to short fire return intervals as they are extirpated from sites when fire return intervals are shorter than the time required for reproductive maturation, evident in both empirical (Zedler *et al.* 1983; Pate *et al.* 1991; Gill & Bradstock 1995; Keeley 2006a; Pausas 2006) and modeling studies (Keeley 1986; Keeley & Swift 1995; Enright *et al.* 1998; Regan *et al.* 2010).

Species differ in mode of storage, with most accumulating seeds in the soil and others by the accumulation of seeds in serotinous fruiting structures (a type of bradyspory) retained on the parent plant. The selective trade-offs are considered in more detail in Chapter 9.

Soil Seedbanks

Many species that disperse seeds at maturity have innate barriers to germination and accumulate dormant soil-stored seedbanks, which are later triggered to germinate by fire-related cues. These include many dominant shrubs and a huge number of ephemeral species that spend most of their lifetime as dormant seedbanks (Pate *et al.* 1985; Keeley 1991; Bell *et al.* 1993; Keeley & Fotheringham 2000). Fire-stimulated germination of soil seedbanks may be triggered by heat or chemicals from the combustion of biomass. These are often considered rather specialized germination mechanisms; however, in the latter case there is evidence that this germination mechanism may be rather ancient, being found in basal taxa within several different lineages (C.J Fotheringham personal communication, 2004; Pausas & Keeley 2009).

Table 3.4 Postfire seeding woody genera in MTC crown fire regimes that depend entirely on seeding (obligate seeders) or couple seeding with resprouting (facultative seeders)

		Number of species			
		Obligate seeders	Resprouters (facultative seeders)	Species with populations of both	Source[a]
Mediterranean Basin					
Cistus	Cistaceae	18	0	2	4
Erica	Ericaceae	1	11	0	4
Genista	Fabaceae	1	9	0	4
Ulex	Fabaceae	1	7	0	4
California					
Arctostaphylos	Ericaceae	48	4	8	1
Ceanothus	Rhamnaceae	23	14	1	3
Chile					
[none]					
South Africa					
Aspalathus	Fabaceae	81	11	1	6
Cliffortia	Rosaceae	60	60	–	[b]
Erica	Ericaceae	390	27	16	5
Leucadendron	Proteaceae	27	3	0	6
Leucospermum	Proteaceae	23	3	0	6
Podalyria	Fabaceae	4	12	2	6
Paranomus	Proteaceae	9	1	0	6
Protea	Proteaceae	28	15	0	6
Southwestern Australia					
Adenanthos	Proteaceae	15	16	0	8
Banksia	Proteaceae	38	20	2	2
Grevillea	Proteaceae	138	62	0	8
Hakea	Proteaceae	77	38	0	7

[a] (1) Parker et al. 2010; (2) Lamont & Markey 1995; (3) Wilken 2010; (4) Paula et al. 2009; (5) Ojeda 1998; (6) Schutte et al. 1995; (7) Young 2006; (8) Bell 2001.
[b] estimate.

Heat-stimulated seeds are known from temperate forests and grasslands, tropical savannas, deserts, MTC shrublands and other ecosystems, in species of Anacardiaceae, Apiaceae, Cistaceae, Convolvulaceae, Ericaceae, Fabaceae, Geraniaceae, Hypericaceae, Lamiaceae, Malvaceae, Portulacaceae, Onagraceae, Restionaceae, Rhamnaceae and Sterculiaceae (Keeley & Fotheringham 2000). These seeds typically exhibit exogenous dormancy imposed by a dense water-impermeable cutinized testa and dense palisade tissue that has led to the term *hardseeded*. Fire disrupts the water-impermeable tissues, allowing imbibition, which typically leads to germination. Hardseeded species with postfire seedling recruitment appear to differ substantially in the intensity and duration of heat shock most stimulatory to germination. Some exhibit optimal germination after brief bursts of high temperatures (e.g. 5 minutes at 105 °C) whereas others have higher germination after a long duration at lower temperature (e.g. 1 hour at 70 °C). The same applies to lethal temperature regimes: for example, large-seeded species survive a short duration at high temperature but are killed by a long duration at lower temperature, whereas very small seeds exhibit the opposite pattern (Keeley 1991). Such differences in stimulation/tolerance regimes may explain some of the variance in microhabitat segregation of postfire floras (e.g. Davis *et al.* 1989).

Since many species are stimulated by a long duration at 70–80 °C, they are not strictly tied to postfire environments as such conditions may be encountered by seeds exposed to direct sun rays on open sites. Thus, heat-shock-stimulated germination does not limit recruitment to burned sites, rather such species can establish in gaps created by other types of disturbance as well. Also, since unburned landscapes often comprise a heterogeneous collection of suitable and unsuitable recruitment sites it is not surprising that most heat-stimulated species have polymorphic seed pools (Keeley 1991, 1995a). Thus, while the bulk of the seedbank may be deeply dormant, a portion may germinate readily and establish in the absence of fire.

It has only recently become evident that in some fire-prone environments the majority of species that recruit after fires lack heat-stimulated germination, but rather chemical products of biomass combustion contained in charred wood or smoke provide the cue that triggers germination and heat plays no role in their germination (Box 3.1). This response is present in many of species from MTC ecosystems in southern Australia, South Africa, California and the Mediterranean Basin (Wicklow 1977; Keeley 1987, 1991; de Lange & Boucher 1990; Brown 1993; Dixon *et al.* 1995; Moreira *et al.* 2010). The chemical trigger is transferred to seeds either in the smoke aerosol during fire, in aqueous leachate or vapors from smoke or charred wood after fire or even secondarily transferred after adhesion to soil particles (Keeley & Fotheringham 1997), and there is evidence for both inorganic (Keeley & Fotheringham 1997) and organic (Flematti *et al.* 2004) compounds as the trigger. This fire-generated chemical stimulus for germination is found in a wide diversity of plant families (Box 3.1).

Longevity of soil-stored seed is not well documented but circumstantial evidence suggests seeds of many MTV species are very long lived. The longest record for shrublands is that inferred from the study of California chaparral burned after approximately 150 yrs (Keeley *et al.* 2005b). Many of the ephemeral species that are restricted to the immediate postfire years were present in the first growing season after fire in these ancient stands. Colonization from outside was impossible considering the 25 000-ha size of the fire, lack of dispersal capacity and short time interval between fire and establishment, indicating these seeds had persisted for roughly a century and a half as dormant seeds. Some shrubs common in MTC shrublands also play a role as a seral stage following crown fires in forests and in one case there is evidence the seedbanks survived over 200 yrs between fires (Gratkowski 1962).

Box 3.1 Smoke and Charred Wood Stimulated Seed Germination

In many MTC ecosystems there are species with seedling recruitment restricted to the immediate postfire years from soil-stored seedbanks that are triggered to germinate by fire. Historically it was thought these were stimulated by intense heat and indeed many species are, but for a substantial proportion of postfire recruiting species, heat does not trigger germination; rather it is cued by chemicals produced by biomass combustion. These chemicals are present in smoke and charred wood, or can even be generated by heating wood to 175 °C (Keeley & Nitzberg 1984; S. Keeley & Pizzorno 1986). Because of the ease of generation most experimental work has used smoke and so for convenience we will refer to this phenomenon as *smoke-stimulated germination* (Brown 1993; Keeley & Fotheringham 1998).

Perhaps the first to recognize this phenomenon were the California Indians who routinely sowed native tobacco (*Nicotiana* spp.) seeds in postfire ash beds (Harrington 1932). Under natural conditions *Nicotiana attenuata* and *N. quadrivalvis* are known to recruit primarily after fire and germination is triggered almost exclusively by charred wood or smoke (Baldwin *et al.* 1994; Keeley & Fotheringham unpublished data). The first scientific report of this phenomenon was by Wicklow (1977) who demonstrated that germination of the postfire annual *Emmenanthe penduliflora* was dependent on charred wood. Subsequent studies demonstrated that many other California annuals and perennials that restrict recruitment to recently burned sites were triggered to germinate by chemicals produced during the combustion of biomass and not by heat shock (Keeley 1987; Keeley & Keeley 1987). The first demonstration that smoke could trigger germination was made on a South African fynbos species by de Lange & Boucher (1990).

This "smoke"-stimulated germination is now known from many postfire recruiting species in South Africa (Brown 1993), Western Australia (Dixon *et al.* 1995), California (Keeley & Fotheringham 2000) and the Mediterranean Basin (Moreira *et al.* 2010). Many of these species could be described as having *smoke-dependent germination*, as the only environmental cue that will trigger germination is smoke. Some species from the matorral in Chile exhibit some enhanced germination in response to smoke but these are not species that typically recruit seedlings after fire (Gómez-González *et al.* 2008). Smoke-dependent germination in postfire recruiting species is now known from hundreds of species. The plant families (Table B3.1.1) where it has been most convincingly demonstrated are rather different in the northern and southern hemispheres and present an apparent example of convergent evolution (Keeley & Bond 1997; see also Chapter 9). Enhanced germination in response to smoke also has been found in some non-MTC species that recruit seedlings after fire (Jefferson *et al.* 2008; Lindon & Menges 2008; Abella 2009).

Continued

Box 3.1 (cont.)

As a general rule species with smoke-stimulated germination have very different seeds from those with heat-stimulated germination (Keeley & Fotheringham 2000). The latter are characterized by having a water-impermeable seed coat that is ruptured by heat shock. Following such a heat treatment, and given the appropriate thermal, light and moisture regime, these seeds germinate readily. Dormant seeds of smoke-stimulated species have water-permeable seed coats and heat has no detectable effect on permeability. Outer seed coats are highly sculptured and in all cases so far examined they lack a dense palisade layer. Commonly the outer coat comprises loosely packed tissues with a subdermal semi-permeable cuticle. Most of the smoke-stimulated species fully imbibe water during dormancy, indicating smoke-stimulated germination is overcoming endogenous dormancy (Keeley & Fotheringham 1998). However, some smoke-stimulated species can be germinated by mechanical seed coat scarification (Keeley & Fotheringham 1997). In most smoke-stimulated species, dormancy appears to be innate but in some cases it apparently overcomes secondarily induced dormancy (Krock *et al.* 2002).

The transfer of smoke or other combustion chemicals to the seed appears to take several pathways. The exposure of dried seeds to smoke prior to moist incubation is one mode, but aqueous and vapor transfers from smoke-treated substrates to seeds during moist incubation also are potential modes (Fig. B3.1.1).

A large number of experimental studies have been conducted searching for smoke-stimulated germination. A significant number of these have failed to appreciate that even in smoke-stimulated seeds, some concentrations of smoke are lethal and the lethal dose is species specific (Keeley & Fotheringham 1998). Studies that report a lack of smoke-stimulated germination, without conducting the experiment over a large concentration gradient, or demonstrating that the seeds are still viable after smoke treatment, are inconclusive.

Smoke has been demonstrated to stimulate germination in many species that do not recruit seedlings after fire. This includes plants from arid land communities where fire is currently rare as well as a great many agricultural species (Pierce *et al.* 1995; Drewes *et al.* 1995; Taylor & van Staden 1998). It has been suggested that this observation raises questions about the adaptive significance of smoke-stimulated germination in species from fire-prone communities (Pierce *et al.* 1995). We agree that all observed cases of smoke-stimulated germination should not be interpreted as evidence of an adaptive trait. However, for species with seedling recruitment restricted to postfire environments, and with seeds that remain dormant except when exposed to smoke (i.e. smoke-*dependent* germination), it seems inescapable that this is truly an adaptive trait. Whether or not it is a true adaptation that evolved in response to fire is a rather different question (see Chapter 9). It is advisable to distinguish between smoke-stimulated germination and smoke-dependent germination in sorting out explanations of trait selection.

Continued

Box 3.1 (cont.)

Table B3.1.1 Selected plant families from MTC regions with numerous species that recruit seedlings after fire and exhibit smoke-dependent germination

Northern Hemisphere	Southern Hemisphere
Asteraceae	Dilleniaceae
Boraginaceae	Ericaceae (including Epacridaceae)
Caryophyllaceae	Goodeniaceae
Ericaceae	Haemodoraceae
Hydrophyllaceae	Myrtaceae
Lamiaceae	Poaceae
Loasaceae	Proteaceae
Onagraceae	Restionaceae
Papaveraceae	Rutaceae
Polemoniaceae	Thymeleaceae
Scrophulariaceae	
Solanaceae	

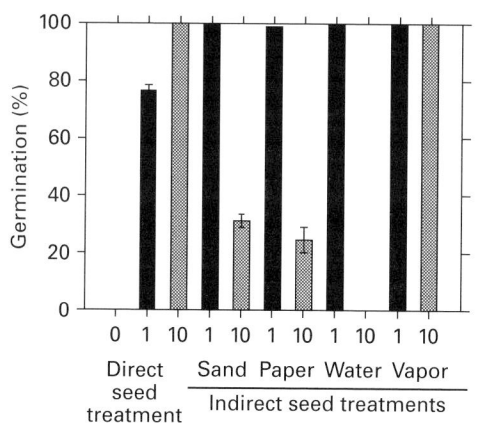

Fig. B3.1.1 *Germination of the California fire endemic annual* Emmenanthe penduliflora *for control (0) and smoke treatments of 1-min or 10-min exposures for direct application of smoke to dry seeds or indirect exposure to seeds during moist incubation. Indirect treatments are Sand: dry sand exposed to smoke and then seeds incubated in this substrate; Paper: dry filter paper exposed to smoke and then seeds incubated on these filter papers; Water: aqueous extracts of smoke applied to seeds during moist incubation and Vapor: seeds exposed to vapors from previously smoke-treated sand. (From Keeley & Fotheringham 1997.)*

Canopy Seedbanks

Serotiny is a type of bradyspory that involves the delayed opening of cones or fruits with fire-triggered synchronous opening and seed dispersal into postfire seedbeds (Fig. 3.5). It is largely restricted to crown fire ecosystems. In the northern

Fig. 3.5 *Serotinous fruiting structures of Gondwana angiosperm shrubs: (a) persistent* Banksia *infructescence (inflorescence of many fruits) opened after previous year's fire in Western*

hemisphere it is a coniferous phenomenon, being limited to species of eastern hemisphere *Cupressus* and western hemisphere *Hesperocyparis* (Cupressaceae), *Pinus* (Pinaceae), *Sequoiadendron* (Cupressaceae) and the monotypic genus *Tetraclinis* (Cupressaceae), which are distributed in a range of habitats that include boreal forests and semi-arid shrublands (Lamont *et al.* 1991; Keeley & Zedler 1998). Although these are seemingly very different ecosystems, they share certain features that contribute to the crown fire regime, including an abbreviated growing season that maintains fuels near the crown, leading to high-intensity crown fires.

In the southern hemisphere, primarily in MTC shrublands of the South African Cape region and Western and South Australia, serotiny is found in about a dozen conifers in three genera of Cupressaceae (*Actinostrobus, Callitris* and *Widdringtonia*), and is very widespread in angiosperm families, including the Asteraceae, Bruniaceae, Casuarinaceae, Ericaceae, Rosaceae, Myrtaceae and Proteaceae (Lamont *et al.* 1991; Oliver & Fellingham 1994). In the last two families are several genera with 40 or more serotinous species: *Banksia, Eucalyptus, Hakea, Leucadendron* and *Protea*. Serotiny is substantially more important in both diversity and abundance in these southern hemisphere ecosystems than any other region of the world.

In all of these taxa, cones or fruits open en masse and disperse seeds within days of being scorched by fire, resulting in a pulse of seedling recruitment that generates stands of even-aged cohorts. Seeds are generally not dormant, are short lived and fire plays little direct role in stimulating germination beyond inducing cone or fruit opening (Lamont *et al.* 1991).

There are intraspecific patterns of varying degrees of serotiny and thus populations may be even aged or uneven aged, possibly tied to fire return intervals (Keeley & Zedler 1998; Ne'eman *et al.* 2004). Retention of seeds is also highly variable across species, ranging from a few months (e.g. *Pinus muricata* cones remain closed through the autumn fire season and disperse seeds in winter, Keeley & Zedler 1998) to many decades (Lamont *et al.* 1991). Delayed dispersal in *Sequoiadendron giganteum* in the Sierra Nevada of California might also be considered a form of serotiny. Although seed dispersal is not strictly tied to fire, cones can remain closed many years after reaching maturity and then shed seeds in mass numbers in the first few days following a surface fire with plumes of convective heat transfer that dries the cones.

There is species-specific variation in retention of seeds in cones in the absence of fire. When seeds are released in unburned stands seedlings typically have a low probability of successful recruitment (Zammit & Westoby 1988). Among the biggest threats to long-term sustainability of serotinous populations are

Caption for Fig. 3.5. (*cont.*) *Australia, (b) unopened* Hakea *fruits from Victoria, Australia, (c) unopened* Erica sessiliflorus *infructescence from Cape region fynbos, South Africa, (d) opened* Protea *infructescences with dispersed seeds and seedlings beneath following recent fire; and serotinous cones of gymnosperm trees in northern hemisphere MTC ecosystems, (e) unopened cone crop in* Pinus halepensis *in southern Spain, and (f)* Hesperocyparis (*formerly* Cupressus) forbesii *after a recent fire with opened serotinous cones dispersing seeds in southern California chaparral. (Photos by Jon Keeley, a–d, f, and Juli Pausas, e.)*

long fire-free periods since seeds are generally released upon death of the parent plant, regardless of fire.

The extraordinary importance of serotiny in South African and Australian shrublands stands in stark contrast to the limited presence of serotiny and nearly complete dependence on soil-stored seedbanks in other MTC ecosystems. Lamont (Lamont *et al.* 1991; Lamont & Enright 2000) points out that this high degree of serotiny in these southern hemisphere systems is correlated with extremely nutrient deficient soils and more predictable rainfall, patterns that will be explored more fully in Chapter 9.

Delayed Postfire Seedling Recruitment from Resprouts

Many woody and herbaceous taxa recover from fire by resprouting with no immediate postfire seedling recruitment and are termed obligate resprouters. Subsequent seedling recruitment is largely a function of growth form. Resprouts from herbaceous perennials, suffrutescents and subshrubs grow rapidly, flower and disperse seeds in the first postfire season. These taxa have limited seed dormancy and many recruit en masse in the second and subsequent years until canopy closure (see obligate resprouting subshrubs in Table 3.3) (Keeley & Keeley 1984; Denham & Auld 2002). Similar patterns are evident with the synchronous flowering after fire in some Australian grasstree *Xanthorrhoea* species (Gill & Ingwerson 1976; Taylor & van Staden 1998). Subsequent seed dispersal largely restricts seedling recruitment to the second and third postfire years in some species (Lamont *et al.* 2004b). Although these are immediate postfire obligate resprouters, they have reproduction that is fire dependent.

Delayed Postfire Seedling Recruitment from Parent Trees

Surface fire regimes have selected for a very different suite of adaptive traits in the dominant overstory species that are largely concerned with ensuring survival of parent seed trees (Keeley & Zedler 1998). Such fires are considered the historical norm for many northern hemisphere conifer forests (Agee 1993; Allen *et al.* 2002), and a common fire regime in certain Australian eucalypt forests (Bell *et al.* 1989; Williams 2000; Gill & Catling 2002).

The strategy of trees in this fire regime is to ensure survivorship of parent seed trees that can disperse seeds into gaps opened up by higher-intensity burning. Two traits that have adaptive value are self-pruning of lower dead branches, which creates a mid-story gap between surface fuels and canopy fuels, and thick bark that reduces lethal effects of surface fires. Conclusions about the selective role of fire in generating these traits requires close examination because they could have evolved in response to other environmental factors. Bark, for example, insulates against fire because of a series of dense cork layers, which could also insulate

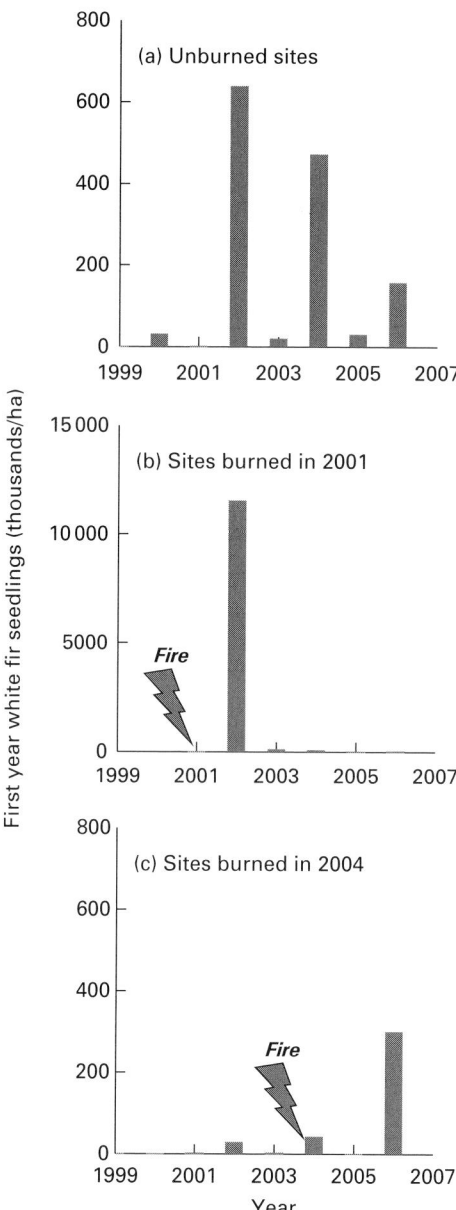

Fig. 3.6 *Interaction between fire and mast cone production illustrated by first-year* Abies concolor *seedlings in 1- or 2-ha sample plots: (a) n = 11 unburned sites, (b) n = 3 sites burned in 2001, one year prior to a mast year, and (c) sites burned in 2004, two years prior to a mast year, in white-fir-dominated forests of Sequoia National Park in the southern Sierra Nevada of western North America. (From Keeley & van Mantgem 2008.)*

against extreme heat or cold. There is one case though where fire would seem to have been the primary selective force. In the genus *Pinus* all species from fire-prone habitats have very thick bark, whereas the non-fire-prone desertic pinyon pines and the timberline white pines, exposed to the hottest and coldest conditions respectively, have the thinnest bark (Keeley & Zedler 1998). A selective basis for thick bark is also supported by its extensive development in savanna oaks compared with the rather thin bark in scrub oaks adapted to crown fire regimes where aboveground survivorship is unlikely (Zedler 1995a).

Regeneration by the dominant species in closed-canopy woodlands and forests with surface fire or mixed fire regimes is commonly restricted to open sites with bare mineral soil created by fires. Regeneration in these species is dependent on the survival of parent trees in proximity to fire-created gaps. When cones or fruits are present at the time of fire, recruitment may occur in the first growing season after fire, but often times it is delayed until subsequent fruiting cycles. Many trees in surface fire regimes have masting cycles of reproduction (Keeley & van Mantgem 2008). If a fire coincides with a mast year there will be abundant seedling recruitment in the first postfire year but following other fires it may be delayed and result in substantially lower recruitment (Fig. 3.6). Thus, there is a certain level of serendipity involved in fires coinciding with seed production. The longer the interval between fire and a mast year the greater the competitive inhibition by understory vegetation, which may dictate very different successional trajectories.

Postfire Colonization

Some fire-sensitive species drastically reduced by fire or species that are transitory on sites are generally absent from recently burned sites and ultimately enter those communities only through colonization. The role of colonization is quite variable between different vegetation types. For example, in California chaparral and in Mediterranean Basin shrublands all dominant shrubs recover endogenously and colonization during early succession typically accounts for less than 10% of the cover (Keeley *et al.* 2005a). In contrast, in semi-arid parts of western North America the dominants in non-MTC woodlands, pinyon (*Pinus* spp.), juniper (*Juniperus* spp.) and sagebrush (*Artemisia tridentata*) all recover slowly from crown fires by recolonization, often from unburned patches resulting from uneven patterns of burning.

Fire-independent Recruitment

In crown fire regimes obligate resprouters, as the name implies, do not recruit seedlings after fires. Comparisons of these postfire obligate resprouters with postfire seeders often contrast these species as representing vegetative vs. sexual

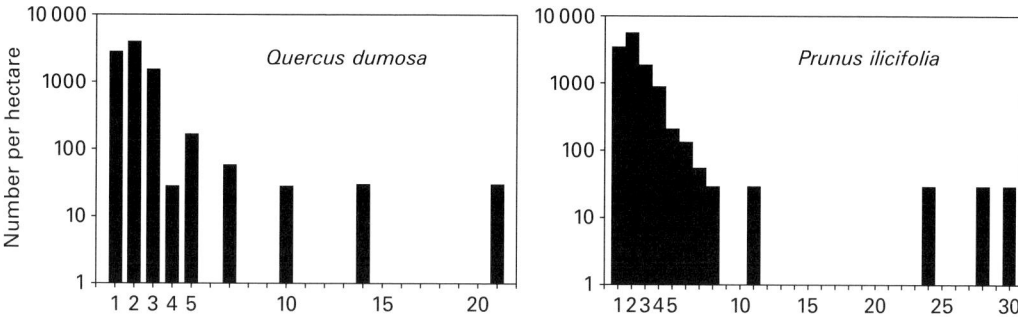

Fig. 3.7 Density of seedlings and saplings in the understory of California chaparral unburned for more than 50 years (Keeley 1992c). In these closed-canopy shrublands saplings generally remain stunted and represent "advanced reproduction" capable of progressing into the canopy only after resprouting following fires.

reproduction. This is a pyrocentric perspective because obligate resprouters do have a sexual cycle with seedling recruitment; however, they have not opted for a strategy of delaying reproduction to a single pulse of recruitment immediately after fire. Obligate resprouting shrubs (Table 3.1) reproduce sporadically during fire-free periods and are best described as having *fire-independent reproduction* (Keeley 1998; García-Fayos & Verdú 1998). In stark contrast to postfire seeders these all have propagules designed for widespread dispersal, including animal-dispersed meaty or fleshy fruits and wind-dispersed plumes (Keeley 1992d; see also Chapter 9). Recruitment is largely restricted to the understory of the shrub canopy and on sites free of fire for extended periods of time (Williams *et al.* 1991; Keeley 1992c) or in gaps of adjacent woodlands (Keeley 1990b). Recruitment appears to be tied to years of high precipitation and in most cases saplings remain stunted in the understory for extended periods of time (Fig. 3.7). In some respects this represents advanced reproduction or seedling bank, as these stunted saplings resprout and grow rapidly after fire, and in some taxa there is a level of fire dependency for successful recruitment into the canopy.

In many forest types there are canopy species that regularly recruit in the understory in the absence of fire. Typically these taxa are more shade tolerant than taxa that specialize in recruitment in gaps after fires. Such species also are important components in some fire-prone landscapes, particularly *Abies* (Pinaceae) in the northern hemisphere and *Eucalyptus* (Myrtaceae) in the southern hemisphere. Some taxa have very flexible recruitment requirements with respect to fire. For example, the North American white fir, *Abies concolor*, dominates forests with a surface fire regime. It regularly recruits in the shady understory between fires but it also capitalizes on postfire environments with massive recruitment (Fig. 3.6). These taxa illustrate that the often-used generalization that species with shade-tolerant understory recruitment are *fire sensitive* is clearly not always true.

Conclusions

Fire regimes constrain the options for postfire regeneration of plant communities. Regeneration after crown fire regimes in sclerophyllous vegetation, both inside and outside MTC regions, is dominated by a combination of resprouting and postfire seedling recruitment. Forest trees in surface fire regimes commonly survive fire by a combination of traits that limit fire spread to dead surface fuels and protect trunks from overheating. Recruitment is dependent on survival of parent trees, and seedlings are dependent on patches of high-intensity burning as safe sites for seedling recruitment.

Section II

Regional patterns

Mediterranean-type climates (MTC) comprise a diverse array of woody, herbaceous, and even succulent vegetation types in each of the five regions. Shrublands are universal across all MTC regions and have been the main focus of past comparative studies; however, woodlands and forests are of great interest in understanding drivers in these ecosystems. In Section II we examine in detail the intraregional patterns of variation in vegetation types and the ecological role fire plays in each of the five regions. We are concerned with the degree to which similar environments have converged on similar fire regimes and those factors responsible for divergence. A boiler-plate approach of topics covered in each region is not the appropriate metaphor here as each region presents very different problems associated with different landscape histories and fire responses. The nominate region, the Mediterranean Basin, has such an extraordinarily long and intensive human history that landscapes exist as a palimpsest with ecological patterns overriding signatures from earlier uses. Crown fires dominate all MTC regions but California is unique among these in having substantial forests of tall conifers prone to surface fire regimes, with different trajectories of trait evolution and fire management impacts. Central Chile exhibits patterns best interpreted as a fading fire regime, once quite evident but, like a dying star, it has been extinguished by the Late Miocene completion of the Andean uplift that now blocks ignition sources due to lightning. The Western Cape of South Africa is considered to have the climate and geology sufficient to support forests, but these are extremely limited and restricted to narrow refugia from fire. Southern Australia is phenomenal in that sclerophyllous-leaved mediterranean-type vegetation (MTV) covers much of the southern third of the continent, despite the lack of a MTC from middle Victoria to New South Wales in the southeastern corner of the continent.

4 Fire in the Mediterranean Basin

The Mediterranean Basin is a meeting point of three continents, Europe, Asia and Africa, and this is responsible for the great diversity of plants, animals and cultures that formed the cradle of Western civilization. It is considered one of the biodiversity hotspots (Myers *et al.* 2000) because of its high species richness and high proportion of endemisms (Thompson 2005). The total area showing a mediterranean-type climate (MTC) is about 2.3 million km^2, with transitions toward temperate forest ecosystems (in the European mountains) and toward arid ecosystems (in North Africa and the Near East). It is not only the largest of the five MTC regions, but also the most geographically complex (with more than 40 000 km of rough coast in different peninsulas and islands) as well as the most socio-economically, culturally and politically varied. Elevations range up to 3756 m in the east (the highest peak in the Taurus mountains, Turkey) and up to 4167 m in the west (the highest peak in the Atlas mountains, Morocco). There are many volcanoes in Italy and the Aegean Islands, with frequent minor eruptions and rare major explosions. The MTC region of the basin corresponds to a narrow rim around the Mediterranean Sea (Fig. 4.1), and includes: (1) in southern Europe, most of the Iberian peninsula (Portugal and Spain), south of France, most of Italy and Greece, the coast of Croatia, Montenegro and Albania; (2) in southwest Asia (the Near East), Cyprus, Lebanon, Palestine, Israel, most of Turkey, and the coast of Syria; and (3) in North Africa (the Magreb), the north of Tunisia, Algeria, Morocco and small coastal areas of Libya. It also includes all the islands in the Mediterranean Sea.

In general terms, summers are hot and dry and winters are mild and relatively wet; winters may be cold in the interior areas with a continental climate influence (e.g. central Spain and central Turkey). The configuration of seas, peninsulas and islands, and the topographic complexity of the area, produce a great regional variety of weather and climate. Rainfall ranges from semi-arid conditions (<300 mm) up to over 2000 mm, and peaks in autumn and spring (in the west) and in autumn and winter (in the east). Because of the air masses' trajectories, the wettest parts of the basin are typically the western parts of the peninsulas (Iberian, Italian and Balkan peninsulas). There are also clear gradients from the colder and wetter northwest (southern France and northern Iberia) to the hotter and more arid south and southeast parts of the basin (North Africa and the Near East). The

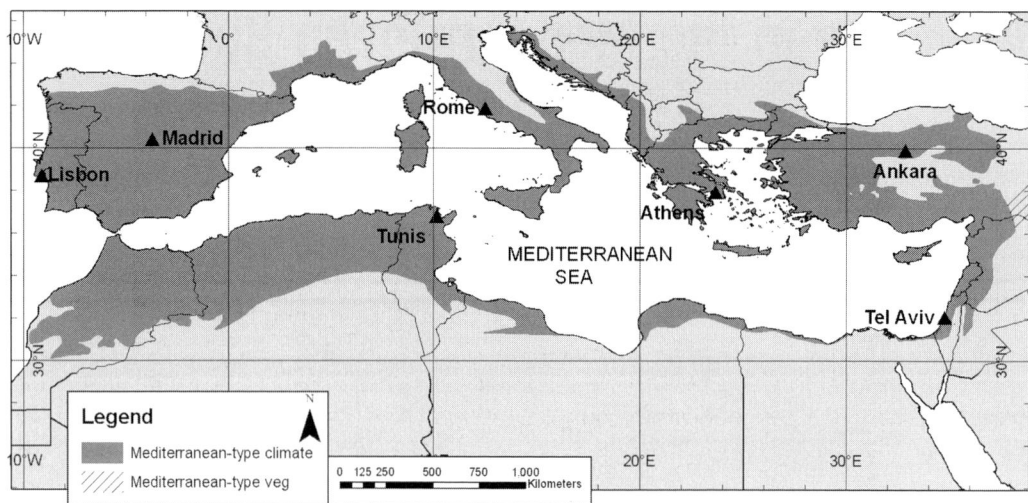

Fig. 4.1 *Distribution of the mediterranean-type climate (MTC) in the Mediterranean Basin (dark shading). (Based on Quézel & Médail 2003.)*

temperature-moderating effect of the sea is highest in the west (Atlantic coast) and lessens toward the east (water temperatures rise from west to east).

Current landscapes in the Mediterranean Basin are a product of long and intense land use, and often comprise mosaics of agricultural land, old-fields and wildland, with urbanization increasingly encroaching on these landscapes. Vegetation types include a wide range of shrublands (e.g. see Fig. 1.6a) depending on the land use history. In the more preserved sites we find tall and dense shrublands (maquis) dominated by evergreen broadleaf sclerophyllous species (e.g. *Rhamnus*, *Phillyrea*, *Arbutus*, *Olea*, *Myrtus*, *Osyris*, *Pistacia* and *Quercus*). In abandoned old-fields, in pine woodlands or shrublands repeatedly burned, and in frequently grazed (sheep, goats) areas, shrublands are mainly dominated by species with small leaves (e.g. *Ulex*, *Genista*, *Fumana* and *Helianthemum*), often aromatic species (e.g. *Thymus*, *Rosmarinus*, *Teucrium*, *Lavandula* and *Satureja*) or by malacophyllous species (*Cistus*, *Phlomis*; see Box 1.2). Grasslands occur mainly as a transition state in recently abandoned fields, and as a steady state in semi-arid zones (e.g. alfa steppes, *Stipa tenacissima*). Trees are usually low in stature (< 20 m); they include species of *Ceratonia*, *Olea*, *Quercus* and *Pinus* that may occur sparsely in shrublands or form dense forests and woodlands (e.g. see Fig. 1.6b). The main forest trees are evergreen sclerophyllous oaks (*Q. ilex*, *Q. suber* in the west, and *Q. ilex*, *Q. calliprinos* in the east) and pines (often favored by extensive plantations; e.g. *P. halepensis*, *P. pinaster*, *P. brutia*, among others), with the species varying depending on the climatic, fire and soil characteristics. These forests usually have a dense shrubby understory. Winter-deciduous and semi-deciduous trees and shrubs also occur (*Quercus*, *Acer*, *Fraxinus*, *Platanus*, *Populus*, *Pyrus*, *Ulmus*, *Amelanchier*, *Lonicera*, *Vitex*, *Rhus*, *Salix*, etc.) on moist sites (e.g. pole-facing slopes and ravines) or at higher elevations in the mountains.

Landscapes Strongly Shaped by Humans

The most striking difference between the Mediterranean Basin and the other MTC regions is its substantially longer history of intensive and widespread land use. Humans first appeared in the African savanna, but they soon spread downstream along the Nile River to the Mediterranean Basin (Stringer & McKie 1996). They have been present in this region longer than in any other MTC region or anywhere else in Europe (Carbonell *et al.* 1995, 2008). It is thought that Paleolithic people (~2.5 Ma–10 000 BP) already burned deliberately for hunting and food gathering (Stewart 1956; Goren-Inbar *et al.* 2004), although the earliest evidences of land and fire management in the Mediterranean Basin are related to the spread of agriculture and domestic grazing during the Neolithic (~10 000–4500 BP) (Naveh 1975). Several cave settlements from ~400 000 BP provide the earliest evidence of at least limited use of fire. Fire use continued to increase, and during the Bronze Age (~4500 BP) it was a general practice widely applied for purposes of deforestation and pasture improvement and occasionally as a war strategy (Thirgood 1981; Pons & Thinon 1987; Grove & Rackham 2001; Kaniewski *et al.* 2008). By the late Bronze Age terraces and irrigation work were being carried out and a large part of the region already presented a "humanized" landscape dominated by crops of winter wheat, barley, vines and olives, and livestock of cattle, sheep, goats and pigs (Huston 1964). Romans and Muslims improved terracing and irrigation and introduced new crops; additional crops were introduced after the Spanish conquest of America.

The Mediterranean Basin has been the theater for the birth, blooming and collapse of many cultures and civilizations (e.g. Sumerian and Mesopotamian, Ancient Egyptian, Phoenician, Jewish, Ancient Greek, Persian, Arab, Roman, Ottoman, among others), and for many political conflicts (wars, changes in land ownership, population movements) that resulted in numerous socio-economic and land use changes. The dominance of each of these cultures was mainly based on the increased commercial and military power provided by having a larger fleet of ships than the previous culture. Thus, the shipping industry together with increased agricultural land are the most important causes of mediterranean deforestation. As an example, the fleet of Arab war galleys that unsuccessfully laid siege to Constantinople in 717 is estimated to have been composed of 1800 ships, all made from local wood. All these facts imply millenia of severe pressure on the land, resulting in clearing, terracing, and cultivating all arable areas and burning, cutting and grazing non-arable areas. From very early times (and especially under the Roman Empire), mediterranean landscapes have been heavily terraced for cultivation, implying the clearing, including uprooting and soil movements, of most of the landscape (Fig. 4.2).

The concomitant increase in human populations and Holocene drying has raised the debate as to whether major modifications to mediterranean vegetation in the last 6000–7000 yrs are more the result of human activities than they are of

Fig. 4.2 *Terraced slopes are a dominant feature of the landscapes in the Mediterranean Basin and illustrate the strong human pressure in the region and the current old-field dynamics. Note that fields are in use at the bottom, and are abandoned (and recolonized by shrubs) at the mid and top portions of the slope (Vall d'Albaida, Valencia, Spain). (Photo by Juli Pausas.)*

climatic changes. Similarly, it is difficult to distinguish between the anthropogenic fire contribution and the natural (lightning) source in the Holocene fire regime changes (Gil-Romera *et al.* 2010; Turner *et al.* 2010). It is often assumed that although climate change drove vegetation change until ~10 000 BP, human activities (e.g. agriculture, forest clearing) have since become the determinant factor influencing vegetation changes, beginning in the east and moving west (Quézel & Médail 2003; Kaniewski *et al.* 2008). Grove & Rackham (2001) argued, however, that many of the current open landscapes are not the product of human activities but rather are driven by climatic factors. How these landscapes might have looked with no human impact is very difficult if not impossible to know because both climate-driven and human-driven changes have occurred simultaneously, and no natural reference landscape exists in the contemporary climate. What is clear is that ancient descriptions of many landscapes bear little relationship to current landscapes and such comparisons suggest intense land use and deforestation over the past few millenia (Blondel & Aronson 1999). Parallel to anthropogenic and climatic changes there have been changes in the fire regime, although the pattern of changes may vary through the basin. For instance, the period with lower fire activity since the Last Glacial Maximum 18 kyr ago coincided with the most arid period in dry areas, perhaps due to fuel limitations

(Turner *et al.* 2008; Linstädter & Zielhofer 2010), and with a moist period in wetter areas, suggesting fuel moisture limitations (Carrión 2002; Sadori & Giardini 2007).

The consequence of the long human pressure on this landscape is that little if any of the landscape has escaped human impacts and most landscapes are very far from their natural state. This fact has important implications for fire ecology and management, as well as restoration ecology, in particular the lack of a reference ecosystem (Vallejo *et al.* 2006).

Recent Socio-economic, Land Use and Fire Changes

The longest available fire history for a Mediterranean Basin site (130 yrs) suggests that there has been a relatively recent shift in fire regime. This was related to socio-economic changes in the 1970s and resulted in an increase in the frequency of large fires (Pausas & Fernández-Muñoz 2011). This shift is specially observed in the northern rim (European) countries where with industrialization came rural depopulation, abandonment of farmland and a reduction in the livestock grazing pressure (without replacement by natural grazers). This process resulted in increases in the amount of fuel, especially in early succession species (many of which are very flammable, Baeza *et al.* 2011), and changes in the landscape pattern and the fire regime (Moreira *et al.* 2001; Lloret *et al.* 2002; Pausas 2004; Bajocco *et al.* 2010; Pausas & Fernández-Muñoz, 2011). Furthermore, many of these old-fields have been extensively planted with pines and *Eucalyptus* during the last few decades (Pausas *et al.* 2004a, 2008), resulting in even higher fuel loads and fuel continuity. Thus, at present, the main ecological process occurring in the Mediterranean, especially on its northern rim, is land abandonment and its impact on fire (Fig. 4.2).

At the same time increased development along coastal areas as a result of both the escalating tourism pressure (holiday resorts and retirement homes) and life style changes (houses instead of flats) has had indirect effects on ecological processes and fire regimes. The spread of new urban and semi-urban populations in rural areas is covering many old-field areas. The consequences of these trends are mainly (1) an increase in water demand due to the reduction of aquifers, (2) habitat fragmentation and (3) an increasing wildland–urban interface, which contributes to increased fire ignitions.

All of these processes are important drivers behind the increase in number of wildfires and area burned in recent decades. In fact, the number of ignitions increased (exponentially) from the 1960s to the 1980s in most countries (Fig. 4.3), and there is a correlation between population density and fire ignitions (Terradas *et al.* 1998; Catry *et al.* 2009). The area burned also increased exponentially on the northern rim of the Mediterranean Basin during this period (Pausas 2004; Fig. 4.4). However, the influence of climatic changes cannot be denied (Piñol *et al.* 1998; Pausas 2004). In fact, fire and climate are strongly correlated during recent decades, but not for the period previous to the fire regime shift of the 1970s

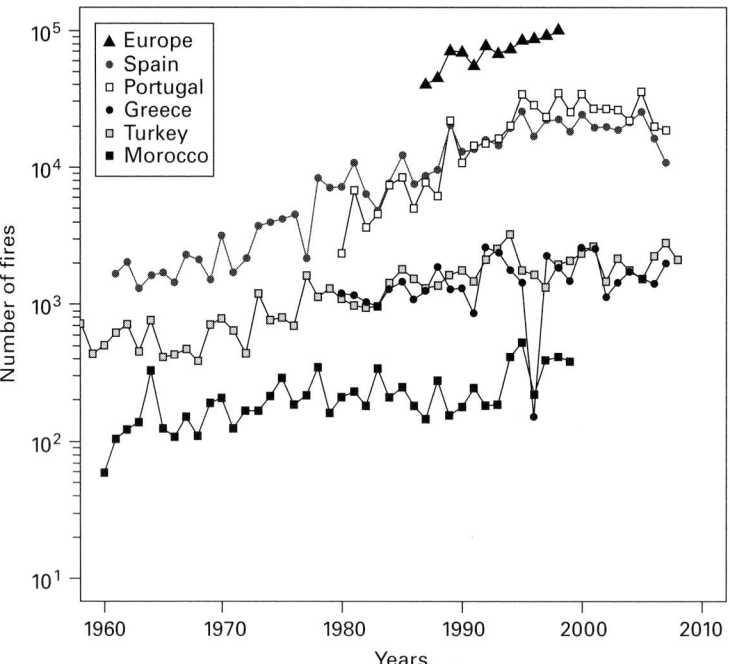

Fig. 4.3 *Progression of the number of fires (log scale) during the last several decades in different Mediterranean Basin regions (Spain, Portugal, Greece, Morocco, Turkey) and in Europe. (Data from national agencies of each country, modified from Pausas 2004 and Pausas et al. 2008.)*

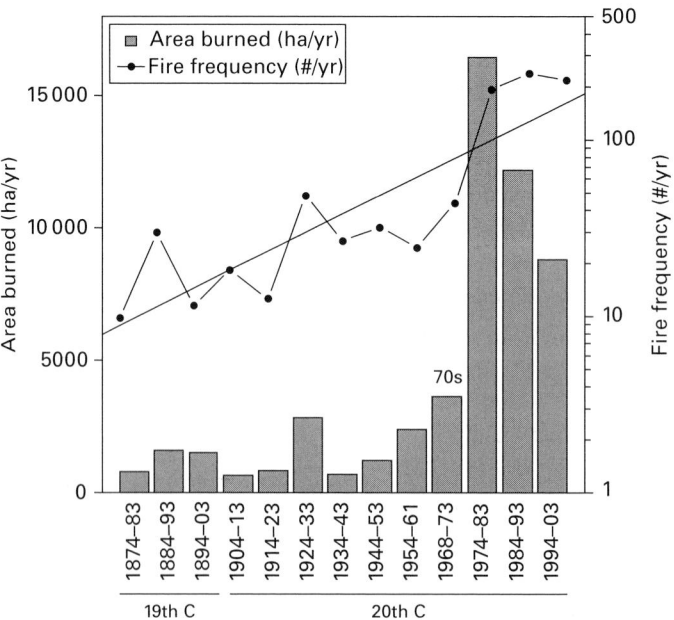

Fig. 4.4 *Average annual area burned and annual number of fires (log scale), from 1874 to 2003 by decades (except for three periods, 1954–1961 and 1968–1973, in which information was available for a shorter period only) for Valencia province (1 080 600 ha, 54% of which is wildland area). (Modified from Pausas 2004 and Pausas et al. 2008; see Pausas & Fernández-Muñoz 2011 for further analysis of this data.)*

(Pausas & Fernández-Muñoz 2011). Prior to this shift, mediterranean landscapes were shaped by agriculture, livestock and other land uses that maintained low levels of highly fragmented fuels. Progressive changes in land use switched from a fuel-limited fire regime to a drought-driven fire regime (Pausas & Fernández-Muñoz 2011). Indeed, recent large fires in the Mediterranean Basin have been related to extreme warm and dry weather (Pausas 2004; Trigo et al. 2006; Founda & Giannakopoulos 2009).

In the Mediterranean Basin southern rim countries, as well as in some eastern rim countries such as Turkey, a large proportion of the population continues to be rural, and heavy use of the land (e.g. overgrazing) continues to be a dominant force behind many land degradation and desertification problems (Mairota et al. 1998; Camci Çetin 2007). Consequently, in these areas, fuels are maintained at low levels and the annual area burned has remained fairly constant over the last few decades. Industrialization and rural depopulation are just beginning in North Africa, where we might expect a similar shift in fire regimes as observed in the northern rim countries. Such a shift is already occurring in Algeria as there is a current depopulation of rural areas for safety reasons, thus fuels are increasing and fires are becoming more prevalent.

Fuel Patterns and Structure

Most vegetation types (shrublands and woodlands) are sufficiently dense to sustain active crown fires. Surface fires are currently rare but they sometimes occur in mountain areas with coniferous forests. It is quite probable that surface fire regimes were more common in forests in the distant past, but recent forest management activities, including fire suppression, have made these ecosystems vulnerable to low-frequency crown fires. Some low shrubs, arid grasslands and open woodlands rarely burn because of the low amount and continuity of fuel loads. In open oak woodlands the low fuel values are the result of the grazing pressure by livestock or the other management actions (e.g. shrub clearing for cork extraction). In subshrubs and arid grasslands the low fuel volume appears to be the result of millenia of overgrazing pressure; thus it may be that some of these landscapes had more frequent fires in the past.

Fuel amounts vary with vegetation type (see Table 2.1). They also vary with fire history, with climatic and topographic conditions, and, in old-fields, with the time since abandonment. For instance, in old-fields of eastern Iberia (Spain), biomass (live and dead, Mg ha^{-1}) varies from about 1.5 (1 year postfire) to 18 (unburned) in *Brachypodium* grasslands (Caturla et al. 2000; see also Table 2.1), and from 13–59 Mg ha^{-1} from the third to the seventeenth postfire year in *Ulex parviflorus* (gorse) shrubland (de Luis et al. 2004; Baeza et al. 2006). Fuel loads vary not only with the time since last fire but also with the fire history previous to the last fire, as has been demonstrated in *Quercus coccifera* garrigue (Delitti et al. 2005). In the western part of Iberia, where rainfall is high but there is still a summer drought,

Fig. 4.5 Ulex parviflorus *(Fabaceae), a postfire obligate seeding species that accumulates a large amount of fine and dead fuel. Although not apparent in black and white, most foliage is dead and live biomass is restricted to the branch tips. (Photo by Juli Pausas.)*

fuel loads may be very high, reaching over 100 Mg ha^{-1} in some Portuguese shrublands (Viegas *et al.* 2002). In contrast, in the eastern part of the basin, we find communities overgrazed for centuries and with a long annual dry period; consequently they accumulate much less biomass (e.g. *Sarcopoterium spinosum* or *Phlomis fruticosa* shrubland; Dimitrakopoulos 2002). In rugged and protected areas, where evergreen oak forest is still present, biomass accumulation is often over 100 Mg ha^{-1} and can reach over 200 Mg ha^{-1} (*Quercus ilex* forest, including stems; Lledó *et al.* 1992; Ibáñez *et al.* 1999), although only a small fraction is consumed by wildfires.

Some of the species in the shrublands accumulate a large proportion of dead biomass (which has profound impacts on fire behavior, see Chapter 2). For instance, most of the biomass in *Ulex parviflorus* occurring in coastal areas of Spain and southern France corresponds to very fine fuel; in a 9-yr-old community (Fig. 4.5), standing dead biomass may account for more than 50% of its total biomass (Baeza *et al.* 2006, 2011). Similarly, *Erica–Cistus* shrublands of central Spain include *c.* 40% of dead biomass, and 30% of the green biomass is fine fuel (diameter < 0.60 cm; Moreno *et al.* 2004). In this community, *Cistus ladanifer* is the dominant species, and also the one with the highest proportion of dead and fine fuel. *Sarcopoterium spinosum*, in the eastern part of the basin, also accumulates large proportions of standing dead fuel (Seligman & Henkin 2000).

The Mediterranean Basin flora is also rich in species with a high volatile oil content, and these oils increase flammability. Examples include many Lamiaceae (*Thymus, Rosmarinus, Salvia, Lavandula*) and some Cistaceae (*Cistus ladanifer, C. populifolius*).

Ignition Patterns and Fire Behavior

Natural fires occur as a consequence of lightning. Especially in summer, dry storms can be an important fire ignition source (Soriano *et al.* 2005; Fig. 4.6), and foehn winds in different parts of the basin (see Box 1.3) greatly facilitate the spread of fires. Currently only a small proportion of the annual area burned (< 10%) is derived from lightning-ignited fires. It is probable that in the past (before the spread of agriculture), when native vegetation covered larger and more continuous areas and fire suppression did not extinguish these fires,

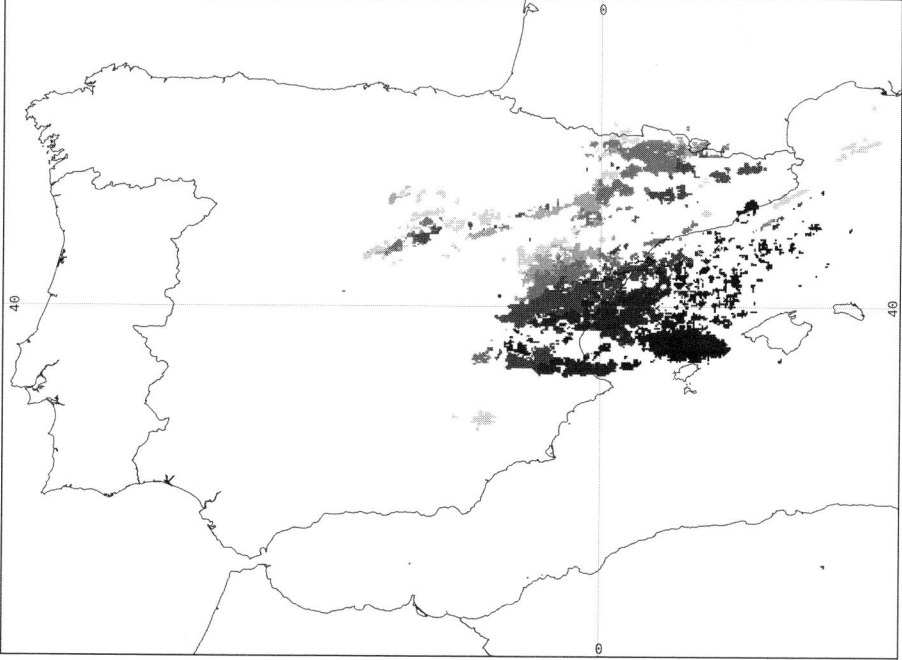

Fig. 4.6 *Image of the Iberian peninsula (Portugal and Spain) on August 1, 2005 (mid summer) showing the location of the 15 134 lightning strikes that were registered in 12 hours. Different shadings indicate 2-hr periods, from 14:00 (light gray, left) to 24:00 (dark, right). (Image from Agencia Estatal de Meteorología, Spain; www.aemet.es)*

Table 4.1 Percentage of the total number of fires and area burned in different parts of the Mediterranean Basin, including the Iberian Peninsula (Portugal in the west; Valencia in the east; and Catalonia in the northeast), Greece (Mount Parnitha National Park, Athens) and southwest Turkey (Datça-Marmaris)
Only fires greater than 1 ha are included. Percentages are based on wildlands in the area of interest. In bold face are the fire size classes that account for most (>50%) fires and area burned (elaborated from data provided by local forest administrations in Portugal, Valencia and Catalonia, from M. Arianoutsou for Mt Parnitha, and Ç. Tavsanoglu for Datça-Marmaris).

	Portugal[a]		Valencia		Catalonia		Mt. Parnitha		Datça-Marmaris	
Fire size classes (ha)	# fires	Area	# fires	Area	# fires	Area	# fires	Area	# fires	Area
$10^0 - 10^1$	25.4	1.6	**68**	2.5	**80.6**	4.4	**52.2**	1.4	**57.2**	0.8
$10^1 - 10^2$	**57.7**	15.8	22	8	14.6	7.4	26.1	7.0	28.9	4.6
$10^2 - 10^3$	15.2	**36.2**	8.2	25.6	3.9	19.5	17.4	41.2	10.6	17
$10^3 - 10^4$	1.6	**31.5**	1.7	**41.34**	0.78	**38.8**	4.3	**50.3**	2.8	**38.5**
$10^4 - 10^5$	0.06	14.9	0.12	**22.37**	0.12	**29.9**			0.6	**39.1**
Period	1990–2005		1968–2007		1987–2002		1959–1996		1968–2008	
Study area (ha)	5 310 900		1 209 264		1 956 791		26 000		138 000	

[a] Small fires in Portugal are underestimated (compared with the other areas) because for some years the data include only fires larger than 15 ha (1992), 50 ha (1993) and 10 ha (2005).

lightning was a more important fire ignition source. Most current fires are due to anthropogenic ignition, either by negligence or intentionally. Human- and lightning-ignited fires tend to have different geographical distributions; the former are more concentrated in coastal areas, while the latter occur in inland areas at higher elevation (Vázquez & Moreno 1998), a feature shared with other MTC regions.

Most fires, including those from lightning, are concentrated in summer when fuels are driest, although fires caused by pasture burning (typically < 5% of the fires, Moreno et al. 1998) are distributed throughout the year. The interannual variability in area burned is correlated with summer rainfall in several different ways (Pausas 2004). During wet summers the area burned is lower than in dry summers, undoubtedly due to higher fuel moisture. Summer rainfall is also positively correlated with area burned for a time lag of 2 yrs, suggesting that high rainfall may increase fuel loads that burn the subsequent 2 yrs.

Due to the fragmented landscapes in the Mediterranean Basin, fires are relatively small (Table 4.1). Large and intense fires occur only in summer, and they leave relatively few unburned patches. The few fires that occur in other seasons of the year are less severe and leave abundant unburned patches (Pausas et al. 2003). Fires are especially intense in forest plantations, particularly in areas that combine summer drought with high rainfall during the rest of the year. As with most fire-prone landscapes, the distribution of the number of fires vs. the burned area is very skewed, with most of the burning being accounted for by a few very large fires (Table 4.1).

Mediterranean Basin Flora and Postfire Strategies

The flora may be characterized as comprising two distinctive character syndromes: (1) species with broad sclerophyllous evergreen leaves, small flowers with a reduced perianth, large non-refractory seeds dispersed by vertebrates, and postfire obligate resprouters, and (2) species with reduced or malacophyllous leaves, with larger flowers and smaller passively dispersed dormant seeds and postfire seeders (Herrera 1992; Verdú 2000; Pausas & Verdú 2005). Species from the first group are from lineages that appeared earlier in the fossil record than those from the second group (Herrera 1992). However, whether or not this represents an earlier evolution or bias in the fossil record is a matter of some dispute (see Chapter 10). Phylogenetic analyses suggest that the acquisition of the capacity to form a dormant seedbank and recruit after fire (Syndrome 2) occurred in relatively few lineages and they mostly lack resprouting ability (see Table 3.4). This contrasts with other MTC ecosystems where reseeding is much more widespread taxonomically and is often found in species that both resprout and reseed (see Chapter 9). These differences suggest a somewhat different evolutionary history of fire and plant trait evolution than in other MTC regions (Pausas et al. 2006b).

All evergreen species in the Mediterranean Basin are able to resprout after fire or other disturbances from basal stem buds, roots or rhizomes. A few have distinctive lignotubers (see Fig. 3.1), such as in *Arbutus unedo*, *Quercus suber*, *Olea europaea*, *Tetraclinis articulata*, and some *Erica* and *Phillyrea* species (see Table 3.2). Some species that resprout epicormically have specialized protected stem buds such as in *Quercus suber* (Pausas 1997).

Heat-stimulated germination is mainly in Cistaceae (see Fig. 11.2a) and Fabaceae species that are hardseeded and water impermeable until fire cracks the outer seed coat layers (González-Rabanal & Casal 1995; Herranz et al. 1998, 1999; Paula & Pausas 2008; Moreira et al. 2010). Smoke-stimulated germination (see Box 3.1) is observed in postfire seeders with water-permeable seed coats like Lamiaceae and Ericaceae, among others (Moreira et al. 2010; Table 4.2). Serotiny occurs only in conifers such as *Pinus halepensis* (see Fig. 3.5e) or *P. brutia*, and to a variable extent in *P. pinaster* (Tapias et al. 2001, 2004). *Cupressus sempervirens* also has serotinous cones (Lev-Yadun 1995; Battisti et al. 2003).

Geophytes, that is, plants in which only the underground storage organ (bulb or rhizome) survives the unfavorable period, are common in the Mediterranean Basin. These include both monocots (*Urginea*, *Gladiolus*, *Crocus*, *Tulipa*, *Iris*, and many *Orchis* and *Orphys* species) and dicots (*Cyclamen*). They are little affected by fire as their buds are well protected by the soil and the plants are normally dormant during the fire season. However, the opening of the canopy by fire often stimulates flowering and increases seed production and seedling recruitment.

Annual plants are very common in the region and are more important in postfire dynamics in the eastern part of the Basin (Kazanis & Arianoutsou 2004;

Table 4.2 Smoke-stimulated shrub seed germination for selected Mediterranean Basin species that exhibit postfire recruitment

Family	Species	Percentage germination[a]		Significance (P)
		Control	Smoke	
Ericaceae	*Erica multiflora*	82	94	< 0.01
Ericaceae	*Erica terminalis*	33	60	< 0.001
Ericaceae	*Erica umbellata*	2	56	< 0.001
Lamiaceae	*Lavandula latifolia*	29	59	< 0.001
Lamiaceae	*Lavandula stoechas*	50	100	< 0.001
Lamiaceae	*Rosmarinus officinalis*	26	40	< 0.05
Lamiaceae	*Thymus vulgaris*	88	100	< 0.001
Primulaceae	*Coris monspeliensis*	9	43	< 0.001

[a] Control conditions were seeds immersed in distilled water and smoke treatments were seeds immersed in liquid smoke solution.
Source: From Moreira et al. (2010).

Kavgaci et al. 2010) than in the western part. This is likely tied to climate but whether it is due to the eastern basin having a higher predictability of winter rainfall or much more limited precipitation in other seasons (see Table B1.1.1) is unknown. Annuals are also more abundant in acidic soils (Pausas et al. 1999). In general, the importance of annuals in postfire dynamics in the Mediterranean Basin seems to be lower than in some other MTC regions such as California (see Table 5.2). However, in the Mediterranean Basin, annuals are common colonizers of abandoned agricultural fields, and in semi-arid areas in mixtures with perennial grasses. Many Mediterranean Basin species that are invasive in other parts of the world are winter annuals (*Bromus*, *Hordeum* and *Avena*). In contrast, in the Mediterranean Basin, postfire exotic invasive plants are a less important factor than in other MTC regions, at least currently (see Chapter 12).

Vegetation Patterns in Response to Climate, Geology and Land Use

Due to the size and geographical complexity of the region, the MTC area of the basin (Fig. 4.1) is often divided into different bioclimatic zones defined by temperature and also related to elevation (Quézel & Médail 2003). Each zone has the potential to grow different vegetation types. The lowlands or *thermo-mediterranean* zone are warm and dry, and closed forests are rare; broadleaf evergreen shrublands (garrigue, maquis) are the dominant vegetation. In the low-elevation *meso-mediterranean* zone, evergreen oak woodlands are the most significant type. In the mid-elevation *supra-mediterranean* or *sub-mediterranean* zone, winter deciduous or semi-deciduous woodlands are common, while at the high-elevation *mountain-mediterranean* zone, coniferous forests are common. At the top of the mountains the *oro-mediterranean* zone vegetation is sparse with some conifers and spiny shrublands. Most of these vegetation types are susceptible to active crown fires (see Chapter 2), except in semi-arid conditions where the low fuel load and continuity limits fire spread. Understory fires, although they are currently relatively rare, may occur in mountain coniferous forests.

However, these climatic patterns interact with past land use and geology in determining fire response (Pausas *et al.* 1999). Many areas with rocky bedrocks such as hard limestones and karsts were not terraced in the past because they have very shallow and decarbonated soils with abundant rock outcrops. They do have abundant cracks that allow deep-rooted plants to access water and, thus, deep-rooted postfire obligate resprouters dominate these soils. On these substrates broadleaf evergreen garrigue and maquis are common on drier sites and evergreen oak woodlands on more mesic sites with deeper soils. Previous land uses were related to gathering forest products, including wood, charcoal, cork, mushrooms, kermes and acorns, and livestock grazing.

In contrast, landscapes with relatively soft bedrocks (e.g. marl–limestone colluviums) were extensively cleared and terraced or intensively grazed in the past. Many of these croplands and grasslands have been abandoned and old-fields are a dominant feature of such landscapes (Fig. 4.2). Under these conditions are shrublands dominated by small-leaved and/or malacophyllous species, which are frequently non-resprouters, and early colonizers; evergreen resprouting species appear later in the succession (see below). Mixed shrublands, with both broadleaf evergreen and small-leaved species also occur. All of these shrublands may have pines (*P. halepensis*, *P. pinaster*, *P. brutia*) with variable densities. In most Mediterranean Basin countries, massive plantations, especially with native conifers (mainly pines) but also with non-native conifers and *Eucalyptus*, have been planted on these substrates during the last century, and many of these afforestations were conducted in old-fields (Pausas *et al.* 2004a, 2008).

In brief, at low elevations landscapes often comprise a mixture of: (1) never-terraced broadleaf evergreen shrublands and forest, (2) old-fields of variable age, often dominated by small-leaved shrubs, and sometimes afforested with pines, and (3) low shrublands resulting from long-term overgrazing or in arid conditions. In the first and second cases, crown fires are common, while in the third they are rare due to low fuel values. Transitions and intermixed mosaics are frequent.

Broadleaf Evergreen Shrublands (Maquis and Garrigue)

Broadleaf evergreen sclerophyllous shrubs and small trees dominate areas that were never terraced, typically hard limestone landscapes at low elevations. This maquis shrubland includes species of *Pistacia, Phillyrea, Olea, Ceratonia, Quercus, Rhamnus, Arbutus, Myrtus* and *Viburnum* among others (Table 4.3). All are strong postfire obligate resprouters and belong to lineages whose origin is tied to the Tertiary Madrean–Tethyan sclerophyllous vegetation that was broadly distributed in North America and Eurasia (Laurasia, see Chapter 10). The most

Table 4.3 Growth form characteristics of woody obligate postfire resprouters in the Mediterranean Basin

Genera	Common species	Family	Fruit type	Leaf type	Leaf size	Growth form
Arbutus	*A. unedo, A. andrachne*	Ericaceae	fleshy	broad	large	large shrubs
Asparagus	*A. acutifolius, A. aphyllus, A. stipularis*	Asparagaceae	fleshy	linear	very small	lianas
Clematis	*C. vitalba, C. flammula, C. cirrhosa*	Ranunculaceae	dry	broad	medium	lianas
Daphne	*D. gnidium*	Thymelaeaceae	fleshy	linear	small	shrubs
Hedera	*H. helix*	Araliaceae	fleshy	broad	large	liana
Laurus	*L. nobilis*	Lauraceae	fleshy	broad	medium–large	tree
Lonicera	*L. implexa, L. periclymenum* (d), *L. etrusca* (d)	Caprifoliaceae	fleshy	broad	medium	lianas, shrubs
Myrtus	*M. communis*	Myrtaceae	fleshy	broad	medium	shrubs
Olea	*O. europaea*	Oleaceae	fleshy	broad	medium	small trees
Phillyrea	*P. angustifolia, P. latifolia, P. media*	Oleaceae	fleshy	broad	small–medium	large shrubs
Pistacia	*P. lentiscus, P. atlantica* (d), *P. terebinthus* (d), *P. palaestina* (d)	Anacardiaceae	fleshy	broad	medium	large shrubs
Quercus	*Q. ilex, Q. coccifera, Q. suber, Q. calliprinos,* etc.	Fagaceae	dry (acorn)	broad	medium	trees, shrubs
Rhamnus	*R. alaternus, R. lycioides, R. palaestina, R. saxatilis* (d)	Rhamnaceae	fleshy	broad	small–medium	shrubs
Rubia	*R. peregrina, R. tenuifolia*	Rubiaceae	fleshy	broad	small	lianas
Ruscus	*R. aculeatus, R. hypophyllum*	Liliaceae	fleshy	broad[a]	medium–large[a]	small shrubs
Smilax	*S. aspera*	Smilacaceae	fleshy	broad	medium	liana
Styrax	*S. officinalis* (d)	Styracaceae	fleshy	broad	medium–large	large shrubs
Viburnum	*V. tinus, V. lantana* (d)	Caprifoliaceae	fleshy	broad	large	shrubs

All are evergreen sclerophyllous-leaved shrubs or trees, except those indicated with (d) are winter deciduous and (m) evergreen malacophyllous species. Leaf size is: very small (<25 mm^2), small (25–225 mm^2), medium (225–2025 mm^2) and large (2025–4550 mm^2).
[a]phylloclades.
Source: Data from the BROT database (Paula *et al.* 2009; Paula & Pausas 2009).

typical maquis would be dominated by the trees *Ceratonia siliqua* and *Olea europaea* var. *sylvestris* and the shrub (sometimes a small tree) *Pistacia lentiscus*. The fact that some of these communities can form woodlands (e.g. in some North African localities) suggests the current shrubland formations may be the result of past land use. Evergreen postfire resprouting lianas such as species of *Lonicera*, *Clematis*, *Smilax* and *Rubia* are also common, although these taxa occur in evergreen oak woodlands as well.

These maquis communities are almost exclusively dominated by postfire obligate resprouters and the proportion of woody species that are killed and fail to resprout after fire is generally very low. As a consequence this vegetation regenerates rapidly after fire. Most of these shrubs have vertebrate-dispersed fruits, often fleshy (Table 4.3), and consequently they are dispersed relatively long distances (Jordano *et al.* 2007; Pons & Pausas 2007). These obligate resprouting species avoid the stressful summer drought by means of deep roots that gain access to water through soil cracks. Thus, many of the seedlings are dependent on establishment in favorable mesic microsites (see Chapter 9) and they often make an early investment in roots (Lloret *et al.* 1999b; Paula & Pausas 2011).

Perhaps one of the most typical broadleaf shrubs in the Mediterranean Basin is *Quercus coccifera*, known as Kermes oak, which is an evergreen species that resprouts vigorously from rhizomes after fire (Malanson & Trabaud 1988). Formations dominated by this shrub are locally termed *garrigue* (Box 4.1). Although this species is resilient to frequent fire, there is a reduction in productivity when subjected to very short fire intervals (Delitti *et al.* 2005). The floristic composition of these communities also changes very little with different fire regimes (Trabaud & Lepart 1980, 1981). In the western part of the basin, *Q. coccifera* tends to be a multistemmed shrub, often forming large and imbricate carpets, but occasionally forming low woodlands on moister sites, for example in North Africa (Charco 1999). In the eastern part of the basin, *Q. coccifera* is very often a woodland tree and this form is often considered a distinct species, *Q. calliprinos*.

A few widespread species in these shrublands have highly reduced leaves and photosynthetic stems, sometimes flattened to form phylloclades. These include species of *Ruscus* (with broad sclerophyllous phylloclades) and species of *Asparagus* (with thin, sometimes thorny, phylloclades). These taxa have fleshy fruits and are obligate resprouters after fire. The native dwarf palm *Chamaerops humilis* may also appear in these western shrublands; its leaves protect the apical meristem from fire and thus it survives crown fires. Indeed, it may become dominant in zones with very high fire recurrence, and also under very high grazing pressure. The only other palm native to the region, *Phoenix theophrasti*, also survives fire but it occurs on very few sites in the eastern part of the basin (Boydak 1985; Barrow 1998).

Intermixed with the broadleaf shrubs, some small-leaved species may also occur, for example mostly resprouting *Erica* species or *Juniperus oxycedrus*, which has both resprouting and non-resprouting populations (Pausas *et al.* 2008); the non-resprouting *Juniperus phoenicea* appears only on sites that have had long fire-free periods or that are fire protected, such as rocky outcrops and cliffs. Other small-leaved species that may appear together with broadleaf shrubs, especially on open microsites, are species of Lamiaceae, Cistaceae and Fabaceae. Many of these have the capacity to recruit seedlings after fire but are much more dominant in other shrubland formations.

Box 4.1 Mediterranean Basin Shrublands

The most common names used for shrublands in the international (English) literature.

Maquis: tall and dense, dominated by a diversity of broadleaf evergreen shrubs. Most maquis species resprout vigorously after fire. The name of this vegetation type varies with different Romance languages like the Corsican (*machja*), French (*maquis*), Catalan (*maquia*) and Italian (*macchia*).

Matorral: Spanish word referring to shrubland (of any type). It can be of any sort: short matorral, tall matorral, etc. In Spain, other names may be applied depending on the dominant species; for example, *aulagar* (*Ulex*), *brezal* (*Erica*), *escobonal* or *piornal* (*Cytisus*), *jaral* (*Cistus*), *romeral* (*Rosmarinus*), *coscojar* (*Quercus*), *tomillar* (*Thymus*), *lentiscar* (*Pistacia*), etc. The word *matorral* (also used in Chile, see Chapter 6) is equivalent to the Portuguese *mato*.

Garrigue: shrubland dominated by *Quercus coccifera* (broadleaf evergreen and sclerophyllous shrub). Original name probably from the Occitan and Catalan languages, currently used also in French. The Spanish name is *coscojar* and the Greek name *prinones*. However, in some eastern parts of the basin (Israel) *Quercus coccifera* grows as a tree (sometimes named *Q. calliprinos*), and the word garrigue may be used for a shrubland dominated by *Calicotome villosa*, a summer deciduous thorny shrub (see also *phrygana*, below). Some have traditionally linked the term garrigue to shrublands on calcareous substrates and the term maquis to those on non-calcareous soils; however, this can lead to confusion as the dominant garrigue shrub *Q. coccifera* does form communities on non-calcareous soils.

Batha: short shrubland dominated by *Sarcopoterium spinosum*, small-leaved dwarf shrubs typical of dry areas of the Near East. Fires are infrequent due to fuel limitations. Name from the Hebrew and Arabic languages. It may be structurally similar to the Spanish *tomillar* and the Greek phrygana (see below).

Phrygana: Greek word referring to open shrublands, dominated by small-leaved (often spiny or aromatic) and/or malacophyllous species (often with seasonal dimorphic leaves). In Israel they use garrigue for these communities, which can lead to some confusion since that term is restricted to a very different vegetation in the western part of the basin (see above).

Tomillar: Spanish word for short shrubland (scrubland), often with low cover, where *Thymus* is abundant, together with other small-leaved shrubs. The Catalan word is *timoneda*. Structurally it may be very similar to the batha and phrygana (see above), and fires are rare due to a low amount of fuel.

Evergreen Oak Woodlands

On moister sites where human impact has been relatively low such as rugged landscapes and steep mountains, dense evergreen sclerophyllous oak woodlands occur. Previous uses were related to charcoal production, gathering forest products and livestock grazing. On previously cleared landscapes following land abandonment oak forests may recolonize (Pons & Pausas 2006, 2007). The most abundant oak tree is *Quercus ilex* known as holm oak (Rodà *et al.* 1999), but others include *Q. calliprinos* or Palestine oak in the eastern basin and *Q. suber* or cork oak on carbonate-free soils in the western basin (Aronson *et al.* 2009). All of these oaks have the capacity to resprout after fire from either basal buds or epicormically along the stem (see Chapter 3). Because of this strong resprouting capacity, repeated coppicing for wood harvesting has been carried out since ancient times and continues on a diminished scale as an important oak woodland management method. The result has been widespread replacement of single-stemmed trees with multiple-stemmed trees, and many current oak woodlands still show the characteristic clusters of trunks from recent or older coppicing. Such coppicing also alters the understory light regime and species composition in these communities. Pollarding (i.e. cutting the tree branches) has been traditionally used in oak woodlands for gathering firewood and improving acorn production to fatten hogs.

Because of their characteristic density, with high vertical and horizontal fuel continuity, oak woodlands usually burn as crown fires. The thick and insulating bark of *Q. suber* protects the stem buds and this species resprouts epicormically (see Fig. 3.3a) after fire (Pausas 1997; Pausas *et al.* 2009; Catry *et al.* 2010). Furthermore, in contrast to other oaks, this cork oak possesses underground dormant buds on a lignotuber (Molinas & Verdaguer 1993). Although cork oak is not very combustible itself, it grows among flammable grasses and shrublands that promote crown fires. Postfire recovery is positively related to tree diameter and bark thickness, such that small diameter (< 12 cm) trees generally resprout only from basal buds (Pausas 1997). Variations in bark thickness are due not only to tree size or age but also to the number of years since the last bark-stripping harvest for cork production (Aronson *et al.* 2009). In fact, bark stripping clearly increases the susceptibility of this species to fire, and failure to resprout has been observed in old, recurrently stripped trees (Moreira *et al.* 2007). The fact that cork oak can quickly regenerate after fire from stem buds gives this species a competitive advantage over other coexisting woody plants. For instance, a mixed cork oak and pine (maritime pine *Pinus pinaster* or stone pine *P. pinea*) forest has traditionally been encouraged for production of both cork from the oak and wood or pine nuts from the pine. However, the pines are sensitive to repeated fires and under such conditions these forests may convert to monospecific cork oak woodlands, a process often facilitated by the cork industry.

The understory of evergreen oak woodlands may be quite dense, particularly on moist sites, comprising most of the same shrubs and lianas that appear in maquis shrublands. Cork oaks very often have an open crown and on these well-lit understory sites the composition may be more typical of phrygana shrublands. Oak seedlings that recruit in the understory often remain suppressed for decades; however, they do resprout after fire, and it may be that a disturbance is required for their successful emergence into the canopy, as is seen with oak recruitment into pine woodlands (Pons & Pausas 2006).

Shrublands Dominated by Small-leaved and/or Malacophyllous Species (Phrygana)

Phrygana includes a wide range of shrublands, dominated by species with small or narrow leaves, often very reduced or even absent (e.g. with photosynthetic stems), or by species with malacophyllous leaves, that are soft, somewhat fleshy and pliable (Table 4.4). Broadleaf malacophyllous shrubs are often semi-deciduous and shed most leaves (sometimes even small branches) under water stress, and some are seasonally dimorphic, replacing their large winter/spring leaves with smaller, thicker summer leaves (in species of *Sarcopoterium*, *Cistus*, *Phlomis* and *Euphorbia*; Margaris 1977; Christodoulakis 1989; Orshan 1989; Aronne & De Micco 2001). All these shrublands are often very rich in suffrutescents (chamaephytes), such as species of *Helianthemum*, *Fumana*, *Thymus* and *Teucrium*, which may become dominant toward arid zones where they form scrublands with very low amounts of fuel.

Many of these shrubs and subshrubs are shallow-rooted non-resprouters, and they counterbalance their lower root allocation through (1) leaf traits that confer higher drought resistance, e.g. high leaf mass per area, even in comparison with sclerophyllous broadleaf evergreen resprouter species (Paula & Pausas 2006), and (2) a root structure that allows them to better explore the upper soil layer and to transport water more efficiently than resprouter species (Paula & Pausas, 2011).

These associations have many local names depending on the dominant species and the language (Box 4.1). Phrygana is used in Greece and other eastern basin countries (Margaris 1976; Arianoutsou-Faraggitaki & Margaris 1982) and tomillar is a term more common in the west (Dufour-Dror 2002). This association often occurs on disturbed sites, such as those that were terraced in the past and then abandoned, and in woodlands that were cleared, overgrazed and later abandoned (Diamantopoulos *et al.* 1994; Pérez *et al.* 2003; Baeza *et al.* 2006). They also tend to occur on sites with acidic, low-nutrient soils forming typical heathland (e.g. Ojeda *et al.* 1996, 2010) but may appear as gap species in woodlands. Short small-leaved subshrubs also occur in drier areas, forming a transition toward arid or cold steppes.

Shrubs or subshrubs in these communities are shorter lived and smaller than broadleaf sclerophyllous shrubs. Many of the dominant woody species recruit after fire (Table 4.4) from dormant seeds with heat- or smoke-stimulated germination (see Chapter 3). They include many legumes (species of *Ulex*, *Spartium*, *Calicotome*, *Cytisus*, *Genista*, *Chamaespartium* and *Anthyllis*), Cistaceae

Table 4.4 Characteristics of common Mediterranean Basin woody species with postfire seedling recruitment arising from dormant soil-stored seedbanks

Genera	Common species	Family	Resprouting	Fruit type	Leaf type	Leaf size	Growth form
Anthyllis	A. cytisoides, A. hermanniae	Fabaceae	yes	dry	broad	small–medium	shrub
Calicotome	C. villosa, C. spinosa, C. intermedia	Fabaceae	yes	dry	broad	small	shrub
Cistus	C. albidus, C. clusii, C. salviifolius, C. ladanifer, C. creticus, C. mospeliensis, C. crispus, C. laurifolius, C. incanus, etc.	Cistaceae	no	dry	broad	small–medium	shrub
Cytisus	C. striatus, C. scoparius, C. multiflorus, C. balansae, C. patens, C. reverchonii	Fabaceae	yes	dry	variable	small–very small	shrub
Dorycnium	D. pentaphyllum	Fabaceae	yes	dry	linear	medium	small shrub
Erica	E. australis (R+), E. manipuliflora (R+), E. vagans (R+), E. umbellata (R−)	Ericaceae	variable	dry	linear	very small	shrub
Fumana	F. ericoides, F. thymifolia	Cistaceae	no	dry	linear	very small	small shrub
Genista	G. florida (R+), G. berberidea (R+), G. scorpius (R+), G. acanthoclada (R+), G. triacanthos (R−)	Fabaceae	variable	dry	variable	variable	shrub
Halimium	H. viscosum, H. alyssoides, H. halimifolium, H. ocymoides	Cistaceae	no	dry	variable	small–medium	shrub
Helianthemum	H. apenninum, H. syriacum, H. marifolium, etc.	Cistaceae	no	dry	linear	small	small shrub
Pinus	P. halepensis, P. brutia, P. pinaster	Pinaceae	no	dry	needle	small	tree
Rhus	R. coriaria	Anacardiaceae	yes	fleshy	broad	large	shrub
Spartium	S. junceum	Fabaceae	yes	dry	small	small	large shrub
Sarcopoterium	S. spinosum	Rosaceae	yes	fleshy	broad	small	shrub
Rosmarinus	R. officinalis	Lamiaceae	no	dry	linear	small	shrub
Thymus	T. vulgaris (R−), T. piperella (R+)	Lamiaceae	variable	dry	broad	small–very small	small shrub
Ulex	U. parviflorus (R−), U. minor (R+), U. europaeus (R+)	Fabaceae	variable	dry	spines	very small	shrub

Resprouting indicates whether the genus is composed of facultative seeders (yes) or obligate seeders (no) and "variable" means that different responses have been observed between species in the genus or for the same species in different populations (when genera are variable, individual species are noted as "R+" if they resprout and "R−" if not. See Table 4.3 for the legend of leaf size.
Source: Data from the BROT database (Paula et al. 2009; Paula & Pausas 2009).

(some species of *Fumana*, *Helianthemum* and *Cistus*), and aromatic Lamiaceae (species of *Rosmarinus*, *Thymus*, *Lavandula* and *Salvia*). The germination of Fabaceae and Cistaceae is primarily stimulated by heat, whereas the germination cue for Lamiaceae and Ericaceae is mainly smoke (Moreira *et al.* 2010; Table 4.2). Resprouting is variable as some are non-resprouting obligate seeders and others have the capacity to both resprout and recruit after fire (facultative seeders). There is a tendency for non-resprouters to have seeds with a greater heat tolerance and a greater heat-stimulated germination than facultative seeders have (Paula & Pausas 2008).

These communities are highly flammable because communities have a large proportion of their biomass in fine fuels with high risk of ignition. Also, some species are highly aromatic containing flammable oils, plus many are relatively short lived and due to site aridity they accumulate a lot of dead standing biomass. Thus, these communities tend to burn more readily than shrublands dominated by broadleaf plants (Ojeda *et al.* 2010). However, at the drier end of the moisture gradient or where sites have been overgrazed, fuel loads are low and discontinuous, which inhibits fire spread. The regeneration of these communities is often very dependent on the seedbank, and thus they may be more sensitive to the length of the inter-fire period than the shrublands dominated by postfire resprouting shrubs.

Erica is one of the common genera forming small-leaved shrublands. Its center of diversity is in the Cape region of South Africa (see Chapter 7), where most are postfire obligate seeders (Ojeda 1998). In contrast, most species in fire-prone ecosystems of the Mediterranean Basin resprout from lignotubers after fire or clipping (e.g. *E. arborea*, *E. multiflora*, *E. australis*, *E. scoparia*; Canadell *et al.* 1991; Lloret & López-Soria 1993; Vilà & Terradas 1995; Cruz & Moreno 2001a) and very few do not (e.g. *E. umbellata*; Quintana *et al.* 2004). Many of these *Erica* species have dormant seedbanks stimulated to germinate by smoke and thus they recruit well after fire (Table 4.2; Moreira *et al.* 2010).

Another fine-leaved resprouter in the western part of the Basin is the tussock grass *Ampelodesmos mauritanica*. This species accumulates large amounts of fine, dead fuel (Vilà *et al.* 2001) in a loose airy structure that is very susceptible to burning. Fire does not penetrate the tightly packed base of the tussock, which survives and the fuel quickly regenerates. Furthermore, this species flowers and produces abundant seeds after fire, resulting in heavy recruitment in subsequent postfire years (Lloret *et al.* 2003).

The mediterranean old-field includes a range of vegetation combinations that depend on environmental conditions: age since abandonment, previous crop and interactions with grazing (Debussche *et al.* 1996; Bonet & Pausas 2004, 2007). Recently abandoned fields are quickly occupied by herbaceous vegetation. A dominant perennial grass in many old-fields is the rhizomatous *Brachypodium retusum*, which resprouts quickly after fire. This species appears in most western plant communities (woodlands, shrublands, and in semi-arid conditions) but it shows its maximum occurrence in early stages of old-field succession (Caturla *et al.* 2000).

In the eastern part of the basin, a good colonizer of old-fields is the thorny dwarf shrub *Sarcopoterium spinosum*, which can form large and continuous monospecific formations, especially in overgrazed areas. It has clonal growth and a strong postfire resprouting capacity, and recruits profusely, both after fire and in the absence of fire (Henkin *et al.* 1999; Seligman & Henkin 2000).

One of the early shrubs appearing in old-fields in the western Mediterranean Basin is gorse or *Ulex parviflorus*, a very flammable shrub with fine fuels and retaining a large proportion of dead biomass (Fig. 4.5; Baeza *et al.* 2006, 2011). It is a non-resprouter and the density of this obligate seeder can be very high, making the old-field a very flammable community that rapidly replaces itself after fire. Other obligate seeder species may coexist with *Ulex parviflorus*, such as some *Cistus* species. Postfire recruitment of *Rosmarinus officinalis* is typically lower than that of *Cistus* and *Ulex*, but it lives longer and may become abundant in later stages of old-field succession or on colder inland sites.

Because most broadleaf evergreen shrubs are more shade tolerant and longer lived than these small leaf, malacophyllous species, they are often functionally late-successional species. Thus, with increasing time since abandonment, resprouters and facultative seeders may colonize and become abundant. It is interesting that resprouting species tend to appear earlier in abandoned tree crops (e.g. olive or carob tree crops) than in herbaceous crops due to the fact that previous tree crops act as perches for birds (Debussche *et al.* 1982; Pausas *et al.* 2006a) and many resprouters have bird-dispersed fleshy fruits (Table 4.3; Pausas *et al.* 2004b) and recruit in the absence of fire (see Chapter 9). Pines may also colonize these old-fields, depending on the distance from a seed source tree.

Pines of the Mediterranean Basin

Pines (*Pinus* spp.) grow naturally in many places in the region, but they have also been extensively planted (Pausas *et al.* 2004a, 2008), and not always from local provenances. Current pine communities are highly variable, and most understory species are the same ones in surrounding shrublands and woodlands. Pines occur at low or moderate densities, or they may form relatively dense monotypic forests. They include pines from both crown fire regimes and surface fire regimes. They are highly flammable and as a consequence of widespread planting have contributed to large, high-intensity crown fires (Pausas *et al.* 2008). Comparing bark thickness and bud tolerance to heating, they can be ranked from the most resistant to the least resistant to fire as follows: *P. pinaster*, *P. pinea*, *P. nigra*, *P. halepensis*, *P. brutia*, *P. sylvestris* and *P. uncinata* (Fernandes *et al.* 2008).

Pinus halepensis (Aleppo pine) and *P. brutia* (Turkish pine) are typical of pines in crown fire regimes (Keeley & Zedler 1998). They are thin barked with branches from the base and are killed by most fires and do not resprout. Serotinous cones open after fire and seed viability in the soil is short lived (Daskalakou & Thanos 1996; Pausas *et al.* 2004a); thus, they recruit profusely in the first growing season (Fig. 4.7; Trabaud *et al.* 1985; Moravec 1990; Thanos *et al.* 1996;

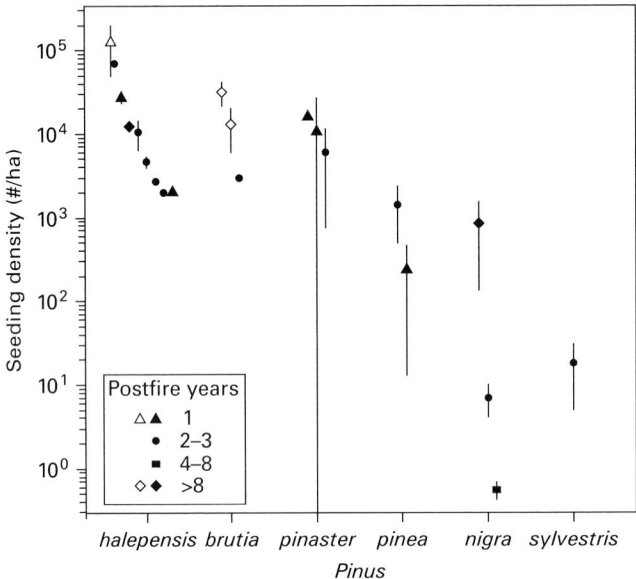

Fig. 4.7 *Seedling density after crown fire (log scale) for Mediterranean Basin* Pinus *species in different localities and with different postfire ages. Position of the symbols indicates the mean values, and type of symbol indicates the postfire age (time since fire); vertical lines are standard deviation (in solid symbols) or range values (in open symbols). (Compiled from different sources by Pausas et al.* 2008.)

Herranz *et al.* 1997; Ne'eman 1997; Tsitsoni 1997; Arianoutsou & Ne'eman 2000; Leone *et al.* 2000; Pausas *et al.* 2004c). The degree of serotiny is quite variable and populations arising from colonization have a lower degree of serotiny than postfire populations (Goubitz *et al.* 2004), illustrating fire selection for serotinous cones. The degree of canopy combustion affects the temperatures experienced by the cones and, thus, seed mortality (Vega *et al.* 2010). The role of pine nut consumers in shaping serotiny (e.g. Mezquida and Benkman 2004) remains to be explored in these species. As wildfires become more frequent, the probability of immaturity risk *sensu* Zedler (1995a) increases. Both *P. halepensis* and *P. brutia* produce cones at a relatively early age (<10 yrs, Thanos & Daskalakou 2000), but they do not produce a significant canopy seedbank before the age of about 10–15 yrs and, thus, inter-fire periods shorter than this can result in local extirpation (Pausas 1999; Arianoutsou *et al.* 2002).

A positive relation between *P. halepensis* seedling mortality and grass cover suggests competition is a factor in seedling recruitment (Pausas *et al.* 2003). Pine seedling growth and survivorship are also higher within the canopy of the pine skeleton (Ne'eman *et al.* 1992; Ne'eman 2000) and this is likely tied to limited competition and enhanced resources. The higher fire intensity in the thick leaf litter beneath the pines eliminates most competitors and increases soil fertility, an effect also observed in North American serotinous pines and cypress (Ne'eman *et al.* 1999; Keeley & Zedler 1998). Low-intensity fires may not completely combust the needle layer and affect soil conditions for postfire germination (Pausas *et al.* 2008). In eastern Spain, Pausas *et al.* (2002, 2003) found no differences in seedling densities between different fire severities measured by crown scorch, but both seedling height and biomass were higher on high fire severity sites where most litter was consumed. Fire severity was also correlated with higher postfire soil phosphorus content, as a consequence of increased microbial activity (Fierro *et al.* 2007).

In North Africa, *P. halepensis* may coexist with another conifer, *Tetraclinis articulata* (closely related to the Australian *Callitris*). It is believed that in the past *Tetraclinis* was a very common tree throughout the area, but the high quality of its wood (it is both aromatic and resistant to rotting) made it valuable to past civilizations. In spite of this, we can still find open woodlands of *Tetraclinis*, thanks to its resprouting from a lignotuber after cutting and fire (Naveh 1975). However, even these burls, which become massive after recurrent coppicing and burning, have been extensively collected for their quality and beauty (Charco 1999).

Pinus pinaster (maritime pine) is a highly variable species and some of this variation is likely tied to past fire regimes. North African populations are highly serotinous and have thin bark typical of pines from crown fire regimes, whereas Atlantic coast populations have low serotiny and thick bark suggesting a surface fire regime (Tapias *et al.* 2004). Unlike *P. halepensis*, *P. pinaster* can survive with a high proportion of the crown scorched (Botelho *et al.* 1998). Under surface fire regimes dendrochronology studies record a mean fire interval of 14.5 yrs for *P. pinaster* over the last 180 yrs (Vega 2000). Given that *P. pinaster* is used for wood production, some Atlantic stands are maintained at very high tree densities and under a fire exclusion policy, so that when burned, they produce high-intensity crown fires. Even in these cases, recruitment seems to be relatively good (P. Fernandes personal communication), although Pérez *et al.* (1997) observed almost no regeneration of *P. pinaster* after a single fire in the subhumid central part of Spain, and attributed this to competition with the shrub *Cytisus eriocarpus*.

Pinus pinea (stone pine) has characteristics of pines from a surface fire regime. It has thick bark and no branches in the lower portion of the tree due to self-pruning of dead branches. It readily survives surface fires, and can survive crown fires with more than 80% of the crown volume scorched (Rigolot 2004; Catry *et al.* 2010). As is the case with other pines from surface fire regimes it does not have serotinous cones (Keeley & Zedler 1998). It has some immediate postfire regeneration (Fig. 4.7) due to the annual seed crop being protected in thick cones and by a dense seed coat (Escudero *et al.* 1999). Due to the fact that this pine produces edible nuts, it has been planted for a very long time, and thus its natural distribution and habitat are uncertain. It seems to have been originally limited to poor sandy soils in dune ecosystems where understory fuels were low. However, its current distribution includes dense forests that sustain crown fires. In such conditions a general decline has been observed in some areas (Rodrigo *et al.* 2007).

Fig. 4.8 *Two examples of surface fire regimes: (a)* Pinus nigra *forest with fire-scarred trees (El Turmell, Baix Maestrat, Spain; photo: Juli Pausas) and (b) cross section of* Abies pinsapo *with dated fire scars (Sierra Bermeja, Málaga, Spain; photo by J.A. Vega; see more details of this forest in Vega 2000.)*

Pinus nigra (black pine) is a thick-barked, non-serotinous, long-lived pine that grows at higher elevations in cooler environments, and its seeds are sensitive to the high temperatures produced during wildfires (Escudero *et al.* 1999; Habrouk *et al.* 1999; Núñez & Calvo 2000). It can survive low severity surface fires, and fire scars on living trees illustrate a history of frequent surface fires (Fig 4.8; Fulé *et al.* 2008). Dense plantations and fire suppression have changed the fire regime toward less frequent crown fires, extirpating this species after large high-intensity crown fires (Trabaud & Campant 1991; Retana *et al.* 2002; Rodrigo *et al.* 2004; Pausas *et al.* 2008).

Other montane pines are *P. sylvestris* (Scots pine) and *P. uncinata* (mountain pine), which are thin-barked, non-serotinous pines that cannot survive crown fires. They grow in environments where fires are currently rare and small, and recolonize burned areas from the edges or from refugia such as rocky outcrops or ridges, similar to some California pines (Schwilk & Keeley 2006). However, the fire history of these pine forests is poorly known and they may have historically suffered more frequent fires as in other European pine woodlands (Stahli *et al.* 2006; Niklasson *et al.* 2010).

Although the volcanic Canary Islands on the north Atlantic coast of Africa are not within the Mediterranean Basin there is a mid-elevational zone with a MTC and a uniquely fire-adapted pine. *Pinus canariensis* (Canary Island pine) is a long-lived tree with thick bark and serotinous cones, and a strong resprouting capacity from epicormic buds (Climent *et al.* 2004), a trait that is rather rare in pines and other gymnosperms (Keeley & Zedler 1998). Fires are frequent due to lightning and possibly volcanic activity and this pine has the capacity to regenerate its crown rapidly following crown fires, not unlike many Australian eucalypts (see Chapter 3).

Other Mediterranean Basin Vegetation Types that Rarely Burn

One type of oak woodland that rarely burns is the *dehesa* (Spain) or *montado* (Portugal) on the Iberian Peninsula. These are mainly open oak woodlands or savannas of anthropogenic origin (Klein 1920; Stevenson & Harrison 1992). They are dominated mainly by evergreen oaks (*Q. ilex* ssp. *ballota*, *Q. suber*) although deciduous oaks may also be common (*Q. pyrenaica*, *Q. faginea*). These woodlands were traditionally used for raising livestock such as cows, sheep and pigs, and for cultivation of cereal crops. The original natural vegetation is, in most cases, oak woodlands that were modified by thinning, leaving trees for acorn production and shade for livestock, or for cork production in *Q. suber dehesas*. The consequence of this agroforestry system is that the oaks in *dehesas* are aging with a lack of regeneration (Pulido *et al.* 2003; Pausas *et al.* 2009). In these ecosystems intense grazing reduces the herbaceous surface fuels required to carry fire. When grazing is excluded, these open woodlands tend towards more close oak forests, and then they can sustain crown fires. Some other species, such as *Olea europaea* var. *sylvestris*, *Arbutus unedo*, *Pistacia atlantica*, *Castanea sativa* or *Ceratonia siliqua*, may also form *dehesas* in some places. *Argania spinosa* (the only member of the Sapotaceae in the Mediterranean Basin) also forms *dehesa*-type woodlands in western Morocco, in very warm areas with very low rainfall but high air moisture from the Atlantic. The tree is grazed by goats, and the fruits are used as a source of cooking oil and for cosmetics.

In the mountains and high plateaus, with cooler summers and colder winters, occur weakly flammable woodlands dominated by conifers. These include *Juniperus phoenicea* (throughout the basin), *J. thurifera* (western basin), *J. excelsa* and *J. drupacea* (eastern basin), *Pinus sylvestris* and *P. uncinata* (Iberian mountains), and other conifer species of restricted distributions like *Cedrus* (North Africa and the Near East) and *Abies* (North Africa, Greece and southern Spain). These conifers have no mechanism to persist after crown fires, and thus may have historically suffered a very low frequency of crown fires with subsequent colonization from the edges (Stahli *et al.* 2006). In the event fires are limited to understory fuels, some of these species can survive such surface fires (Fig. 4.8; Vega 2000). Many of these conifers have been heavily logged for millenia, to be used in the ship industry and house building, and their current distribution is dramatically reduced.

Other vegetation types that rarely burn because of low fuel levels and low fuel continuity are the scrublands that appear in semi-arid continental MTCs, forming a transition toward the steppe (e.g. central Iberian and Anatolian peninsulas). These low shrublands have been traditionally overgrazed and are thorny formations (*Genista*, *Sarcopoterium*), often with abundant aromatic plants (e.g. *Thymus*, *Rosmarinus*, *Salvia*). Toward the southern and eastern regions, with low rainfall, vegetation is dominated by perennial grasslands such as *Stipa tennacissima* (tussock grass), together with some small shrubs (e.g. *Artemisia*). Cover is typically low, and the bare soil between tussocks inhibits the spread of fires. Where aridity is combined with high water table, salt bushes (e.g. *Atriplex*, *Salsola*, *Suaeda*) appear extensively. Another semi-arid community found in northern Africa is the woodland dominated by *Acacia gummifera*, which yields abundant gum marketed locally and is the only species of this genus that can be considered to reach the Mediterranean Basin, in a transition toward desert vegetation.

In Mediterranean Basin mountains a number of winter semi-deciduous (marcescent) oaks are common that are within the MTC and form a transition to temperate forests. These may coexist with evergreen oaks in moist gullies and pole-facing slopes or with deciduous oak woodlands. Examples are *Quercus faginea*, *Q. cerrioides*, *Q. pyrenaica* and *Q. canariensis* in the west, and *Q. ithaburensis* and *Q. infectoria* in the east. Typical/fully winter-deciduous species also occur in the basin, but are restricted to especially mesic sites and include oak woodlands of *Q. pubescens* (= *Q. humilis*), *Q. frainetto*, *Q. trojana*, *Q. petrea* and communities of *Ostrya carpinifolia*, *Carpinus orientalis*, *Fraxinus ornus*, *Sorbus*, *Acer*, *Aesculus*. Sweet chestnut or *Castanea sativa* woodlands also occur in sub-mediterranean environments; this deciduous species is considered indigenous to the Balkan peninsula and northern Turkey but it is widespread and naturalized throughout southern Europe. Most of these deciduous species are resprouters and have been coppiced for a long time and fires are very rare. The Italian chestnut and pasture landscapes are the product of a historical management based on burning the dead leaves on the floor to promote pasture under trees (Grove & Rackham 2001).

At the tops of mountains and ridges, where wind is a limiting factor, together with grazing, spiny cushion-shaped dwarf shrubs with *Erinacea* species and *Juniperus sabina* form oro-mediterranean communities. Fuels are low and discontinuous and fires are small and infrequent.

Riparian vegetation in the north and at high altitudes, with constant water flows, may seldom be subjected to fires and these sites sometimes act as refugia for temperate deciduous trees at their southernmost boundaries or Tertiary (see Fig. 9.1) relict species like the *Liquidambar orientalis* (Altingiaceae, Turkish sweetgum). These riparian ecosystems also harbor high biodiversity because they provide refugia for species near the edge of their distribution (Hampe & Petit 2005). Under drier conditions in the lowlands and further south, water flow is seasonal and may dry sufficiently to carry fire. As with woody riparian species throughout the world these trees are obligate resprouters and include *Nerium oleander*, species of *Salix*, *Populus* and *Ulmus*. Resprouting in these lineages may be as much a response to winter channel-scouring damage as to fire. Seedling recruitment is not tied to fire but rather to the annual flooding cycle.

Marshes, wetlands and coastal lagoons used to be frequent all along the Mediterranean coast in many countries of the region. In the past many were drained for agricultural purposes and more recently for building houses and tourist resorts. Remnants of these ecosystems are still very important biodiversity spots and for the ecosystem functions they provide such as maintaining the local water cycle (Millán 2002). Coastal dunes, with *Pinus pinea* and *Juniperus* species, have also been dramatically reduced. Although under the right conditions these ecosystems may be flammable, they do not regularly burn.

Fire Management

Fire has been used as a management tool for thousands of years (Pausas & Keeley 2009). Shepherds around the Mediterranean Basin have traditionally burned shrublands and grasslands to promote palatable plants and stimulate growth. Indeed, fire has often been considered as necessary to shepherding as ploughing is to farming. Even in some deciduous woodlands, leaf-litter fires were lit to improve understory pastures. However, at least in mediterranean Europe, the current cultural framework suggests that dense forests are the best possible vegetation for any landscape, and that shrublands are the product of degraded forests. In this framework fires are viewed by land managers and society in general as bad, and land management is largely focused on reducing fires and increasing forests (Seijo 2009). This is mainly a cultural parading, not always based on scientific grounds, and is strongly influenced by the northern European forest tradition. The management of MTC ecosystems requires a paradigm shift, whereby ecosystems are viewed in their ecological, historical and biogeographical context and where fire regimes are an important ecosystem process (Pausas & Vallejo 2008).

The consequences of this current land management framework, together with the recent socio-economic changes, is that fire use has declined and most places it has been outlawed. This has produced changes in landscape-level fuel loads and a shift from frequent light fires to infrequent high-intensity wildfires in some mountain coniferous forests. Indeed, surface fires are currently very rare in the Mediterranean Basin whereas they were probably common in the past in mountain coniferous forests (Vega 2000; Fulé *et al.* 2008). The combination of the abandonment of fire as a management tool, the reduction of grazing, interest in maximizing wood production, and efficient fire-prevention and fire-suppression measures, have worked in concert to decrease fire frequency and increase fuels. Thus, when a fire occurs in these ecosystems, it is in the form of a crown fire. Most mountain coniferous trees do not have traits to regenerate after a crown fire, and thus these fires are threatening forest structure and biodiversity. There is increasing evidence that periodic fires or grazing would help to maintain surface fires and this would contribute to fewer crown fires and greater conservation of these mountain forests (Vega 2000).

A very different story is occurring in the lowlands (garrigues, heathlands, maquis, and other chaparral-like vegetation), where vegetation is able to successfully regenerate quickly after repeated fires, and thus fires do not normally pose a threat to biodiversity, except where extremely high fire frequencies stress the tolerance of some species. The major fire problem in the coastal regions is the large human population density that is expanding into highly flammable watersheds and putting people and associated infrastructures at risk to fire. Current mediterranean societies have failed to adapt to living in the midst of such flammable ecosystems, and most current urban planning does not adequately consider the fire risks in planning decisions. Given the increasing population density in coastal areas, the most critical issue for fire management is acting at the wildland–urban interface (WUI) to protect properties and lives. Specific regulations for living in fire-prone landscapes such as regulations for building, urban planning, gardening, etc. are lacking in many Mediterranean Basin countries or seldom followed. In fact, many settlements in the WUI are illegal or have been legalized after building (i.e. without adequate planning, and without following appropriate regulations). In this sense, currently, the most advanced country is France, where there are specific regulations and recommendation guides for living in forested areas. The limited consideration of recurrent fires when planning homes at the WUIs may have catastrophic consequences when the inevitable fires occur. This has been seen during recent European heat waves with large fires reaching extensive WUI areas. For instance, the wildfires in Greece during summer 2007 destroyed 2850 homes and caused 78 fatalities (Xanthopoulos 2007).

The current and most widespread fire management strategy in the Mediterranean Basin combines fire prevention through fuel treatments with fire suppression. The objective is to minimize the area burned, especially near urban and suburban areas. Typical fuel treatments consist of fuel break networks of different widths, maintained mechanically or more rarely by grazing/browsing. Fuel breaks require frequent maintenance, otherwise early successional and highly flammable grasses and small shrubs invade and diminish their value as barriers to fire spread. There is evidence that under certain extreme weather conditions fuel breaks, including major highways, are ineffective, while in other cases they have successfully protected properties. However, a detailed cost-effectiveness analysis remains to be done (Rigolot *et al.* 1999). On complex landscapes (e.g. Fig. 4.2), cultivation of less flammable crops such as vineyards and olive groves may act as fuel breaks, although under severe fire weather, such as the 2007 Greek fires (Xanthopoulos 2007), even these landscapes will burn.

Prescribed burning is only used locally or sporadically in France, Portugal and Spain, but it is not even allowed in Greece, Turkey and most of Italy. The technique is well established only in France, where it is used for fuel reduction and for reducing shrubland colonization of open pastures in order to improve vegetation quality for grazing and wildlife habitat (Rigolot *et al.* 1998). In Portugal, fuel-reduction burning is being used in productive *Pinus pinaster* stands (Rego *et al.* 1987; Fernandes *et al.* 1999), and in Turkey, understory burns have been tested for reducing competition and improving seed beds for regeneration of target tree species (*Cedrus libani*; Boydak *et al.* 1998).

Modeling is a useful tool for testing alternative fire management options. Simulations applied to the Iberian peninsula suggest that fire exclusion scenarios slightly enhance large fires, whereas prescribed fire scenarios reduce them; however, the total area burned did not vary with the two scenarios (Piñol *et al.* 2005). Furthermore, even without changes in the total area burned, different fire management scenarios produce different spatial distribution of local fire regimes, and thus they have different consequences on the persistence of the different plant types (Pausas 1999; Pausas & Lloret 2007).

A prominent effort is now focused on using remote sensing information and GIS tools to assess fire management issues that involve prefire, during fire and postfire assessments (Chuvieco 1999, 2009). This includes fuel mapping (Riaño *et al.* 2002), assessment of fire risk by calibrating satellite data with fuel moisture and then providing fire risk maps (García *et al.* 2008), forecasting fire danger from meteorological data plus GIS information (topography, vegetation, fire history, etc.) (Chuvieco *et al.* 2004), detecting fires and following their growth using real-time satellite data (Martin *et al.* 1999), estimating and mapping burned areas and burned severity using spectral signatures (Mitri & Gitas 2008; De Santis *et al.* 2009), and monitoring postfire regeneration from vegetation indices (Díaz-Delgado *et al.* 2002; Abdel Malak & Pausas 2006). Although all these tools are currently quite well developed, they are seldom routinely used for land management by forest administrations.

Because many mediterranean landscapes have been continuously modified for a long time (terracing, plantations, etc.), recurrent high-intensity fires may lead to soil losses and degradation problems (for a review on erosion in the region, see Pausas *et al.* 2008). In addition, there is a cultural framework where dense forests are the target for many landscapes in the region. The traditional approach was to create extensive plantations in burn areas, mainly with conifers, sometimes with non-native species (Pausas *et al.* 2004a, 2008). The objective was mainly watershed protection, and sometimes wood production. With the new social demands for conservation of biodiversity and ecosystem services, restoration actions are increasingly focused on: (1) soil and water conservation, (2) improving resistance and resilience of ecosystems, (3) increasing mature woody formation, both shrublands and forests, and (4) promoting biodiversity (Vallejo *et al.* 2006). The techniques are mostly based on rapid mulching and/or seeding after fire as well as plantations of resprouting shrubs and trees. Although there is now sufficient knowledge for applying these restoration methods, there is still a need for better understanding of microsite preferences for species and for long-term monitoring to evaluate effectiveness.

Conclusions

The most striking difference between the Mediterranean Basin and the other MTC regions is its substantially longer history of intensive and widespread land use. Little of this landscape has escaped human impacts and this has implications for fire ecology and management. An appropriate metaphor is that of a palimpsest, where modern plant communities are overlaid on landscapes still containing markings of earlier uses. All evergreen shrub species are obligate postfire resprouters and a few have well-developed lignotubers. Resprouting success is very high and maquis shrublands regenerate rapidly after fire. Thus, the small postfire gaps have selected against postfire seeders, which are less common in maquis shrublands. These obligate resprouting shrubs and small trees avoid the stressful summer drought by means of deep roots and thus seedlings are dependent on finding favorable mesic microsites. This is enhanced by production of fleshy, vertebrate-dispersed fruits with long-distance dispersal. On more arid sites shorter-lived semi-deciduous shrubs or subshrubs exhibit a very different fire response in that most have specialized reproduction by delaying it until the postfire environment. Many leaf and stem traits of species on these sites contribute to making these communities highly flammable. Site aridity coupled with a high frequency of high-intensity fires contribute to the abundance of postfire gaps and selection for postfire seeders. The majority of communities are crown fire ecosystems although a few montane pines historically burned in surface fire regimes. Land use changes during the latter quarter of the twentieth century have resulted in a shift in fire regime. This is due to rural depopulation, abandonment of farmland and recolonization of sites by woody species that greatly add to the fuel load and increase the size and intensity of wildfires. This coupled with urban expansion is creating a wildland–urban interface fire problem not previously seen in this region.

5   Fire in California

On the west coast of North America lies the state of California, USA (Fig. 5.1), the bulk of which is dominated by a mediterranean-type climate (MTC). Elevations range from sea level to over 4000 m. Mountain ranges are largely oriented north to south with a major valley between the coastal ranges and the interior Sierra Nevada range. In the rain shadow east of the interior mountain ranges the climate is more continental with much colder winters and increasing proportion of summer precipitation eastward. This easternmost part of the state has steppe climates in the northern portion and desert climates in the south. In Arizona and a few other parts of southwestern USA and northeastern Mexico are disjunct patches of sclerophyllous-leaved vegetation that closely resembles California MTC vegetation. These include evergreen shrublands, broadleaf woodlands and conifer forests and represent mediterranean-type vegetation (MTV) under non-MTCs. Further east at similar latitudes but under different climates are sclerophyll forests with many similarities to MTC conifer forests.

The California Floristic Province (Raven & Axelrod 1978) essentially circumscribes the MTC vegetation of North America and extends across the latitudinal range of the state. On the western slopes of the major mountain ranges is a rich diversity of vegetation types that change along the elevational gradient. Ascending the coastal mountains the main vegetation types sort out along gradients of decreasing aridity in the following order: grasslands, semi-deciduous woody sage scrub, evergreen chaparral shrublands, oak woodlands and conifer forests. A similar pattern is evident on the west side of the interior Sierra Nevada except for the absence of sage scrub. These vegetation types exhibit marked differences in fire regime and tolerance to disturbance tied to the different patterns of fuel structure resulting from changes in dominant growth forms along the elevational gradient. Along this gradient there is an interaction between fires and aridity such that lower fire frequency is required to displace shrubland associations with grasslands and other herbaceous vegetation on xeric than on mesic landscapes (Keeley 2002b). Consequently there are complex local mosaics due to differences in aspect and fire history (see Fig. 1.6c).

Associated with these vegetation types are gradients in primary productivity, seasonality and ignitions that result in very different ecosystem roles for fire. For example, the fuel structure in chaparral and sage scrub shrublands maintains biomass throughout the shrub profile resulting in all fires burning as

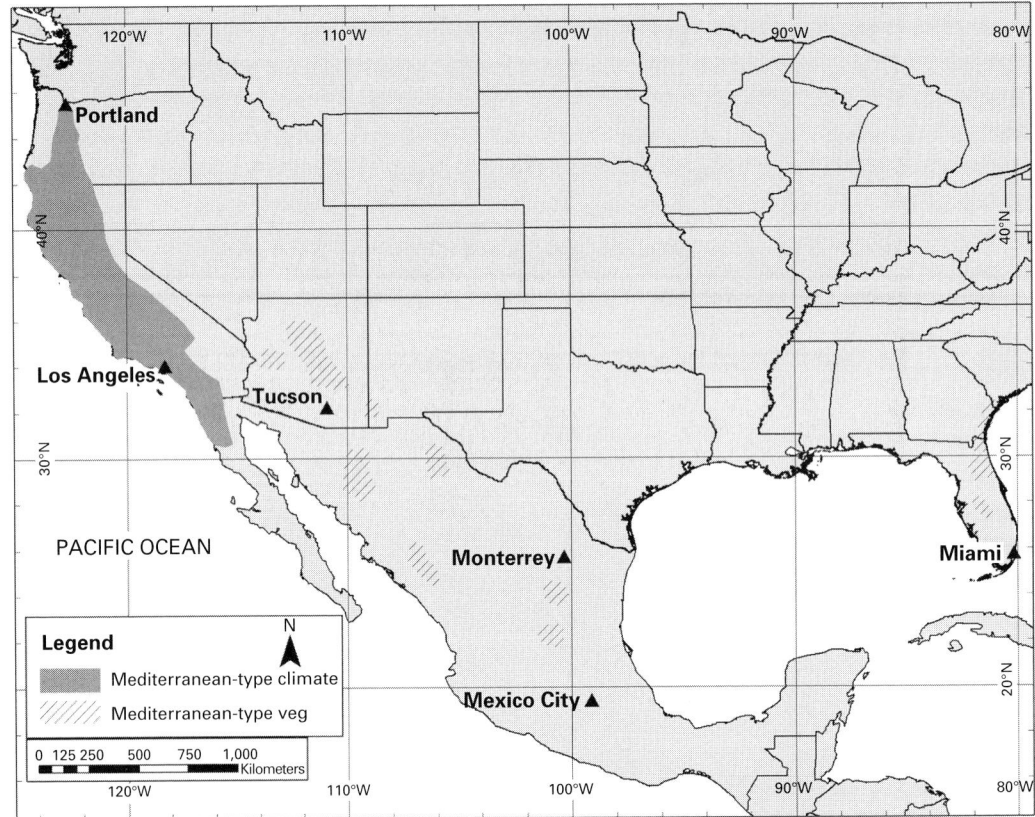

Fig. 5.1 *Distribution of the mediterranean-type climate (MTC) in California and adjacent parts of North America (dark shading) and shrublands dominated by mediterranean-type vegetation (MTV) outside the MTC region (hatched).*

high-intensity crown fires. In contrast, at the more mesic end of the elevational gradient, montane mixed conifer forests accumulate dead surface fuels (see Table 2.1), and because of higher growth rates maintain their canopies far above the surface fuels; this separation of surface and canopy fuels is conducive to surface fires. Other surface fire regimes include the lower-elevation ponderosa pine forest and blue oak savanna ecosystems that are more open and driven by herbaceous surface fuels. Here we illustrate the range of fire regimes in California by contrasting the crown fire regime in chaparral with the surface fire regime in conifer forests.

Chaparral Crown Fire Regime

The taller-stature evergreen chaparral replaces sage scrub on mesic north-facing slopes at low elevation (Fig. 5.2), and on most slopes between about 600 m and

Fig. 5.2 *Chaparral and sage scrub mosaic in southern California. (Photo by Jon Keeley.)*

1500 m. Although there are fine-grained differences in plant associations related to slope aspects, fires generally burn in a coarse-grained manner (Keeley 2006a).

Fuel Patterns

Mature shrublands typically range from 1–5 m in height and form a dense closed canopy that excludes most herbaceous surface fuels. Dead surface fuels accumulate slowly because many species retain a substantial proportion of dead branches in the canopy (Schwilk 2003). This fuel structure results in a fire regime dominated by active crown fires with relatively little surface fire.

On most sites of moderate fertility a postfire ephemeral flora of annuals and short-lived perennials germinate from seeds that survive fire and develop dense stands during the early seral stage of postfire recovery. This ephemeral flora produces a significant load of fine fuels that die back each summer (Fig. 5.3) and, coupled with dead skeletons from the last fire, comprise fuel loads often in excess of 10 Mg ha^{-1} (see Table 2.1). These fine fuels are easily ignited and are sufficient to carry fire (Fig. 5.4), but when they occur within the first 10 yrs they are generally detrimental to the recovery of the shrub dominants.

Typically within the first decade after fire the shrub canopy closes and the ephemeral flora remains primarily as dormant seedbanks. Over the subsequent decade, as shrub canopies expand, the ratio of live to dead fuel may be too high to carry fire except under severe weather conditions. Fuel loads tend to increase with stand age although there is substantial species-specific variation in the relationship between stand age and live/dead ratio that may be an important determinant of flammability under all but the most extreme conditions.

Fig. 5.3 *Seral stage chaparral 5 years after fire dominated by resprouting* Adenostoma fasciculatum *and a dense layer of ephemeral subshrubs arising from dormant seedbanks (primarily* Lotus scoparius*) illustrating the high potential flammability of postfire chaparral during the first decade. This scene is the site of the 2002 Bouquet Canyon in northern Los Angeles County and 2700 ha reburned in the 2007 Buckweed Fire. (Photo by Jon Keeley.)*

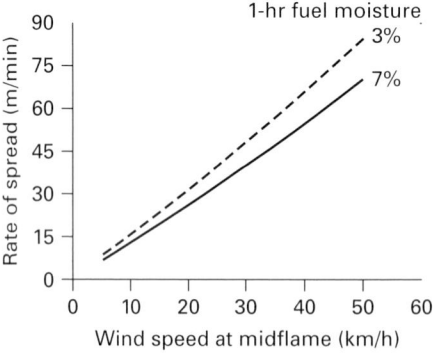

Fig. 5.4 *Behave model results for rate of fire spread for the young chaparral illustrated in Fig. 5.3 demonstrating that, contrary to some assertions, young fuels in postfire chaparral can carry fire (from Keeley & Zedler 2009). Such reburns are a major contributor to type conversion of native shrubland to alien annual herbaceous associations (see Chapter 12).*

As a consequence of these different successional stages, these communities go through a change from being highly vulnerable to fires during the first 5 yrs or so, then less susceptible for a decade or more (Schoenberg *et al.* 2003) until dead fuels accumulate in the shrub canopies (Fig. 5.5). In general, there is about a 30% live/dead ratio for mature stands that varies according to shrub species composition and stand age (see Table 2.1). Shrub species that are fire dependent for

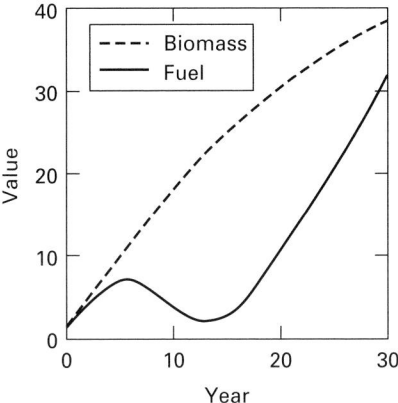

Fig. 5.5 *Schematic model of temporal changes in biomass and fuels in chaparral. In the early years the herbaceous ephemeral postfire flora dries completely during the summer drought and contributes to substantial fuels during the early years, but as the shrub canopy closes in this understory flora is shaded out by less flammable green shrub foliage.*

seedling recruitment (*Adenostoma*, *Arctostaphylos* and *Ceanothus* species) have a marked tendency to retain dead branches in the canopy and this has important effects on subsequent fire intensity (Schwilk 2003). Other shrubs that recruit in the absence of fire such as species of *Quercus*, *Prunus*, *Rhamnus* and *Rhus* self-prune dead branches, and are expected to have a higher live to dead ratio, although this has not been well documented. The retention of dead branches is hypothesized to be part of a character syndrome with evolutionary implications as discussed below.

Landscape patterns of fuels exhibit spatial variation as well. From south to north, shrublands dominate a decreasing proportion of the landscape and thus shrubland fires tend to be the largest in the southern half of the state. Even within this region there are marked differences in fuel patterns that affect fire size (Keeley & Zedler 2009). The largest fires (> 50 000 ha) have mostly occurred either in San Diego County or further north in Santa Barbara/Ventura Counties because the topography of both regions supports large contiguous east–west swaths of shrubland fuels where both offshore and onshore wind flows can drive fire over very long distances. An important exception is the 2009 Station Fire, the largest fire in Los Angeles County. Unlike most large fires it was not driven by Santa Ana winds (see Box 1.3) and this likely contributed to its extraordinary size. Fires driven by these winds generally have a northeast to southwest trajectory and in Los Angeles County this constrains ultimate fire size (Keeley & Zedler 2009). Due to the lack of Santa Ana winds, the Station Fire was a plume-driven fire (see Fig. 1.2), which created intense downdrafts that spread the fire along multiple fire fronts. Although fuels were not outside the historical range of variability for this type, about half of the landscape was older than typically the case in Los Angeles County, and this, coupled with very rugged terrain and multiple years of intense drought, contributed to this large fire event.

Table 5.1 Lightning- and human-ignited fires in selected sites in California

	Total fires (#/million ha/yr)	Percentage due to lightning
Coastal southern California	4290	< 1
Coastal central California	117	17
Coastal northern California	507	3
Interior southern California	2803	6
Interior central California	117	17
Interior northern California	456	55

Source: Based on data from Keeley (1982).

Ignition Patterns

In California humans have been a source of ignitions for only slightly more than 10 000 yrs, but the early Holocene populations may not have been large enough to have had a very widespread influence on fire regimes. However, very little is known about these early stages of New World colonization. By the mid Holocene the expanded utilization of seeds, in particular acorns from oak trees, was coupled with increased populations. For the past ~5000 yrs Native Americans potentially affected significant portions of the California landscape (Erlandson & Glassow 1997), and on many shrubland landscapes at low elevation they undoubtedly increased fire frequency over that due to lightning alone (Timbrook et al. 1982; Keeley 2002b).

Prior to the colonization by Europeans and Americans in the nineteenth century, lightning would have been the primary source of ignitions over vast stretches of rugged and uninhabited parts of the chaparral region. Even today lightning contributes to some significant fires in remote areas. Lightning-ignited fires vary spatially because summer thunderstorms are rare near the coast and most frequent at higher elevations in the interior (Keeley 1982, 2006a; Greenlee & Langenheim 1990; van Wagtendonk 1993). In coastal southern California lightning-ignited fires are uncommon (Table 5.1) and tend to be concentrated in the late summer just prior to the Santa Ana wind season (see Box 1.3). In the northern California San Francisco Bay area, lightning fires are quite rare (four lightning fires per decade per 100 000 ha, Keeley 2005).

Today most ignitions in coastal California are started by people and commonly near the wildland–urban interface. These ignitions increased in frequency during the twentieth century, concomitant with population growth (Keeley & Fotheringham 2003a). However, toward the latter part of the century ignitions in many parts of the state apparently have reached a threshold and leveled off or even declined. This change has been ascribed to patterns of housing development in the wildland–urban interface, particularly as the early development stages that form intermixes with wildland areas transformed to more classical interface zones (Syphard et al. 2007).

Fire Regime

The continuity of shrub canopies and lack of surface fuels ensures that these shrublands burn in stand-replacing crown fires. Commonly crown fires will burn large portions (i.e. 10^3–10^4 ha) of the landscape, leaving few unburned patches.

Historically the fire cycle was rather long and likely limited more by ignitions than by fuels (Box 5.1). In coastal mountain ranges this frequency was perhaps as low as once or twice a century, but in interior mountain ranges the higher lightning frequency would have promoted higher fire frequencies, perhaps several fires a century (Keeley 2006a). The peak season for fires was in the summer and these often remained small and of low intensity due to the higher humidity and less severe winds at this time of the year (Minnich 1987a). Further, the peak for lightning is in late summer and early autumn and it nearly overlaps with the autumn Santa Ana wind season; thus it is inconceivable these landscapes would have escaped large fire events (Keeley 2006a; Keeley & Zedler 2009). This model only requires persistence of lightning-ignited fires for a few weeks to ensure being caught by the annual gale-force Santa Ana winds. Thus, the historical fire regime would have been a regime of summer lightning fires punctuated periodically by huge Santa Ana wind-driven fires (see Fig. B1.3.1). In all likelihood the bulk of the burning on these landscapes occurred under such conditions.

This historical fire regime is supported by the observation that summer fires often continued for months and consumed no more than a few thousand hectares (Minnich 1987a) and at this rate it would have taken many centuries to burn this landscape (Keeley & Zedler 2009). However, Santa Ana wind-driven fires often consume more than 10 000 ha in a single day (Keeley et al. 2004, 2009b) and thus when an occasional lightning ignition carried over until the autumn the total area burned would have increased by orders of magnitude. Fire scar dendrochronology studies of one of our lowest elevation conifers, *Pseudotsuga macrocarpa* (see Foothill coniferous trees section below), which is often juxtaposed with mid- to high-elevation chaparral, reinforces this conclusion that large landscape fires were historically commonplace on these landscapes (Lombardo et al. 2009). These dendrochronology studies showed that since at least the seventeenth century, fires covering 100 000 ha or more occurred roughly every 35–75 yrs on the Los Padres National Forest in southern California.

In shrublands across the California MTC region there are large differences in fire behavior related to climate, weather, topography and human demography. These patterns translate into differences in fire hazard (Moritz et al. 2004). During the latter half of the twentieth century most southern California counties have had a fire rotation interval of 30–40 yrs, the central coastal region 40–80 yrs (Keeley et al. 1999a), the east bay area of San Francisco approximately 100 yrs (Keeley 2005), and the foothill chaparral of the southern Sierra Nevada perhaps even longer (Keeley et al. 2005b). The short fire rotation in southern California is due to a longer period of annual drought, coupled with the annual autumn foehn winds known locally as Santa Ana winds (see Box 1.3), plus the high population

> **Box 5.1** Comparison of Fire Regimes between Southern California and Northwestern Mexico (Baja California)
>
> Dodge (1975) postulated that burning patterns were likely different in chaparral and coniferous forests on the two sides of the USA/Mexican border because of different management practices. North of the border fire prevention and fire suppression were policy, whereas south of the border neither were widely practiced; indeed, in rural Mexico it was customary to set fire to chaparral, sage scrub and grasslands as soon as sufficient dead material had accumulated to carry a fire. Dodge found that the most significant difference between the vegetation of these two regions was in land use practices; in particular the intensive grazing and browsing by cows and horses south of the border limited fuel continuity. He suggested that vegetation patterns and fire behavior in Baja California were the result of overgrazing and extensive anthropogenic burning.
>
> Minnich (1983) used Landsat remote imagery to compare patterns of burning in southern California shrublands between 1972 and 1980 on both sides of the border. He concluded that during this 9-year period fires were larger north of the border; however, fire size was not compared statistically and critics have contended that further analysis shows no difference in fire size (Strauss *et al.* 1989). Keeley & Fotheringham (2001a) pointed out that the inclusion by Minnich (1983) of two huge fires north of the border, which were outside the study period, biased the conclusions because they were based on U.S. Forest Service historical records and such records are not available in Mexico.
>
> In an attempt to compensate for the lack of written records south of the border, Minnich (Minnich & Dezzani 1991; Minnich 1995, 1998; Minnich & Chou 1997) used historical aerial photographs to prove that large fires were absent from northern Baja California. This conclusion was based on photos from three time periods over an 80-year period, and these studies have been criticized because of the lack of demonstration that this photo series, with a 16–18-yr gap between photos, could capture all large historical fires (Keeley & Fotheringham 2001a, 2001b). Further challenges to the notion that Baja California did not historically have large fires are based on written accounts by early explorers that described in detail massive wildfires in northern Baja California (Keeley & Zedler 2009).
>
> The conclusion from the Baja studies is that the smaller fires south of the border are reflective of the natural southern California fire regime and larger fires north of the border are a modern artifact of fire suppression. An important management conclusion from these studies is that widespread prescription burning is needed in southern California to return the landscape to its natural condition with fuel loads insufficient to carry massive fires. In the late twentieth century this model was readily accepted by fire managers in southern California as this seemed like a reasonable extension of the western conifer model of fire suppression causing fuel accumulation which has led to large contemporary
>
> *Continued*

Box 5.1 (cont.)

fires in those forests. However, the fire history data fail to support the idea that fire suppression has excluded fires in shrubland crown fire ecosystems in southern California (Keeley *et al.* 1999a; Mensing *et al.* 1999; Keeley & Fotheringham 2001a, 2001b). In fact rather than this landscape having massive stands of anomalously old vegetation, quite the opposite is true (see Fig. 5.10). Further evidence against the hypothesis that large fires are due to contemporary fire management practices is the demonstration that large chaparral fires were frequent prior to the modern fire suppression era (Keeley & Zedler 2009; Lombardo *et al.* 2009).

Although there is clear evidence that fire suppression has not caused anomalous changes in southern California fuels, Minnich (2001; Minnich & Chou 1997) has insisted that there is a marked change in the pattern of burning immediately south of the border (Fig. B5.1.1) and this must be viewed as support for their model of how different fire management policies have changed fire behavior. However, one of the alternative explanations for the differences in fire size north and south of the USA/Mexican border identifies the very different demographic patterns and rural land use (Keeley & Fotheringham 2001a, 2001b). North of the border most of the population is concentrated in coastal cities and much more of the interior landscape is largely protected wildlands. However, south of the border, population centers are distributed much further inland and there is a striking change in social patterns immediately south of the border (Fig. B5.1.2). Intensive rural development has greatly fragmented fuels, with consequent impacts on fire patterns. In addition, the warmer and drier conditions south of the border, coupled with more intensive land use, contribute to more extensive type conversion to flashy fuels of weedy alien annuals (see for example vegetation maps in Minnich & Franco-Vizcaino 1998), which, along with limited attention to fire prevention, accounts for the much higher fire frequency in Baja California.

In short, the model that burning patterns in Baja represent the natural pattern in southern California is not supported by multiple lines of evidence:

(1) Large fires in southern California predate modern fire suppression.
(2) Fire suppression policy in this crown fire system has been unable to exclude fire.
(3) Since the arrival of Euro-Americans, southern California landscapes have burned more frequently than historically was the case; thus fuel loads are lower than historically.
(4) Most large fires are driven by extreme Santa Ana winds (see Box 1.3) and fuel age has limited control over fire spread.
(5) A mosaic of young age class chaparral fuels is incapable of stopping the spread of many large fires.

Continued

Box 5.1 (cont.)

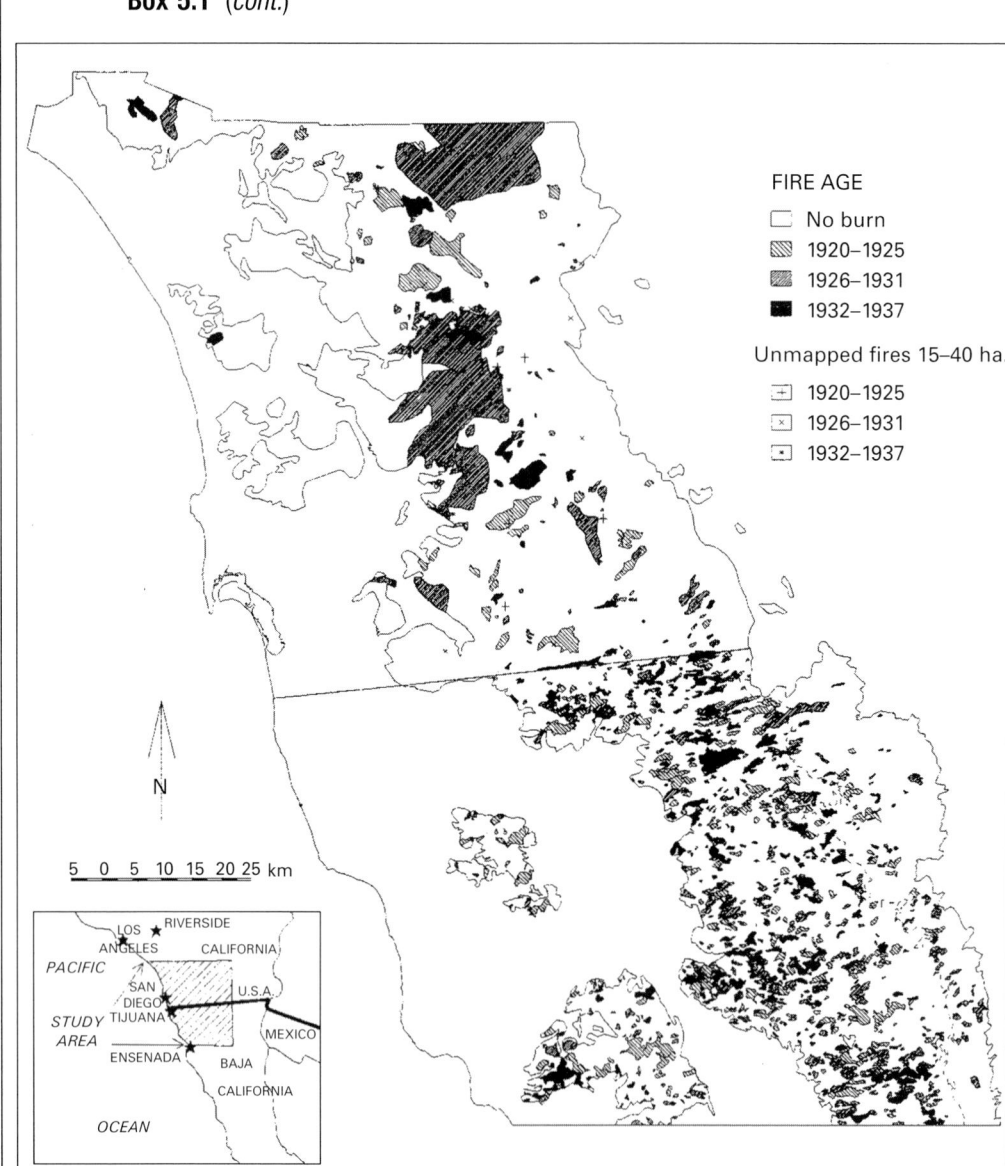

Fig. B5.1.1 *Fire sizes north and south of the USA/Mexican border over a 17-yr period. (From Minnich & Chou 1997.)*

Despite the body of scientific evidence against the Baja model, it is still used to justify managing southern California wildland fires by use of prescription burning (e.g. Minnich 2003; Chastain 2007). Some agencies insist this is a debate and contend that since there isn't a consensus they take the "we have

Continued

Box 5.1 (*cont.*)

Fig. B5.1.2 *Aerial image of the interior landscapes (50–75 km inland), adjacent to the USA/ Mexico border (arrowed). The immediate change south of the border is evident with the large population center of Tecate, Mexico, which contributes to extensive rural development along much of the southern edge of the border, something not observed along the northern edge of the border.*

our expert, you have yours" stand. However, management has never relied on consensus in decision making as most debates in science have some dissenting opinions; witness for example "climate change" science. Best management practices require accepting the preponderance of evidence and in the case of fires in southern California, it is blatantly clear that age of fuels is not the primary determinant of catastrophic fire losses. The primary problem with ignoring this evidence is that it distracts from real solutions to fire problems in the region, which are not tied to fuel treatments in the wildlands but rather on concentrated effort at the wildland–urban interface. In the twenty-first century most agencies in the region have abandoned the idea of trying to create mosaics of fuel age classes as a means of controlling wildland fires (see Chapter 13).

A similar Baja model has been used to find support for hypothesized impacts of fire suppression in California forests with a surface fire regime. This model is based on comparisons of fire regimes and forest structure between southern California and northern Baja California forests (Savage 1997; Minnich & Everett 2002; Stephens *et al.* 2003). These studies assume the only significant difference between forests north and south of the border is their different history of fire management. However, there are important climatic differences between these regions (Minnich 1987b; Keeley & Fotheringham 2001a) that could explain differences in fire regime and forest structure. Regardless of the value of these studies in demonstrating fire management impacts, the observed climatic differences may ultimately prove to be valuable in investigating future

Continued

Box 5.1 (cont.)

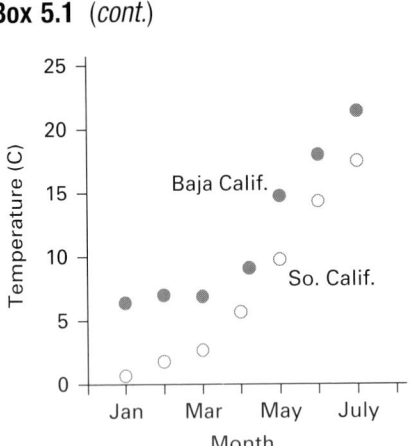

Fig. B5.1.3 *Monthly average temperatures at approximately 2000 m in the San Pedro Mártir of Baja California (from Alvarez & Maisterrena 1977) and in the San Bernardino Mountains of southern California (at Big Bear Lake; NOAA 1998).*

climate change impacts on both forest structure and fire regimes. At comparable elevations the Baja California San Pedro Mártir is substantially warmer than sites north of the USA–Mexico border (Fig. B5.1.3). The degree of global warming that Westerling et al. (2006) suggested would result in major changes in fire regime throughout the western United States is far less than the present difference between Baja and southern California forests. Since the greatest temperature differences between Baja California and southern California occur in winter, this likely has profound impacts on snowpack, a factor that potentially has substantial impacts on fire regimes (Westerling et al. 2006). Some climate forecasts predict changes in seasonal distribution of precipitation for California that may impact fire regimes (Fried et al. 2008) and the substantial differences in seasonal distribution between Baja California forests, which receive substantial summer rains, and southern California forests may also prove of some value as a potential model system for understanding future climate change impacts in California.

density (>150 people km^{-2}) and extensive road infrastructure, which leads to frequent and widespread ignitions (Table 5.1). The much lower fire frequency in the San Francisco east bay area is tied to the higher rainfall and shorter droughts, which have contributed to a more limited window of time when shrublands provide available fuels for burning. In the southern half of the Sierra Nevada the lower population density has likely played some role in limiting fire ignitions. In addition, the lack of passes and the high peaks to the east appear to hold foehn winds aloft, and thus the chaparral-dominated foothills are not subjected to this extreme fire weather. Historically lightning-ignited fires likely originated upslope in conifer-dominated forests and during the twentieth century these fires have been successfully suppressed at a small size (see Coniferous Forest section below).

Postfire Community Response

Chaparral and sage scrub shrublands are highly resilient to fires at intervals of 30–150+ yrs. Within this range these communities return to prefire functional states rapidly with little or no loss of species. The process has been described as "auto-successional" because all components of the prefire state are present after fire and colonization plays a limited role in returning these systems to their prefire state (Hanes 1971). In other words, plant regeneration is almost entirely endogenous, involving dormant seedbanks that germinate after fire and resprouting from persistent roots and basal lignotubers. As a consequence, the spatial extent of shrubland area that burns has relatively little impact on success of vegetative recovery.

Species diversity is substantially higher in the first postfire year (see Chapter 11), unless there is a significant rainfall deficit, in which case diversity may peak in the second postfire year (Keeley et al. 2005c). Diversity declines in later years although this trend may be reversed by very high rainfall in early seral stages (Keeley et al. 2005a).

Detailed postfire studies have shown that at a community scale (0.1 ha), roughly 90% of the cover in the fifth year after fire was from species present in the first postfire year, and because of the timing of fires and phenology of dispersal it is certain all of this is due to endogenous regeneration (Keeley et al. 2005a, 2006b). Of particular interest though is the observation that this represented only about 50% of the species richness present in year 1. Thus, it would appear that diversity in early seral stage shrublands is strongly affected by colonization. However, essentially all species present in the fifth postfire year were present immediately after fire somewhere within the fire perimeter; thus, rather than colonization of the burned site, this pattern is more an example of mass action effects (Shmida & Wilson 1985). In other words many species had limited populations immediately after fire and these populations expanded throughout the burned area during the first five years after fire.

As the shrub canopy returns, herbaceous species persist largely as dormant seedbanks. The intensity of herbivory, competition and resource limitations are primary factors that limit herb growth in these mature stands (Swank & Oechel 1991). These factors have selected for deeply dormant seeds that persist in the soil seedbank until triggered to germinate by fire (Keeley & Fotheringham 2000).

Ephemeral Postfire Flora

The one feature that separates California shrublands from all other North American ecosystems is the unique and highly diverse postfire flora comprising many species endemic to only postfire environments (see Fig. 11.2c). The brief pulses of resources present after fire are utilized by many short-lived species that vary in their recruitment patterns (Table 5.2). Some species are so strictly tied to postfire

Table 5.2 Examples of postfire annuals in California chaparral and yearly distribution of total population recorded during the first 5 yrs

Species	Family	Percentage by year				
		1	2	3	4	5
Native postfire endemics						
Allophyllum gilioides	(Polemoniaceae)	50	47	1	0	2
Calandrinia ciliata	(Portulacaceae)	35	64	1	0	0
Emmenanthe penduliflora	(Hydrophyllaceae)	49	49	1	0	1
Guillenia lasiophylla	(Brassicaceae)	98	1	0	0	1
Lupinus succulentus	(Fabaceae)	70	21	2	2	5
Menzelia micrantha	(Loasaceae)	87	12	0	0	1
Nicotiana attenuata	(Solanaceae)	99	1	0	0	0
Papaver californicum	(Papaveraceae)	37	63	0	0	0
Phacelia brachyloba	(Hydrophyllaceae)	84	16	0	0	0
Silene multinervia	(Caryophyllaceae)	43	56	1	0	0
Native postfire specialists						
Antirrhinum coulterianum	(Scrophulariaceae)	18	66	5	10	1
Calyptridium monandrum	(Portulacaceae)	44	35	0	0	21
Camissonia bistorta	(Onagraceae)	2	77	0	1	20
Chaenactis artemisiifolia	(Polygonaceae)	6	75	8	9	2
Eucrypta chrysanthemifolia	(Hydrophyllaceae)	60	22	4	5	10
Gilia capitata	(Polemoniaceae)	8	74	4	14	0
Lupinus bicolor	(Fabaceae)	11	48	10	19	12
Nemacladus ramosissimus	(Campanulaceae)	35	57	1	0	7
Phacelia cicutaria	(Hydrophyllaceae)	59	21	6	8	6
Salvia columbariae	(Fabaceae)	1	55	12	19	13
Native postfire opportunists						
Amsinckia menziesii	(Boraginaceae)	6	18	32	10	34
Antirrhinum kelloggii	(Scrophulariaceae)	7	32	20	2	39
Apiastrum angustifolium	(Apiaceae)	5	40	12	13	30
Cryptantha clevelandii	(Boraginaceae)	3	62	10	6	19
Lepidium nitidum	(Brassicaceae)	25	48	9	2	16
Lotus hamatus	(Fabaceae)	13	54	19	5	9
Pterostegia drymarioides	(Polygonaceae)	2	41	28	14	15
Rafinesquia californica	(Asteraceae)	9	49	9	4	29
Triodanus perfoliata	(Campanulaceae)	2	6	79	0	13
Vulpia octoflora	(Poaceae)	1	23	17	22	37
Native late successional increasers						
Bowlesia incana	(Apiaceae)	0	0	0	0	100
Claytonia perfoliata	(Portulacaceae)	9	17	5	24	45
Collinsia parryi	(Scrophulariaceae)	1	42	0	1	57
Crassula connata	(Crassulaceae)	7	9	9	9	66
Eriastrum sapphirhinum	(Polemoniaceae)	5	1	2	2	91
Galium aparine	(Rubiaceae)	18	1	35	0	46
Pectocarya linearis	(Boraginaceae)	19	5	13	0	63
Plagiobothrys canescens	(Boraginaceae)	0	80	0	0	20
Trifolium microcephalum	(Fabaceae)	0	6	31	18	45

From Keeley *et al.* (2006b).

conditions they are known as *pyro-endemics*; others are more opportunistic and are most abundant after fire but persist long after fire in gaps and other disturbances. A few annuals and herbaceous perennials seem to thrive best in low numbers in mature chaparral.

Annuals dominate the postfire flora, typically comprising more than 50% of the cover and diversity in the first few postfire years. The dormant seedbanks of these species are triggered by either intense heat shock or combustion products from smoke or charred wood (Keeley 1991; Keeley & Fotheringham 1998). Heat-stimulated germination is found in hard-seeded species that have water-impermeable seeds until the outer seed coat is scarified and include families such as Fabaceae and Convolvulaceae. Chemically stimulated species have water-permeable seeds that imbibe water but remain dormant until cued by combustion products from smoke or charred wood (see Box 3.1). These species include members of the Hydrophyllaceae, Lamiaceae, Polemoniaceae, and Scrophulariaceae, among others (Keeley & Fotheringham 2000).

Shrub Canopy Recovery

Postfire regeneration of the shrub dominants is by either resprouting and/or seedling recruitment, and the relative importance of each is a function of (1) innate characteristics of species, (2) fire intensity and (3) postfire drought.

The majority of shrub taxa have the capacity to resprout from adventitious buds in underground stems or basal lignotubers (see Fig. 3.1). For some species the lignotuber is an ontogenetic stage of development that is evident by the first year of growth (see Table 3.2). These include *Adenostoma fasciculatum* (Rosaceae), and many species in the two largest shrub genera, *Arctostaphylos* (Ericaceae) and *Ceanothus* (Rhamnaceae). Others, such as species of *Quercus* (Fagaceae) will produce basal swellings following repeated decapitation from fire or browsing. Most other resprouters generally lack any sort of basal burl.

A postfire pulse of seedling recruitment is observed in some taxa but not all. *Adenostoma fasciculatum*, and all species of *Arctostaphylos* and *Ceanothus*, produce seeds on an annual to biennial basis and the bulk remain dormant in the soil for decades or even centuries until germination is triggered by fire. These postfire recruiters are most common on more xeric sites, such as low elevations and south-facing exposures (Keeley 1986; Meentemeyer & Moody 2002).

Many species resprout after fire and lack seedling recruitment from dormant seedbanks. These postfire obligate resprouters have not been selected to delay reproduction to a single postfire pulse of recruitment, rather they recruit in the inter-fire interval (see Table 3.1). These include species of *Prunus*, *Quercus*, *Rhamnus*, and *Cercocarpus*. Other vigorous resprouters that have a low level of recruitment after fire include *Fremontodendron* spp., *Rhus* spp. and *Garrya* spp. It has been hypothesized that these different fire responses are tied to character syndromes that include drought tolerance response, dispersal and past history (see Table 9.1).

Animal Communities

Other than a few burrowing species, most animals in chaparral shrublands flee from fires or perish (Cook 1959; Lawrence 1966; Quinn 1979; Fox et al. 1985). As a consequence recovery is dependent on recolonization and thus metapopulation patterns play a critical role, although this has not been well studied. On contemporary landscapes large fires may pose problems for recovery due to habitat fragmentation and loss of natural corridors.

Coniferous Forest Surface Fire Regime

In mountains above 2000 m in southern California (latitude $33°-34°$) and progressively lower further northward in the state, conifer forests replace shrublands and associated woodlands (see Fig. 2.2a). These forests are dominated by a small number of species: in the Pinaceae, *Pinus ponderosa, P. jeffreyi, P. lambertiana, Abies concolor* and *Pseudotsuga menziesii*, and in the Cupressaceae, *Calocedrus decurrens*, all of which reach 50–70 m in height (Rundel et al., 1977; Thorne 1977; Minnich 2007; Fites-Kaufman et al. 2007). Half of these are endemics to the MTC region and three, *P. ponderosa, A. concolor* and *Ps. menziesii*, are widespread outside this climatic region (Fowells 1965). These species assemble in diverse patterns both locally and regionally, with much variation in landscape mosaics and fire regime characteristics (Sugihara et al. 2006).

Throughout the interior mountain ranges of California two forest types predominate. Forests dominated by *Pinus ponderosa* often form open-canopy communities with herbaceous understories at lower elevations, and higher up on equator-facing exposures. The closed-canopy, *Abies concolor* dominated, mixed conifer forests are typically at higher elevations, and on pole-facing slopes at lower elevations. Much of this landscape is topographically heterogeneous and thus clusters of closed-canopy forest often form a mosaic with persistent gaps in which relatively little establishes, as well as mosaics with persistent shrub thickets (Conard & Radosevich 1982; North et al. 2002). Both forest types are characterized by a surface fire regime or a mixed surface and crown fire regime.

Fuel Structure

High productivity of trees in these forests allows for rapid growth rates and the capacity to outgrow surface fuels. As a consequence both forest types typically maintain a separation between the tree canopy fuels and surface fuels, and the taller the trees the greater the potential separation (Fig. 5.6a). This separation of fuels is enhanced by traits such as self-pruning of dead branches, thick bark and deep roots (Keeley & Zedler 1998).

The character of the surface fuels varies as a function of canopy closure. On the drier ponderosa pine forest sites, wide spacing of trees results in a more open forest

Fig. 5.6 *Ponderosa pine* (Pinus ponderosa) *(a) illustrating self-pruning and thick bark, tall stature with understory of* Chamaebatia foliolosa, *(b) with grass understory. (Photos by Jon Keeley.)*

with high surface radiation that supports herbaceous understory fuels (Fig. 5.6b), usually perennial grasses. On more mesic *A. concolor* dominated mixed conifer sites, nearly complete canopy closure reduces light levels at the soil surface and limits herbaceous growth. It has been suggested that this lack of a grass understory is the result of fire suppression that has allowed the canopy to close in over the past century and shade out grasses. However, this is not supported by historical studies showing a lack of grass phytoliths in the soils of mixed conifer forests (Evett *et al.* 2006, 2007). It is also not supported by historical fire responses to antecedent rainfall (see Fig. 2.8). Surface fuels in the mixed conifer forest comprise dead needles and branches, which accumulate rapidly due to the high productivity of these forests (Fig. 5.7).

Grasses in ponderosa pine forests maintain a continuous fuel cover that enhances fire spread. In addition, these vertical fuels produce a high surface area to volume ratio, and this packing ratio, plus the loose packing of long-needle *P. ponderosa* litter, enhances combustion. Since these forests have a relatively open canopy, surface fuels are dry for longer periods beginning in late spring, they can carry fire at short intervals, of 5 yrs or less.

In closed-canopy mixed conifer forests, dead surface fuels of needles and branches are typically oriented horizontally and this, plus the very short needles of *A. concolor*, results in a low packing ratio that inhibits combustion. In addition,

Fig. 5.7 *Surface fuels of dead branches and other litter in the understory of a mixed conifer forest. (Photo by Jon Keeley.)*

the closed canopy may limit the length of the season of drying that converts litter to available fuels, and fuel distribution is often more heterogeneous, which may inhibit fire spread, all of which contribute to longer intervals between fires. This longer fire interval is a factor in greater fuel accumulation and this, coupled with the uneven distribution of fuels, leads to mixed fire behavior of low-severity surface spread punctuated in spots with higher-intensity passive crown fire. In mixed conifer and broadleaf forests pine generally have a selective advantage over oaks (*Quercus* spp.) in that conifer fuels promote higher temperatures that inhibit oak regeneration (Williamson & Black 1981; Knapp & Keeley 2006).

Fuel type also affects fire season. On drier open sites herbaceous fuels dry sufficiently to carry fire from late spring to autumn. On more mesic closed-canopy sites, dead surface fuels dry later in the year, and combined with the snowpack in these higher elevation forests leads to a later fire season.

In both forest types, when sites fail to burn for extended periods, high tree sapling density adds an additional fuel layer. These fuels reduce the degree of separation between surface and canopy fuels and form what is often referred to as *ladder fuels*. These fuels are frequently not included in estimates of forest fuel volume. The spatial pattern and extent of these ladder fuels have likely changed under fire suppression management as discussed below.

Ignition Patterns

On the Pacific slope of North America, lightning strike density generally increases with elevation and thus it is no surprise that these montane forests have a much higher incidence of lightning-ignited fires than the lower chaparral-dominated foothills (Show & Kotok 1923a; Court 1960; Keeley 1982). Today these fires still represent the bulk of fire starts in some regions, and a significant source of burning throughout most forested regions (Table 5.1). Lightning-ignited fires are concentrated in summer months of June, July and August and arise from monsoonal storms that are concentrated in the southwestern USA, but often affect interior mountain ranges in the interior of California.

Lightning ignitions in California are spatially quite variable. Throughout the region lightning strikes increase with elevation but exhibit little preference for slope aspect. Prior to human occupation fires in the lower foothills may have been ignition limited, whereas in the higher mountains forests were saturated with natural lightning ignitions (van Wagtendonk & Cayan 2008) and under conditions of saturating ignitions, fuels play a larger role in controlling fire frequency. In the higher mountain ranges lightning ignitions typically peak at mid-elevations (Vankat 1985).

Fire Regime

Fires in these forests typically spread by low-intensity surface burning but may include patches of high-intensity passive crown fire. This mixed surface and crown fire regime varies in the proportion of these two modes of burning both spatially and temporally. Factors that affect the proportion of surface to crown burning include climate, weather, topography and past management practices. In forests with herbaceous fuels, crowning is often dependent on ladder fuels, whereas in other forests localized accumulations of dead branches and leaves may be sufficient to cause fire behavior to switch from surface to crown fire.

Cross sections of trees with fire-scarred bases (see Fig. 4.8a) typically reveal numerous embedded fire scars that can be accurately dated by association with annual growth rings (see Figure 4.8b). These studies show that historically these forests were subjected to frequent fires, and since the trees survived (and have no resprouting capacity), these were obviously low-intensity surface fires (Kilgore & Taylor 1979). Fire-scar dendrochronology studies have been done in forests throughout California and the western half of the USA and reveal a common pattern of frequent fires prior to the twentieth century, but nearly total fire exclusion over the past century (Skinner & Chang 1996). As a consequence, this subsection will focus on the historical fire regimes, and in a later subsection we will discuss how these patterns have been perturbed by land management practices of the twentieth century.

The historical fire frequency varied spatially and temporally at different scales. In the Sierra Nevada Range the lower-elevation *Pinus ponderosa* forests had a higher fire frequency than the upper-elevation *Abies concolor* dominated mixed conifer forests (Swetnam & Baisan 2003). Also, within the mixed conifer forest at the same elevation, equator-facing slopes had a higher fire frequency than more mesic pole-facing slopes (Caprio 2004). It is unlikely that ignitions were a factor in these patterns due to the characteristics of lightning distribution noted above. Fuel structure and fuel moisture are the primary drivers behind these fire regime differences and these can often vary over very short distances (e.g. Stephens 2001).

There are many other sources of variation in historical fire frequency. At a local scale fire frequency is often highest on ridgetops, followed by mid-slope forests and then lower slope forests (Skinner *et al.* 2006). There is also substantial regional variation. Drier southern California mixed conifer forests have substantially longer fire intervals (Everett 2003) than Sierra Nevada forests, due to lower primary productivity and slower fuel accumulation. More mesic forests in the northern part of the state also have longer intervals than Sierra Nevada forests, due to a combination of a shorter growing season for fuel production and a shorter fire season when fuels are available to burn (Agee 1991; Taylor & Skinner 1998).

In addition to spatial variation there is a clear record of temporal variation in response to climate. Not surprisingly, drier years are associated with more widespread burning (Taylor & Beatty 2005). Utilizing the 2000-yr record in *Sequoiadendron giganteum*, Swetnam (1993) showed that fire frequency increased during warmer periods and was reduced during cooler periods. These patterns also affected fire intensity since the longer fire-free intervals resulted in greater fuel volumes that contributed to higher fire intensity.

On the basis of these fire-scar records, the frequency of low-intensity surface fires is the best-understood fire regime parameter in these forests. Patches of high-intensity crown fire that kill even the largest trees and create gaps in the forest canopy have occurred in the past, but we know relatively little about the size distribution of such gaps and how they have varied relative to historical climate changes (Stephenson *et al.* 1991). The size of these gaps may vary from an area the size of a single tree to hundreds or thousands of hectares (Show & Kotok 1923b). There is some evidence that different forest types are prone to different sized gaps (Collins & Stephens 2010). Because all trees in gaps are killed in high-intensity crown fires, there is little record of the past distribution of gap sizes. Past gaps can be inferred from localized patches of similar aged trees but extending such an analysis over broad areas has not been done. This is an important area of research for understanding historical fire regimes and because gap size plays a critical role in the recruitment of many tree species.

Another fire regime parameter relatively well documented by fire scar records is seasonality. Studies show that most fires occurred late in the growing season, perhaps August–September (Swetnam & Baisan 2003). This interval is offset from the peak lightning-ignited fire season so it seems likely that this seasonal distribution was controlled by the state of the surface fuels, in particular fuel moisture.

Postfire Community Response

Tree mortality from surface fires or mixed surface and crown fires varies largely by tree size and fuel accumulation and less by species. Seedlings and saplings are often killed by low-intensity surface fires but high-intensity crown fires may kill even mature individuals. Where high surface fuel loads and or ladder fuels have accumulated, surface fire will spread to passive crown fires. These fires result in mortality, because none of these conifer species have the capacity to resprout, either from the base or epicormically. These fire-caused gaps play an important role in both the maintenance of community and landscape diversity of flora and fauna but also in the regeneration of forest trees (Keeley & Stephenson 2000; Keeley & van Mantgem 2008).

Tree Regeneration
None of the dominants in these forests have serotinous cones, nor do they accumulate dormant soil seedbanks. Regeneration is heavily concentrated in open gaps created by patches of high-intensity burning. Species such as *Pinus ponderosa* are dependent on gaps for seedling recruitment, with high light at the soil surface and absence of litter, both of which are needed for successful establishment (Fowells 1965). Other species such as *Abies concolor* are more opportunistic, as they readily recruit seedlings in the understory of unburned forests, but also exhibit massive recruitment on burned sites (Tappeiner & Helms 1971; Mutch & Parsons 1998; North *et al.* 2005a; van Mantgem *et al.* 2006).

Size of fire-caused gaps plays an important role in tree regeneration. Seedling recruitment is dependent on parent seed trees dispersing seeds into the gaps; thus as gap size increases the seed rain declines (McDonald 1980; Greene & Johnson 2000). Gaps on the order of 0.01–10 ha generally have successful recruitment, but much larger gaps pose a problem for regeneration. In general the very smallest gaps have abundant seed fall but seedling growth rate may be inhibited by shading, whereas larger gaps sometimes favor higher seedling growth rates if seeds can reach them (McDonald & Abbott 1994).

Although mineral soil and high light environments characteristic of gaps play key roles in seedling establishment, another feature of gaps that enhances seedling recruitment is that they retain higher soil moisture longer into the summer drought (Ziemer 1964). Another factor critical to recruitment is the vulnerability of young trees to even low intensity surface fires. A key attribute of gaps is the absence of overstory trees, and thus a slower rate of surface fuel accumulation and a greater probability of subsequent fires missing them until saplings reach sufficient size and bark thickness to survive fires (Keeley & Stephenson 2000). This pattern is reflected in the fact that most fire-scar dendrochronology studies show that the interval between establishment to the first fire scar is nearly always longer than subsequent fire intervals.

Unlike crown fire shrublands where many dominant species recruit in a pulse of seedlings from dormant soil-stored seedbanks, none of the forest tree dominants exhibit this trait. Forest regeneration is dependent on parent seed trees surviving fire, which disperse seeds into fire-generated gaps. All of the dominant tree species have periodic masting cycles; *Abies concolor* commonly produce massive seed crops on a 2–3 yr cycle and *Pinus ponderosa* and other species have a longer periodicity (Fowells 1965; Krannitz & Duralia 2004). As a consequence, recruitment is largely a function of the coincidence of fire and a masting cycle (Keeley & van Mantgem 2008; see Fig. 3.6). In addition, successful recruitment will often depend on a year of high precipitation as well as the seasonal timing of rain (North et al. 2005a; League & Veblen 2006). Not surprisingly, the pattern of gap formation relative to wind currents (Gordon 1970) and seed predation also affect regeneration (Shearer & Schmidt 1970).

Thus, patterns of postfire recruitment are expected to be complex due to species-specific differences in response to the coincidence of fire, masting and precipitation. Additionally, surface fuel type will play a significant role because in ponderosa pine forests grass regrowth can inhibit pine regeneration (Pearson 1942), particularly if seed fall is delayed very long after fire. In mixed conifer forests shrub recruitment from dormant seedbanks may have a similar inhibitory effect (McDonald & Fiddler 1990), but this is often very patchy, and sometimes shrubs have positive nurse-plant effects (Keyes et al. 2001, 2007). On top of all this, there are marked elevational differences in response to rapid climatic changes along steep mountain slopes (van Mantgem et al. 2006; Gworek et al. 2007).

Understory Recovery

Understory species vary with forest type and region (van Wagtendonk & Fites-Kaufman 2006). Most perennial herbs in the understory survive low-intensity surface fires and resprout, but are killed by high-intensity fires (Knapp et al. 2007; Rocca 2009). In ponderosa pine forests the rhizomatous subshrub *Chamaebatia foliolosa* is persistent even after fairly high intensity fires (Rundel et al. 1981). Similarly, fire intensity affects survivorship of soil seedbanks and thus the extent of postfire annuals. Some annuals that are restricted to gaps in mature forests have dormant seedbanks in the understory that germinate in response to fire; however, strict pyro-endemics, which are only found on burned sites, are rare in these forests (Keeley et al. 2003). Postfire recovery patterns have a deterministic component resulting from species-specific life history components and a stochastic component resulting from site history and timing of disturbance (Halpern 1989) and surface fuel load (Rocca 2009). Persistence of understory species is dependent on microhabitat variation in light and moisture (Naumburg & DeWald 1999; North et al. 2005b).

Several shrub species, *Ceanothus integerrimus*, *C. cordulatus* and *Arctostaphylos patula*, are widespread throughout the conifer region in California, and establish after fire from soil seedbanks (Kauffman & Martin 1991). These shrubs form an early seral stage (Knapp et al. 2007) and, depending on soil depth, may outcompete tree regeneration for water resources (Royce & Barbour 2001). After several decades shrubs may be shaded out by overstory trees (Conard & Radosevich 1982), but seedbanks appear to persist for hundreds of years (Quick 1961). In permanent shrub patches, shrubs such as *Quercus vaccinifolia*, *Chrysolepis sempervirens*, *Ceanothus cordulatus* and *Ribes* spp. resprout after fire and, due to severe substrate conditions, are not displaced by forest trees (Kauffman & Martin 1990; Nagel & Taylor 2005).

Other California Vegetation Types

Crown fire chaparral shrublands and surface fire conifer forests provide a useful contrast illustrating the range of fire regimes present in the California MTC region. These two ecosystems dominate the bulk of the landscape, yet there is a rich diversity of other vegetation types, each with its own unique fire regime characteristics. Some of that variation is reflected in these examples.

Grasslands

Most grasslands in California are disturbed communities that lack much structural diversity and are dominated by non-native annual grasses and forbs (Huenneke & Mooney 1989; Hamilton 1997). These annual grasslands appear to be derived from several different origins. In the central coast ranges and southern California extensive burning of native shrublands by Native Americans replaced these perennial systems with annual herbs that were susceptible to alien invasion upon European colonization (Keeley 1990a). Cooper (1922) cited remnant stands of *Adenostoma fasciculatum* in the northern part of the Great Central Valley as evidence of a former wider distribution across the valley that had been devastated by Native American burning and this is affirmed by more recent research (Bloom & Watson 2006).

Others have long contended that throughout the Central Valley annual grasslands have replaced perennial grasslands due to intensive livestock grazing and extreme droughts during the nineteenth century (Heady 1977). In the southern part of the Central Valley and foothills of the Sierra Nevada there is evidence that extensive "grasslands" were originally dominated by native annual forbs, and these communities have been heavily invaded by non-native annual grasses (Hoover 1936; Hamilton 1997). Native forbs in genera such as *Amsinckia*, *Cryptantha*, *Lupinus*, *Madia* and *Plagiobothrys* still persist in these annual grasslands (Keeley 1990a; Schiffman 2007; Minnich 2008).

Native grasslands today are limited to widely disjunct and isolated fragments. Structurally they are more interesting than annual grasslands with diverse growth forms dominated by bunchgrasses, geophytes, rhizomatous perennials and annual forbs (Corbin *et al.* 2007). The most widespread native bunchgrass, *Nassella pulchra*, is endemic to the California MTC region, as are many of the geophytes such as species of *Bloomeria*, *Brodiaea*, *Calochortus*, *Fritillaria* and *Triteleia*. Historically it is believed that native grasslands dominated large portions of the northern Central Valley (Bartolome *et al.* 2007). Farther south, and throughout the central and southern coastal ranges, native grasslands are thought to have been distributed in small patches determined by soil and drainage characteristics (Huenneke 1989; Keeley 1993a).

All grasslands are relatively resilient to frequent surface fires (Reiner 2007). Although most annual species have transient seedbanks, the annual seed rain appears sufficient for postfire regeneration of annual grasses and forbs. Also, the relatively low fire intensities resulting from burning these herbaceous fuels lead to high seed survivorship. Native perennial bunchgrasses, such as *Nassella pulchra*, *Poa secunda*, *Koeleria cristata* and numerous geophytes, resprout after fire. Since the aboveground parts of these species die back each year regardless of fire, this response is indistinguishable from the normal seasonal resprouting. Enhanced flowering of herbaceous perennial resprouts after fire generates substantial seed crops that recruit in the second and subsequent postfire years (Tyler & Borchert 2002; Keeley *et al.* 2006b). Numerous native annuals share their distribution between grasslands and chaparral, but have markedly different germination and establishment patterns, suggesting ecotypic differentiation (Keeley & Davis 2007). This includes species of *Gilia*, *Lotus*, *Lupinus*, *Phacelia*, *Silene*, *Trifolium* and others. In closed-canopy chaparral soil, seedbanks of these annuals are deeply dormant until germination is triggered by either heat shock or smoke. In grasslands most native annuals establish every year, although populations exhibit extraordinary annual fluctuations. Some of this is tied to years of high rainfall but other factors may play a role. For example, extremely dry years are commonly followed by years of high annual presence and this may be tied to the reduction of thatch during dry years and its potential for opening up larger gaps in the following year.

California Sage Scrub

Sage scrub shrubland is of smaller stature than chaparral and tends to include mostly summer deciduous subshrubs (Fig. 5.2) and is distributed from Baja California to central California (Rundel 2007). Sage scrub is found on more arid sites, usually at lower elevation, than chaparral on both coastal and interior sites, but is absent from the foothills of interior ranges such as the Sierra Nevada.

The fire regime of sage scrub is remarkably similar to chaparral. Crown fires kill nearly all aboveground biomass. It is resilient to a wider range of fire frequencies than chaparral, as it will tolerate shorter fire intervals. Coastal sites dominated by this vegetation were likely burned at frequent intervals by Native Americans (Timbrook *et al.* 1982), but in the absence of human ignitions fires likely spread from associated shrublands and grasslands in the interior and would have occurred at much longer intervals.

Regeneration is also much like chaparral. Most shrubs are facultative seeders, both resprouting and recruiting from seed. The soil-stored seedbank is often polymorphic with some seed recruiting between fires and other seeds more deeply dormant and stimulated to germinate by fire (Keeley 1991). Resprouting is commonly more successful in younger plants and it has been suggested that this reflects the herbaceous ancestry of most of these subshrubs (Keeley 2006c). The ephemeral postfire flora is largely indistinguishable from chaparral (Keeley *et al.* 2006b).

Riparian

Riparian communities comprise mixtures of hardwood trees and shrubs. Distribution of riparian communities is controlled by drainage patterns as these are restricted to stream courses and often the same plant assemblage is distributed across a large elevational gradient (Harris 1987; Holland & Keil 1995; Rundel & Sturmer 1998). Indeed, although the taxa may change, the same genera dominate riparian habitats far outside MTC California: *Acer*, *Alnus*, *Baccharis*, *Fraxinus*, *Platanus*, *Populus*, *Salix*, *Pluchia* and others. Almost all of these taxa are winter deciduous, which stands out in stark contrast to the evergreen flora that dominates the surrounding upland landscape. This seasonal pattern reflects the trade-offs in carbon gain during summer when water availability is sufficient to maintain large leaves but winter day length and temperature is insufficient to maintain a viable carbon balance (e.g. Mooney 1972).

The annual flooding cycles have been a far greater selective force than periodic fires (Bendix 1997). Essentially all woody species in these communities have the capacity to resprout after top-kill from either flooding or fire (Davis *et al.* 1988; Busch 1995). In most years, regardless of the aboveground water flow, these woody species will have access to underground water for much of the dry summer season and thus live fuel moisture will have a dampening effect on fires. At the level of individual plants, access to soil moisture will enhance their capacity to survive and resprout after fire. At the community level the higher fuel moisture will act as a heat sink, often preventing fire spread across stream courses (Dwire & Kauffman 2003). During extreme droughts low fuel moisture will reduce the capacity of this vegetation to dampen fire intensity and thus result in severe crown fires with complete crown scorch and less resprouting.

Seedling recruitment in these riparian habitats is highly influenced by annual flooding cycles that scour substrates and select against long-term soil seed storage. Recruitment after fire will be a function of flowering and seeding behavior of mature trees and resprouts that survive fire. In most woody species seed dispersal occurs in late winter and seeds must germinate quickly as viability is extremely short, on the order of weeks (Vaghti & Greco 2007). This recruitment pattern occurs annually regardless of fire. Following high-intensity fires, locally available seed sources may be lost and the regeneration delayed until seeds disperse in from other riparian sites (Dwire & Kauffman 2003).

Oak Savannas and Woodlands

Savannas with a grass understory are dominated by one of several oak species, including winter deciduous *Quercus lobata* and *Q. douglasii*, evergreen *Q. agrifolia* and semi-evergreen *Q. engelmannii* (Barbour 1987). Surface fires are carried by herbaceous fuels, and trees survive fire because of thick bark and also deep roots that maintain high foliage fuel moisture, which acts as a heat sink and dampens fires. During extreme droughts fuel moisture declines, making these trees vulnerable to fires. Also, invasion by some understory alien species may act as ladder fuels, increasing the potential for crown fires (see Fig. 12.1).

Following fire, mature oaks resprout epicormically if fires are not too intense, and some resprout from the base when immature (Allen-Diaz et al. 2007). Oak species differ in canopy architecture, bark thickness, self-pruning and various other traits, but there has been little comparative study of how these differences affect fire regimes and ecosystem responses. For example, the savanna blue oak *Q. douglasii* is something of an anomaly because it retains a substantial amount of dead branches, and this lack of self-pruning would seem to be disadvantageous in the current surface fire regime. However, Cooper (1922) suggested that the current savanna habitat is a modern anomaly and this oak was originally part of a shrubland assemblage with crown fires. In this context, lack of self-pruning may have selective value in a crown fire ecosystem where localized higher fire intensity could eliminate potential competitors (e.g. Bond & Midgley 1995). Consistent with Cooper's idea of a blue oak shrubland community is the observation that seedling recruitment of *Q. douglasii* is facilitated by shrub nurse plants (Callaway et al. 1991). In a similar vein, Wells (1962) proposed that prior to increased anthropogenic burning in the Holocene, the understory of *Q. agrifolia* savanna was not grass, but rather sage scrub shrubland. Thus, many of the current oak savannas may be human artifacts derived from oak shrubland associations.

Below the conifer belt on more mesic slopes, or sites free of fire for long periods of time, broadleaf evergreen trees form closed-canopy woodlands. Common tree taxa include *Quercus wizlizennii*, *Q. chrysolepis*, *Lithocarpus densiflora*, *Arbutus menziesii* and *Umbellularia californica*, and these often merge with riparian woodlands (Sawyer et al. 1977). In southern California, foothill woodlands are dominated by winter-deciduous *Juglans californica* in association with several obligate resprouting evergreen chaparral species (*Heteromeles arbutifolia*, *Prunus ilicifolia* and *Rhamnus crocea*) that persist as gap-phase arborescent shrubs (Keeley 1990b). At higher elevations throughout the state the dominant oak is the winter-deciduous *Q. kelloggii*, and it merges with conifer forests. With one exception, all of these species are endemic to the MTC California Floristic Province; *Q. chrysolepis* is disjunct to Arizona. In average rainfall years fuel moisture remains high and fires dampen down and are largely extinguished by evergreen woodlands, but in drier years these woodlands are subject to crown fires. All tree species resprout epicormically if not burned too intensively, and some resprout from the base, often from massive burls.

None of the dominants in these vegetation types have postfire seedling recruitment; rather they produce transient seedbanks with little seed carry-over from year to year. Successful seedling recruitment is largely restricted to fire-free intervals and is often dependent on a sequence of high-rainfall years that promote seed production in one year and seedling establishment in the following year.

Foothill Coniferous Trees

Included here are trees that occur in small patches juxtaposed with chaparral, including: *Pinus attenuata*, *P. coulteri*, *P. muricata*, *P. radiata*, *P. sabiniana*,

P. torreyana, *Pseudotsuga macrocarpa* and *Torreya californica* in the Pinaceae family and several cypress species in the genus *Hesperocyparis* (formerly *Cupressus*) in the Cupressaceae family (Vogl *et al.* 1977; Minnich 1978; Zedler 1986; Barbour 2007). Nearly all of these are endemic to the California Floristic Province. Those in the latter family seldom exceed 10 m height whereas the others range from 20 m to 40 m in height. All occur in relatively small pure populations distributed amongst a mosaic of chaparral, grassland, woodland and forest.

The range of responses to fire is similar to those in chaparral. Some taxa are obligate resprouters and others seeders. *Torreya californica* persists by resprouting from the base but does not have postfire seedling recruitment, whereas most of the pine and cypress species are non-resprouting serotinous trees. These serotinous species have many attributes for persistence in a crown fire regime. Due to their low stature, branching near the ground and limited self-pruning, they form dense localized populations that burn in high-intensity crown fires, factors leading to the evolution of serotiny (Keeley & Zedler 1998). Cones typically remain closed on the tree for periods from a few years to a few decades, depending on the species. Soon after fire, cones open and disperse seeds (see Fig. 3.5). Seedling recruitment is restricted to a single postfire pulse of seedlings, and in this respect they resemble the crown fire chaparral response. Axelrod (1980), who assumed fire was largely a recent anthropogenic phenomenon, hypothesized that serotiny in these pines was not an adaptation to fire but rather to drought. However, we believe fire is the selective force for the origin and maintenance of serotiny in these species. Fire has been a potential selective force throughout the evolution of land plants, and life history analysis would not predict drought to select for delayed reproduction to a single postfire year (see Chapter 9).

A few tree species, such as *Pinus sabiniana* and *Pseudotsuga macrocarpa*, have a unique fire response that involves metapopulations with dynamic fire-related fluctuations (Keeley 2006a; Schwilk & Keeley 2006). Both are often embedded in chaparral and subject to periodic crown fires. Neither is serotinous nor has a persistent soil seedbank. Although the latter species resprouts epicormically, on sites with heavy fuels both species are vulnerable to local extirpation. However, these trees persist in refugia, such as open alluvial plains, grasslands, rock outcrops or extremely steep slopes with limited fuel loads, where they persist in the face of high-intensity crown fires in the surrounding landscape. Under long fire-free periods they spread into chaparral sites and persist until a high-intensity crown fire once again restricts them to a more limited number of safe sites.

Coast Redwood and Other Mesic Forests

The northwest region of the California Floristic Province is moist with moderate winters and the main vegetation is dense forests dominated by *Pseudotsuga menziesii* (Pinaceae), *Lithocarpus densiflora* (Fagaceae) and localized populations of *Sequoia sempervirens* (Taxodiaceae). There is a long history of high-intensity crown fires in these forests (Veirs 1979, 1982; Stuart 1987), but fire-scar dendrochronology studies show many of these forests have a history of frequent low-intensity surface fires (Finney & Martin 1992; Brown & Swetnam 1994). Undoubtedly this latter pattern is the result of human interference resulting from Native American land management (Stephens & Fry 2005). This is supported by the fact that the region has a very low lightning-ignited fire frequency (Keeley 2005), and historically it had one of the highest Native American densities in North America (Heizer 1978).

Prior to the entry of people into this region in the late Pleistocene, the fire regime in these forests was undoubtedly characterized by long-interval crown fires driven by unusually severe and widespread droughts at rare intervals. Such conditions would be needed to adequately dry understory fuels in these closed-canopy forests as well as enhance the probability of fire spread from surrounding lightning-rich areas in the interior. Consistent with this crown fire regime is the fact that redwood (*S. sempervirens*), the tallest tree in the world, is a vigorous resprouter after high-intensity crown fires (Stuart 1987), an unusual trait in conifers and not to be expected in a surface fire regime (Keeley & Zedler 1998). Under these extended fire-free periods, other disturbance events such as tree falls likely played a very important role in structuring these forests (Hunter & Parker 1993).

Mediterranean-type Vegetation outside the Mediterranean-climate Region

Mediterranean-type vegetation in other climatic zones may be viewed at two levels: (1) California Floristic Province taxa that range far beyond the MTC and (2) communities outside this climate that are composed of different taxa that are structurally similar to those that comprise our MTC communities.

Chaparral Outside the Mediterranean-type Climate

Chaparral is the most widespread vegetation within the MTC region of California. Components of chaparral vegetation are also found east of the California MTC (Table 5.3), in Arizona with a bimodal rainfall pattern and northeastern mainland Mexico with a summer drought/winter rain climate (see Fig. B1.1.1). The primary feature these MTVs have in common with chaparral is in the predominance of sclerophyllous-leaved evergreen shrubs, and they do share a few shrub species (Muller 1939, 1947; Shreve 1939). In addition, because they form dense shrublands in seasonal environments, they are all subject to periodic crown fires.

The vast majority of California chaparral shrub species are endemic to the MTC region, although there are noteworthy exceptions. Shrubs such as *Malosma laurina* are distributed south along watercourses into arid deserts of Baja California (Shreve 1936) and *Heteromeles arbutifolia* is disjunct from southern California to the tip of Baja California in a winter drought/summer rain climate (Phipps 1992).

Table 5.3 Comparison of North American regions with mediterranean-type vegetation comprising chaparral shrublands, in California with a MTC, and Arizona and northeastern mainland Mexico with non-MTCs

	California	Arizona	Sierra Madre Oriental, Mexico
Total extent (1000 ha)	7565	1214	~100
Latitude	30–40N	~30–34N	16–20N
Longitude	~117–123W	~108–115W	~97–101W
Elevation range (m)	100–2000	900–2400	2000–2800
Mean daily temperature range (°C)	7–27	0–33	11–17
Annual precipitation (mm)	275–675	250–650	380–640
Precipitation distribution	Winter to early spring	Winter to early spring and mid to late summer	Summer
Water deficit (evap > ppt)	Late spring to early autumn	Late spring to early summer and mid to late autumn	Winter to spring
Percentage shrubs evergreen	>95%	~75%	~50%
Number of shrub species in common with California	–	11	4
Woody plant diversity	Two highly diverse genera	No highly speciated woody genera	No highly speciated woody genera
Annuals	Highly diverse	Pretty diverse	Little diversity
C_4 grasses	Non-existent	Very diverse	Very diverse
Postfire shrub recovery	Sprouting and seedlings	Sprouting and seedlings	Sprouting and seedlings
Herbaceous postfire	Diverse spring postfire annual flora	Postfire ephemeral flora comprises two seasonal floras: spring and autumn	Only a summer postfire flora

Arizona chaparral shares many shrub species with California, including *Arctostaphylos pringlei, A. pungens, Ceanothus greggii, Cercocarpus betuloides, Eriogonum fasciculatum, Garrya flavescens, Quercus turbinella, Rhamnus californica, R. crocea, Rhus ovata*, and *R. trilobata*. The islands of chaparral in the Sierra Madre Oriental of northeastern Mexico share only four of these species but also have closely related species in several other genera.

Despite the sharing of shrub dominants, these non-MTC chaparrals comprise very different plant communities. This largely derives from the difference in growing seasons. In contrast to California, Mexican chaparral has a warm growing season and as a result the herbaceous understory and postfire flora comprise fewer annuals and many C_4 perennial grasses (Shreve 1939). Arizona chaparral, with a bimodal pattern of rainfall, comprises elements of both Californian and northeastern Mexican chaparral. This disparity is particularly evident after fire in Arizona, where the postfire ephemeral flora comprises a spring flora of annuals and an autumn flora dominated by herbaceous perennials (Fotheringham 2009).

Due to these two seasonal floras, the total flora in the first postfire year in Arizona chaparral sites is substantially higher than observed in California chaparral.

The similarities between California and northeastern Mexico chaparral include patterns of postfire shrub resprouting and seedling recruitment. Lloret *et al.* (1999a) concluded that because the mexical vegetation is burned infrequently, yet possesses many of the fire traits evident in frequently burned California chaparral, such traits are pre-adaptations to fire. However, the current fire regime in California is anthropogenic. In the absence of a constant bombardment of human ignitions in densely populated southern California (Keeley 1982, 2006a), fire regimes would be much more similar (Rodríguez-Trejo 2008). Thus, the similar postfire responses are not unexpected and can't be used to infer selective differences between regions.

In general, chaparral vegetation is widely distributed in both MTC California as well as non-MTC Arizona, with bimodal rainfall, and northeastern Mexico, with winter drought and summer rains. Other components of these communities are radically different and thus these "chaparrals" do not represent the same plant ecosystems. This is interesting in terms of paleoecology because shrub leaves are the most likely plant parts to be preserved, and one might expect that fossil floras of these three regions would tend to point toward the same ecological communities, when in fact they represent very different ecosystems.

Coniferous Forests outside the Mediterranean-type Climate

Sclerophyllous-leaved MTV conifer forests with surface fire regimes are not restricted to MTC California. Indeed, the two most widespread forest types, ponderosa pine forest and the mixed conifer forests, occur outside the MTC region. The two dominants of these forests, *Pinus ponderosa* in the former and *Abies concolor* in the latter forest type, are also dominants in similar forest types under non-MTCs in other parts of the western USA.

Pinus ponderosa has roots back to the mid Tertiary (see Fig. 9.1) and throughout this period climates have shifted and the taxon has migrated in step. The present distribution dates to only the last glacial episode (Anderson 1989). *Pinus ponderosa* is one of the dominants of the MTC mountains, throughout California and into Oregon, Washington and Idaho (Fig. 5.8), all of which have some semblance of a MTC with winter rains and summer droughts. However, this species is also widespread throughout the interior non-MTC Rocky Mountains. The primary climatic difference between regions is not as much in temperature as in the seasonal distribution of rainfall (Pearson 1951; Norris *et al.* 2006). Interior populations of *P. ponderosa* occur under a continental climate with summer rains; however, as with MTC populations, their distribution is characterized as occurring on sites exposed to prolonged soil moisture deficit (Barrett *et al.* 1980), and numerous factors affect soil moisture, including precipitation, degree of evapotranspiration, and soil structure (Stephenson 1998).

Mediterranean-type Vegetation outside the Mediterranean-climate Region

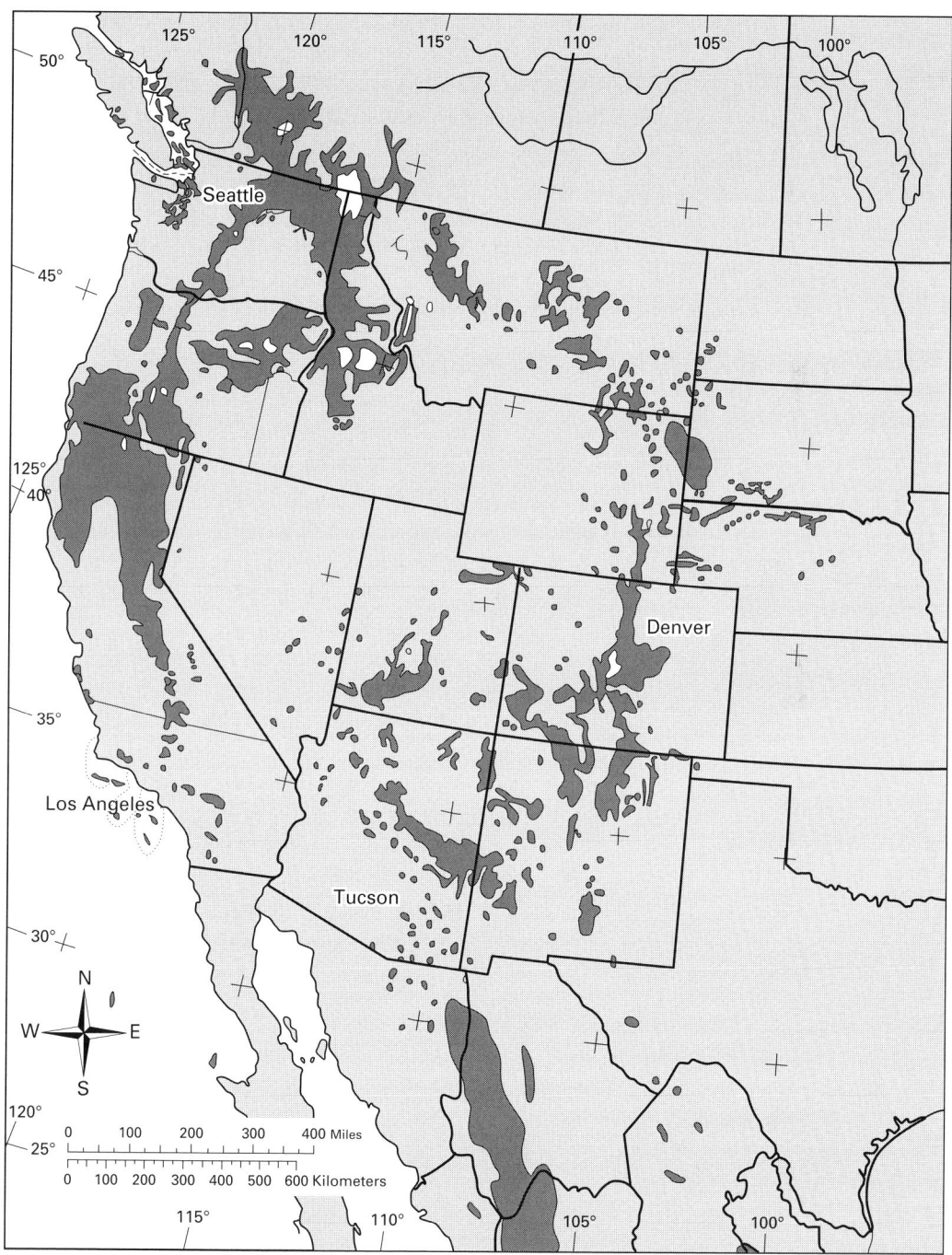

Fig. 5.8 *Distribution of ponderosa pine* (Pinus ponderosa) *in the MTC region of the Pacific Coast, and the non-MTC interior of western North America. (Based on Steele 1988.)*

If one takes a climate-centric view of species' distribution patterns then it would appear that *P. ponderosa* is widely distributed under quite different climate conditions as a result of ecological sorting processes, and evolutionary adaptation is not a major factor determining its current distribution (e.g. Ackerly 2004a). An alternative hypothesis is that the most relevant selective factors have not changed, in particular cold winters and seasonal soil moisture deficit, as well as susceptibility to a surface fire regime – factors common throughout the current range of ponderosa pine and likely the selective factors responsible for the origin of this species.

This hypothesis of course does not rule out the possibility of continuing trait selection in response to subtle differences in climate (Sorensen *et al.* 2001) and fire regime (Parker 1987). There is evidence of evolutionary fine tuning within regions at the microhabitat scale of different slope faces, elevation and soil types (Linhart 1988; Zhang & Cregg 2005).

Despite the widespread distribution of tree dominants outside the MTC, the assemblage of species in Californian conifer forests, both in terms of codominants and understory species are relatively unique. For example, in ponderosa forests codominants include endemics to the California Floristic Province such as *Quercus kelloggii* and *Calocedrus decurrens*. The understory also includes endemics in all main growth forms – shrubs, herbaceous perennials and annuals – although many of the genera may be shared across different regions (Steele 1988). Understory species within California forests are distinctly different from similar forest types outside California; however, within California these species do not exhibit high fidelity to particular forest types (Mellmann-Brown & Barbour 1995). The mixed conifer forest exhibits a similar pattern in that the assemblage in California contains many unique elements, not the least of which is the giant sequoia, *Sequoiadendron giganteum*. There is also much intraregional variation based on substrate characteristics (Abella & Covington 2006). Thus, while the distribution of tree dominants may be tied more to fire regime than to climate, there appears to be a clear MTC signal in the current assemblages.

Fire Management

Native American Management

When Spaniards first settled the California coastline late in the eighteenth century they inherited a landscape that Native Americans had long managed with fire. However, compared with other MTC regions, the New World landscapes in California and Chile have had the shortest period of human land management. The oldest reliable archeological records are about 15 000 yrs before present (BP). North America and South America together have relatively few sites in this range and there is some contention about their validity. Considering the number of scientists searching for such sites it seems almost certain that even if humans were here during that early period they were not a dominant presence on these landscapes.

Approximately 13 000 BP represents the age at which there is nearly complete agreement on human occupation and rapid expansion throughout North and South America. Between 13.5 and 13 cal kBP (thousands of calendar/calibrated years before present), 17 or more genera of North American megafauna (animals $\geq$ 100 kg) became extinct (Martin & Klein 1984). The timing of this event was coeval with Paleoindian expansion, implicating human involvement, but the extinctions also coincided with significant climatic changes, making it difficult to parse out the relative importance of each. However, this megafauna survived previous Pleistocene interglacial episodes of sharp climatic warming; thus the arrival of humans would seem to be an important factor driving their extinction.

It is hypothesized that this period of time marked a change in fire regimes in California and other parts of the western USA, driven by novel anthropogenic sources of ignitions, altered fuel structure related to megaherbivore decline, and warming climates (Pinter *et al.* 2011). On coastal and foothill landscapes the arrival of human ignitions resulted in an immediate increase in fire frequency and subsequent changes in vegetation such as localized replacement of chaparral with herbaceous vegetation. In the higher interior mountains where Native American populations were scarce, and there were abundant natural ignition sources from lightning, forest structure was probably little affected by humans.

This model contradicts the climate-only model of Marlon *et al.* (2009) because it infers a multifactorial explanation in which humans, herbivores, climate and fire sort out along landscape gradients. It is consistent with Marlon *et al.*'s data because that study sampled largely from high-elevation landscapes where natural lightning ignitions were not limiting, and human settlements were absent. However, for those landscapes such as the lower elevations in California where natural ignitions were limiting, humans had the potential for greater impact on fire regimes, both from increased ignitions and from altered landscape fuel patterns due to the demise of megaherbivores.

The mid Holocene marks a change in food economy of the California Indians with increased emphasis on natural seed crops as the main dietary staple, and populations increased throughout all regions within the California Floristic Province (Erlandson & Glassow 1997). Although agriculture had already begun to flourish in parts of the Americas, it was basically absent in California. Nonetheless, Native Americans did manage the California landscape, often very intensely through the use of fire (Timbrook *et al.* 1982; Keeley 2002b). As was the case over much of the globe, the primary management goal was to displace less useful woody vegetation with high seed output annual forbs and grasses, and fire was a reliable means of doing this. The primary focus of burning was to break up dense shrublands into mosaics of herbaceous and woody species (Cooper 1922; Bauer 1930). Many of the typical postfire species in these shrublands are high seed producers and were heavily exploited by Indians as food sources. Fire was used for other purposes, for example for herding animals during hunting or killing invertebrates such as fleas that would invest homes, weevils that would attack acorn crops, or grasshoppers, which were too hard to capture without fire. Landscape-level impacts were produced by their management goal of wanting to open up, or entirely type convert, dense shrublands to annual forbs and grasses. This landscape change increased seed resources and created a major food source in many parts of the region. This change also improved travel and reduced attacks from grizzly bears and competing Native Americans. Ideally, total elimination of shrublands may not have been a goal since retention of some wood resources would have been desirable for cooking fires. However, these native shrublands by and large are not resilient to the continuous onslaught of fires and eventually are type converted to annual herbaceous associations.

Although use of repeated burning to convert shrublands to grasslands was a well-developed management technique in Spain, there are relatively few reports of this practice by European colonists in California. Possibly the prior burning by Native Americans had opened up much of the landscape to grasslands and thus there was less incentive for further type conversion of shrublands and woodlands. However, this eventually changed as resource competition by the burgeoning California population experienced increasing limitations to livestock grazing and need for more rangeland in the late nineteenth century (Burcham 1957).

Contemporary Management

The chaparral crown fire regime and the montane forest surface fire regime have very different characteristics and require very different fire management approaches. Despite the fact that both have been managed under the same fire suppression policy for the last century, the outcome has been radically different. Fire suppression policy has been highly successful at excluding fire in forests but largely unsuccessful at excluding fire in chaparral.

Fire suppression has been highly effective at excluding fire in forests because the surface fire regime is conducive to rapid fire suppression. Surface fuels generate shorter flame lengths that can be effectively suppressed by fire fighters. Many of these fires are ignited by lightning during moderate weather conditions not conducive to rapid fire spread and the wind resistance created by dense forests reduces rates of spread. As a consequence, fire management agencies have been extraordinarily successful at excluding fires from these environments for over a century. One consequence of fire suppression is that surface fuels as well as ladder fuels have accumulated to levels far in excess of historical conditions. The primary concern is that higher fuel loads will spread fire from the surface to the canopy and will lead to crown fires of much greater extent than historically occurred. In other words the gap size to be expected by mixed surface and crown fires will become much larger as a result of historic fire suppression and buildup of fuels (Fig. 5.9). This not only increases fire hazard but has negative resource impacts such as the reduction of parent seed trees, making regeneration more precarious.

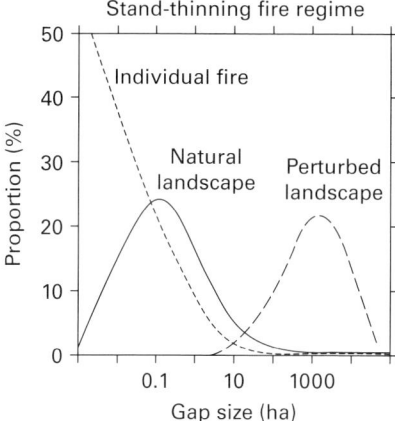

Fig. 5.9 *Projected gap size in mixed conifer forests of the Sierra Nevada under historical conditions and perturbed conditions due to a century of fire suppression policy that has excluded fire. (From Keeley & Stephenson 2000.)*

These forests require management that restores historical fire regimes or utilizes mechanical thinning to reduce current fuel loads (Franklin et al. 2006). These approaches sound like simple alternatives but there are untold problems behind both. Prescription burning is a viable option in some parts of California, but due to human demographics, it is not widely applicable in many areas highly fragmented by development. Additionally, atmospheric circulation in most of these ranges is closely linked with foothill and valley communities and as a consequence air quality constraints greatly limit the window of opportunity for prescription burning. Mechanical thinning (i.e. logging) is perhaps the only option in many forests, but it is costly, particularly if the prescriptions focus on removing those tree sizes most responsible for fire hazard. Mechanical thinning projects can pay for themselves, but usually only if large trees are harvested. This scenario creates a dilemma because removing large trees promotes further in-growth of saplings and may exacerbate the fire hazard problem, requiring future mechanical thinning at shorter intervals.

In chaparral, fire suppression has not been effective at eliminating fires. On the basis of our understanding of historical burning patterns it is apparent that contemporary fire regimes are not qualitatively different from historical fire regimes (Keeley 2006a). However, fire frequency has increased with demographic growth and the human subsidy of fires has greatly increased fire frequency. Thus, the lower foothills and coastal plains have not experienced fire exclusion and in fact the landscape burns far more frequently now than historically was the case (Keeley et al. 1999a). This does not imply that fire suppression has had no impact. Indeed, without suppression over the past century this landscape would have burned at a much greater frequency than was the case historically. Fire management of this landscape is particularly difficult because it is also the region of greatest human development. The primary management dilemma is

148 Fire in California

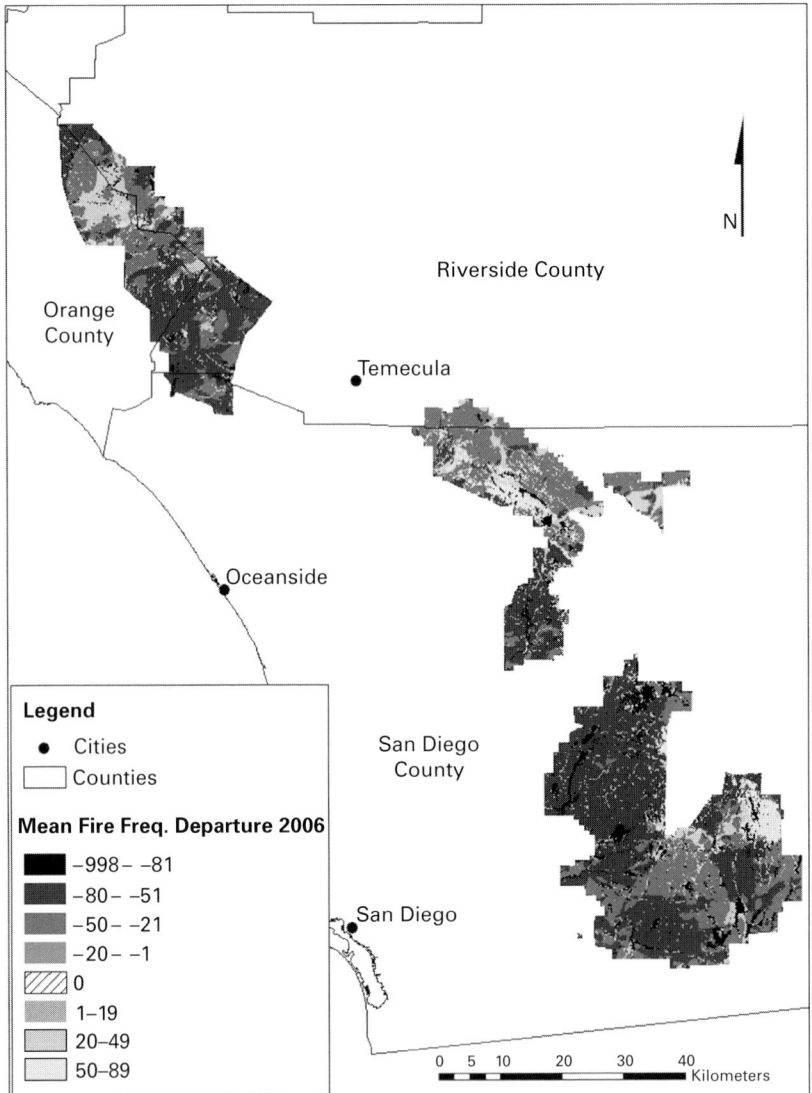

Fig. 5.10 *Fire departure map for the Cleveland National Forest in southern California, USA. Negative indices indicate an excess of fires during the past century higher than historical levels and are typical of the lower foothills dominated by chaparral, sage scrub and grassland. Positive indices indicate missed fire cycles and are in higher elevation coniferous forests. (Data from Hugh Safford, U.S. Forest Service, 2008.)*

how to reduce the vulnerability of communities to wildfire risk and manage natural resources for sustainability.

Because much of this landscape burns at a frequency greater than historical levels (Fig. 5.10), excess fires are a major resource concern in much of the current chaparral landscape. Many species in these shrublands are extirpated when fire frequency exceeds more than a single fire in a decade (Keeley 2006a). Because so much of the landscape has greatly exceeded the historical fire frequency there has already been substantial loss of native habitat and replacement with alien grasslands (see Chapter 12). This resource concern often comes into conflict with fire management when fire hazard reduction treatments threaten to increase disturbance on an already overly disturbed landscape.

Conclusions

The California MTC landscape is dominated by shrublands, oak woodlands and conifer forests, each with very different fire regimes. The sclerophyllous-leaved MTV that dominates these communities is highly fire-prone and species exhibit numerous traits adaptive under different fire regimes. Although the composition of these plant communities is unique to California, many of the shrub and tree dominants have a limited presence in other non-MTC semi-arid plant communities in western North America. Periodic droughts and fires are common selective factors present throughout the range of distribution of this MTV.

6 Fire in Chile

The Mediterranean-type Climate Region of Chile

The mediterranean-type climate (MTC) in Chile (Fig. 6.1) is distributed from La Serena (30° S; Región IV, see Appendix 6.1) in the north to Concepción (37° S; Región X) in the south. It is constrained to the west side of the Andean mountain range, although as the height of this range decreases in the south, a MTC is observed at least as far eastward as Bariloche, Argentina. Although a pattern of winter rains and summer droughts extends northward into the Atacama Desert, this area falls outside our definition of MTC because winter evaporation exceeds rainfall in these areas of extremely low precipitation. The northern border of the MTC region is the transition from desert communities to shrubby matorral, while the southern border corresponds to the point of transition from sclerophyll woodlands to Valdivian evergreen forests (Gajardo 1994; Amigo & Ramirez 1998; Rundel et al. 2007).

The landforms of central Chile can be divided into three north–south trending geomorphic zones (Fig. 6.1): the Coastal Cordillera, the Central Valley, and the high Andean Cordillera (Armesto et al. 2007). The Coastal Cordillera rises relatively sharply from the coast, with little extent of coastal terraces, and reaches elevations as high as 2222 m at Cerro Roble and 1880 m at Cerro Campana, which lie between Valparaíso (Región V) on the coast and Santiago (Región Metropolitana) at the base of the Andes. The Central Valley is a structural basin filled to great depth by sediments from the surrounding mountains. North of Santiago, spurs from the Andes extend west across the valley and connect with the Coastal Cordillera, separating individual river basins such as that of the Río Aconcagua (Región V). From Santiago to the south, however, the valley extends uninterrupted for a distance of 900 km to Puerto Montt (Región X), with typical elevations of 400–700 m. The Andean Cordillera marks the eastern boundary of the MTC zone of central Chile. It is the product of complex tectonic activity beginning in the Cretaceous (see Fig. 9.1), but with major uplift in the late Tertiary. Paleobotanical evidence suggests that the central Andes had not attained more than half their current elevation by 10 Ma (Gregory-Wodzicki 2000), but was close to the present height by the end of the Miocene (Reynolds et al. 1990). Elevations reach from 4000 m to nearly 7000 m. To the north of Santiago the Andes are largely composed of metamorphosed sedimentary rock, but a major volcanic zone extends from south of Santiago through the Lake District (Los Lagos Región X).

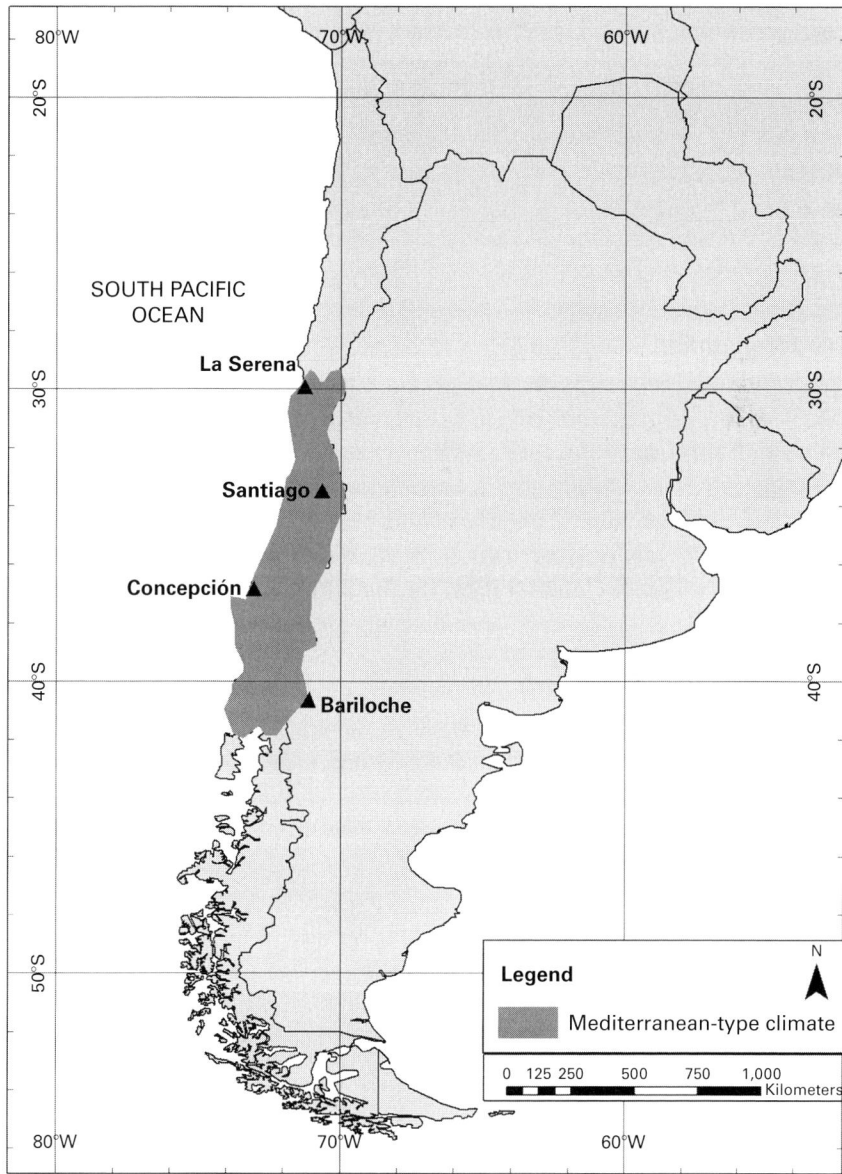

Fig. 6.1 *Distribution of the mediterranean-type climate (MTC) in Chile and adjacent parts of South America (dark shading).*

Valparaíso on the central coast receives about 390 mm of annual precipitation, while Santiago inland at the base of the Andes receives about 350 mm. La Serena at the northern margin of the mediterranean zone receives only about 125 mm of annual precipitation, with a wide variability between years. To the south, Concepción along the coast at the traditional southern margin of the region receives about 1275 mm annually. Higher levels of mean annual precipitation reaching 2000–5000 mm or more are present in the forested areas of the Andean Cordillera in Los Lagos Región. Eastward there are many pockets of drier sites with a MTC, for example, Bariloche, Argentina (41°06′ S/71°10′ W) receives only about 10% of its annual 717 mm precipitation in the summer (Müller 1982).

The cold Humboldt Current flowing northward along the Pacific coast of Chile moderates coastal temperatures. Mean maximum summer temperatures in January are typically 20–23 °C along the central coast, while inland cities in the Central Valley reach 27–30 °C. Mean winter minimum temperatures in July reach below freezing only in the high mountains and some of the steep river valleys below the Andes where cold air drainage dominates, although snow occurs occasionally in Santiago.

Vegetation

Matorral shrublands (see Fig. 1.6d) form the characteristic evergreen sclerophyllous shrublands of central Chile (Rundel 1981a; Armesto et al. 2007). These communities exhibit mixed dominance of many shrub species (Table 6.1), with the most common including *Lithraea caustica*, *Kageneckia oblonga*, *Schinus polygamus*, *Escallonia pulverulenta*, *Trevoa trinervis* and *Retanilla ephedra*. Also present are a variety of semi-woody shrubs with semi-deciduous or deciduous leaves (Table 6.1), which may dominate on more arid sites. These subshrubs include genera such as *Baccharis* and *Satureja*, which are also found in California shrublands (Parsons 1976). In contrast to California chaparral, herb cover is relatively abundant in mature matorral, often covering up to 40% or more of the ground surface (Montenegro et al. 1978). Much of this herb cover is composed of herbaceous perennials, including a rich flora of geophytes. Dry equator-facing slopes in the matorral exhibit a different vegetation structure, with open stands of the arborescent cactus *Echinopsis chilensis* and *Puya* species, with *L. caustica* and *Colliguaya odorifera* as associated shrubs. Coastal sites are often dominated by a smaller-stature scrub vegetation with species of *Puya* well represented (Fig. 6.2).

Fig. 6.2 *Coastal scrub with flowering stalks of* Puya *in central Chile. (Photo by Jon Keeley.)*

Table 6.1 Characteristic species of Chilean shrublands, woodlands and transitional forests

Species	Family	Growth form	Phenology
Matorral (interior and coastal)			
Acacia caven	Fabaceae	shrub/tree	deciduous
Baccharis linearis	Asteraceae	shrub	evergreen
Baccharis rosmarinifolius	Asteraceae	shrub	evergreen
Colletia spinosissima	Rhamnaceae	shrub	evergreen
Colliguaya odorifera	Euphorbiaceae	shrub	evergreen
Echinopsis chiloensis	Cactaceae	succulent	evergreen
Ephedra andina	Ephedraceae	shrub	evergreen
Escallonia pulverulenta	Escalloniaceae	shrub	evergreen
Flourensia thurifera	Asteraceae	shrub	evergreen
Kageneckia oblonga	Rosaceae	shrub	evergreen
Lithraea caustica	Anacardiaceae	shrub	evergreen
Lobelia salicifolia	Campanulaceae	shrub	evergreen
Podanthus mitiqui	Asteraceae	shrub	deciduous
Proustia cuneifolia	Asteraceae	shrub	deciduous
Puya berteroniana	Bromeliaceae	shrub	evergreen
Puya chilensis	Bromeliaceae	rosette	evergreen
Retanilla ephedra	Rhamnaceae	shrub	evergreen
Schinus polygamus	Anacardiaceae	shrub	evergreen
Talquenia quinquenervia	Rhamnaceae	shrub	evergreen
Teucrium bicolor	Lamiaceae	shrub	deciduous
Trevoa trinervis	Rhamnaceae	shrub	deciduous
Sclerophyll and mesic woodland			
Beilschmiedia miersii	Lauraceae	tree	evergreen
Crinodendron patagua	Elaeocarpaceae	tree	evergreen
Cryptocarya odorifera	Lauraceae	tree	evergreen
Drimys winteri	Winteraceae	tree	deciduous
Jubaea chilensis	Arecaceae	tree	evergreen
Nothofagus macrocarpa	Fagaceae	tree	deciduous
Persea lingue	Lauraceae	tree	evergreen
Peumus boldo	Monimiaceae	tree	evergreen
Quillaja saponaria	Quillajaceae	tree	evergreen
Transitional forest			
Aextoxicon punctatum	Aextoxicaceae	tree	evergreen
Beilschmiedia berteroana	Lauraceae	tree	evergreen
Eucryphia glutinosa	Cunoniaceae	tree	evergreen
Gomortega keule	Gomortegaceae	tree	evergreen
Myrceugenia obtusa	Myrtaceae	tree	evergreen
Nothofagus allessandri	Fagaceae	tree	deciduous
Nothofagus alpina	Fagaceae	tree	deciduous
Nothofagus glauca	Fagaceae	tree	deciduous
Nothofagus leonii	Fagaceae	tree	deciduous

Fig. 6.3 Cryptocarya alba *woodland in central Chile. (Photo by Jon Keeley.)*

Sclerophyll woodland (*bosque esclerófilo*) is widespread locally on mesic pole-facing slopes of the Coastal Cordillera and in moist valleys (San Martin 2005). Common trees or tree-like shrubs include *Quillaja saponaria, Cryptocarya alba* (Fig. 6.3), *Lithraea caustica, Beilschmiedia miersii* and *Peumus boldus* (Table 6.1) These stands can reach 10–15 m or more in height. On mesic sites these forests have been termed hygrophilous forest, with species more characteristic of austral forest to the south, such as *Aextoxicon punctatum, Aristotelia chilensis, Crinodendron patagua, Persea lingue,* and *Drimys winteri*.

An open savanna community known as *espinal* covers low-lying areas of the Central Valley (Fig. 6.4) and eastern slopes of the Coastal Cordillera (*secano interior*) from about 32° S in the north to the Laja River valley in the south at 36° S (Ovalle *et al.* 1990; Fuentes *et al.* 1990). This community largely owes its origin to historical opening of matorral and sclerophyll woodland vegetation by fire, woodcutting and cattle grazing, leading to an open savanna of *Acacia caven* in a matrix of alien annual grassland (Bahre 1979; Fuentes *et al.* 1989).

Unlike the situation in California and the Mediterranean Basin where evergreen shrublands give way to coniferous forests above about 1200–1500 m elevation, forests are largely lacking from much of the Andean Cordillera in central Chile. Only in the southern portions of central Chile does forest vegetation begin to occur. For most of the central and northern Andean slopes of the region, matorral gives way at these elevations to a low and open evergreen shrubland termed montane matorral. This community extends to elevations of about 2000–2200 m where it is replaced by subalpine and alpine communities of cushion plants and low herbaceous perennials (Armesto *et al.* 1980).

Fig. 6.4 Acacia caven *savanna in the Central Valley of central Chile, in some parts replacing degraded matorral. (Photo by Jon Keeley.)*

The northernmost occurrence of conifers in the Chilean Andes are pockets of *Austrocedrus chilensis*, which include remnant stands just north of Santiago to more extensive stands in south-central Chile. More extensive are the forests of deciduous species of *Nothofagus* which cover large areas of the southern MTC region of Chile. *Nothofagus obliqua* var. *macrocarpa* reaches its northern limit in small stands around Cerro Campana and Cerro Roble in the Coastal Cordillera (Región V), and forms relatively pure forests at elevations of about 1000–2000 m on moist south-facing slopes of the Andes from about 34°30′ to 36°50′ S (Hinojosa *et al.* 2006). Similar forests exist in both the Coastal Cordillera and Andean Cordillera in a transition from the traditional mediterranean region of central Chile to the Valdivian forests to the south (Regiones VI and VII), with typical rainfall levels around 700–1200 mm or more (San Martin 2005). Forest dominants here include *Nothofagus glauca*, *N. allessandri*, *N. alpina* and *N. leonii*, along with evergreen trees such as *Beilschmiedia berteroana*, *Eucryphia glutinosa*, *Myrceugenia obtusa* and the rare endemic *Gomortega keule* (Table 6.1; Hoffmann *et al.* 2001).

High-elevation forests dominated by deciduous *Nothofagus pumilio* typically form the Andean tree line extent in south-central and southern Chile, extending from about 36° S latitude in the mediterranean region to as far as 56° S in Patagonia. These forests are little affected by fire (Veblen *et al.* 1979; Lara *et al.* 2005), although seedlings of *Nothofagus alpina* do become established following fire (Palfner *et al* 2008).

Excluding the heavily modified areas of the Central Valley that are largely agricultural today, various shrublands comprise over two thirds of the land cover of the MTC region of Chile (see Table 1.1). Evergreen sclerophyllous forests and woodlands make up 20% of the area. Further south, winter-deciduous forests with *Nothofagus* dominate the southern portions of the country.

The austral Valdivian rainforests of the Lake District of south-central Chile (39°30′–43°30′ S; Regiones VIII and IX) lie beyond the traditional definition used in Chile of the mediterranean zone, but retain a strong MTC of winter precipitation with dry summers comparable to that of western Oregon in North America. Native forests today cover about half of this region, with much of this concentrated on the western slopes of the Andes. The lower coastal ranges in this area have been heavily deforested, as has the Central Valley. The most widespread forest community is the Valdivian rainforest with a moderately diverse assemblage of evergreen broadleaf trees including *Nothofagus dombeyi*, *Laureliopsis philippiana*, *Eucryphia cordifolia*, *Aextoxicon punctatum* and other species (Donoso 1993, Ramírez & San Martin 2005). Here, landslides and volcanic eruptions are the most important natural disturbances (Veblen *et al.* 1979, 1980), with fire much less important. Higher-elevation montane forests are dominated by deciduous species of *Nothofagus*. Forest stands with a major canopy cover of conifers are rare in these austral forests. The exceptions are seral stands of *Araucaria araucana* (see Fig. 9.2) in the northern margin of the Lake District, small areas of *Austrocedrus chilensis* in the Andes, and scattered forests of *Fitzroya cupressoides* in both the Coastal Cordillera and Andes. Large areas of the coastal ranges now support plantation stands of *Pinus radiata* (Estades & Escobar 2005).

Much of the coastal region at this latitude is very wet but eastward there are many pockets of drier landscapes with a clear MTC, for example as around Bariloche, Argentina (41°06′ S/71°10′ W). Associated with this MTC are matorral shrublands that are subject to fire. Although the strong summer dry conditions of MTCs do not extend into the Patagonia region of southern Chile, patches of fire-prone matorral occur in the region. These are often dominated by a resprouting shrub form of *Nothofagus antarctica*, forming shrublands as well as the understory of *Araucaria* forests (Steinke *et al.* 2008). Large areas of these shrublands have burned at irregular intervals (Cerrillo *et al.* 2008).

Fire Regimes

Throughout the MTC region of Chile summer wildfires are widespread (Montenegro *et al.* 2004). The fire season extends from October through April, with more than three fourths of all fires occurring during the summer months of December (13% of fires), January (26% of fires), and February (39% of fires). On average, some 5400 fires are recorded each year, affecting an average of 52 400 ha. These averages, however, do not indicate the ongoing trends in increased fire frequency

Table 6.2 Annual average occurrence and area burned by forest fires in Chile by five-year periods from 1963 to 2003

Five-year period	Number of fires per yr	Average area burned per yr (ha)	Average area per fire (ha)
1963–1968	383	26 767	70
1968–1973	883	39 682	45
1973–1978	2207	21 891	10
1978–1983	4239	40 679	10
1983–1988	5451	69 853	13
1988–1993	5090	47 617	9
1993–1998	5653	53 269	9
1998–2003	6346	52 348	8

Source: Soto (1995) and CONAF (2003).

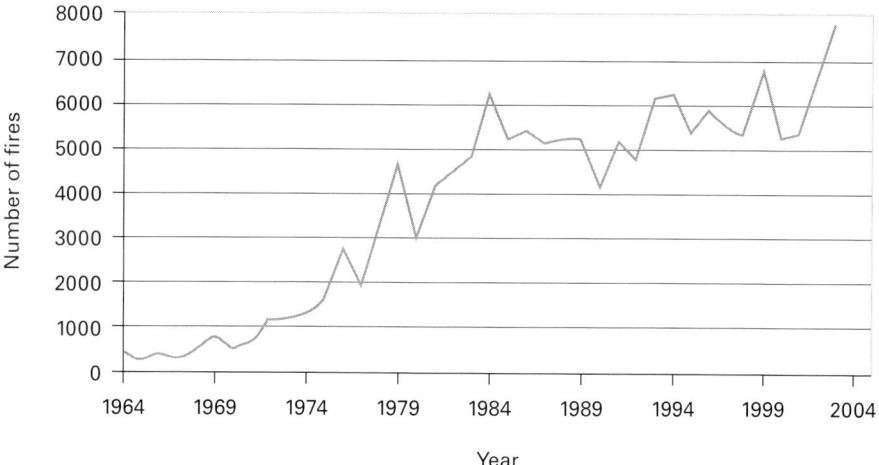

Fig. 6.5 *Total number of fires per annual fire season for Chile. Each season stretches from October to April of the following year. The value indicated for the season corresponds to the year the season ended. (Data from Soto 1995; CONAF 2003.)*

(Table 6.2), from fewer than 1000 fires annually in the early 1970s to a high of 7500 fires in the 2003 fire season (Fig. 6.5).

The annual area consumed by fire is highly variable between years (Fig. 6.6), with much of this variation explained by summer drought (Montenegro et al. 2004). For example, more than 100 000 ha burned in 1999, and both 1998 and 1999 were severe drought years due to La Niña effects (Quintana 2000). Two years later under an El Niño influence of high rainfall, the 2001 season experienced less than 11 000 ha burned. Most fires are small, with

Fire in Chile

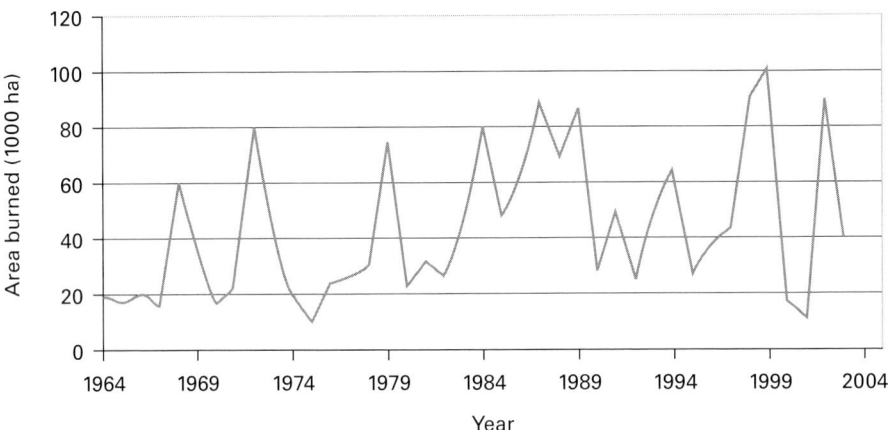

Fig. 6.6 *Total surface area of fires in Chile by annual fire season. Each season stretches from October to April of the following year. The value indicated for the season corresponds to the year the season ended. (Data from Soto 1995; CONAF 2003.)*

only about 1% extending over large areas and causing significant economic damage (Soto 1995; CONAF 2003). Data collected since the 1960s indicate that the average size of fires has dropped from >50 ha in the 1960s and 1970s to <10 ha in the 1990s (Table 6.2). There are few large fires: 87% of the fires are less than 5 ha and more than 90% are under 10 ha. However, large fires up to 50 000 ha have occurred.

It is clear that the historical fire frequency in the Valdivian forests of Los Lagos Región (Región X) was dramatically increased after 1880 with the settlement of the area by European colonists. Fire was widely used in this period to clear land for agriculture and grazing, and tree ring records clearly show this increase in fires (González *et al*. 2005), with an estimated 400 000 ha of forest burned in 1853 alone (Cavelier & Tecklin 2005). More than 11 million ha of forest were burned overall with the expansion of agriculture and ranching in this region (Haltenhoff 1991). Detailed fire records have been kept by the Corporación Nacional Forestal (CONAF) since 1979. Virtually all fires from Los Lagos Región over this period were attributed to accidental or intentional human actions, although both lightning and vulcanism have been documented as natural sources of fire for the region (Lara *et al*. 1999; Veblen *et al*. 1999). Annual climate cycles and seasonality show a strong relationship to the size of fires. Three of these dry summer regimes with large fires were associated with strong ENSO (see Box 1.2) conditions: 1983, 1987 and 1997. However, other years with dry summers have not had large fires, suggesting the timing of human ignitions are a factor.

Half of the fires in Chile occur within the MTC area of central Chile, from north of Valparaíso to near Concepción (Regiones V, VI and VII). Fires are more limited northward because the open shrublands and semi-desert scrub lack the continuity of fuels to carry fire. In the south a large increase in numbers of major fires has occurred in recent years in Los Lagos Región of south-central Chile, often in areas of secondary vegetation or in tree plantations. About 35% of the fires occur in native forests and 15% in plantations of Monterey pine (*Pinus radiata*) and *Eucalyptus*.

Humans are the cause of virtually all fires in central Chile. The high Andean Cordillera to the east prevents summer convective thunderstorms from moving westward from Argentina (see Fig. 10.4) and, as a result, lightning is rare to virtually non-existent over much of the MTC region (Armesto & Gutiérrez 1978; Montenegro *et al.* 2003). On a geological timescale this is a relatively recent effect and lightning-ignited fires may have been relatively common in central Chile through much of the Miocene, until the Late Miocene completion of the Andean uplift formed an effective barrier to westward storms (see Chapter 10).

Historically the primary cause of fires in Chile has been accidental ignitions associated with human traffic and transport through wildland areas, but arson fires have become the primary cause in recent years. This increase in intentional fires has been particularly evident in the forest lands and plantations of south-central Chile.

As in other MTC regions of the world, fires in Chile include both stand-replacement crown fires and lower-intensity ground fires. Fires in matorral shrubland, sclerophyll woodland, and young Monterey pine and *Eucalyptus* plantations are characteristically intense stand-replacement fires that consume the majority of aboveground biomass. Fire in the humid temperate forests of *Nothofagus*, mixed evergreen trees and conifers are generally of low to moderate intensity and consume understory vegetation without killing the forest canopy trees.

Fire in Matorral Shrublands

Natural disturbance regimes in the matorral have been altered greatly over the last four and a half centuries since European settlement, due to land clearance, fire, grazing, charcoal production and invasive herbivores such as the European rabbit (Armesto *et al.* 2010).

As a consequence of the absence of natural fire as a major disturbance factor in matorral, the flora shows fewer life history specializations to fire than found in the other four MTC regions (Montenegro *et al.* 2003). For example, none of the Chilean matorral shrubs have strict postfire seedling recruitment and there is no postfire ephemeral flora as in other systems. However, fire-adaptive traits are not entirely lacking either. There are a few annuals that do germinate in profusion after fire (see *Loasa* sp. under the shrub skeleton in Fig. 11.2d), although not strictly tied to fire. All shrubs resprout after fire (see Table 3.1; Araya & Ávila 1981, Armesto & Pickett 1985; Ginnochio *et al.* 1994; Montenegro *et al.* 2003). Although resprouting *per se* is not unique to fire-prone environments, lignotubers (see Fig. 3.1 and Table 3.2), which are closely associated with fire-prone environments (see Chapter 9), are present as ontogenetic traits that develop early in seedling growth in several woody Chilean matorral species (Hoffmann & Kummerow

1978; Montenegro *et al.* 1983). The vigor of resprouting is typically high in woody shrub and arborescent species such as *Lithraea caustica*, *Kageneckia oblonga*, *Quillaja saponaria*, *Azara dentata* and *Cryptocarya alba*, but is relatively weak in colonizing and semi-woody species such as *Baccharis linearis*, *Muehlenbeckia hastulata* and *Colliguaya odorifera* (Segura *et al.* 1998).

All Chilean matorral shrubs establish seedlings in the absence of fire (Fuentes *et al.* 1984, 1989, 1994; Fuentes & Espinoza 1986). This fire-independent recruitment syndrome is typically associated with particular fruit characteristics that enhance dispersal (see Chapter 9), and for species with this syndrome (i.e. obligate resprouters) there are marked similarities between Chile and California (Hoffmann *et al.* 1989). In Chile, most woody shrubs have animal-dispersed propagules, and isolated shrubs provide suitable perching sites (Jiménez & Armesto 1992). This dispersal pattern may account for the preponderance of seedling recruitment being associated with shrub clumps (Fuentes *et al.* 1984), although factors during recruitment also appear to play a role (Holmgren *et al.* 2000b).

A small number of matorral shrubs accumulate soil seedbanks between fires, and may also produce a small pulse of seedling recruitment in the first growing season after fire. However, these are from hard-coated seeds in families such as the Fabaceae and Rhamnaceae, and they germinate under a range of disturbance conditions. Examples of shrub species that may establish abundant seedlings after low-intensity fires are *Acacia caven*, *Trevoa trinervis*, and *Muehlenbeckia hastulata* (Muñoz and Fuentes 1989; Fuentes *et al.* 1994; Segura *et al.* 1998). Canopy storage of dormant seeds in serotinous cones or fruits, as occurs in the South African and Australian MTC regions, is not present in any of the Chilean matorral species.

Although fires promote the germination and establishment of some herbaceous species in matorral (S. Keeley & Johnson 1977; Ávila *et al.* 1981), most seeds in matorral soils are killed by the heat of a fire (Muñoz & Fuentes 1989; Segura *et al.* 1998). The widespread annual flora of open matorral stands largely consists of generalist species, as well as an abundance of non-native European annual grasses and forbs, with transient seedbanks (Figueroa & Jaksic 2004). Several studies have suggested that anthropogenic fires in central Chile may promote the invasion of alien plants with fire-adaptive traits not present in the native flora (Muñoz & Fuentes 1989; Segura *et al.* 1998; Holmgren *et al.* 2000a, 2000b; Figueroa *et al.* 2009; Gómez-González & Cavieres 2009). Field studies in matorral have found that fire has little or no influence on the composition of herb floras and herb abundance (Holmgren *et al.* 2000a), although fire appears to promote the invasion of alien annual species under some conditions (Ávila *et al.* 1988). Other studies have shown that high-intensity fires negatively affect the seedling emergence of both native and alien species, but more strongly so in native species. Low-intensity fires do not significantly affect the emergence of native herbs but did lead to increased alien species richness (Gómez-González & Cavieres 2009). Fire-following annuals with dormancy broken by chemical cues from ash have not been recorded from Chile.

An indication that fire has not been an important ecological disturbance factor in some parts of central Chile is the widespread occurrence of fire-sensitive arborescent cacti such as *Echinopsis chilensis* on dry equator-facing slopes in matorral communities. In contrast, in the North American California/Baja California MTC region, fire has apparently eliminated this growth form of cactus from shrublands and restricted it to steep canyons where it cannot be reached by the frequent fires in the region.

Frequent fires on the matorral landscape since the arrival of Euro-Chileans have undoubtedly had some impact on the structure and composition of matorral communities. There is no question that matorral stands in proximity to urban areas exhibit much more open structure in a mosaic of shrubland and non-native annual grassland. As with other MTC shrublands, fires in matorral communities burn with varying intensities, often leaving a patchy distribution of lightly burned stems and intense fire areas with little aboveground biomass remaining. It is reasonable to assume that intense fires offer a relative advantage to vigorously resprouting species over colonizing species, while low-intensity fires would not favor either group (Segura *et al.* 1998). The limiting factor for shrub seedling establishment after fire may be more an effect of seed availability due to lack of seed dormancy. When seedlings do occur after fire they survive better and grow faster than in areas cleared by hand or in competition with herb growth (Holmgren *et al.* 2000b). Without more detailed and extensive studies of fire in matorral ecosystems it is difficult to fully understand causality in impacts on shrub establishment and growth given the confounding effects of four centuries of wood gathering for charcoal production and intensive browsing by domestic goats.

We hypothesize that fire was a factor in Chile through much of the Miocene and has been lost for only a few million years since the Andes reached near their current height at the end of the Miocene (see Chapter 10). This hypothesis leads to the conclusion that the presence of fire-adaptive traits such as lignotubers in contemporary matorral are relictual traits that have persisted in the absence of fire until the late Pleistocene arrival of humans reintroduced fire (Armesto *et al.* 2010). The virtual lack of fire in the region for the last few million years accounts for the lack of fire-dependent reproduction in either woody or herbaceous species as delaying reproduction would be rapidly selected against in the absence of fire. This hypothesis answers the quandary posed by Fuentes *et al.* (1994) of how to account for fire-adapted traits in the absence of a contemporary lightning regime. They hypothesized that this was the result of volcanic activity that has provided a reliable ignition source for matorral fires (Fuentes & Espinosa 1986). However, it seems doubtful this would have been predictable enough in time and space to drive trait evolution, particularly in most of the MTC region. Another putative source of natural ignitions are fires thought by some to be spontaneously ignited as drops of sticky nectar on *Puya* leaves create a lens that focuses solar heat, though this has never been rigorously tested.

Fig. 6.7 *Recently burned matorral near Bariloche, Argentina: (a) dominated by resprouting* Nothofagus antarctica *(b) with well-developed lignotuber and (c) postfire annuals not evident in nearby unburned matorral, (d) including a species of* Phacelia *(erect with white flowers), one of the dominant postfire annual genera in California chaparral. (Photos by Jon Keeley.)*

Disjunct patches of matorral shrublands occur under a MTC on the eastern side of the Andes around Bariloche, Argentina, at around 41° S latitude and provide some insight into the role of lightning. Some of these patches are dominated by shrubs with well-developed lignotubers (Fig. 6.7a,b) and resprout after fire. More importantly though, these communities have a number of annual forbs that appear to be restricted to postfire sites (Fig. 6.7c,d), not unlike the ephemeral postfire herbaceous flora in California chaparral (see Chapter 5); including many of the same families: Hydrophyllaceae, Boraginaceae, Scrophulariaceae, and Portulacaceae. This postfire flora is in striking contrast to the lack of such a fire response in central Chile. It may be tied to the fact that the Andes are much lower in this region and thus it is subjected to predictable natural lightning-ignited fires, and that these shrublands are particularly prone to regular high-intensity crown fires (Mermoz et al. 2005). Thus, these shrublands have had a far longer fire regime than the anthropogenic fire regime of central Chile.

Fire in MTC Woodlands and Forests

The mesic MTC woodlands and forests of *Nothofagus* in south-central Chile (Regiones VIII and IX) remain poorly known with respect to fire regimes and fire history. Forests of *N. glauca* in the Coastal Cordillera experience a low incidence of lightning and thus fire as likely as not has been a factor in natural disturbance regimes. However, extensive clearing of these forests in recent decades for the establishment of Monterey pine plantations, often utilizing burning, has led to an increased occurrence of fire with destructive impacts on vegetation structure and soil characteristics (San Martin & Donoso 1995; Litton & Santilices 2002, 2003).

Fire in Austral Forests

Fire has long been a natural disturbance regime in many austral forests (Regiones IX and X), especially in areas dominated by coniferous species (Aravena *et al.* 2003), although with a geographic gradient in the frequency of fire from west to east. There is evidence of fires dating back to the beginning of the Holocene (Moreno 2000; Abarzúa & Moreno 2008; Markgraf *et al.* 2009). Much of what is known about fire history comes from studies not in Chile but across the Andes in adjacent areas of Argentina where lightning-ignited fires from summer convective storms are relatively common compared with fire frequency to the west of the Andes (Kitzberger *et al.* 1997; Veblen *et al.* 1999, 2003, 2008; Mermoz *et al.* 2005). Nevertheless, there was widespread clearing and burning of forests over much of the austral forest area by European settlers in the late nineteenth century (Armesto *et al.* 2010). These fires have had very profound effects on forest distribution and composition. Under reduced fire frequency over the past century there has been a shift in dominance from short-lived resprouting species (mostly shrubs) toward longer-lived species such as *Austrocedrus chilensis* and *Nothofagus dombeyi*. Due to limited seed dispersal of these tree species, the spatial configuration of remnant forest patches has played a significant role in influencing current landscape patterns (Kitzberger & Veblen 1999, 2003).

Fire is clearly an important component of the disturbance regime of *Austrocedrus chilensis* on the eastern slopes of the Andes in Argentina. Regeneration of *Austrocedrus* in burned areas is very slow, and may depend on a series of environmental and biotic factors, among which the availability of seeds could be extremely important because of a variable soil seedbank that is greatly influenced by the degree of fire disturbance (Urretavizcaya & Defosse 2004; Urretavizcaya *et al.* 2006). Fire is certainly much less common in stands of *Austrocedrus* on the west face of the Andes in Chile.

Historical records and tree ring reconstructions have been used to establish a 550-yr chronology of fire history of *Austrocedrus* stands on the east slope of the Andes in Argentina (Kiztberger *et al.* 1997; Kitzberger & Veblen 1997; Veblen *et al.* 1999, 2003, 2008) Long-term records suggest that El Niño–Southern Oscillation

(ENSO) events have been associated with a regional pattern of years with widespread fires. There was increasing fire frequency during the latter half of the nineteenth century coinciding with increased Native American occupation of the area, followed by a sharp decline in fire frequency following the demise of the Native American population in the late 1800s. Fire frequencies peaked again with the arrival of Euro-Chilean settlers (Torrejon *et al.* 2004; Bustos-Schindler *et al.* 2010), but fire suppression was widely practiced and effective during the twentieth century.

Fire in austral forest regions of the Los Lagos Región is heavily centered on conifer-dominated ecosystems (Lara *et al.* 2003). The emergent conifer *Araucaria araucana* grows with an understory of the shrubby *Nothofagus antarctica* in south-central Chile and Argentina. Both species are well adapted to survive fire (Burns 1993; Aagesen 2004). *Araucaria araucana* possesses thick bark, readily sprouts from epicormic buds, and protects terminal buds on high branches. *Nothofagus antarctica* resprouts vigorously from root crowns after fire. Large *Araucaria* trees survive fire without harm, and readily establish seedlings below the resprouting canopies of *N. antarctica*. In the absence of fire, the shade-intolerant *N. antarctica* is eventually eliminated as canopies of *Araucaria* and arboreal species of *Nothofagus* close over. Thus, fire acts as a medium for species coexistence between the vigorously sprouting, shade-intolerant species (*N. antarctica*), and the more shade-tolerant and fire-tolerant species (*A. araucana*).

Araucaria araucana forms mixed stands with another tree, *Nothofagus pumilio*, in the Andean cordillera of south-central Chile. Here fires are of mixed severity and often high-intensity patches produce gaps that are filled with similar age cohorts of one or the other species (González *et al.* 2010). The occurrence of large lightning-ignited fires covering thousands of hectares of *Araucaria* forest in the summer of 2002 were the first historical fires over much of this region. However, stands of *Araucaria araucana* studied near Lago Villarica have shown a 300-yr chronology with a much more frequent fire regime (González *et al.* 2005; González & Veblen 2006, 2007), indicating that fire has been an important disturbance regime influencing these communities. The tree ring chronologies suggest that the fire regime has included a mixture of mild surface fires and catastrophic crown fires in past centuries. High-severity widespread fires were relatively infrequent (e.g. 1827, 1909 and 1944). The mean fire interval over this period varied from 7 yrs for all fires to 62 yrs for intense fire events where $> 25\%$ of recorder trees were scarred. The spatial extent of fires ranged from small patchy events to those that burned more than 40% of the entire landscape in the study area (i.e. > 1500 ha).

Dendrochronological studies with the long-lived *Fitzroya cupressoides* have established a 600-yr chronology of fire occurrence in the coast range and central valley of Los Lagos Región. These studies show that *Fitzroya* is able to withstand infrequent low-intensity fires. There was also evidence of even-aged stands resulting from seedlings established after catastrophic stand-replacing fires (Lara *et al.* 1999; Silla *et al.* 2002).

Agroforestry

Tree plantations of Monterey pine (*Pinus radiata*) and *Eucalyptus* cover an estimated 2.5 million ha in Chile, with Monterey pine representing 80% of the total. From a relatively small industry in the 1960s and early 1970s with about 6000 ha of plantations added annually (Lara & Veblen 1993), a massive expansion of this industry now adds an average of over 70 000 ha of new plantation area annually. Exports of wood chips and pulp from these plantations provide approximately 13% of the total Chilean exports. Chile is the world's third largest exporter of wood chips and sixth largest of pulp.

About 60% of this area of tree plantations lies at the southern end of the MTC region of Chile (Regiones VII and VIII). Most of the original areas of these plantations lie in the coastal ranges, which had previously been deforested and degraded by erosion (Estades & Escobar 2005). As the total area of plantations has increased, however, native sclerophyll forests have been cleared to provide suitable plantation sites, which can produce more than 20 m^3 ha^{-1} yr^{-1} of wood biomass.

Fire has been widely used in clearing native forests for plantations and as a management tool in plantation forestry due to its low cost and ease of implementation. Such fires have been shown to significantly affect some soil chemical and physical properties, as well as plant species composition (Litton & Santelices 2002). The intensive silviculture utilized in managing Chilean plantations predisposes them to forest fires (Lara & Veblen 1993). These are single species plantations (monocultures) planted in high density, and carrying a large biomass of flammable dead fuels. Moreover, the coastal mountain ranges are characterized by irregular topography with gullies and canyons that can act as convective chimneys and greatly increase rates of fire spread.

Fire Management

Although it seems that fire has not played an important role in the recent evolutionary history of the matorral shrublands of central Chile, fire has been widely applied to the landscape for more than four centuries as an important tool in agricultural and forest clearance (Armesto *et al.* 2010). Widespread traditional views of land management using fire clearance and burning of agricultural residues and logging slash has brought significant economic gains with little attention given to possible environmental consequences. With the increased numbers of fires in recent years (Table 6.2) and concern about their ecological impacts, there has been an evolving program of improved fire management objectives and protocols.

Over the last few decades there have been new laws enacted to control and regulate the use of fire in forestry and agricultural activities, with significant penalties for infractions. Controlled burning to eliminate vegetation cover is limited to specific criteria and protocols to keep fire under control. In addition, these laws have defined suppression responsibilities and the obligations of fire protection by private forest owners. While prescribed burning is widely utilized, alternative methods of mechanical brush clearance are encouraged.

Fire management in Chile is a responsibility of the national government through CONAF, which is the equivalent of the U.S. Forest Service and National Park Service combined, an organization created in 1970 and attached to the Ministry of Agriculture. It is responsible for the promotion of agroforestry and the timber industry in native forests, as well as the establishment and management of national parks. These duel responsibilities have produced inherent conflicts. CONAF carries out an organized fire management program through actions of prevention, presuppression fuel treatments, and suppression throughout the country, and implements a single national standard in forest firefighting operations. The major objectives of the CONAF fire management program have been: (1) to reduce the frequency and extent of large fires that cause damage to structures and natural resources; and, (2) enlist regional government agencies and private landowners in the prevention, control and mitigation of fire damage. This latter co-operative work with private landowners, most notably the private forestry sector, is particularly important since 68% of land in Chile is privately owned. Many large and medium-sized forestry firms have adopted and implemented their own fire protection programs.

Conclusions

The high Andean Cordillera to the east of central Chile effectively prevents summer convective thunderstorms from moving westward from Argentina, and as a result, lightning is rare to virtually non-existent over much of the MTC region. However, on a geological timescale this is a relatively recent phenomenon and lightning-ignited fires may have been relatively common in central Chile through much of the Miocene. Indeed, even as recently as 10 Ma the central Andes had only reached half their present height. Since the end of the Miocene Andean uplift has formed an effective barrier to westward storms and natural sources of fires. The Chilean matorral exhibits fire-adaptive traits such as lignotubers that could have been selected for during an earlier fire-prone Miocene landscape. The virtual lack of fire in the region for the last few million years undoubtedly accounts for the lack of fire-dependent reproduction in either woody or herbaceous species.

Appendix 6.1

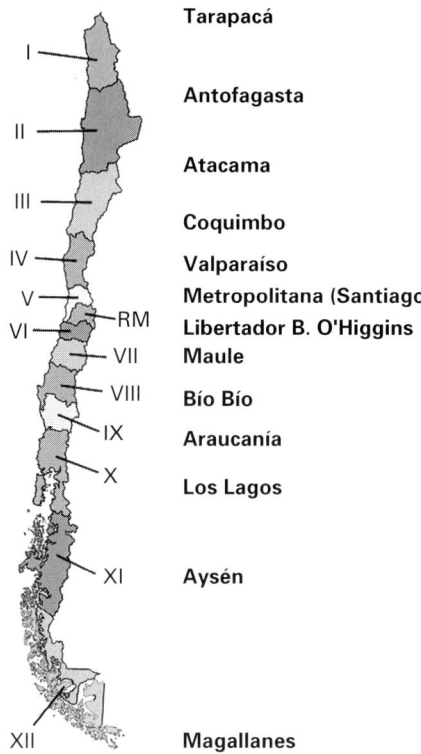

Fig. A6.1.1 *Región map of Chile.*

7 Fire in the Cape Region of South Africa

South Africa's mediterranean-type climate (MTC) region is the smallest of the five MTC regions, centered in the southwestern corner of the Western Cape Province (Fig. 7.1). This Cape region is dominated by fynbos shrublands (see Fig. 1.6e) but this fynbos biome continues eastward far outside the MTC. The Cape region is unusual in that shrublands dominate under climate regimes that also support forests. Entire landscapes can support alternative ecosystem states. Even the semi-arid areas can support entirely different vegetation: fire-prone shrublands or fire-resistant broadleaf thickets. Perhaps more than any other MTC region, fire plays a central role in determining major vegetation patterns of winter rainfall regions of South Africa. Soils are also thought to be of major importance since much of the Cape's MTC region is on nutrient-poor sandy soils (see Fig. 1.5). A complex interplay between soils, fire and climate and, in the east, large mammal herbivory, determines boundaries of major biomes. The Cape Floristic Region is extremely rich in species with very high levels of endemism (Linder 2003). It is the world's richest temperate flora and is largely restricted to fire-prone ecosystems (Cowling *et al.* 1996; Linder 2003). So, contrary to the widely held popular belief that fires are an anthropogenic disturbance (e.g. Pillans 1924; Axelrod 1980), or merely incidental to this formation (Hopper 2009), a rich endemic flora has evolved in the Cape whose members are overwhelmingly fire dependent, implying a long history of natural fires as a selective force.

Major Vegetation Patterns

This chapter discusses fire regimes in the Cape region, what little is known of their determinants, and how they influence major vegetation patterns in the region. Though a large number of studies have explored plant responses to fire (reviewed by Bond 1997; Cowling *et al.* 1997a), these are heavily biased toward fynbos shrublands, the dominant vegetation cover of the region (Fig. 7.2). Fire responses of species belonging to other vegetation types are poorly known. Yet the existence of these other vegetation types is one of the central conundrums of the Cape region. It implies failure of climate alone to explain apparent convergence with

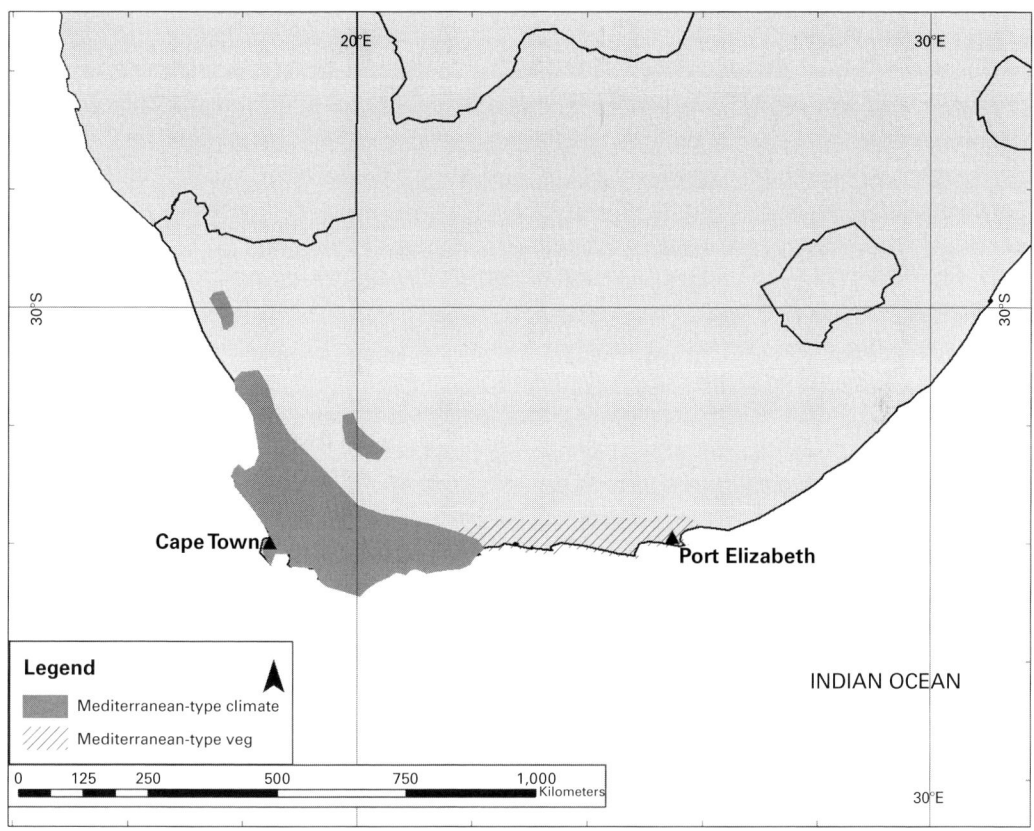

Fig. 7.1 *Distribution of the mediterranean-type climate (MTC) in the Western Cape Province of South Africa (dark shading) and shrublands dominated by mediterranean-type vegetation (MTV) outside the MTC region (hatched).*

other MTC regions (Chapter 1). For example, low shrublands would be expected in deserts replaced, as rainfall progressively increases, by taller shrublands, woodlands and then forests. But this is clearly not the case in the Cape region. The dominant fynbos vegetation shows very little variation in aboveground biomass from arid desert fringes (mean annual precipitation ~250 mm) to rain-drenched high-altitude heathlands (> 3000 mm) (Fig. 7.3). Yet across the entire rainfall gradient fynbos co-occurs with alternative ecosystems with much greater woody biomass. These broadleaf thickets and forests have an entirely different floristic and functional composition and often are restricted to isolated fire-protected refugia (Fig. 7.2; Taylor 1978; Kruger 1979; Cowling et al. 2005; Rebelo et al. 2006). The implication is that apparent convergence of shrubby fynbos growth forms with other MTC plant communities cannot be understood in terms of climate alone and that one needs to think in terms of the climate, fire, geology filter (see Fig. 1.4).

Fig. 7.2 *Fynbos-dominated landscape (foreground and distant slopes) with forests restricted to ravines, Saasveld, Western Cape, South Africa. (Photo by Jon Keeley.)*

Alternatively, since climate has limited explanatory power it is apparent that fire or some other factor is necessary to account for Cape landscape patterns. Figure 7.4 illustrates how the major vegetation types of the region are distributed in relation to fire, rainfall and soils while Table 7.1 shows characteristic taxa of each type. Cape fynbos is the most prominent fire-prone formation occurring over a wide rainfall gradient with rather little structural change across the gradient. It is usually associated with nutrient-poor sandy soils. In mountain fynbos these are derived from very pure quartzites. Lowland fynbos has more varied geology but is typically restricted to nutrient-poor sandy, or limestone-derived soils. It is replaced by a second major fire-prone shrubland, renosterveld, on clay-rich soils. Renosterveld is dominated by shrubby members of the Asteraceae, particularly *Elytropappus rhinocerotis*, which can dominate in monotypic stands (Taylor 1978; Boucher and Moll 1981). Renosterveld often occurs on granite and shale-derived soils in valleys and as an apron on the lower slopes of mountain fynbos. As rainfall

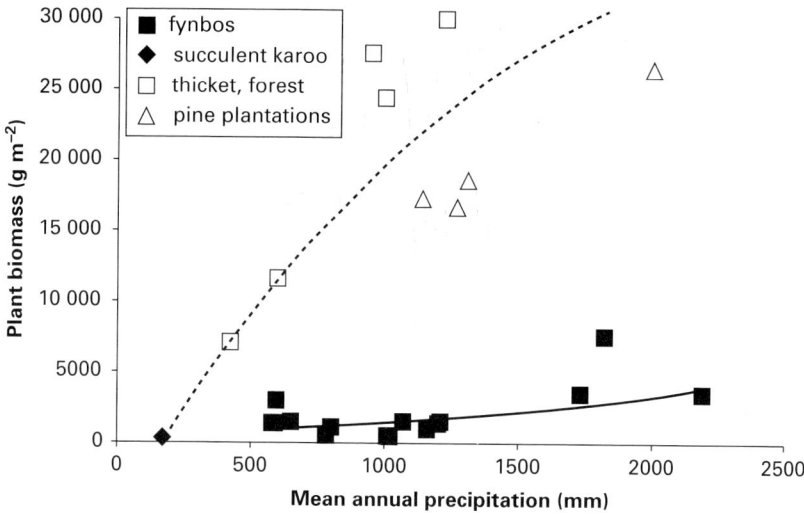

Fig. 7.3 *Aboveground biomass of Cape ecosystems along a precipitation gradient. Fynbos shrublands show little variation in biomass along the gradient (solid line). Conifer plantations, grown in the same fynbos landscapes, produce tenfold higher biomass than fynbos. Native non-flammable thicket and forest have much higher biomass for a given rainfall than fynbos (dashed line). (Data from: Kruger 1977; Le Maitre et al. 1996; Midgley & Seydack 2006; Mills & Cowling 2006; van Laar 1982; van Wilgen 1982.)*

increases (> 600 mm), renosterveld is replaced by fynbos on clay-rich soils (Kruger 1979).

Succulent karoo shrublands are by far the most extensive fire-resistant vegetation type, replacing fire-prone fynbos and renosterveld in more arid sites. They are dominated by leaf succulents of the Mesembryanthemaceae and Crassulaceae, along with stem-succulent Euphorbiaceae and evergreen shrubs of the Asteraceae (Table 7.1; Taylor 1978; Rebelo et al. 2006). Thickets are broadleaf plant communities that contain a mix of evergreen, deciduous and succulent shrubs, many of which are spiny (Table 7.1). Thickets are common along the coast where they are referred to as *strandveld*, but they also occur in mountain fynbos and intermontane valleys. With increasing rainfall, thicket stature and plant cover increases and the succulent and deciduous components diminish. Above ~750 mm, the thickets are replaced by closed forests with a high proportion of evergreen species (Table 7.1). Small patches of forest or scrub forest are common in fire refugia such as on scree slopes, at the foot of cliffs, and along stream banks, in most mesic fynbos landscapes. These fire-resistant alternative ecosystem states of thickets and forests occur on all soil types, nutrient-rich and nutrient-poor, across the entire rainfall gradient (Kruger 1979). Within a landscape, there are often clear edaphic patterns with fynbos on poorer sandy soils and thicket, scrub or forest on deeper, more nutrient-rich soils, but the latter are also found on the nutrient-poor substrates (Vlok et al. 2003; Cowling et al. 2005).

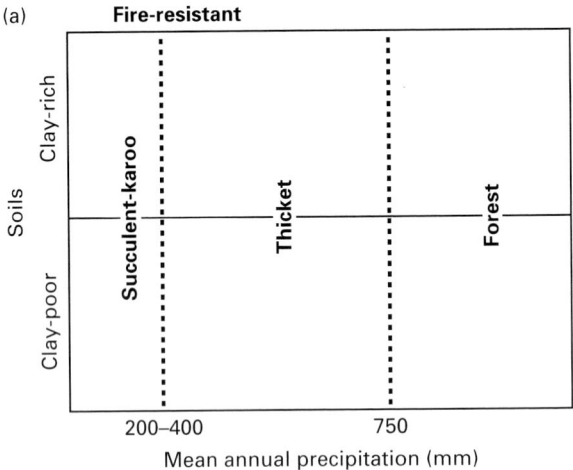

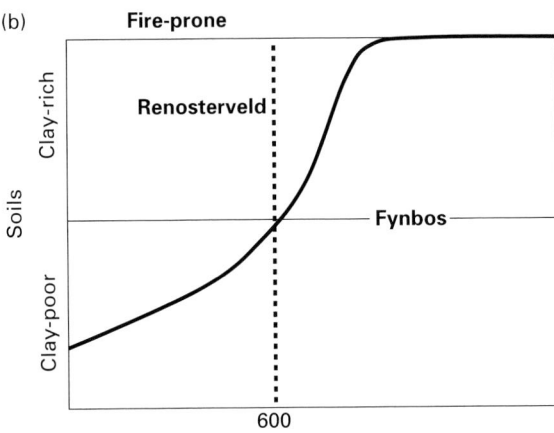

Fig. 7.4 *The major vegetation types of the Cape in relation to fire, rainfall and soils. (a) Fire-resistant ecosystems seldom burn and occur on both clay-rich (granite and shale-derived soils) and clay-poor (quartzites, aeolian sands, limestone) soils. (b) Fire-prone formations are alternative ecosystem states to fire-resistant vegetation and occupy the same environmental space. Fynbos occurs over a wide rainfall gradient but is restricted to sandy (nutrient-poor) soils in more arid sites. Renosterveld generally occurs on more clay-rich soils and in more arid sites, often on the lower slopes of fynbos-clad mountains.*

Thus the Cape region has alternative fire-prone and fire-resistant ecosystem states distributed across wide rainfall gradients and diverse soils. The fire-prone types dominate in the higher-rainfall regions with succulent shrublands dominating the arid regions. The fire-resistant thickets and forest form tiny patches too small to map on most scales. Yet their existence is deeply unsettling for ideas on what determines vegetation pattern:

Table 7.1 Some typical taxa of the biomes of the Cape region of South Africa

FYNBOS	RENOSTERVELD	THICKET	FOREST	SUCCULENT KAROO
Woody plants	**Woody plants**	**Woody plants**	**Woody plants**	**Succulents**
Bruniaceae	Asteraceae:	Anacardiaceae:	Podocarpaceae:	Aizoaceae (incl. Mesembr-yanthemaceae): *Ruschia, Conophytum, Drosanthemum* etc.
Ericaceae: *Erica*	*Elytropappus,*	*Heeria, Loxostylis,*	*Podocarpus*	
Fabaceae: *Aspalathus, Podalyria, Psoralea*	*Eriocephalus, Felicia, Helichrysum, Pteronia, Relhania*	*Rhus*	Celastraceae: *Cassine, Pterocelastrus*	
		Apocynaceae: *Carissa*		
		Araliaceae: *Cussonia*		
		Asteraceae:	Cornaceae: *Curtisia*	
Geraniaceae: *Pelargonium*	Rubiaceae: *Anthospermum*	*Brachylaena, Tarchonanthus*	Cunoniaceae: *Cunonia*	Asteraceae: *Senecio, Othonna*
Geissolomataceae	**Herbaceous**	Boraginaceae: *Ehretia*	Ebenaceae: *Diospyros*	Crassulaceae: *Crassula, Cotyledon*
Grubbiaceae	Poaceae: West: *Ehrharta, Merxmuellera, Pentaschistis*	Celastraceae: *Cassine, Elaeodendron, Gymnosporia, Maurocenia, Maytenus, Pterocelastrus, Putterlickia*	Icacinaceae: *Apodytes*	
Peneaceae				
Polygalaceae: *Muraltia*			Lauraceae: *Ocotea*	Euphorbiaceae: *Euphorbia*
Proteaceae: *Aulax, Diastella, Protea, Leucadendron, Leucospermum, Mimetes, Serruria, Spatalla*	Poaceae: South, East: *Themeda, Hyparrhenia, Eragrostis*		Myrsinaceae: *Myrsine, Rapanea*	Geraniaceae: *Pelargonium*
		Ebenaceae: *Diospyros, Euclea*		**Woody**
		Fabaceae: *Indigofera, Schotia*	Myrtaceae: *Metrosideros*	Asteraceae: *Pteronia, Eriocephalus*
Rhamnaceae: *Phylica*		Malvaceae: *Grewia*	Oleaceae: *Olea*	
Rosaceae: *Cliffortia*		Oleaceae: *Olea*	Oliniaceae: *Olinia*	Sterculiaceae: *Hermannia*
		Polygalaceae: *Polygala*		
Rutaceae: Diosmeae, Stilbaceae (+ Retziaceae)		Rubiaceae: *Canthium*	Rubiaceae: *Burchellia, Canthium*	**Herbaceous**
		Sapindaceae: *Dodonea, Hippobromus*		Amaryllidaceae
				Asclepiadaceae
Herbaceous		**Succulents**		Eriospermaceae
Cyperaceae: *Ficinia, Tetraria*		Asphodelaceae: *Aloe*		Hyacinthaceae
		Didiereaceae: *Portulacaria*		Oxalidaceae
Iridaceae: Irideae, Ixioideae				Poaceae
Orchidaceae: Disineae, Coryciinae		Euphorbiaceae: *Euphorbia*		
Poaceae: *Ehrharta, Pentaschistis*				
Restionaceae: African clades				

Source: See Linder (2003) and Mucina & Rutherford (2006).

(1) They imply that MTC shrublands are not at equilibrium with climate and that features of forests and thickets, together with those of the shrublands, have to be compared to identify which plant traits are common to MTCs
(2) Similarly for low nutrient soils: what traits, if any, are convergent in both fire-prone and fire-resistant vegetation growing on the same nutrient-poor substrates or differ depending on their disturbance regime?
(3) What traits are convergent because of shared responses to fire or mammal herbivory (e.g. spinescence), rather than to climate and soils?

This chapter focuses on fynbos shrublands since these are by far the most studied. However, we do discuss the other fire-prone and fire-resistant vegetation types with a view to understanding the major vegetation patterns in the region. Within the wider African context, the Cape region is a tiny enclave of shrubby vegetation relative to the vast expanse of flammable C_4 grassland ecosystems. The fynbos and renosterveld crown fire regimes are very different from the surface fires of grassland and savanna ecosystems and have selected for entirely different reproductive and persistence traits. Indeed the most distinctive feature of MTCs in the Cape may be that they select for crown fire regimes with return intervals much longer than the frequent surface fires fueled by C_4 grasses.

Fynbos Fire Regimes

Fynbos has a crown fire regime (see Chapter 2). Fynbos fires come in a large range of sizes from a maximum of 58 000 ha (a lightning fire in the Cedarberg in 1988) to less than a hectare. A database of digitized fire maps for the Cape region has recently become available derived from fire scars sketched onto 1:50 000 maps since as early as the 1920s (Scientific Services, Capenature, Stellenbosch). The database has more than 2000 fires, mostly from protected areas in mountain fynbos across the biome, and is most consistent and reliable from the 1970s. It has been used in several fire regime analyses and shows that large fires (> 1000 ha) accounted for only 22% of the total number of fires but 85% of the area burned (Forsyth & van Wilgen 2007, 2008; Seydack et al. 2007; Southey 2009; Wilson et al. 2010).

Most fynbos fires occur at intervals of 10–30 yrs (Seydack et al. 2007; Forsyth & van Wilgen 2008; Southey 2009; Wilson et al. 2010). Fires seldom burn in stands younger than 7 yrs and few stands survive more than 40–50 yrs without burning. Fires can burn in any season, including winter, though the largest number of fires, and total area burned, is in summer (Fig. 7.5). There is some geographic variation in fire season with winter burns being common in the humid coastal ranges in the eastern mountains of the region (van Wilgen 1984; Southey 2009). These are generally associated with long periods of warm dry foehn-like winds, called *bergwinds* (see Box 1.3). Under experimental conditions, fynbos fire intensities range from 500 to ~20 000 kW m^{-1} with flame lengths of 2–7 m and rates of

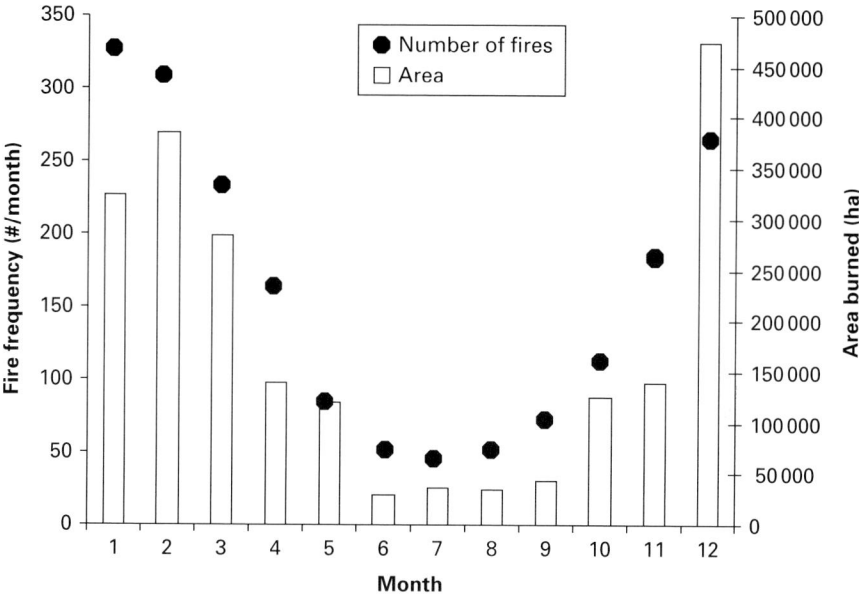

Fig. 7.5 *Number of fires and area burned in protected areas across the Cape region. Data ex Cape Nature and includes montane and lowland fynbos (total area ~1 100 000 ha, total n = 1969 fires, data shown from 1927 to 2006).*

spread ranging from 0.04 to 0.89 m s^{-1} (van Wilgen et al. 1985). However this range of values does not include fires burning under extreme conditions when fire intensity is likely to be considerably greater and when large fires are more likely to occur. Comparison of fynbos fuel loads with other MTC ecosystems (see Table 2.1) shows they are higher than Australian heathlands and comparable to younger stands of Mediterranean Basin matorral and California chaparral.

Postfire Responses

Natural History of Fire Responses

Reproduction

One of the most striking features of a postfire stand of fynbos is the abundance of seedlings contrasting with unburned stands where there are usually none. Curiously, the phenomenon was not recognized by botanists early in the twentieth century who argued forcefully for fire suppression (e.g. Pillans 1924). Though no one has counted them all, it seems probable that > 90% of the ~6200 plant species of the fynbos biome recruit seedlings in the first year or two after fire (Le Maitre & Midgley 1992; Cowling et al. 1997a). Fire-stimulated recruitment is strong evidence for the importance of fire in plant life histories.

Fire-stimulated recruitment implies the presence of persistent propagules in which the population locally persists in propagule form after 100% scorch by fires (Pausas et al. 2004b). Propagules that survive fire must have fire-stimulated cues to produce seedling cohorts after a fire, including fire-stimulated flowering, heat- or smoke-stimulated seed germination or seed release from serotinous cones (see Chapter 3). Fire-stimulated flowering is particularly common in geophytes and graminoids, including grasses. Flowering is also stimulated by fire in many resprouting shrubs. Keeley (1993b) showed that flowering in *Cyrtanthus*, a fynbos geophyte, is smoke stimulated, but for many species, the cue may be increased resources in the postburn environment rather than specific fire-related cues (e.g. Verboom *et al.* 2002 for *Ehrharta calycina*, a grass). Fire stimulates seed germination by physical or chemical cues. In fynbos, heat shock stimulates germination in members of several families including Fabaceae, Rhamnaceae (*Phylica*) and Sterculiaceae (*Hermannia*). Brits *et al.* (1993) described a novel cue, heat desiccation, for ant-dispersed (myrmecochorous) seeds of several Proteaceae. In California chaparral Wicklow (1977) first reported that chemicals leached from charred wood triggered germination of fire-dependent species and later smoke was found to do the same (see Box 3.1). Smoke-stimulated germination was first reported for a fynbos species by de Lange & Boucher (1990) for *Audoinia*, a member of the endemic family Bruniaceae. Smoke-stimulated seed germination has since been reported for many fynbos species in diverse families (reviewed by Brown *et al.* 2003).

Keeley & Bond (1997) noted that fire-stimulated germination cues of heat or chemicals from charred wood and smoke were convergent traits in chaparral and fynbos and that different cues were associated with different growth forms. Annuals were most commonly smoke stimulated whereas shrubs were relatively evenly spread among smoke and heat-shock stimuli. Species with non-refractory seeds were quite common in both systems, strikingly so in plants with other modes of postfire regeneration. Thus geophytes, most of which have fire-stimulated flowering, had non-refractory seeds, as do serotinous species, which release seeds en masse after the plant is burned (Keeley & Bond 1997; Brown *et al.* 2003). Although the discovery of smoke-stimulated germination in fynbos species led to an explosion of new plants for horticulture (Brown *et al.* 1995), it is less certain how many fynbos species have an obligate dependence on smoke. Smoke stimulates germination in a wide variety of plants from diverse biomes (Pierce *et al.* 1995; Brown *et al.* 2003) and qualitative or quantitative differences in response in species from fire-prone vs. fire-resistant ecosystems have seldom been explored. Pierce *et al.* (1995) found that smoke stimulated germination in members of the Mesembryanthemaceae from both fire-resistant succulent shrublands and fire-prone fynbos and suggested that their results cast some doubt on the ecological significance of smoke as a fire-related cue. While constituents of smoke may be a general germination cue in angiosperms, the specific importance of smoke as a germination cue for plants sharing similar fire life histories has been shown in comparative studies of seed germination in Californian chaparral (Keeley & Fotheringham 1998), Mediterranean Basin matorral (Moreira *et al.* 2010) and fynbos (Keeley & Bond 1997).

Fire-stimulated seed release from serotinous cones is common in many fynbos and Australian flowering shrubs (see Fig. 3.5a–d), as well as a few coniferous gymnosperms. However, serotiny is restricted to just a few conifers in California and the Mediterranean Basin (Bond 1985; Lamont *et al.*, 1991). It is particularly common in the proteoid shrubs that form the overstory of many fynbos communities (species of *Protea, Leucadendron, Aulax* in the Proteaceae; Box 7.1). There are also serotinous members of the Asteraceae, Bruniaceae, Ericaceae, Rosaceae

---

Box 7.1 Proteaceae

This family is almost exclusively in the southern hemisphere and comprises about 80 genera and 2000 species of mostly woody evergreen plants. It provides a striking example of adaptive radiation. There are several genera with more than 50 species and some with more than 350 species (*Grevillea*). Many species produce very showy inflorescences, consisting of many small flowers densely packed into a compact head or spike, and have been used extensively for ornamental purposes (Fig. B7.1.1). Flowers are usually pollinated by insects and birds, and more rarely by bats and small marsupials (Myerscough *et al.* 2000). *Banksia* and *Protea* are iconic species of Australia and South Africa, which are the main centers of diversification (Fig. B7.1.2), although there are marked differences in diversity patterns of this family between these regions (Cowling & Lamont 1998).

Proteaceae occur on virtually every land mass in the southern hemisphere (including New Zealand, New Caledonia, New Guinea and South America), reflecting an ancient Gondwana origin. The traditional vicariance biogeographical view was that Proteaceae taxa drifted along with the Gondwana fragments and thus became distributed throughout all ancient Gondwana lands. However, recent phylogenetic analyses suggest that some nodes associated with transoceanic disjunctions postdate the breakup of Gondwana (Barker *et al.* 2007). If correct, then the current distribution of Proteaceae is explained by a combination of vicariance and long-distance dispersal between continents.

Proteaceae are well adapted to the two prominent characteristics of Australian and South African MTC landscapes: fire and infertile soils. Many species (with the exception of the most basal genera, e.g. *Agastachys, Symphionema*, and the subfamily Pesoonioideae; Fig. B7.1.2) possess a characteristic root structure, called proteoid roots, especially efficient in poor-nutrient soils. Proteaceae also shows a large variety of adaptive fire traits. Serotiny (see Fig. 3.5) appears in species with fruits that are formed of woody follicles (e.g. *Banksia, Hakea, Lambertia, Xylomelum*). It also appears in species with small one-seeded achenes that are held between persistent bracts tightly clustered in oval or globular heads (e.g. *Petrophile, Isopogon*). Indeed, Proteaceae is the family that includes most of the serotinous species worldwide. Seed dormancy and fire-stimulated germination is observed in some species of *Conospermum, Grevillea, Persoonia* with soil-stored seeds (Myerscough *et al.* 2000). Fire-stimulated flowering has

Continued

Box 7.1 (*cont.*)

(a)

(b)

Fig. B7.1.1 *The two major genera of Proteaceae. (a)* Protea *is common in fynbos shrublands of South Africa; depicted is* P. cynaroides, *king protea, and (b)* Banksia, *represented here by* B. menziesii, *a dominant in many heathlands and woodlands of southwestern Australia. (Photos by Jon Keeley.)*

Continued

Box 7.1 (*cont.*)

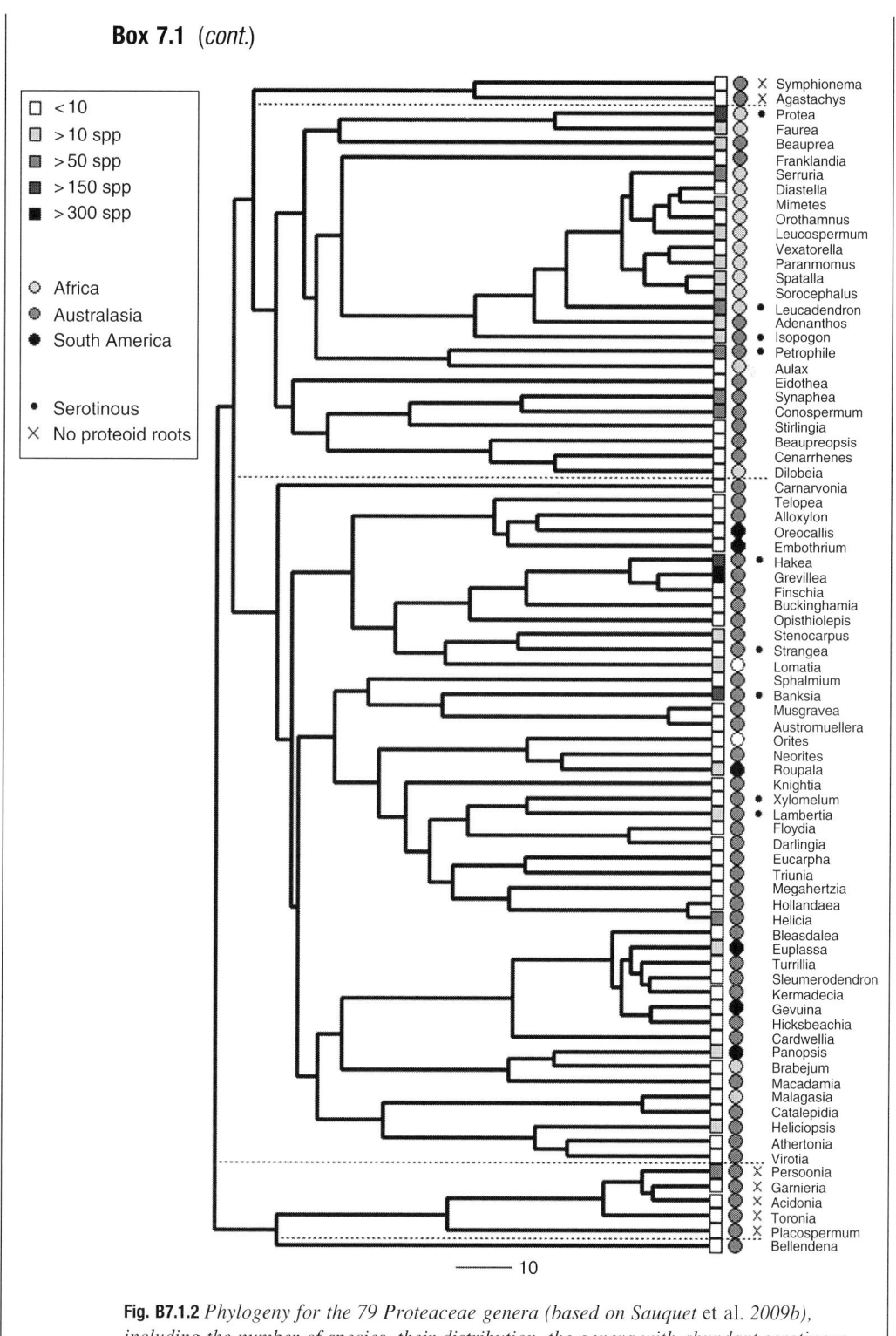

Fig. B7.1.2 *Phylogeny for the 79 Proteaceae genera (based on Sauquet et al. 2009b), including the number of species, their distribution, the genera with abundant serotinous*

Continued

> **Box 7.1** (cont.)
>
> **Caption for Fig. B7.1.2** (cont.) *species and the genera that lack proteoid roots. Scale bar represents 10 Ma. The different sections of the tree marked by dotted lines refer to subfamilies (from top to bottom: Symphionematoideae, Proteoideae, Grevilleoideae, Persoonioideae and Bellendenoideae; Weston & Barker 2006). The genus* Banksia *sensu stricto is paraphyletic and so here it includes* Dryandra *(Mast et al. 2005; Mast and Thiele 2007). Some genera of Australasia distribution also reach southeast Asia (e.g.* Helicia, Macadamia*). Most African genera are confined to the Cape region.* Lomatia *and* Orites *are native to both Australia and South America, but are more species rich in Australia. Genera native to South America are not generally found in MTC ecosystems.*
>
> been reported in a few species such as *Lomatia silaifolia* (Denham & Whelan 2000) and *Telopea speciosissima* (Pyke & Patton 1983; Bradstock 1995). Another fire-adaptive trait is resprouting, which is also widespread in the family, including many species with lignotubers (see Table 3.2).

(*Cliffortia*), Cupressaceae, and even a strange monotypic genus in the Anacardiaceae, *Laurophyllus capensis*. Unlike serotinous cones in some conifers that remain closed with viable seeds for decades, serotiny in most fynbos taxa is relatively weak, with seeds typically being held for only a few years and then being released from old fruits even in the absence of fire (Midgley & Enright 2000). However, successful recruitment between fires is very rare. Serotiny has not been recorded in South African savannas, even where *Protea* spp. form the dominant woody plants. Thus, it is not a fire adaptation *per se* but an adaptive response to a particular fire regime characterized by crown fires that recur on the order of decades in fynbos (see Chapter 3).

Persistence
There are several ways in which plant species survive fires, either by resprouting vegetatively or by seedling recruitment. Many fynbos shrubs resprout after fire and most have seedling recruitment after fire as well and are termed facultative seeders. A significant number of these resprout from specialized lignotubers (see Fig. 3.1 and Table 3.2).

In contrast, many shrubs do not survive fire and are known as non-resprouters or *obligate seeders* since their ability to persist on burned sites is entirely due to seedling recruitment from dormant seedbanks (see Table 3.4). A few taller proteoids are usually obligate seeders, but may survive fire if the apical buds are not damaged (e.g. *Leucospermum conocarpodendron*, *Mimetes fimbriifolius*, *Leucadendron argenteum*). Such species have thicker bark then most other fynbos species, suggesting adaptation to a somewhat different fire regime. Many plants are only top killed by fire and belowground parts survive and resprout from vegetative

Fig. 7.6 *One of the typical fynbos graminoids in the Restionaceae family on nutrient-poor quartzite soil. (Photo by Jon Keeley.)*

tissues or root suckering. The incidence of root suckering is poorly known but does occur in several shrubby species such as *Cliffortia ruscifolia* where a single clone may cover several hectares. Clonal spread by underground rhizomes is also a distinctive feature of several aggressive graminoids, including species of Restionaceae (Fig. 7.6). Finally, a few woody species resprout from epicormic buds on branches that survive the fire (e.g. *Protea nitida*) but this is very rare in fynbos, in contrast to Australian *Banksia*. Fynbos grasses have remarkably diverse vegetative morphologies, linked to different fire responses, with greater diversity of forms than in grassland or savanna biomes (Linder & Ellis 1990). They include fire ephemerals, geophytes and tall bamboo-like shrubby forms able to outcompete woody species.

Obligate seeding life histories are very common in the fynbos flora with about 2000 species reported (Cowling *et al.* 1997a). Most of these are ericoid shrubs; indeed, the genus *Erica* has over 600 species and the majority are obligate seeding shrubs endemic to the Cape Floristic Region (Ojeda 1998). Obligate seeding proteoid shrubs form the overstory and dominate the biomass of many fynbos communities. Obligate seeding graminoids also occur, making up an estimated 20% of the 550 species of grasses, restios and sedges (Cowling *et al.* 1997a). Most (> 80%) fire ephemerals, which usually die out before the next fire, do not resprout even if they survive long enough to burn.

Obligate resprouters are shrubs that regenerate after fire solely by resprouting and do not recruit seedlings (see Chapter 9), a growth form dominant in Mediterranean Basin and Chilean matorral and frequent in California chaparral, but very rare in the fynbos with only 100 species (< 2% of the flora). However, these obligate resprouters are common in Cape ecosystems, such as strandveld, thicket and forest, which do not burn as readily as fynbos. As in these other MTC regions, obligate resprouters are nearly all broadleaf species with fleshy fruits dispersed by birds and have forest/thicket affinities (e.g. *Rhus, Diospyros, Maytenus, Heeria, Myrsine*). They do not recruit after fire but establish between fires, often under perches frequented by birds (Manders *et al.* 1992; Cowling *et al.* 1997b).

Fire and Community Dynamics

Which species or functional groups occur together in communities and how do different community assemblies respond to perturbations? The answers have generally been sought in differing abilities to compete for resources or to escape competitors by dispersal. However in fire-dependent ecosystems, the emphasis has been on the ability of a species to persist in the face of fires at varying intervals, intensities and seasons and competitive structuring of communities has received less attention. In fynbos, as in other MTC ecosystems, there have been numerous studies on the mode of recovery from burning, survival of propagules from one fire to the next, and fire-stimulated recruitment. The mix of species in a community has been viewed as a function of the individual attributes of a species and whether these would allow them to persist in a given fire regime. The implicit assumption is that the vital attributes of a species would remain the same regardless of the species mix. This view has been amended since the early 1990s following the recognition of competitive interactions between different functional groups, particularly overstory vs. understory plants, or between different growth forms in the understory or overstory (Cowling & Gxaba 1990; Moore & Noble 1990; Keith & Bradstock 1994; Vlok & Yeaton 1999; Bond & Ladd 2001; Keith *et al.* 2007). For example, a non-resprouting shrub might persist over a range of fire frequencies but would be unable to do so in the presence of fast-growing overstory shrubs which suppress its growth and flowering.

The great majority of fynbos species have some form of fire-stimulated recruitment. They are also mostly very intolerant of shading. This is readily apparent when fynbos stands are invaded by alien trees, such as northern hemisphere conifers and Australian acacias and hakeas (see Chapter 12). Most fynbos species die under dense alien trees while some persist but fail to flower. A small group of species tolerate shade, mostly the obligate resprouters that are distinguished by their broad leaves and fleshy fruits and affinities to closed-forest lineages, but also several graminoids, especially Cyperaceae, and some geophytes such as *Oxalis* spp. (Holmes & Cowling 1997a). Shading by tall native species also suppresses the understory and can markedly influence community dynamics (Cowling & Gxaba 1990; Vlok & Yeaton 1999, 2000). Tall proteoid shrubs dominate many fynbos communities and shade out understory plants. Most of these overstory species are serotinous obligate seeders, killed after every fire, so that the shading effect is cumulative, increasing with age since fire. Resprouting graminoids that spread vegetatively after fire are also important understory competitors, suppressing seedlings that emerge after a fire. Vlok & Yeaton (1999, 2000) have argued that, in the absence of shading by dense proteoids, competition from graminoids can reduce alpha diversity by as much as 50%. They argue that the presence of a dense, but transitory, proteoid cover is a key factor maintaining high alpha diversity in fynbos communities. Effectively, proteas create regeneration gaps by shading out the vigorously resprouting graminoid layer, thereby providing safe sites for seedling establishment after fire.

It is interesting to contrast this interactive view of fynbos dynamics, where the species mix matters, with one based purely on the interaction between fire and vital attributes of species (see also Keith *et al.* 2007 for Australian heathland examples). For example, van Wilgen & Forsyth (1992) used Noble and Slatyer's (1980) vital attributes to classify 210 fynbos species in the Swartboschkloof area of the south-western Cape, focusing particularly on the timing of critical life history events such as age to first flowering, life span, and longevity of the seedbank. They used this trait analysis to evaluate how species would respond to changes in fire regime. They concluded that only four species would be eliminated by frequent fires (5-year intervals): two serotinous proteas, and two forest trees. The two protea species would also be vulnerable to long intervals between fires (due to death of adults and simultaneous death of seeds, a common hazard for serotinous species, see Chapter 9) with another 16 species possibly vulnerable depending on the longevity of their seedbank. Thus, with the exception of serotinous proteas, fynbos species appear to be remarkably resilient to short fire intervals because they survive vegetatively, or have persistent seedbanks, and/or have short maturation times. Studies of recovery of stands cleared of alien trees, after different periods of invasion, reveal that many fynbos species have sufficiently long-lived seedbanks to survive long (> 30 yr) intervals between fires (Holmes & Cowling 1997b; Holmes 2002). However, seedbank persistence varies among sites, with more persistent seeds in mountain fynbos, while in lowland fynbos long-lived obligate seeders had transitory seedbanks and may therefore be extirpated by long fire return intervals (Holmes 2002). Holmes & Newton (2004) studied patterns of seed persistence across diverse growth forms in a seed burial experiment. They found surprisingly short-lived seedbanks (< 2 yrs) in most of the small-seeded species, which suggests either very high rates of dispersal into burned patches or experimental conditions that do not mimic natural cues for promoting seed dormancy. The latter seems more likely.

Event-dependent effects on postfire recovery
Most attempts to define plant functional types for predicting fire responses have focused on predicting the consequences of varying fire frequencies. Thus, age to first flowering, plant life span, and longevity of seedbanks are key attributes for predicting how plants will respond to different fire intervals. However, event-dependent effects, such as the season and intensity of a fire, also influence postfire recovery (Bond & van Wilgen 1996; Keeley *et al.* 2005a for chaparral). Fynbos fires can occur in any season. Anthropogenic fires have a broader seasonal range than lightning fires, which typically peak in summer. The season of fire can have major impacts on postfire recruitment, especially of serotinous proteas.

Recruitment from seedlings is generally poorest after winter and spring fires and best after summer and autumn fires, perhaps due to increased predation following early season fires (Bond et al. 1984; van Wilgen & Viviers 1985; Le Maitre 1988). Resprouters would be expected to recover best from summer and autumn fires, when roots are likely to have accumulated reserves, and to recover poorly from winter and spring fires when root reserves are mobilized for new shoot growth (see Chapter 3). *Watsonia pyramidata*, a fynbos geophyte, shows striking differences in flowering depending on fire season, with sparse flowering after spring fires and abundant flowering and corm division after autumn fires (Le Maitre 1984). This, and related species, were used as food by hunter–gatherers in the fynbos region and it has been suggested that fires were deliberately set in autumn to promote such corm production (Deacon 1992).

Most prescribed fires are conducted under weather conditions that minimize fire intensity and the risk of fire escapes. In contrast, wildfires burn large areas under extreme weather conditions when fire intensities would be much higher. These very intense fires may kill resprouting species (Richardson & van Wilgen 1986), resulting in long-lasting changes in the community because resprouters are generally poor seedling recruiters. However, intense fires promote recruitment of large-seeded myrmecochores and other species requiring a heat pulse for germination (Bond et al. 1990, 1999). Intense fires kill seeds in the surface layers of the soil so that only deeply buried, large seeds are able to emerge (Bond et al. 1999). These effects seem restricted to deep soils with more even heat penetration. On rocky, mountain fynbos soils, there is greater fine-scale heterogeneity in soil heating and no marked effects of fire intensity on seedling emergence have been observed (Holmes & Foden 2001).

Proteoid shrubs
Proteas (*Protea* spp; family Proteaceae) form the overstory in many fynbos communities and contribute the bulk of the biomass. By far the most common species are serotinous obligate seeders. There have been many studies of population biology and fire responses of this functional group in the Cape region with strong parallels to studies in Australia. Seeds are stored in the canopy of the plant in woody cone-like infructescences which remain closed for several years. After the death of the plant, seeds are released en masse. Seeds released between fires very seldom recruit successfully so populations are typically even aged and date from the last fire. Serotinous proteas generally have longer youth periods than other fynbos elements, taking 3–12 yrs or more (longer on the more arid sites) to flower for the first time. Most species have relatively short life spans and few survive longer than ~40 yrs (longer with decreasing precipitation). Thus protea populations can be extirpated by short fire return intervals and also by very long ones. Protea recruitment is also sensitive to season of fire with very poor recruitment after winter fires (e.g. less than one seedling for every ten parent plants; Bond et al. 1984), poor recruitment after spring fires, and best recruitment after summer and, especially, autumn fires (> 40 seedlings per parent). Seasonal effects on

recruitment success appear to vary geographically, with marked effects in the western and southern Cape but no apparent effects in the eastern Cape under a non-MTC (Heelemann et al. 2008). Postfire recruitment may also be poorer after very intense fires that appear to incinerate seeds, but little data are available.

Some protea species also show intrinsic variation in recruitment success due to density-dependent effects. Whereas the "law of constant yield" states that most plant populations produce constant seed output across a wide range of densities (Harper 1977), some dense protea stands show a decline in stand-level seed output. Protea flower heads are large structures borne on terminal shoots in many species. Dense crowding may lead to etiolation of shoots, reduced numbers of thick shoots, and therefore reduced developmental potential to produce the flower heads (Bond et al. 1995). For candidate species, fires in mature (12–20 year old) stands, and in the most favorable season for seedlings, can produce very dense populations that, after the next fire, have very poor seedling recruitment. Repeated fires in the most favorable season for seedlings would lead to oscillating populations, which is a rare example of chaotic dynamics in plants (Bond et al. 1995).

Unlike northern hemisphere MTC shrubland species, serotinous protea populations often fluctuate greatly from one fire to the next, whether from strong density dependence or from unfavorable fires. Local extinction is not uncommon after a particularly unfavorable fire. Thus, fynbos proteas not only produce transient overstories in space (being killed by fire) but also in time where an entire stand may be eliminated by a single fire. They are rare examples of plant metapopulations with quite frequent local extinction of subpopulations in some species but rapid recolonization. Whereas most fynbos species have very limited dispersal (1–100 m), serotinous proteas disperse hundreds to thousands of meters in the few days after a fire. They do so primarily by seeds rolling over the soil surface, with the distance traveled dependent on seed attributes, wind velocity and terrain roughness (Bond 1988; Schurr et al. 2005). The combination of high vulnerability to local extinction and remarkable dispersal ability makes for highly labile populations of serotinous proteoids, the dominant component of many fynbos communities.

Serotiny contrasts strongly with myrmecochory, a dispersal syndrome characterized by specialized seeds that attract ant dispersers. Ants move seeds only a few meters but seed life span in the soil is long. Some myrmecochorous proteas have disappeared only to reappear decades later after fire from long-lived seedbanks. The combination of seed dispersal by ants and long-lived dormant seeds is thus an alternative strategy of surviving highly variable fynbos fire regimes (Christian & Stanton 2004; see also Chapter 9). *Mimetes stokoei* is a spectacular example (Slingsby & Johns 2009). A single specimen of this very rare 3-m-tall protea was first discovered in 1922. A second population of just five plants was found in 1925. The populations regenerated after several fires until the last flowering plant died in 1950 (it set no seed). Following disturbance of the ground for a planned planting of proteas, in 1965 a single seedling appeared but died before flowering. No plants were seen again, despite several subsequent fires, until an intense fire in 1999 when a few dozen plants emerged in the same two populations. Apparently the seeds had survived since at least 1944, when the last flowering plants had set seed. Surprisingly, the seeds had also survived three or four subsequent fires without germinating. Given their importance in shading the understory, and creating regeneration gaps for understory species, obligate seeding proteas, either serotinous or myrmecochorous, create an unusually dynamic shrubland community contributing to high alpha diversity in fynbos (Vlok & Yeaton 1999, 2000).

Fire in Other Vegetation Types

Renosterveld – the Other Fire-prone Shrubland

Renosterveld is a fire-prone shrubland occurring on relatively clay rich soils and at lower rainfall than fynbos (Fig. 7.7). It has been considered analogous to sage scrub shrublands in California and garrigue in the Mediterranean Basin (Taylor 1978). Renosterveld occurs across the entire Cape Floristic Region (Fig. 7.1). The community is dominated by the renosterbos, *Elytropappus rhinocerotis*, a shrub in the Asteraceae. Other members of this family are also prominent (e.g. *Relhania*, *Felicia*, *Pteronia*). Renosterveld differs from fynbos in the near absence of Proteaceae and Restionaceae (Taylor 1978; Boucher & Moll 1981; Rebelo *et al.* 2006). Many other characteristic fynbos families and genera are absent, including all the Cape region endemic families. Taller broadleaf shrubs, with forest and thicket affinities (e.g. *Rhus*, *Olea*, *Diospyros*) occur sporadically. Grasses are the most prominent graminoids, with C_3 grasses in the west and C_4 grasses increasing to the

Fig. 7.7 *A patch of renosterveld in the Western Cape, South Africa. (Photo by Jon Keeley.)*

east with increasing summer rainfall (Cowling 1983a). Geophytes are common and diverse with many species endemic to renosterveld (Proches *et al.* 2006). Diversity patterns are poorly known but high alpha diversity has been reported in western regions, with lower diversity and less turnover of shrub species in eastern communities (see Chapter 11).

Renosterveld is poorly studied relative to fynbos. The vegetation has been highly fragmented by agriculture (> 90% transformed in the Cape region) with the most extensive remnants on steep mountain slopes (Kemper *et al.* 2000). Large areas of renosterveld, especially in the east, may have been grasslands with *Elytropappus rhinocerotis* invading depending on grazing and fire history. The shrubland has been converted to C_4 grasslands in southern and eastern examples by frequent burning to promote grazing (Cowling *et al.* 1986). Similar human-caused type conversion has occurred in parts of the Cape region as well (Kraaij 2010).

The fire ecology of renosterveld is poorly known. The vegetation supports crown fires at intervals of several decades, with median frequencies longer than fynbos in the southern Cape (20–30 yrs; Seydack *et al.* 2007). Cape region renosterveld, unlike many fynbos communities, resembles California sage scrub in that the fuel is made up of a continuous shrub layer with a discontinuous graminoid layer, and similar low fuel loads (see Table 2.1).

Elytropappus, the dominant shrub, has a finely branched canopy with very small cupressoid leaves. It accumulates dead branches and is highly flammable. Indeed the flammable properties of this single species may account for why renosterveld replaces succulent Karoo vegetation over much of its range – a plausible example of *niche construction* by evolution of flammability in a single species (Bond & Midgley 1995; Kerr *et al.* 1999). *Elytropappus* is usually killed by fires, though some resprouting populations are known from the eastern parts of its range. Recruitment is fire stimulated with shade-intolerant seedlings (Levyns 1929). As in fynbos, the large broadleaf shrubs in renosterveld become more prominent in thicket formations and all are obligate resprouters (Taylor 1978). The fire responses of other shrubs seem to vary along geographic gradients with fire-stimulated obligate seeders in mountain renosterveld in the Western Cape (e.g. *Aspalathus* spp.), but mostly resprouting members of the Asteraceae (*Relhania*, *Pteronia*) in the eastern distribution of the formation.

Many of the associated geophytes in renosterveld have fire-stimulated flowering but there is no information on whether fire is an obligate cue. Fire ephemerals also occur and some have smoke-stimulated germination (C.J. Fotheringham personal communication) suggesting that, in at least part of its range, renosterveld has fire-dependent elements. Despite its aridity relative to fynbos, the more mesic stands of renosterveld are prone to invasion by alien trees, usually conifers and acacias. Local patches of broadleaf forest dominated by *Olea africana* and *Rhus* spp. occur in fire-protected sites, indicating a potential for shifting to an alternative ecosystem state in the absence of burning (Boucher & Moll 1981).

Fig. 7.8 *Strandveld vegetation along the coast of the Western Cape, South Africa. (Photo by Jon Keeley.)*

Less Fire-tolerant Ecosystems: Strandveld, Thicket, Forest

As discussed above, the fire-prone shrublands of the Cape region, fynbos and renosterveld, generally have much lower biomass than expected for the level of annual precipitation. These fire-prone formations often abut taller scrub and forest vegetation that occur in fire refugia such as river valleys, deep ravines, scree slopes (Fig. 7.4; Geldenhuys 1994) and adjacent to the ocean as with strandveld (Fig. 7.8) There are striking differences in floristic composition, structure, and functional attributes between the fire-prone and less flammable formations. The latter are dominated by broadleaf species that, unlike fynbos shrubs, cast dense shade. Also unlike fynbos, vertebrate-dispersed fruits are common and fire-stimulated recruitment traits are absent (Manders *et al.* 1992). Seeds are short lived, there are no serotinous species, and no fire-stimulated flowering. Analogous vegetation is known as *vine thicket* or *rainforest* in Australia (Bowman 2000). Analogues are less obvious in other MTC regions but would include closed forest formations such as temperate forests where fire is an irregular event and unimportant for recruitment. These formations would have been considered *climax* vegetation in older literature because they are self-maintaining, at equilibrium with the climate, and cannot be invaded except after catastrophic disturbance (Phillips 1931).

Forest and thicket patches do occasionally burn but fires seldom penetrate more than 50–100 m. However, under extreme weather conditions, fires may burn larger forest patches. Following weeks of bergwinds (see Box 1.3), an extreme fire event occurred in the late 1860s and is said to have destroyed extensive forest patches in the southern Cape region, the largest forested area in South Africa (Phillips 1931).

Van Wilgen et al. (1990b) have compared fuel properties of fynbos and forest. Fuel mass in fire-prone fynbos was less than half that in the closed forest. However there was a pronounced separation of litter and canopy fuels in the forest and the latter were discontinuous and patchy relative to fynbos. Fuel moisture contents of the forest fuels were 50–100% higher than fynbos, and dried out less in the dry summer months, but heat yields and fat content of forest and fynbos fuels were similar. Thus, the physical features of fynbos fuels promote flammability whereas forests are much less flammable.

Although forests do not burn easily, forest margins do burn. Two species of *Virgilia*, hard-seeded legumes with long-lived seedbanks, are fire-stimulated obligate seeders that occur only on forest margins abutting fynbos. No other forest trees have been reported to have fire-stimulated recruitment. Rare fires may influence forest composition for decades by selecting for species capable of resprouting. Thus, forest margins are often dominated by multistemmed trees (*Rapanea*, *Cassine*, *Pterocelastrus*) indicating past disturbance (Kruger et al. 1997). Tall single-stemmed individuals of these species, and of fire-intolerant non-resprouting species (e.g. *Podocarpus latifolius*, *P. falcatus*, *Olea capensis* ssp. *macrocarpa*, *Curtisia dentata*), occur deeper in the forest.

Forest has been observed colonizing fynbos in several localities where fire has been excluded for several decades (Masson & Moll 1987; Manders 1990; Manders et al. 1992; Luger & Moll 1993; Cowling et al. 1997b). Many colonizing tree species have bird-dispersed fruits and establish under perch sites on taller shrubs (Manders et al. 1992; Cowling et al. 1997b). Forest saplings may take many years to become fire proof: that is, with the capacity to resprout after burning. Manders (1990) noted that all saplings of *Cunonia capensis*, a forest species colonizing fynbos unburned for ~30 yrs, died after being burned regardless of plant size. Manders et al. (1992) concluded that forest could potentially occur in most fynbos sites with > 650 mm precipitation but were prevented from doing so by the combined effects of frequent fynbos fires and slow sapling growth rates.

Determinants of Fire Regimes

It is interesting to contrast fynbos fire regimes with those of other MTC ecosystems and with the other major fire-prone biomes in South Africa, which include grasslands and savannas dominated by C_4 grasses. Fires burn more frequently in fynbos than in California chaparral (see Chapter 5) but less frequently than kwongan shrublands of Western Australia (see Chapter 8). Fynbos burns much less frequently than mesic (> 750 mm annual precipitation) C_4 grassy ecosystems, much of which burns annually or biennially (Balfour & Howison 2002). Fires in semi-arid savannas are strongly limited by the continuity of grass fuels, which depends on preceding rainfall (van Wilgen et al. 2004). Though mesic fynbos and savannas are both fire-prone and fire-dependent ecosystems, they have strikingly different woody plant life histories reflecting their very different fire regimes

(Le Maitre & Midgley 1992; Bond & Keeley 2005). Fynbos is rich in species with fire-stimulated recruitment but this is extremely rare or absent in savannas. Many woody fynbos plants are obligate seeders (see Chapter 3), that is they lack resprouting capacity after fire and seedling recruitment is restricted to a single postfire pulse. In contrast, all South African savanna trees and shrubs have some capacity to resprout after fire and none restrict seedling recruitment to a single postfire burst. Thus, though both fire-prone biomes experience predictable fire, the different fire regimes in the two biomes have selected for entirely different life history traits.

Fynbos-like shrublands dominate MTC landscapes and also occur in isolated patches throughout the mesic grassy biomes of summer rainfall areas (O'Connor & Bredenkamp 1997). A climate-only hypothesis might propose that the extent of drought-prone habitat favoring fynbos shrubs is greater in MTC landscapes and the patchy occurrence in summer rainfall regimes is due to a more limited distribution of appropriate habitat. However, landscapes in both of these climatic regimes are fire-prone, but with very different regimes that could be a factor in determining shrubland dominance in these two regions. These communities have likely assembled due to interactions between climate, fire and perhaps geology (see Fig. 1.4). In this light we might hypothesize that the high fire frequency maintained by C_4 grasses is sufficient to exclude sclerophyllous fynbos shrubs over much of the summer rainfall landscape. This model involves a close integration between climate and fire, as very high fire frequency is the proximal factor excluding fynbos shrubs in C_4 grasslands, but this is ultimately a response to the monsoon climate selecting for rapid fuel production capability of C_4 grasses and high fire frequency dependent on frequent lightning storms. Geology also plays a role in that fuel continuity and fire spread are enhanced by the broad open plains in a manner similar to the North American prairies where woodlands are restricted to escarpments by frequent fires (Wells 1965).

Fire exclusion experiments in the summer rain grassland biome support this model with repeated instances of invasion of woody sclerophyllous MTV where fire has been excluded for a decade or more (Titshall *et al.* 2000; Bond *et al.* 2003). Indeed, shrubby Ericaceae, Rosaceae (*Cliffortia*) and Asteraceae (*Stoebe*) with fynbos affinities have been perceived as woody invasives by livestock farmers, and fire regimes have been developed to eliminate them and promote grasslands (Trollope 1973). In the absence of fire, both fynbos and the C_4 grasslands in more humid regions have the climate potential to support forest (Bond *et al.* 2003). It seems probable that both fire-prone biomes exist because of fire, and differ in the traits of dominant plants in part because of different fire regimes.

If fire regime determines the boundaries of fynbos and grassy systems, why do they have such different fire regimes? Why are fires so much less frequent in fynbos than in C_4 grasslands? Fynbos was much more extensive in South Africa in the last glacial period extending into areas that are now pure C_4 grasslands or savannas (Chase and Meadows 2007). How might climate change potentially change the fire regime and therefore the distribution of MTC ecosystems?

Scientific understanding of the determinants of fire regimes is far from complete. Gill et al. (2002) have provided a very useful conceptual framework. They note that large fires, though small in number, account for most of the area burned and therefore the fire regime. Large fires occur when there is a high degree of connectivity of fuel at a landscape scale (Gill et al. 2002). Key factors that affect fuel connectivity are landscape features that act as fuel breaks, sufficiently continuous vegetation cover to support fire, and the coincidence of drought and extreme weather conditions that enhance flammability. Gill et al. (2002) argue that fire regimes are strongly structured by fires that coincide with high connectivity events. The frequency of such events, together with the coincidence of ignitions that ignite the fire, will determine the frequency of large fires. Their frequency, in turn, will be a major determinant of the mix of plant functional types that can occur in a fire-prone landscape.

These concepts are useful for analyzing determinants of fynbos fire regimes and why they differ so strikingly from C_4 grassy ecosystems. The continuity of plant cover as fuel varies with productivity and the growth form mix. Rainfall is a major determinant of productivity in both shrublands and grasslands. Most studies of fynbos as fuel have been in mesic fynbos with > 1000 mm precipitation, which equates with mesic C_4 grassland and savanna (> 750 mm). The discussion that follows refers to mesic fynbos and grasslands. Kruger (1977) recorded biomass accumulation in fynbos of ~2.5 Mg ha^{-1} (ranging from 1 to 4 Mg ha^{-1}) for the first 2–3 yrs after a fire but growth rates decline rapidly after that. Fynbos stands < 7 yrs are often effective in stopping fires burning in older stands (Fig. 7.9).

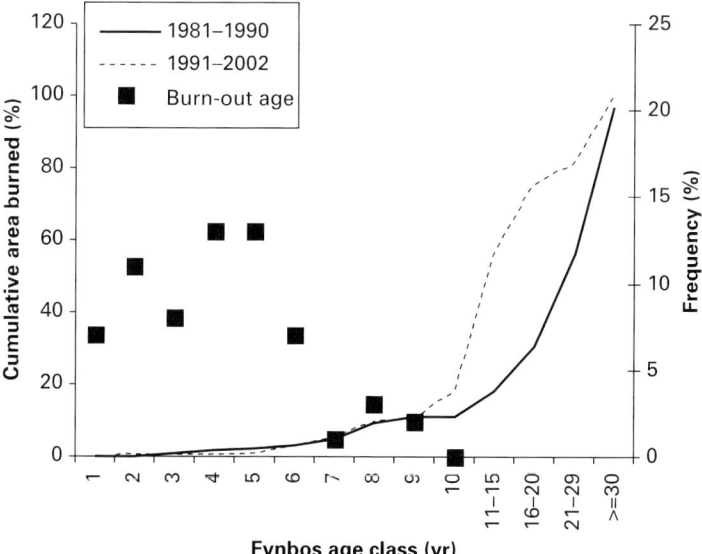

Fig. 7.9 *Effect of postfire age on cumulative area burned (%), and the frequency of ages against which fires stopped burning (burn-out age, squares), Swartberg, Southern Cape. A natural burning policy was in place from 1986 to 2002. (Based on data from Seydack et al. 2007.)*

Unlike California chaparral, postfire fynbos has a low cover of herbaceous fire ephemerals and therefore lacks the dry flashy fuels that can cause young chaparral stands to burn (Zedler et al. 1983; Keeley & Zedler 2009). However, fynbos differs from all MTC ecosystems in having a significant component of perennial graminoid growth forms, primarily restios (Restionaceae) and sedges (Cyperaceae). In many fynbos communities, perennial graminoids produce a savanna-like physiognomy with a more or less continuous grass layer interspersed with ericoid shrubs and emergent proteoid shrubs. Kruger (1977) suggested that the presence of a graminoid layer, together with fine-leaved ericoid shrubs, accounted for the shorter fire interval in fynbos relative to California chaparral and Mediterranean Basin matorral, and both lack a significant perennial herbaceous component (see also van Wilgen 1982; van Wilgen & van Hensbergen 1992). However, Australian heathlands have a higher fire frequency than fynbos and they generally lack a dense restio cover, but like fynbos they do have a more or less continuous layer of fine fuels and a discontinuous layer of taller emergent proteoid shrubs.

It is interesting to note that postfire fuel accumulation rates of fynbos are not dissimilar to montane C_4 grasslands of South Africa, which also produce ~2.5 Mg ha^{-1} yr^{-1} with 1000 mm precipitation (O'Connor & Bredenkamp 1997) yet typically burn at intervals of 1–2 yrs. It is not the biomass alone, but the type of biomass that influences the fire regime. C_4 grassy fuels are not only more continuous than a young postfire fynbos stand but also cure in the winter dry season, producing highly flammable fuel. Fynbos graminoids are evergreen and have high moisture content in the initial postfire recovery phases. The implication is that the capacity to carry fire is dependent on dead fuel accumulation rates after the first few postfire years.

Given sufficient biomass to burn, weather conditions strongly influence flammability by altering the moisture content of fuels. Van Wilgen & Burgan (1984) used a burning index based on the U.S. National Fire Danger Rating System, to analyze climate and weather effects on the occurrence of fynbos fires. The burning index peaked in the summer months, November to March, for most of their weather stations but not for the coastal ranges in the eastern regions where the burning index was relatively consistent across the year, with bimodal peaks in winter and summer. The winter peak is associated with dry, hot bergwind conditions. The number of fires, and the area burned, in different months closely matched the climate-derived burning index peaking in summer in most stations but occurring all year round in the coastal mountains. Thus, fire season in Cape fynbos seems largely determined by weather. Moreover, van Wilgen & Burgan (1984) also found a strong relationship between area burned (large fires) and extreme values of the burning index, suggesting that weather conditions are an important contributor to the high fuel continuity conditions that trigger large fires (see also Forsyth & van Wilgen 2008; Wilson et al. 2010). We do not yet know the extent to which the frequency of such weather conditions determines fire frequencies in the Cape region. To do that, we would need to know both the frequency of the extreme weather events and how often they were associated with large fires, and this analysis has not been done.

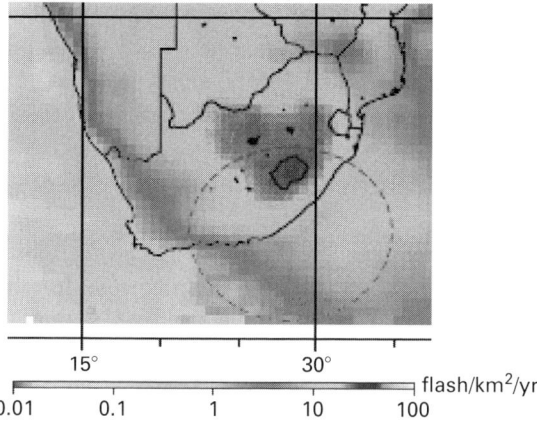

Fig. 7.10 *Annualized lightning flash rate density over South Africa for the period January 1, 1999 to December 31, 2004, collected from Lightning Image Sensor data on the TRMM satellite. Lightning flash densities vary from ~1 km^{-2} yr^{-1} in the southwest to ~10 flashes km^{-2} yr^{-1} in the eastern part of the Cape Floristic Region. Peak lightning activity is 26.4 flashes km^{-2} yr^{-1} over the Drakensberg escarpment and the highveld plateau. (Figure from Collier et al. 2006.)*

Even if there is sufficient fuel, and weather conditions are suitable for burning, fires have to be ignited. Natural sources of ignition in the Cape region include lightning and rockfalls. Fires ignited by rock falls are rare, igniting only 4% of fires of natural origin over a 50-yr period in the Swartberg mountains (Horne 1981). Within South Africa, ground-based measures of lightning showed that the fynbos and karoo (arid shrubland) biomes have the lowest average lightning frequencies, with savannas having intermediate values and montane grasslands the highest (Manry & Knight 1986). The entire western Cape region had a lightning flash density of <1 flash km^{-2} yr^{-1}. In contrast, the C_4 montane grasslands of the summer rainfall region had the highest lightning incidence in the country with 10–14 flashes km^{-2} y^{-1}. In the Swartberg mountains to the east of the Cape region, 10 to 20 thunderstorms occur per year. Lightning days build up from a low of 1.2 in winter, 4.4 in spring, peaking in summer (6.2) and autumn (5.9) (Seydack *et al.* 2007). Satellite-based studies show a gradient of increasing lightning strikes from lowest in the west (~1 km^{-2} yr^{-1}), increasing in the southern Cape with highest flash densities in the eastern Cape (Collier *et al.* 2006). Maximum flash densities in South Africa (26.4 km^{-2} yr^{-1}) were recorded over the Drakensberg mountains and the highveld plateau of the interior (Fig. 7.10), which are mostly C_4 grasslands. The relationship between flash densities and fire starts has not been studied in South Africa.

Despite the low incidence of thunderstorms relative to the rest of the country, lightning is a common source of ignition of fynbos fires, especially in mountain fynbos. In the Swartberg mountains, fires lit by natural causes (mostly lightning) accounted for 78.6% of the total area burned between 1986 and 2002, a period when all anthropogenic fires were suppressed. This contrasts with 44.7% during the preceding period of prescribed burning (1975–1985) and 16.6% during a period of active fire suppression (1951–1974). The mean area burned during these three periods was similar (7907 ha natural, 9015 ha prescribed, 5612 suppression). Thus, at least for this region of the fynbos biome, lightning regularly causes fires and lightning-ignited fires account for similar areas burned to those ignited by people (Seydack et al. 2007).

In summary, we have an incomplete knowledge of what determines fire regimes in fynbos. Fire return intervals are usually longer than the fuel accumulation threshold of ~ 4–6 yrs before fires will spread under typical summer conditions of wind and humidity. Large fires are associated with prolonged summer droughts but these are quite rare, occurring at mean intervals of 5–8 yrs in one study (Richardson & Kruger 1990). Given the combined constraints of sufficient fuel, and weather conditions promoting high fuel continuity, return intervals for large fires would be a decade or more, depending on ignition. Fynbos has a much lower fire frequency than more mesic C_4 grassy biomes. The different fire frequencies cannot be attributed to slow biomass accumulation – the rates are similar at 2.5 Mg^{-1} ha^{-1} yr^{-1} (Kruger 1977; O'Connor & Bredenkamp 1997). But fuel structure (see Chapter 2) is very different. Annual retention of green foliage and slow accumulation of dead fuels in fynbos limits fire starts in early seral stages and the nutrient-poor soils further limit fuels in early stages due to the near lack of annual species. By comparison, fuel structure is very different in C_4 grasslands, where the annual turnover of living foliage in herbaceous perennial grasses generates highly flammable fuel loads on an annual basis.

Human Impacts

Humans evolved in Africa. The oldest human artifacts in the Cape region are those of the earlier Stone Age dating back more than a million years (Deacon 1992). Some of the earliest anatomically modern human fossils, dating back ~120 000 yrs BP, have been collected from the south coast of the Cape Floristic Region (Deacon 1992). There are abundant hearths in human occupation sites starting from the Late Pleistocene (120 000 to 60 000 yrs BP) implying that, from this time, humans could make fire at will. Yet, though humans learned the use of fire in Africa, the archaeological record of early human use of fire is fragmentary and very poorly known, especially relative to Australia and North America. It has been speculated that "fire-stick" farming may have been used to manage fynbos for food resources for ~100 000 yrs (Deacon 1992). Corms of geophytes (*Watsonia, Gladiolus, Ixia, Moraea*) dominate the food debris of archaeological sites throughout the fynbos region and these shrublands may have been burned frequently in autumn to promote this food resource (Deacon 1992; Le Maitre 1984). However, the extent of hunter–gatherer impacts on fire regimes of the region during the Stone Age period are probably slight and localized, with the total population for southern Africa estimated as less than 300 000 at the time of European settlement (Hoffmann 1997).

The first pastoralists appeared in the region 2000 yrs ago (Hoffmann 1997). They were initially sheep herders acquiring cattle about 1300 yrs BP. These Kho-Khoi pastoralists displaced hunter–gatherers to the mountains and more arid regions of the interior, which may have led to more pronounced human impacts on mountain fynbos fire regimes in the last two millennia. The pastoralists utilized the coastal forelands in the west and south with extensive transhumance patterns of herd movement. The Kho-Khoi population in the southwestern Cape region was estimated at 50 000 in the mid seventeenth century with stock ratios of 5–10 cattle and 20 sheep per individual, indicating thousands of livestock in the region. This was in addition to large herds of African ungulates. The implication is that extensive grazing lands occurred in the region. Most of these must have occurred in renosterveld, or on the floodplain fringes of estuaries. Since more than 90% of lowland renosterveld has been converted to crops in the western and southwestern Cape, it is very difficult to reconstruct the nature of the vegetation that supported these vast herds of grazers. The clear implication is that grass was present in abundance in contrast to the dominance of shrubs in contemporary renosterveld. Presumably the shrubs such as the common renosterbos, *Elytropappus rhinocerotis*, were kept at bay with frequent fires, as is still the case (Kraaij 2010). Fynbos makes very poor forage, even when frequently burned, so that extensive areas of lowland and montane fynbos may have been relatively little affected by fires set by Kho-Khoi pastoralists.

The first crop farmers reached South Africa by ~300 AD bringing with them the use of iron. These Iron Age farmers were also pastoralists. Their crops originated from West Africa and included C_4 cereals (sorghum, millet, finger millet) with maize first being planted in the eighteenth century (Hoffmann 1997). The winter rainfall climate of the Cape region was not suitable for the cultivation of the African cereals and may explain why Iron Age people did not settle in the region and therefore had little impact on vegetation and fire regimes. However, it is interesting to note that Scott (2002) was unable to find changes in a long charcoal record in the savanna regions following Iron Age settlement, suggesting negligible effects of these farmers on fire frequency, at least at his site.

Europeans settled in the Cape region in the mid 1600s. By 1806, the human population in the southwestern Cape was still only 75 000 people with half of those in Cape Town and its vicinity (Deacon 1992). The effects of European settlement were, and remain, greatest on the coastal forelands, which had soils suitable for cultivating temperate crops. Wildfires threatened people and property and were, at least in theory, suppressed by the colonial authority. In practice, farmers burned pasture, learning their management techniques from the Kho-Khoi. Since fynbos occurs on nutrient-poor soils, unsuitable for agriculture or pastoralism, extensive areas may have been relatively lightly impacted by the intensified land use. Rapid extirpation of the African megafauna in the Cape would also have had relatively little impact on fynbos vegetation because of the poor forage quality for all but small, specialized mammal herbivores. Renosterveld, in contrast, was heavily impacted, presumably first by loss of megafauna, and then by cultivation (Boucher & Moll 1981). The largest impacts on fynbos in the past few centuries have been indirect and the consequences of introductions of woody plants from distant lands. Conifers were introduced for timber, Australian acacias for stabilizing sand dunes, and *Hakea* spp. for hedges. Many of these species are invasive and pose the most direct threat to the continued existence of fynbos (Richardson & van Wilgen 1986). Extensive conifer plantations were established in higher rainfall regions in the southwestern and southern Cape. The first intensive efforts at fire management of fynbos were implemented to protect the plantations from wildfires. Firebelts, sometimes extending more than a kilometre, are burned at frequent intervals ($\leq$ 6 yrs) and adjacent fynbos stands are often burned to afford additional protection. Increasing urbanization, as in all MTC regions, has fragmented fynbos so that large fires are now more likely to be contained. However, fire sizes on the Cape Peninsula mountain chain, surrounded by the city of Cape Town, have not changed significantly between the periods 1970–1988 and 1989–2007, while fire return intervals have apparently increased (Forsyth & van Wilgen 2008).

The human impact on fire regimes thus has a long history in the Cape region. It is likely to have been relatively slight in the mountains because of their remoteness and, with the exception of afforestation (planting forests), low utility for farming. The second major mediterranean-type shrubland of the region, renosterveld, has been so heavily converted that our knowledge of its ecology is fragmentary and hopelessly incomplete.

Fire Management

The use of fire in managing fynbos has changed over the last century as a result of changes in societal values and needs, scientific knowledge, and resources. In the first few decades of the twentieth century, fires were burned to promote grazing. This was followed by a period of fire suppression policy until the 1970s when prescribed burning programs were initiated in the mountain catchment areas. Since the mid 1980s, the areas burned in prescribed fires have declined, linked to changes in state agencies managing the land (from forestry to conservation agencies) and a decline in financial and human resources for managing fires. In more remote areas, prescribed burning has been replaced by natural fire zones in which attempts are made to suppress anthropogenic fires while allowing naturally ignited fires to burn to their full extent (Seydack *et al.* 2007). Prescribed burning has also become more difficult, especially near the wildland–urban interface (WUI), because of legal and financial liabilities should fires burn beyond landowner boundaries.

The principal management objective for prescribed burning in mountain catchment areas has been to promote streamflow by reducing vegetation biomass. Closely associated with this objective, fire is also used to manage alien invasive trees and shrubs, especially conifers, but also Australian *Hakea* and *Acacia* species. The Working for Water program, launched in 1995, has made a major contribution to controlling invasive woody plants in the Cape region (van Wilgen *et al.* 1996). Water is a key constraint on urban and rural development in the Cape and the initial motivation for the program was to reduce the impacts of invasive trees on streamflow (Le Maitre *et al.* 1996). The success of the program, and the large level of state support, is owed mostly to its success in creating work opportunities for the unemployed (van Wilgen *et al.* 1996) but recent analyses indicate that it is proving to be cost effective for promoting streamflow (Marais & Wannenburgh 2008). Fires are also burned for biodiversity objectives following widespread recognition, since the 1980s, that fynbos is a fire-dependent vegetation. On the WUI fringe, the main objective is protection from fire damage to property. This is achieved by clearing fire breaks (firebelts) and by early suppression of fires, primarily by aerial methods (helicopters with fire buckets).

Prescribed burning from the 1970s was initially based on fixed fire return intervals of ~ 12 yrs in spring. The spring fire season was based on the idea that soils should be moist when burned to reduce soil organic matter losses. Fire regimes became more flexible with increasing knowledge, especially of serotinous Proteaceae, with a shift to longer fire return intervals where maturation rates were slow and with a shift to autumn burns because of the negative effects of spring burns on protea recruitment observed in the southern and southwestern Cape (van Wilgen *et al.* 1992a, 1994; Kraaij 2010). In the last decade, more variable fire regimes have been favored, partly because of management constraints on burning at fixed seasons and frequencies, but also because of the realization that different species and functional groups are promoted by different fire intervals and events (season and intensity). However, variability in frequency, season and intensity of fires in a particular area are monitored within a framework of *thresholds of potential concern* (Rogers & Biggs 1999; Bond & Archibald 2003). These are upper and lower thresholds set for fire frequency, season, intensity, area, and so on. For example, if a fire burns a large area of fynbos or renosterveld at ≤ 6 yr, then this activates a threshold of potential concern because most individuals in serotinous protea populations would not yet have developed seedbanks, resulting in poor recruitment. On reaching such a threshold, managers either re-evaluate the basis for this threshold or intervene; in this example by preventing the area from burning again before proteas have matured. The scope for intervention remains strongly constrained by human and financial resources. Over the last few years, a new organization, Working on Fire (spawned from the very successful Working for Water program) has been initiated with trained firefighting teams that can be sent to different parts of the country (www.workingonfire.org). This promises a more proactive fire management than the reactive mode of the preceding decade when resources were often too limited to influence fires. It is also providing many job opportunities and work skills.

Fire Policies and their Impact on Fire Regimes

Among the various approaches to fire management in fynbos is a *natural fire policy* conducted in *natural fire zones* (Seydack et al. 2007). The natural fire zones are an interesting management experiment in the Cape region. A comparison of fire regimes under different management policies, including 20 yrs under natural fire management, has recently been reported by Seydack et al. (2007) in the Swartberg Mountain range of the eastern inland ranges of the Cape region. The Swartberg is a narrow lozenge-shaped mountain rising to 2000 m elevation from arid shrublands in the lowlands. Because fires in the mountains pose little threat to adjoining properties, a natural fire management regime was implemented in the 1980s and has been maintained to the present. Fires ignited by lightning and rockfalls were allowed to burn without interference, whereas human-ignited fires were prevented from spreading wherever practical (Seydack et al. 2007). This regime contrasts with periods of fire suppression and prescribed burning at different periods of the twentieth century. The experiment provides a rare perspective on determinants of fynbos fire regimes with and without human intervention.

The *natural burning zone* management regime was successful in greatly reducing the area burned by anthropogenic fires (to < 25% per year). Initially the annual area burned (by lightning-ignited fires) was much lower than the previous era of prescribed burning, suggesting that natural fires occurred at much longer intervals than the usual assumption of ~15 yrs. However, the impact of fire policies on the fire regime has to be evaluated against longer-term cycles in area burned. Seydack et al. (2007) analysed the data in 5-yr periods from 1941 to 2000. There were four 5-yr periods when the mean area burned annually equalled or exceeded 10 000 ha. These peak burning periods recurred at intervals of 15, 25 and 15 yrs, more or less the mean fire return interval for fynbos (Fig. 7.11). Seydack et al. (2007) suggest that the peaks in fire activities are associated with periods of more active summer convectional storms bringing more summer rain and also lightning activity. However, periods of high fire activity followed long periods of low activity under cool moist conditions, during which fuel accumulated. This suggests that both fuel and weather conditions influenced natural fynbos fire regimes and the frequency and extent of large fires. Multiyear periodicity of fire activity also means that the effects of different management policies on fire regimes are difficult to assess unless maintained for many decades.

For the available data, fires were significantly smaller in eras of fire suppression (1951–1974) and prescribed block burning (1975–1985) vs. under natural (lightning) fires (1986–2002) (mean and median ha burned: suppression era 761, 84; prescribed era 601, 171; natural era 1430, 276, respectively). The area burned per year in the three eras was 5612, 9015 and 7907 ha respectively but Seydack et al. (2007) attributed the variation largely to multi-annual cycles in climate patterns. The most noticeable differences were changes in fire season, with fires during the natural fire era shifting to the summer months and fewer fires in spring and autumn relative to anthropogenic fires.

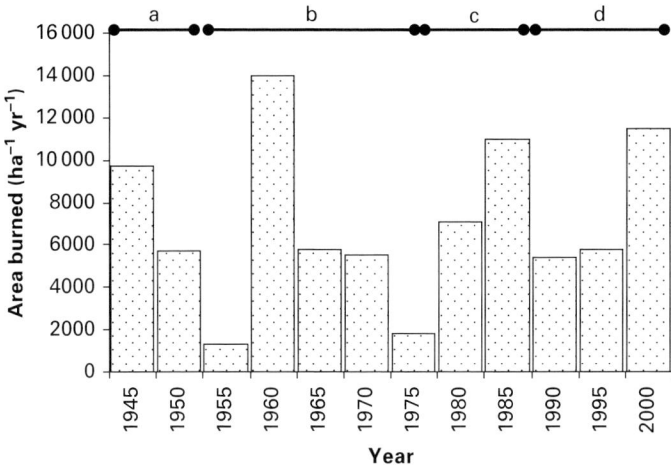

Fig. 7.11 *Mean area burned per annum for 5-yr periods in the Swartberg Mountain range under different management regimes. The year indicated is the last year in the 5-yr period. Management policy: (a) burning for grazing (b) fire exclusion and suppression (c) prescribed burning in "blocks" separated by a network of fire breaks, (d) natural fires. Graph modified from Seydack et al. (2007), who attribute the periodicity in area burned to: less burned area, cool moist periods during which fuel accumulates; more burned area, warm drier periods with convectional (lightning) summer storms. Total area of the Swartberg study region is 170 856 ha.*

The limited degree to which fires can be managed in mountain fynbos is also evident in a recent analysis on the causes of fire in the Cape Nature data set across the Cape region. Southey (2009) analyzed cause of fire for four regions, including the Swartberg, between 1970 and 2007. Prescribed fires ignited by managers accounted for 3 (Swartberg), 6 (Cederberg), 7 (Hottentots-Holland) and 15% (Outeniquas) of the total area burned. Yet despite these low levels of management control over area burned, apparently fynbos fires can be managed quite effectively to protect people and property.

The Urban Interface

Economic damage caused by wildfires at the wildland–urban interface seems to be far less than fires in California and the Mediterranean Basin (e.g. Spain, Greece). For example, two severe fires occurred in the Cape Peninsula in January 2000 on the urban fringe, burning a total of 8000 ha. Homeowners were evacuated and there was general panic. Although these were among the most severe fires that have been recorded at the fynbos WUI, only eight structures were destroyed and a further 51 damaged. No lives were lost. Initial press reports estimated insurance claims of US$500 million and suppression costs in the region of US$3 million (Calvin & Wettlaufer 2000). However, a subsequent government-commissioned report listed insurance claims as ~$5.7 million with suppression costs in the region of $500 000 (Kruger et al. 2000). The relatively low damage to lives and property of fynbos WUI fires probably owes much to the hard urban edges, a legacy of past urban planning, and to the relatively manageable fuel loads in typical fynbos (see Table 2.1). This fire though did initiate a renewed interest in preventive measures for future fires that included control of alien plants and improved management of the urban edge among others (Combrinck *et al.* 2003).

Conclusions

The South African MTC region is unusual in that shrublands dominate under climate regimes that also support forests, and boundaries are determined by fire history. Fynbos shrublands on nutrient-poor soils are extremely rich in species and endemics, as are renosterveld shrublands on clay-rich substrates. The majority of species in fynbos and many in renosterveld are fire dependent in one way or another, which implies a long history of natural fires as a selective force. Most fynbos stands burn at intervals of 10–30 yrs and few stands survive more than 40 yrs without burning. Most shrubs are postfire seeders. Some also resprout and these facultative seeders sometimes have much lower postfire seedling recruitment than obligate seeders that have no capacity for postfire resprouting. Dormant seedbanks are maintained in the soil and in serotinous cones and the latter are more sensitive to long fire intervals. Obligate resprouters regenerate after fire solely by resprouting and do not recruit seedlings. Although this life history is found in several evergreen sclerophyll shrubs widely distributed in fynbos and renosterveld, it only becomes dominant in associated thicket, strandveld or forest vegetation. When these other vegetation types burn, some woody species are resprouters and others are non-resprouters, but none are postfire seeders. Expanding urban development in the Cape region is creating increasing fire problems. Fire suppression policy was begun more than 60 yrs ago and continues to the present. In more remote areas of the Cape region, prescribed burning has been replaced by natural fire zones in which attempts are made to suppress anthropogenic fires while allowing naturally ignited *veld* fires to burn to their full extent.

8 Fire in Southern Australia

The mediterranean-type climate (MTC) in Australia spans from the southwestern part of Western Australia to include much of South Australia and western Victoria (Fig. 8.1), which covers a longitudinal distance second only to the Mediterranean Basin MTC region. As in other MTC regions, the highly fire-prone evergreen sclerophyllous shrub and tree mediterranean-type vegetation (MTV) extends much further east and north into climatic zones that are not MTC. Australia, however, is distinctly unlike other MTC regions in that fire-prone MTV is extensive across the southern part of the continent and transcends climatic boundaries with relatively subtle changes in community structure and composition. Sclerophyllous MTV dominates both the MTC region of the southwestern corner of the continent as well as the southeastern corner under an aseasonal climate. Both regions share a common fire season of summer to early autumn (McArthur 1972); however, the MTC southwest has a potential fire season every summer whereas in the southeast it is tied to weather anomalies that occur once to several times a decade.

Mediterranean-type Vegetation

Within the southern Australian MTC zone (Fig. 8.1) evergreen sclerophyllous vegetation dominates. Such MTV is sometimes defined as shrub dominated (Specht 1979), and indeed large areas of sclerophyllous heaths (see Fig. 1.6f), shrublands (Fig. 8.2a) and mallee (Fig. 8.3) occur. However, woodlands and forests form integral parts of the MTC biome (Dell *et al.* 1989; Gill 1994), and thus MTV includes shrublands, woodlands and forests, and in southern Australia they dominate both in the MTC region and outside that climatic zone (Fig. 8.1). MTV is found across the southern temperate latitudes of Australia (Table 8.1) in an arc below about 30° latitude, accounting for dominant vegetation types in infertile habitats throughout temperate Australia. We specifically focus on the various heaths, shrublands and dry sclerophyll forests that constitute the most fire prone communities in these temperate landscapes. Although similar fire-prone MTV heathlands occur extensively within the tropics on the northern end of the continent (Keith *et al.* 2002; Russell-Smith & Stanton 2002), here we focus on the temperate MTV, but do consider broader relationships with other

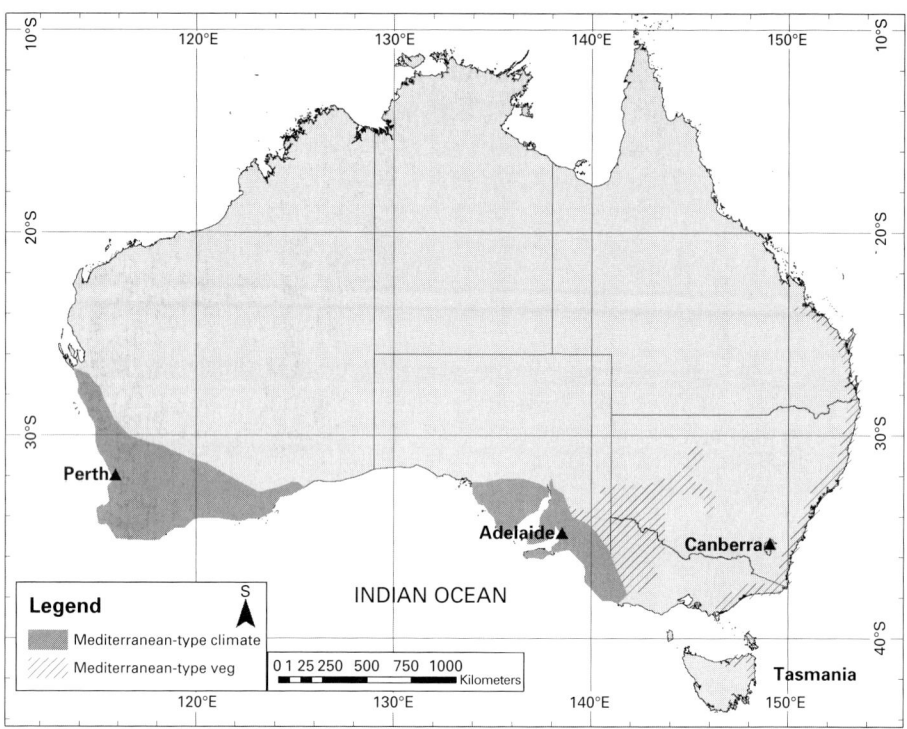

Fig. 8.1 *Distribution of the mediterranean-type climate (MTC) in Western Australia, South Australia and adjacent parts of Victoria (dark shading) and shrublands dominated by mediterranean-type vegetation (MTV) outside the MTC region (hatched).*

(a)

Fig. 8.2 *(a) Treeless heath shrublands, and (b) scattered* Eucalyptus *in grasstree* (Xanthorrhoea) *heathland in the southwestern part of Western Australia. (Photos by Jon Keeley.)*

(b)

Fig. 8.2 (cont.)

Fig. 8.3 *Mallee represents a multistemmed* Eucalyptus *shrubland stunted from repeated fire as illustrated in this mixed stand with the serotinous conifer* Callitris *in the MTC South Australia. (Photo by Jon Keeley.)*

adjoining vegetation types (e.g. wet sclerophyll forests, rainforests, and various arid and semi-arid woodlands and shrublands).

Temperate MTV is important because it (1) constitutes non-tropical enclaves of high diversity and endemism (Barlow 1994; Hopper & Gioia 2004; Hill 2004), (2) is varied in structure and function and forms important habitat for other biota

Table 8.1 Characteristics of major vegetation groups constituting fire-prone sclerophyllous MTV in southern Australia (Fig. 8.1)

Major vegetation group	Climatic regions	Area (km²)	Structure	Dominants	Habitat	Land uses	Current fire regimes
Eucalypt open forests	High rainfall Mediterranean, high aseasonal rainfall, temperate and coastal subtropical (E1, D5, F3–4)	150 000	Trees 10–30 m tall. Shrubby understory on sandy/rocky infertile soils. Grassy understory with shrubs on deeper more fertile soils.	*Eucalyptus, Angophora, Corymbia, Callitris* subdominant.	> 500 mm annual rainfall, wide-ranging soils including lateritic duricrusts in the southwest. Mountainous and undulating terrain. Variations in structure (woodland variants) in rocky, shallow soils.	Significant areas cleared for grazing and agriculture up to early twentieth century. Commercial forestry and extensive conservation reserves. Extensive interfaces with urban and rural development.	Fire intervals 5–20 yrs common. Fire intensity highly variable (500–50 000 kW/m). High fire intensities (crown fire) possible under severe weather. 1–6% p.a. treated by prescribed fire.
Mallee woodlands and shrublands	Mediterranean, semi-arid and aseasonal rainfall areas (e.g. inland NSW). (E1–3,6)	200 000	Mallee *Eucalyptus* spp. are multistemmed, 5–6 m in height.	*Eucalyptus* spp. *Melaleuca, Acacia, Allocasuarina Callitris* and *Hakea*.	< 400 annual mm rainfall on sandy, calcareous and siliceous substrates, including dunefields and sandsheets in coastal and inland environments.	Extensively cleared for cropping until late twentieth century. Large intact expanse on southern coast of WA and SA. Large areas in conservation reserves.	Fire intervals chiefly in 10–50-yr range. Limited suppression access and resources in sparsely inhabited regions. Limited use of prescribed fire for wildfire abatement purposes.
Other shrublands and heathlands	Mediterranean, semi-arid and aseasonal rainfall areas, including coastal subtropical (E13,6, D5, F3–4)	130 000	< 5 m with dense canopies of obligate seeders common. Heaths < 1–2 m.	Highly varied: e.g. *Allocasuarina, Banksia, Bursaria, Dodonaea, Eremophila, Hakea, Grevillea, Kunzea, Leucopogon, Melaleuca*.	Infertile and/or waterlogged sites, on rock or sand in coastal, inland and mountainous environments.	Coastal heath/shrubland extensively cleared for mining, agriculture and urban development. Other communities exploited for grazing in dry inland. Considerable areas in conservation reserves.	Variable fire frequencies in 5–30-yr range. High frequencies adjacent to urban development. Small areas subject to prescribed fire for asset protection.

Vegetation groupings are based on the National Vegetation Information System classification (Dept. of the Environment 2007), with approximate extant area given and bioclimate based on Hutchinson et al. (2005). More detailed descriptions are contained in Beadle (1981) and Groves (1994).

Fig. 8.4 Eucalyptus-*dominated woodland with charred trunks from previous fires on the eastern margin of the MTC in western Victoria. (Photo by Jon Keeley.)*

(Burrows & Friend 1998; Gill & Catling 2002; Burrows & Wardell-Johnson 2003; Yates *et al.* 2003a) and (3) is juxtaposed with contrasting communities, such as rainforest relics (Bowman 2000), dry grassy woodlands, chenopod shrublands and hummock grass/shrublands (Beadle 1981; Groves 1994) on landscapes where differential expression of fire plays a key role in determining landscape patterns. As discussed later, MTV is also important because it forms the economic and social backdrop to the most densely populated regions of the continent, with attendant agricultural and natural-resource-based industries (e.g. forestry).

Eucalypts (collectively used to describe *Eucalyptus*, *Corymbia* and *Angophora* spp. within the Myrtaceae; Hill and Johnson 1995) are a key component of Australian MTV. Eucalypts form the dominant stratum in open woodlands and forests and they are inevitably associated with fire (Fig. 8.4). They also dominate certain shrublands and woodlands in multistemmed *mallee* (Fig. 8.3) and are a variable component of other shrub-dominated plant communities such as heaths, scrub-heaths and thickets (Fig. 8.2b). Even low-growing heathlands such as the southwestern kwongan (Pate & Beard 1984) contain varying proportions of eucalypts in shrub or arborescent forms (Hopkins *et al.* 1983; Beard 1984). Communities not dominated by eucalypts are often azonal enclaves such as rock outcrops, and wet heathlands within a matrix of vegetation containing a prominent component of eucalypts.

MTV refers to woody evergreen sclerophyll plants that dominate systems, but such taxa often occur in radically different ecosystems. For example, MTV often forms an overstory in semi-arid and arid biomes composed of hummock grasslands and extensive chenopod shrublands (Beadle 1981; Groves 1994). Although evergreen woody sclerophyll species are prominent, these represent very different ecosystems, some fire prone, some with C_4 grasses (largely unknown in MTC sclerophyll vegetation), and some with semi-succulent growth forms. As a consequence the term MTV has a narrow definition that refers to the presence of evergreen sclerophylls that are usually very fire prone but says very little about the ecosystem, and for that reason the term mediterranean-type ecosystems needs to be applied with care.

MTV contains prominent centers of high plant diversity, most notably the kwongan shrub-dominated complexes in MTC southwestern Australia (Pate & Beard 1984; Hopper & Gioia 2004). Species richness tends to be lower in structurally similar vegetation in other MTC landscapes in southern and southeastern Australia, particularly at higher levels of scale. Nonetheless, regions containing large tracts of shrubby vegetation have significant diversity in the southeast (Beadle 1981; Rice and Westoby 1983). MTV is nationally and internationally important in this regard.

MTV is strongly associated with fire throughout most of its range. Fire is potentially important in maintaining the composition and character of MTV and its place in Australian landscapes.

Habitats

MTV commonly occurs on soils that are acidic and sandy, or else derived *in situ* from limestone or aeolian deposits of calcareous origin. The latter occur in and around coastal regions in southwestern and southern Australia, plus the lower Murray basin in southeastern Australia (Specht & Moll 1983; Blackburn & Wright 1989). Parent materials are often ancient, particularly in the west (Hopper 2009) and subject to deep *in situ* weathering, as illustrated by lateritic caps and escarpments in the southwest, and with considerable redistribution of surface material via wind occurring in southern dune and sandsheet environments (Wasson 1989). MTV occurs at greater elevations and in more rugged terrain in much of southeastern Australia, where Tertiary and Quaternary (see Fig. 9.1) uplifting has created habitats of steep terrain and skeletal soils, often in high rainfall environments (Wasson 1982). Tertiary volcanic activity in the east may also have diversified soils, both *in situ* and through erosion and deposition across the Murray-Darling basin (Beadle 1981; Specht & Specht 1999).

Many MTV soils are very infertile (low phosphorus and nitrogen) as a result of their antiquity (Specht & Moll 1983; Pate & Dell 1984). A comparison of MTC regions in Fig. 1.5 shows that southern Australia overlaps with the Cape region of South Africa; both are far more infertile than other MTC landscapes. Beadle (1981) invoked nutrients as a key determinant of community distributions and diversity in Australian ecosystems. Specht and Moll (1983) argued that demarcations between tree-dominated vs. treeless communities are governed by interactions between moisture and degree of soil leaching, which is indicative of nutrient availability. Beard (1984) noted a lack of functional hypotheses involving nutrients that would explain distributions of

MTV communities in the southwest. He emphasized the role of soil texture/depth in interaction with rainfall as the overriding determinant of community distributions. Thus, soils with similar texture and depth may yield differing vegetation types under alternative rainfall regimes (Burrough et al. 1977; Beadle 1981; Beard 1984; Sparrow 1989). Recent insights into varying strategies of acquisition of key nutrients in low fertility soils (e.g. mycorrhizae, cluster roots, etc.) suggest that functional specialization could also play a role in fine-scale community patterning in relation to soil variations (Lambers et al. 2008).

Complex mosaics of differing communities can be found throughout MTV regions of southern Australia, which reflect topographic and soil variations (Hopkins & Griffin 1984; Hill 1989; Gill 1994; Hopper & Gioia 2004). Vegetation is often arranged as catenae with major community and edaphic boundaries coinciding (Beard 1984; Keith & Myerscough 1993; Specht & Specht 1999). Water availability, which is a function of soil depth, texture and rainfall, is crucial in determining the balance between trees, shrubs and grasses. Treeless vegetation is typically confined to soils where root growth is impeded due to shallow soils over rocks, which may be seasonally waterlogged, or heavy clay soils, or where water may be periodically unavailable, such as deep sand in drier environments (< 500 mm average annual rainfall). Shrubby mallee woodlands and shrublands are found in lower-rainfall environments (300–500 mm) on deeper sandy soils, often interspersed with grassy woodlands on heavier textured loams. Open forests and woodlands with a shrubby understory are found in higher-rainfall environments on a variety of soils where drainage and waterlogging do not limit tree growth.

Fire Regimes and Land Use

Charcoal has been found in Tertiary sediments in southeastern Australia along with fossils and pollen of MTV taxa (Kershaw et al. 2002; Hill 2004). Charcoal fluctuated greatly through time but abundance generally increased in the late Tertiary and Quaternary (see Fig. 10.9). However, studies are largely biased toward the southeast and there is limited Tertiary data from the MTC regions. The available record does show evidence of a positive association between charcoal, climate and vegetation signals throughout the Quaternary until the late Pleistocene (Kershaw et al. 2002; Lynch et al. 2007). The arrival of humans may have partially decoupled fire activity from climate (Kershaw et al. 2002; Lynch et al. 2007), resulting in more frequent burning. Such trends have intensified with population changes in the late Holocene (Hassell & Dodson 2003; Lynch et al. 2007; Enright & Thomas 2008).

Fire history studies indicate complex trends spanning the European occupation of the continent and the breakdown of indigenous societies in the nineteenth to early twentieth centuries. Interpretations are controversial (Lynch et al. 2007; Bradstock 2008) because of the way signals, such as charcoal in sediments and fire scars from dendrochronology studies, are interpreted. For example, a variety of southeastern Australian studies, spanning spatial scales from local to subcontinental, coastal to alpine environments, and differing methodologies (e.g. charcoal counts, tree rings) indicate that fire activity increased substantially during the early European era (nineteenth to early twentieth century; Banks 1989; Mooney *et al.* 2001; Kershaw *et al.* 2002; Mooney & Maltby 2006; Zylstra 2006). Colonial exploitation involved *laissez-faire* use of fire (Pyne 1991; Collins 2006) and is observed in other MTC regions upon European colonization. Some studies indicate a decline in fire activity in the later twentieth century, due in part to increasingly effective fire suppression, to levels similar to the pre-European era (Kershaw *et al.* 2002).

A unique technique for reconstructing fire history is based on growth bands in the arborescent monocot known as grasstrees *Xanthorrhoea preissi* (Fig. 8.2b) (Ward *et al.* 2001) and has been applied in southwestern Australia. It indicates a very high fire frequency (3–4 fires per decade at a point scale) prior to European colonization and a reduction in fire activity following colonization. Fluctuations in putative European-era fire activity measured by this technique correspond with differing eras of fire and land management policy (Ward *et al.* 2001; Lamont *et al.* 2003). Recent attempts at validation of the technique in kwongan heathlands, however, have indicated major anomalies (Enright *et al.* 2005; Miller *et al.* 2007; Enright & Thomas 2008) that limit the ability of this method to reconstruct fire history prior to the early twentieth century. In all likelihood aboriginal burning was very patchy and potentially not representative of regional fire regimes (Enright & Thomas 2008).

The fire regimes of contemporary MTV communities vary widely according to local climate and ignition patterns (Table 8.1; Bradstock *et al.* 2002). Generally, the point fire return interval in Australian MTVs is decadal to multidecadal (once every few decades or more) (Burrows & Friend 1998; Gill & Catling 2002; Hobbs 2002; Keith *et al.* 2002; Enright & Thomas 2008), though important variations occur above and below this range (Table 8.1). Gill and Moore (1997) and Boer *et al.* (2009) documented recent fire regimes in southwestern Australia open forests (predominantly *Eucalyptus marginata* dominated jarrah forest). They found that under the influence of extensive prescribed burning the average fire return interval was 5–10 yrs. Fire activity is related to rainfall (Clarke 2002a; Pausas & Bradstock 2007; Burrows 2008) with average fire frequency declining toward the drier end of the rainfall gradient due to fuel constraints. In contrast, other MTV such as some forests and wet heaths in high-rainfall areas (> 1000 mm) may have fire activity limited by high fuel moisture and dependent on anomalously dry years, leading to longer multidecadal fire intervals.

Impacts of European land use on fire regimes are varied. Vast areas of shrubland and woodland have been cleared for cereal cropping or utilized for rangeland grazing in drier regions (McIvor & McIntyre 2005). This has resulted in a decline in fire, among other changes to ecosystem functions (Hobbs 2002). There is an emerging role for active restoration of fire in remnant vegetation of this kind for biodiversity conservation purposes (Hobbs 2005; Prober *et al.* 2007). Large areas of intact dry shrublands remain in sparsely inhabited coastal regions of southern Western Australia and South Australia, where lightning ignites occasional large fires (McCaw et al. 1992) resulting in multidecadal fire regimes. Suppression capacity and other management activities in these areas are often minimal.

Mosaics of open forests, woodlands and shrublands in higher-rainfall areas are juxtaposed with densely populated urban centers and intensive rural industries such as plantations, orchards, vineyards, etc. Considerable areas of eucalypt open forest are managed for timber production, though recent, large-scale transfers of such land into conservation reserves have occurred. In many states such reserves now occupy a larger area than natural forests used for timber production. Fire management in these regions has a strong emphasis on protection of urban and rural assets, through active fire prevention involving surveillance or fuel reduction and fire suppression. There is increasing emphasis on formal ecological management of fire using appropriate information and planning systems (Keith et al. 2002; Bradstock & Kenny 2003; Burrows 2008; Gill & Allan 2008). In these reserves both human ignitions from accidents and arson, as well as lightning ignitions, are important (Bradstock & Gill 2001; McCaw & Hanstrum 2003).

Major unplanned fires on the scale of 10^3–10^5 ha occur regularly in southern Australian ecosystems (Ellis et al. 2004; Esplin et al. 2003; Collins 2006). There is an apparent connection with El Niño generated droughts and major fire activity in the southeast (e.g. Verdon et al. 2004; Hennessy et al. 2006) but these linkages require more formal exploration. Major fires are driven by warm, dry air masses generated by southern ocean cold fronts and/or tropical depressions in the southeast, whereas in the southwest such air masses are generated by the eastward passage of strong anticyclones (Luke & McArthur 1978; McCaw & Hanstrum 2003; Hennessy et al. 2006; Hasson et al. 2009). Global change factors may further direct fire regimes toward extremes (Williams et al. 2009; Bradstock 2010).

Diversity in MTV – Resources vs. Fire?

Contemporary MTV has origins in the Cretaceous with diversification throughout the Tertiary and Quaternary (Hopper & Gioia 2004; Hill 2004), encompassing the break-up of Gondwana, opening of the southern ocean, northern drift of the continent and consequent onset of aridity (see Chapter 10). Australian vegetation in the early Tertiary under a relatively warm, wet climate, despite a position of high latitude, consisted of rainforests with gymnosperm and angiosperm dominants. It is postulated that the ancestors of the diverse sclerophyll flora occupied oligotrophic habitats on forest margins, such as shallow, periodically waterlogged soils, that may have shaped the capacity to cope with infertile soils. Families of plants attributed to these origins include Proteaceae (see Box 7.1), Fabaceae, Casuarinaceae, Myrtaceae and Ericaceae. Sclerophyllous vegetation rose in prominence throughout the late Tertiary and Quaternary in conjunction with increasing aridity and evidence of fire (Hill 2004).

The role of fire as an ecological and evolutionary driver of vegetation transition in the Tertiary is difficult to determine, given the paucity of appropriate paleoecological data (Bowman 2000; Kershaw *et al.* 2002). Reviews (e.g. Kershaw *et al.* 2002; Hill 2004; Hopper & Gioia 2004) highlight the prominence of climate, particularly the transition from warm, wet to seasonally dry (and often cool) climatic regimes, as a driver of change. The argument is that fire has followed the effects of climatic-induced changes that result in the opening up of plant cover due to rising aridity (e.g. Kershaw *et al.* 2002). The other putative driver is nutrients because sclerophyllous vegetation is inherently adapted to exploit the low-nutrient soils that typify soils in many parts of southern Australia. Tertiary climates may have contributed to the ongoing impoverishment of soils (Beadle 1981; Barlow 1994; Hopper & Gioia 2004), thus expanding sclerophyll habitat. Fire then accompanied these changes owing to the highly flammable character of evergreen sclerophylls resulting from the high carbon to nitrogen ratio of leaves engendered in dry, infertile habitats (Orians & Milewski 2007).

In this model, fire is seen as being an *emergent property* of sclerophyll vegetation, rather than a force *promoting* its emergence (Martin 1994). For example, Cowling *et al.* (1996) attribute the diversification of species in kwongan plant communities of southwestern Australia to fire regimes, whereas others contend it is a direct result of soils and landscape heterogeneity with an uncertain role for fire (Hopper 2003; Lambers *et al.* 2008). However, in reality the roles of fire and resources (climate, moisture and nutrients) in shaping (drivers of evolution) and maintaining (ecological sorting) of MTV plant diversity are confounded. The influence of a drying climate or nutrient limitations on a plant community assembly may not be readily disentangled from a resultant increase in fire activity.

Patterns of diversity are commonly perceived to be a function of variations in soil moisture and nutrients. Specht & Specht (1999) linked diversity, structure and cover to indices of evaporative potential at a continental scale. In particular they hypothesized that in moist environments, where cover in the dominant stratum will be dense, diversity will be relatively low due to competitive effects on the composition of subdominant strata. Measures of diversity have been correlated with a host of edaphic factors such as nutrients, cations, pH, texture, at both inter- and intra-community scales (e.g. Beadle 1962, 1981; Adam *et al.* 1989; van der Moezel & Bell 1989; Keith & Myerscough 1993; Hahs *et al.* 1999; Le Brocque & Buckney 2003; Clarke *et al.* 2005; Wills & Clarke 2008). The species complement is related to the productive potential of the habitat, with greater numbers of species more likely in less productive than wetter and/or drier MTV habitats (Huston 2003; Pausas & Bradstock 2007). Competitive effects and/or selective influences on functional types are commonly invoked to explain these patterns.

Measures of fire are not usually incorporated into correlative analyses of resources and diversity at intra- and inter-community scales (Le Brocque & Buckney 2003). Enright *et al.* (1994) included time since last fire in their study of MTV plant communities in western Victoria, finding little effect on trends overwhelmingly governed by soil depth and moisture availability. Thus, the potential role of fire in organizing communities appears to be oblique, via effects on soil and geomorphological processes (McIntosh *et al.* 2005; Shakesby *et al.* 2007) and reinforcement of habitat and community distributions and boundaries. As a consequence, fire has been characterized as a recurrent perturbation with minor selective effects on community composition (Florence 1996; Specht & Specht 1999).

The notion that resources and disturbance in MTC ecosystems may have coupled influences on diversity was recognized by Huston (2003) who proposed models of diversity based on discrete and interactive effects of disturbance frequency and productivity, and a classification of ecosystem types. In addition, Huston (2003) linked disturbance frequency in fire-prone systems to flammability, as an outcome of ecosystem productivity (i.e. an endogenous or "bottom-up" model of flammability). The two mechanisms that drive these models are intensity of competition driven by productivity, and disruption of competition by disturbance. Species diversity is hypothesized to exhibit unimodal responses to variation in each of these influences. Huston (2003) characterized the bulk of MTV as being of low productivity and subject to low disturbance frequency. A high level of sensitivity to variations in disturbance frequency and productivity were predicted as a consequence.

This provides a basis for evaluating plant diversity responses to fire regimes at differing levels or scales (i.e. within and between communities), including consequences of habitat variations and spatial variability in fire regimes. Huston (2003) postulated that plant recovery rates from disturbance and therefore competitive pressures will tend to be low due to relatively low productivity. His model predicts that high levels of patch-scale heterogeneity of fire regimes should be due to low rates of regrowth, which implies discontinuity of fuels, and due to effects of fine-scale habitat variability, which implies flammability variations. Both processes are hypothesized to promote high plant richness and diversity at a variety of spatial scales. Does the empirical evidence conform to these predicted plant diversity patterns and responses to fire regimes?

Fire Regimes and Diversity

Patterns of within-community diversity (measured as species richness, see Chapter 11) in Australian MTV are a function of individual species' responses to fire regime components as determined by life history markers and species' vegetative and reproductive characteristics, which include: (1) persistence, (2) seed supply constraints; (3) inhibition of establishment, and (4) risk of senescence. These life history attributes will vary in response to habitat characteristics, thereby influencing rates of recovery from fire and the intensity of competition. Models of community response to fire regime variations can be inferred from an overview of these responses and their patterns of variation. These topics are discussed in order below.

Persistence

Resprouting is common in species in most fire-prone environments, in most growth forms and most taxonomic groupings. Australia stands out amongst MTC regions in having the greatest diversity of shrubs with lignotubers (see Fig. 3.1 and Table 3.2). In MTC Australian shrublands many of the shrub dominants resprout and most also recruit seedlings after fire and are termed facultative seeders. In some MTC communities there are woody species that lack the capacity to resprout and so postfire recovery is entirely dependent on seedling recruitment (see Table 3.4). This life history mode is termed obligate seeding and it is particularly prominent in genera that are dominants of shrublands or as understory shrubs in woodlands and forests: e.g. *Allocasuarina*, *Banksia*, *Dryandra*, *Grevillea* and *Hakea* (Bell 2001). Tree species such as *Eucalyptus*, *Corymbia*, *Angophora* and *Allocasuarina* are usually resprouters (Gill 1997; Burrows & Wardell-Johnson 2003; Nicolle 2006). There are exceptions: for example, low-stature, obligate seeder eucalypts known as mallees occur in some dry southwestern environments. Graminoids and herbs exhibit varying degrees of resprouting. In Restionaceae (e.g. see Fig. 7.6) there are higher proportions of obligate seeders in the MTC southwest than in the aseasonal climate of the southeast (Bell *et al.* 1984; Pate *et al.* 1991; Clarke *et al.* 1996; Benwell 1998; Bradstock *et al.* 1997; Keith *et al.* 2007).

Most resprouters have high resprouting success, although high fire intensities may elevate mortality in some resprouters (Bradstock & Myerscough 1988; Burrows & Wardell-Johnson 2003; Vivian *et al.* 2008). Woody obligate seeders uniformly die when burned and for most species this is not a variable response, as suggested by Vesk & Westoby (2004). A few obligate seeders with a tall growth form and thick bark may survive low-intensity fires, such as some *Allocasuarina* and *Callitris* species (Morrison 1995; Bell & Williams 1997; Bradstock & Cohn 2002b).

There is an overall trend toward higher proportions of resprouters (woody and herbaceous) in moist environments (up to 75%) and lower proportions (~ 50%) in dry environments (Lamont & Markey 1995; Groeneveld *et al.* 2002; Keith *et al.* 2002; Burrows & Wardell-Johnson 2003; Clarke *et al.* 2005; Keith *et al.* 2007; Pausas & Bradstock 2007), a pattern also seen in Californian MTC ecosystems (see Chapter 5). Such a pattern could reflect effects of productivity on availability of space. In moist or more fertile environments, resprouting graminoids and shrubs usurp space soon after fire, thus disadvantaging seedling recruitment by obligate seeders (Yates *et al.* 2003a; Clarke 2002b; Clarke *et al.* 2005; Enright *et al.* 2007; Pausas & Bradstock 2007).

Obligate seeder shrubs dominate at the drier end of the moisture gradient and this life history is commonly associated with narrow or needle leaves that, due to the low surface area to volume ratio, enhance water use efficiency in these habitats (Lamont & Groom 1998; Bell 2001; Pausas *et al.* 2004b; Lamont *et al.* 2007). Obligate seeder shrubs tend to be taller with fewer stems than closely related resprouting taxa, which appear to be allocating more resources to basal lignotubers and other root regenerative structures (Lamont & Markey

1995; Vesk & Westoby 2004; Falster & Westoby 2005; Knox & Clarke 2005). The extent to which these differences affect competitive interactions between obligate seeders and facultative seeders is a matter of some controversy (Keith & Myerscough 1993; Clarke et al. 2005; Wardell-Johnson et al. 2007) and is discussed further in Chapter 9.

Seed Supply Constraints

The bulk of species have persistent seedbanks in which there is some seed carry-over from year to year and seeds are usually available for germination *in situ* following fire (Bell 1999; Auld et al. 2000; Pausas et al. 2004b; Enright et al. 2007; Lamont et al. 2007; Pausas & Bradstock 2007). The formation of persistent *in situ* seedbanks is tied to deep dormancy and relatively short range dispersal for the bulk of the seed pool (long-distance dispersal of some of these is known, Bradstock et al. 1996; Lamont et al. 2007). Such species are not reliant on annually seeking gaps for establishment, but instead are dependent on disturbance *in situ* for the provision of gaps.

Seedbanks stored in the canopy through serotiny (see Fig. 3.5 and Table 9.2), a form of bradyspory, represent about 25% of the postfire seeders in MTC ecosystems in Australia (Lamont et al. 1991, 2007). Soil-stored seedbanks comprise about 70% (Keith et al. 2002; Pausas et al. 2004b). Soil seedbanks of woody species may accumulate more rapidly and have greater longevity than their serotinous counterparts. This is due to several factors. Species with soil seed storage tend to mature earlier (Auld 1987; Bradstock & O'Connell 1988; Keith et al. 2002; Abbott & Burrows 2003; Burrows et al. 2008). Also, soil-stored seedbanks possess dormancy mechanisms that may enable seeds to outlive parent plants (Bell 1999; Auld et al. 2000; Whelan et al. 2002; Wills & Read 2007). In addition, there is some evidence that soil seedbanks are not entirely exhausted by postfire germination and may persist through repeated fires (Whelan et al. 2002; Tozer & Auld 2006); see also Chapter 7.

Canopy storage in obligate seeder species of shrubs generally take 4–10 yrs to be initiated. Slower rates of maturation occur in extreme habitats such as rock outcrops (Clarke 2002a; Yates et al. 2003a, 2003b, 2007; Burrows et al. 2008) and in gymnosperms such as *Callitris* spp. (Bradstock & Cohn 2002b). Canopy storage in facultative resprouters is replenished more rapidly because flowering begins very early in resprouts and thus this secondary juvenile period is much shorter than the primary juvenile period following germination (Bradstock 1990; Lamont et al. 1998, 2007). Compared with congeneric obligate seeders, seed crops of facultative resprouters with canopy storage also tend to be lower (but not always, see Enright et al. 2007) (Zammit & Westoby 1988; Bell 2001; Pausas et al. 2004b; Lamont et al. 2007). Serotinous storage generally reaches maximum levels > 10 yrs postfire, with rate of growth of storage declining at this time, due to losses from predation (vertebrate and invertebrate) and spontaneous release (Gill & McMahon 1986; Bradstock & O'Connell 1988; Bradstock 1990; Lamont & Groom 1998; Lamont et al. 2007).

Seed retention in canopy storage is variable within species, between congeneric species and across families (Cowling & Lamont 1987; Lamont *et al.* 1991; Enright *et al.* 1998; Lamont & Groom 1998). For example, high rates of retention are prominent in Proteaceae (see Box 7.1) and Casuarinacae but not in Myrtaceae (Pannell & Myerscough 1993; Gill 1997; Lamont & Groom 1998; Lamont *et al.* 2007). There is evidence within *Hakea* and *Banksia* of linkages between moisture availability and levels of retention, with high rates of retention favored in drier habitats (Cowling & Lamont 1998; Lamont & Groom 1998; Lamont *et al.* 2007). This may reflect more restricted establishment opportunities or in facultative seeders higher propensity for fires of lethal intensity.

Accumulation of soil seedbanks is less well known than for canopy storage, but modeling studies based on inputs and losses indicate a tendency for an increase and eventual decline with time since fire for *Acacia suaveolens* (Auld 1987) and *Grevillea caleyi* (Regan *et al.* 2003). However, Wills & Read (2007) found that the richness and density of the seed pool did not change significantly up to 26 yrs postfire in a chronosequence study in southeastern Australia. Thus, soil seedbanks of many species may be long lived (> 30 yrs) and longevity of soil seed pools of some taxa, most notably hard-seeded perennial plants, may significantly exceed that of established plants (Auld *et al.* 2000; Wills & Read 2007).

A substantial number of species persist after fire solely by resprouting but utilize the subsequent postfire years for seedling recruitment. This group of pyrogenic flowering plants (up to 25% of species in heathlands) resprout, flower and set seed in the first year and establish seedlings in the second year and subsequent years after fire (Bell *et al.* 1984; Keith *et al.* 2002). This life history is found in many geophytes and other perennial herbs as well as some shrubs and the arborescent monocot *Xanthorrhoea* (Dixon & Barrett 2003; Auld & Denham 2006; Lamont *et al.* 2004b). A resprouting gymnosperm *Podocarpus drouynianus* in Western Australian MTC forests exhibits such a pyrogenic cone production life history (Chalwell & Ladd 2005).

Inhibition of Establishment

In most MTV, seedling recruitment is inhibited in unburned vegetation and occurs in a pulse after fire. Fire-stimulated germination and establishment occur across a wide range of taxa of many different growth forms (Bell 1999; Dixon & Barrett 2003; Merritt *et al.* 2007; Auld & Ooi 2008). Heat and smoke break dormancy in soil-stored seedbanks and typically species respond to one or the other of these triggers (Bell 1999; Auld & Ooi 2008). For some species germination is affected by interactions between heat and smoke but these effects are complex and only partially understood (Thomas *et al.* 2007; Auld & Ooi 2008). See Chapters 3 and 9 for a detailed discussion of heat-stimulated and smoke-stimulated germination.

Effects of heat on subsequent seed germination are complex, given that temperatures in the soil profile are mediated by patterns of fuel structure and fuel consumption. For heat-stimulated species germination is positively correlated with fire intensity, which is correlated with the level of surface fuel consumption

(Bradstock & Auld 1995; Tozer & Auld 2006). Species responses are also variable, with differing optima for release from dormancy apparent among closely related taxa (Auld & O'Connell 1991; Bell 1999; Auld et al. 2000; Thomas et al. 2007).

Some taxa, such as Ericaceae, form dormant soil seedbanks that are sensitive to seasonal temperature fluctuations (Ooi et al. 2006; Auld & Ooi 2008). Release from dormancy is therefore seasonal and not directly related to fire, though the subsequent fate of emergents may be strongly dependent on fire-created gaps. Autumn release from dormancy may be advantageous under a MTC rainfall regime, where timing of germination will regularly coincide with rainfall. Under an aseasonal rainfall regime in the southeast, such a response may be less advantageous (Ooi et al. 2006; Auld & Ooi 2008).

Fire releases seeds from canopy storage, either through direct heating or else indirectly through mortality of stems. Many taxa with canopy storage have non-dormant seeds that are available for germination upon release (Merritt et al. 2007). In some cases (e.g. *Kunzea*, Myrtaceae) there is evidence of the need for further fire-related cues, such as smoke following incorporation of released seeds into the soil (Merritt et al. 2007). The widespread occurrence of serotinous taxa (including both facultative and obligate seeder) across MTV communities (Gill & Catling 2002; Keith et al. 2002; Burrows & Wardell-Johnson 2003; Pausas et al. 2004b) reflects the degree of inhibition to recruitment in the absence of fire, and the regularity of fire in the environment. Serotiny functions to overcome inhibition of establishment in unburned conditions and is favored by decadal intervals of fire with low variance (Lamont et al. 1991; Enright et al. 1998; Pausas et al. 2004b). Rates of seed release and establishment of seedlings of canopy-stored species are positively related to fire intensity (Zammit & Westoby 1988; Enright & Lamont 1989a; Bradstock 1990). There is limited evidence that fire intensity may have negative effects on canopy storage of seeds (Bradstock et al. 1994; Burrows & Wardell-Johnson 2003) but instances of widespread recruitment failure due to mortality of canopy-retained seedbanks are lacking.

Notwithstanding these variations among seedbank types and the nature of mechanisms, the net effects of fire stimuli are positive. Establishment in unburned conditions is infrequent (Gill & McMahon 1986; Whelan et al. 1998; Yates et al. 2003a; Lamont et al. 2007). Germination and establishment may be ongoing in species that lack dormancy but are limited by other factors in unburned conditions. In some species with soil storage there is evidence of significant fractions of seeds without dormancy, which correlates with lower probability of fire (Bradstock 1989; Pausas & Bradstock 2007). Clarke & Dorji (2008) found a greater proportion of fire-cued ephemerals in rock outcrops compared with neighboring forests. This was assumed to reflect reduced postfire competition in rock outcrops due to the lower cover of woody resprouters in outcrops than in forests. Bond & Ladd (2001) noted similar trends across a broader range of taxa in kwongan shrublands and woodlands.

Continous leakage of seeds from serotinous (Lamont & Groom 1998; Lamont et al. 2007) and non-serotinous species such as eucalypts in the absence of fire may ensure that seed supply is not limiting if recruitment opportunities present themselves. One advantage to delaying dispersal until fire is that litter beds form a surface barrier that restricts radicles of germinants from reaching the soil (Bradstock 1989; Tozer & Bradstock 1998). However, disturbances other than fire such as digging and raking of litter by animals may expose mineral soil and present conditions for establishment (Kirkpatrick 1997). Plant cover may inhibit the early growth of seedlings through shading and competition for water (Specht & Morgan 1981; Enright & Lamont 1989a; Wellington 1989; Kirkpatrick 1997), but canopy disturbances from treefalls and other factors often provide recruitment opportunities in the absence of fire.

Predation of seeds and seedlings is a consistent inhibitor to establishment under unburned conditions (Auld 1986; Andersen 1989a, 1989b; Enright & Lamont 1989b; Wellington 1989; Bradstock 1991; Cohn & Bradstock 2000; Hanley & Lamont 2001). Seed supply and seedling establishment is restricted by both invertebrate and vertebrate herbivory, though postfire herbivory can also be severe (e.g. Cohn & Bradstock 2000). Diminished predation of seeds and seedlings in burned compared with unburned conditions may be due to predator satiation (O'Dowd & Gill 1984; Wellington & Noble 1985).

Senescence Risk

Long periods without fire may pose a risk that is dependent on longevity of plants and their seedbanks. Estimates of life spans of perennial woody plants are largely based on extrapolation of survival rates (Auld 1987; Bradstock & Kenny 2003; Pausas *et al.* 2004b). Much of what is known about senescence and mortality in the absence of fire is largely anecdotal and speculative.

Beard (1984) discussed obervations on the fate of kwongan shrublands in the long-term absence of fire. He noted that structure and composition may change through growth, death and the opening of canopies, and that the emergent conifers *Callitris* and *Actinostrobus* sometimes became more prominent. However, wholesale replacement of the community was unlikely and these conclusions are largely endorsed by Lamont *et al.* (2007) for the northern kwongan. Holland (1986) documented little change in the composition of long-unburned mallee in western New South Wales, though there were some structural changes such as a taller overstory and more bare ground. Hopkins & Robinson (1981) hypothesized that woodland in southwestern Australia may be the result of transformation of mallee-heath in the absence of fire. By contrast, Gent & Morgan (2007) hypothesized that in the absence of fire coastal woodlands on dunes in southern Victoria would eventually lose the trees and be replaced by grassland. Eucalypt-dominated woodlands and forests may undergo quasi-successional change to dominance by *Allocasuarina* or *Callitris* in dry sites or rainforest in wet sites if fire is absent for long periods of 50–100 yrs (Close *et al.* 2009).

Decline in species diversity has been documented due to shrub invasion under long-unburned conditions in heaths, shrublands and woodlands in eastern Australia (Bennett 1994; Cheal 1996; McMahon *et al.* 1996; Kirkpatrick 1997; Lunt 1998). In these cases, dominance by long-lived woody shrubs and trees (e.g. *Acacia*, *Allocasuarina*, *Leptospermum*, *Melaleuca*, *Callitris*), with consequent shading of lower-stature shrubs and herbs, has been invoked as the cause of a marked decline in diversity. The rise of *Allocasuarina* as an alternative dominant to eucalypts when charcoal abundances were low during the Quaternary has been documented in various paleoecological studies (Lynch *et al.* 2007), though there are anomalies in this trend.

Callitris species are generally capable of recruitment in the absence of fire in both eastern and western MTV (Bowman & Harris 1994; Enright *et al.* 1994), with these species tending to be prominent in fire refugia in shrubland vegetation (Beard 1984; Enright *et al.* 1994) and off-shore islands (Yates *et al.* 2003a). High abundance of *Callitris* and *Allocasuarina* pollen in the Quaternary paleo record is correllated with low abundance of charcoal (inferred low fire activity, e.g. Thomas *et al.* 2001; Kershaw *et al.* 2002). The relationship between rainforest and sclerophyll vegetation has been extensively debated on the basis of both paleo and contemporary ecological evidence (Bowman 2000; Kershaw *et al.* 2002; Lynch *et al.* 2007). Whether this transition applies in dry, infertile or skeletal edaphic conditions is open to question. *Allocasuarina* and *Callitris* spp. may fulfill this role in these habitats.

Community Responses to Fire

Predicted community responses of Australian MTV to fire regimes reflect the outcome of life history variations, demographic processes, competition and habitat effects. Aboveground plant diversity (measured as species richness, see Chapter 11) tends to recover rapidly, reaching a peak within about a decade after fire, with subsequent stasis or decline (Fig. 8.5; Gill 1999; Bell 2001; Burrows & Wardell-Johnson 2003), conforming to an initial floristic composition model. Gill (1999) noted variations around this basic pattern such as the greater occurrence of peak diversity immediately after fire from *fire ephemerals* in drier MTC environments.

Diversity of MTV communities is often inferred to be a unimodal function of the length of time between fires, known as the fire return interval or inter-fire interval, with a peak at 10–20 yrs and diminution at both shorter and longer intervals. Such inference is based on extrapolation of trait responses via dynamic models of individual species and functional types, in particular the sensitivity of obligate seeder shrubs to length of the fire interval. This model includes the assumption of strong buffering against major variations in fire regimes due to the relatively high but variable proportion of resprouters present. Fire regime variations are therefore assumed to largely affect the woody obligate seeder component of the community (Gill & Bradstock 1995; Gill & Catling 2002; Keith *et al.* 2002; Bradstock & Kenny 2003; Burrows & Wardell-Johnson 2003; Pausas *et al.* 2004b; Lamont *et al.* 2007; Burrows *et al.* 2008; Groeneveld *et al.* 2008), though sustained high frequencies of fire may also deplete some resprouter populations (Watson & Wardell-Johnson 2004; Pausas *et al.* 2004b; Lamont *et al.* 2007). The predicted amplitude of the diversity and fire interval relationship (Fig. 8.6) is therefore mainly a function of the obligate seeder component of MTV communities.

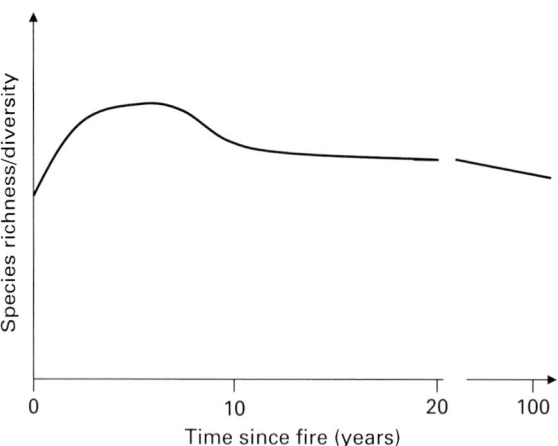

Fig. 8.5 *The characteristic relationship between species richness/diversity and time since last fire in MTV (see Gill 1999).*

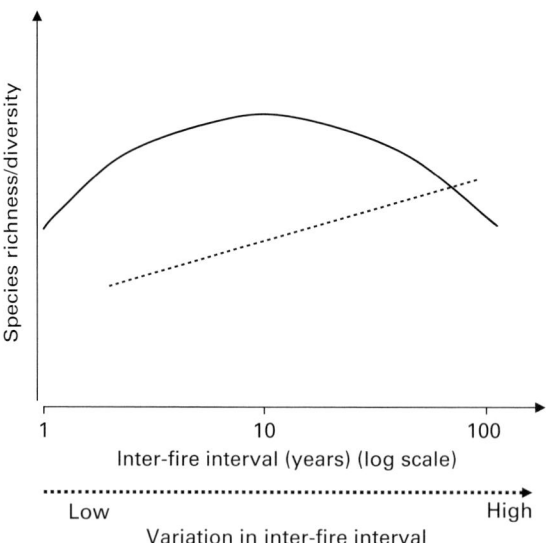

Fig. 8.6 *The predicted relationship between species richness/diversity and mean length of inter-fire interval (i.e. fire return interval) in MTV. The dotted line and scale represent response of MTV communities under relatively high productivity conditions (e.g. wet heaths and open forests). In this case richness and diversity may be more responsive to variability in the inter-fire interval (dotted line and scale).*

Obligate seeders are predicted to have a unimodal response to fire recurrence, principally because establishment is typically tightly keyed to fire. Thus, extirpation is likely under frequent or infrequent fire and persistence favored by intermediate disturbance frequency and low variance in frequency. Shrub-dominated communities in relatively dry habitats are most sensitive to changes in fire interval because of higher proportions of woody obligate seeders (e.g. Clarke et al. 2005).

The length of the juvenile period and patterns of replenishment of seedbanks following fire (e.g. Gill & Nicholls 1989; Burrows et al. 2008) affects the position of the peak in the relationship between diversity and length of inter-fire interval (Fig. 8.6). The domain of highest diversity is a minimum interval of about 10 yrs and this equates to levels of seed accumulation in canopy storage that are sufficient for population replacement of the slowest maturing species (Burrows & Wardell-Johnson 2003; Bradstock & Kenny 2003; Lamont et al. 2007; Burrows et al. 2008). Conversely, declines in serotinous taxa may occur with putative senescence. Similar declines in soil storage are predicted, though longevity may exceed the life span of established plants.

A predicted decline in diversity at longer intervals between fires (Fig. 8.6) also reflects the inhibitory effect of a lack of fire for many species. Release from inhibition is correlated with degree of removal of litter and plant canopies by fire, as a result of direct and indirect stimuli. Accordingly, plant establishment is a general positive function of fire intensity. Where the probability of fire is lower (e.g. semi-arid communities), dependence on direct fire stimuli can be lower. The long-term absence of fire results in a decline in diversity through senescence, lack of recruitment and possible competitive displacement by long-lived dominants that may recruit in the absence of fire (e.g. *Callitris* spp.). Some or all of these mechanisms may apply in differing communities. In a successional sense, MTV is disclimax vegetation, with inherent high richness/diversity sustained by recurrent fire on a decadal to multidecadal cycle. Rainforests, which have potential to replace MTV in the southeast, may contain a completely different assemblage of species, often with lower diversity than neighboring MTV (Clarke et al. 2005).

The fire regime model (Fig. 8.6) corresponds with observed responses of species, functional types and communities to different fire regimes based on opportunistic analyses (i.e. natural experiments) and on meta-analyses (e.g. Pausas et al. 2004b) in shrub-dominated communities. Corroboration of the model via field observations is biased toward studies from southeastern Australia (e.g. Nieuwenhuis 1987; Cary & Morrison 1995; Morrison et al. 1995; Bradstock et al. 1997; Benwell 1998; Ross et al. 2002, 2004; Watson & Wardell-Johnson 2004; Myerscough & Clarke 2007), though empirical observations have been made on individual species' responses to fire regimes in the south and southwest (e.g. Specht 1981; Gill & McMahon 1986; Wooller et al. 2002; Yates et al. 2003b, 2007; Lamont et al. 2007). Such work confirms that obligate seeders (up to 50% of the species) are relatively sensitive to variations in fire interval in particular (i.e. as indicated by the amplitude and modal tendency), but also demonstrate that resprouters exhibit sensitivity in this regard as well (Watson & Wardell-Johnson 2004). In particular the serotinous obligate seeder subcomponent (e.g. usually less than 5% of species, Pausas et al. 2004b) has a pivotal role in shaping this relationship, given high sensitivity to both frequent and infrequent fire.

The response of communities in more productive habitats (i.e. wetter/deeper or more fertile soils) is less sensitive to variation in length of the interval between fires than in drier, shrub-dominated communities. Thus the composition of swamps and wet heaths on the one hand, and open forests on the other may be more sensitive to variance in inter-fire interval length. Reductions in diversity in low-stature herbaceous and shrub growth forms occur under competitive dominance by obligate seeder shrubs in heaths and shrublands (Keith & Bradstock 1994; Bond & Ladd 2001; Tozer & Bradstock 2003; Keith et al. 2007). Diversity may be dependent on high levels of variation in between-fire interval rather than the mean length of the interval (Fig. 8.6, Tozer & Bradstock 2003; see also Jacobsen et al. 2004 for California). Effects may be more acute in moist compared with dry habitats, where greater productivity and cover dominated by resprouters limit understory diversity (Bond & Ladd 2001; Clarke & Dorji 2008).

Pekin et al. (2009) found a positive correlation between diversity measures and fire frequency in MTC jarrah forests, reflecting positive responses of herbaceous functional types to reductions in understory shrub cover. By contrast, Burrows & Wardell-Johnson (2003) found that both frequent and infrequent fire in experimental jarrah forest plots caused declines in abundance of woody taxa in the understory, in accordance with the model in Fig. 8.6. Penman et al. (2008a) found that differing frequencies of experimental burning had little impact on richness in a southeastern, dry sclerophyll forest, though a long-term decline in richness was detected in all treatments. This decline may be due to a failure of seedbanks to be triggered through the absence of occasional moderate or high-intensity fire (Penman & Towerton 2008; Penman et al. 2008a, 2008b).

Varied outcomes of studies of this kind may reflect inherent limitations in measurement of fire regimes and plant responses. Fire history data covering natural experiments is often limited to several decades at most. Manipulations in formal experiments and the range of contrasting fire treatments are similarly limited. Species richness and cover measures are inherently insensitive indicators of composition and may mask major demographic changes within species and functional types (Bradstock et al. 1997; Keith et al. 2007). Thus, the model may respond differently to different functional types and perhaps even to different fire regimes.

In conclusion, a variety of evidence supports the prediction that fire functions as a non-equilibrium determinant of diversity in MTV communities through the disruption of competition. Relatively frequent fire (< 5–10 yr interval) may deplete diversity due to low rates of recovery, though the nature of this effect may vary with productivity. Infrequent fire (> 30 yr interval) may also deplete diversity, through inhibition of recruitment and replacement by dominant species capable of recruiting in the absence of fire. Habitat variations affect these mechanisms via productive potential and the availability of space created by fire as predicted by Huston (2003). MTV spans a wide range of possibilities in this regard. In the least productive communities such as dry, sandy or rocky habitat, competitive effects of shrubs may be reduced in importance due to the ready availability of open space.

In more productive habitats where competition for space is more acute, the variance in inter-fire interval will strongly affect diversity (Fig. 8.6). High variance in fire frequency (e.g. occasional inter-fire intervals < 5 yrs) may be required to maintain diversity in the long term. Such an effect has been well described in eastern high rainfall wet/dry heaths (Keith & Bradstock 1994; Tozer & Bradstock 2003; Keith et al. 2007). Fire and productivity therefore have interactive and even interdependent effects on diversity.

Other mechanisms not accounted for by Huston (2003) contribute to the interplay between fire regimes and diversity. Postfire establishment introduces randomness into plant populations that promotes coexistence. Movements of seeds after fire on soil surfaces may remix species and favor certain microsites in a manner that creates space, allowing coexistence of species with differing competitive abilities (Whelan 1986; Lamont & Groom 1998; Howell et al. 2006; Lamont et al. 2007; Esther et al. 2008). Differential drought tolerances among species may result in alternative recruitment success after different fires, resulting in fluctuating abundances (Lamont et al. 2007). Long-term resilience is enhanced by wide ranging "storage effects" afforded by persistent seedbanks and regenerative organs, common across taxa and growth forms. Abundance fluctuates, thereby preventing dominance by individual species (Lamont et al. 2007). Interactions with other disturbances and drought (e.g. Ross et al. 2004) enhance these effects.

Variations in fire regimes that include fire return interval and fire intensity are also very important. Arguably MTV exhibits greater sensitivity in this regard than other fire-prone Australian communities (e.g. tropical savanna woodlands, arid shrublands). Management of the pivotal obligate seeder component of MTV that underpins this sensitivity may hinge on spatial variability of fire regimes (Groeneveld et al. 2002, 2008; Bradstock & Kenny 2003; Bradstock et al. 2006; Parr & Andersen 2006). Greater emphasis on the interplay between modeling and empirical studies is required to understand both the pure and applied ramifications of the spatial game of fire and habitat that species must play to coexist in high-diversity MTV landscapes. Many aspects of this model likely apply to other MTC regions as well, although with very different timescales.

Flammability and Diversity

General models postulate a unimodal response of fire activity to moisture and productivity (Fig. 8.7; Huston 2003; Bond & Keeley 2005). Such models implicitly integrate the respective effects of moisture on plant growth and the availability of fuel to burn. For example, Pausas & Bradstock (2007) showed that a range of MTV communities in southeastern Australia have a negative relationship between flammability and productivity. Therefore, MTV broadly occupies part of a tail of the fire, moisture and productivity spectrum (Fig. 8.7). How do variations in the flammable characteristics of MTV communities emerge from the combined influence of productive potential of habitats and *in situ* fire weather? Do these variations in these influences lead to high levels of fire regime heterogeneity (landscape

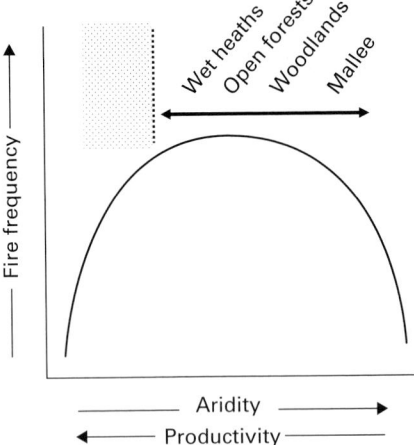

Fig. 8.7 *Relationships between fire frequency, productivity and aridity (from Pausas & Bradstock 2007). The range of MTV is indicated by the solid line and the approximate order of differing communities is indicated. The stippled area represents the domain of dominance by cool temperate rainforest.*

scale), as predicted by Huston (2003)? Does such fire regime heterogeneity reinforce the patterns of varying functional type abundance across habitats?

Fuel Systems, Fire Weather and Flammability

MTV fuel structure leads to very different fire regimes and is determined by several elements. Where shrubs and trees dominate, litter fuels from woody plants are the primary determinant of ignition and fire spread potential, whereas in herbaceous-dominated communities standing dead grasses and forbs are the primary fuel. At the wet end of the MTV spectrum (Fig. 8.7) in graminoid-dominated wet heaths, fire spreads in herbaceous surface fuels with passive crown fires of individual shrubs. At the dry end, in woodland communities herbaceous perennial and annual grasses are similarly important in fire spread. Moisture therefore plays a role in community flammability and patterns of fire spread. In many Australian ecosystems, flammability is heavily affected by the extraordinarily high flammability of eucalypts.

Variations in eucalypt density directly affect ignitability, rate of spread and fire intensity via the spatial continuity and loading of their litter fuels (Catchpole 2002). In woodlands with high densities and greater than 20% canopy cover, eucalypts create a continuous layer of surface litter that is readily ignitable and highly combustible due to its bulk density and energy content. These fuels represent a substantially higher fire hazard than litter produced by other perennial sclerophyllous taxa, particularly those with narrow, needle leaves or other vegetation such as rainforest (Dickinson & Kirkpatrick 1985; Bradstock & Gill 1993; Scarff & Westoby 2006; Plucinski & Anderson 2008). Surface loadings of litter, either in eucalypt-dominated woodlands and open forests or in mixtures such as in heaths and shrublands, are highly variable (see Table 2.1), but typically in the range of 10–20 Mg m^{-2}. Equilibrium surface loadings are usually reached within 10 yrs postfire (Walker 1981; Raison et al. 1983). Accumulation rates and equilibrium loading of surface litter are positively related to rainfall in open forests (Fox et al. 1979; Huston 2003).

Semi-arid shrublands are dominated by narrow-leaf species (Lamont & Groom 1998) that form litter beds difficult to ignite (Scarff & Westoby 2006; Plucinski & Anderson 2008). Ignition and fire spread depends on the arrangement of live and dead foliage in the canopy, and is strongly controlled by wind. Patches of eucalypt litter and other broadleaf species, when dry, can provide the surface energy after ignition that sustains propagation of wind-driven crown fires in shrub-dominated vegetation (McCaw 1997).

Conditional thresholds in fuel moisture, in combination with wind speed, often govern crown fire propagation in many vegetation types (Catchpole 2002; Plucinski & Anderson 2008). In dry communities such as mallee shrublands and woodlands, litter fuels are patchy and clustered around individual plants, which are linked by grasses and ephemeral herbs that carry surface fires. In wet eucalypt forests litter is often too wet to burn in most years but dependent on anomalously dry years.

At the wet end of the MTV spectrum (> 1000 mm rainfall), there are relatively few days of Very High to Extreme fire danger, often less than 10 days per year on average. The number of days increases with decreasing rainfall so that at intermediate levels of rainfall (600–1000 mm) there are 10–20 days per year and at the dry end (300–600 mm) severe fire weather conditions, on average, occur more than 30 days per year (Bradstock 2010).

Fire Regime Heterogeneity and Diversity

The link between moisture, fuel types and fire weather creates the potential for variable flammability ensembles, with different combinations of fire frequency and fire intensity as predicted by Huston (2003). Heterogeneity of fire regimes may lead to the assembly of communities with functional types representing unique flammability characteristics. Pausas & Bradstock (2007) estimated that across a precipitation range of 250–450 mm the average fire cycle varied from 30 to 200 yrs in southeastern shrublands. This variation reflects the coupled influences of fire weather and key functional types. By contrast, in moister open forest at annual rainfall levels > 600 mm the fire cycle is typically < 20 yrs (Bradstock 2010), and in wet areas (> 1000 mm per annum) occupied by tall open forests, the fire cycle may increase to century scales (Gill & Catling 2002).

At the local scale, quantitative evidence of strongly demarcated fire regime variation between MTV communities is sparse, perhaps due in part to lack of appropriate measurement and mapping of fire regimes. Boundaries between heaths, woodlands and forests appear to have limited effects on fire spread, but may have significant effects on fire severity (Chafer et al. 2004;

Hammill & Bradstock 2006; Clarke *et al.* 2007). Anecdotal evidence of fine-scale variation exists, with habitats such as dunes and rock outcrops having a lower probability of burning than adjacent areas (Gill & Bradstock 1995; Bradstock & Cohn 2002a; Clarke 2002a; Lamont *et al.* 2007; Burrows 2008). Such effects are due to lower mass and continuity of fuel in these relatively drier habitats.

Clarke *et al.* (2005) posed a model where community flammability and fire frequency was inversely related to moisture and productivity. This model, developed for eastern, warm temperate vegetation, ranked rock outcrops as having higher flammability than other neighboring communities such as dry sclerophyll forest, wet heath and wet sclerophyll forests. The generality of this model is debatable, as under some circumstances wet heaths may be more ignitable than forests. Rock outcrops often contain highly discontinuous fuel compared with neighboring forests and may be less likely to sustain spreading fire than surrounding vegetation (Clarke & Dorji 2008).

Such differences in probability of burning and its effects on fire frequency and intensity may only partly depend on *in situ* fuel characteristics determined by habitat. The spread of fire, a spatially contagious process, is determined by the combination of fuel and ambient weather characteristics, as represented in fire behavior models. Under severe fire weather, probability of burning in communities with relatively low flammability may increase, particularly if such communities are isolated within a more flammable matrix (e.g. rock outcrop communities within an open forest matrix). Thus, relative flammability may be variable according to weather conditions. In particular, differences in flammability may diminish under extreme weather conditions, which may lessen the influence of fuel characteristics on the probability and rate of spread of fire (Peters *et al.* 2004; Bradstock *et al.* 2010).

Habitat, diversity and resprouting are linked (see discussion above). Hypothesized causes for these links include competition for resources and selection via alternative fire regimes, reflecting differential flammability among habitats (Clarke & Knox 2002; Clarke *et al.* 2005). As noted, however, definitive evidence of close correspondence between fire regime variations, particularly fire frequency, is lacking, but limited data suggest confounded effects of habitats, fire regimes and functional type selection.

An elegant example, involving the spatial partitioning of obligate seeder and facultative seeding (resprouting) shrubs in kwongan heath, has been described by Groeneveld *et al.* (2002) and Lamont *et al.* (2007). In this case, habitats and fire interact spatially to segregate and confine species distributions within the wider confines of their potential habitat range. Pausas & Bradstock (2007) showed that increasing richness, abundance of obligate seeders and decreasing fire return interval were correlated along a subcontinental scale gradient of MTV shrublands. Clarke *et al.* (2005) postulated that habitat influences will have a substantial role in shaping woody species richness and abundance of resprouters. They also predicted that variations in fire regimes will also have a more limited selective influence on these attributes.

The spatially contagious nature of fire and issues of habitat context mean that the scope for tight correspondence between flammable attributes of habitat patches and *in situ* fire regimes is highly variable. There may be scope for widely diverging fire regimes within habitat types, forced by varying combinations of *in situ* and *ex situ* influences. Exploration of the variability in these patterns is required to understand the level to which *in situ* flammable feedback may control functional type composition.

Fire and the Maintenance of MTV

The mix of communities in Australian MTV can be regarded as comprising a diverse, sclerophyll *fire world*. Such an entity is largely self-reinforcing via the coupled effects of fire weather and habitats that emerge from the transitional range of temperate climates in southern Australia. Strong evidence of fire in organizing community distributions and boundaries via differential flammability within MTV is, however, lacking. One possibility is that differences in flammability among functional types are relatively small and largely overwhelmed by the relatively frequent occurrence of severe fire weather across MTV environments (see above). Additionally, the ubiquity of eucalypts results in competitive domination of vegetation structure, composition and fire regimes. The exceptional resilience of eucalypts throughout MTV may also restrict the ability of more flammable functional types to eliminate them. Thus vegetation boundaries and community diversity appear to be primarily determined by major edaphic variations, in turn influenced by moisture. However, the scope for fire to interact with these influences is poorly explored. Little is known of the extent of potential niches of major functional types and the way these are mediated by fire and competition, as demonstrated for arid shrub and grass combinations characteristic of alternative habitats by Nano & Clarke (2010).

Plausible alternative dominants to eucalypts and sclerophyllous shrubs in MTV communities are restricted to fragments in the landscape (e.g. rainforest patches in the east, Bowman 2000), or scattered, low density populations (e.g. *Callitris* spp.) within MTV. These alternatives, with inherently lower flammability, persist in this tenuous manner due to adverse fire regimes that emerge from the characteristic fire weather and wide habitat availability of more flammable competitors under MTV climates in mainland, southern Australia.

The switch to dominance by *Callitris* spp. in sandy, southeastern MTV environments under cooler Holocene climates (Thomas *et al*. 2001) indicates the potential for at least a shift in fire weather, such as a reduction in number of days conducive to fire spread, to alter dominance and composition. Rainforest patches in the continental southeast may also represent fire refugia (e.g. moist gullies, Bowman 2000) where effects of adverse fire regimes are ameliorated through a reduction in conditions conducive to frequent and intense fire (i.e. higher fuel moisture, lower wind speed and exposure to the sun). Patches of this kind in the southeast may also partly represent areas of *core habitat* where rainforest species have an advantage over sclerophyll competitors. Often, such habitat effects may be confounded with fire-refuge effects (e.g. moister gullies may provide a competitive advantage for mesic taxa, along a barrier to spread of intense fires). The disappearance of rainforest in the Pleistocene from southwestern Australia is consistent with this model. The subdued terrain in this region translates into a scarcity of incised canyon refugia and the annual summer drought of this MTC region potentially increases annual fire hazard.

By contrast, under the cool, high rainfall climate of western Tasmania, the nature of these interactions is altered sufficiently through both "top-down"" (very infrequent severe fire weather, low ignition rates) and "bottom-up" (large expanses of moist habitat, impeded drainage, organic soils, etc.) effects on habitats shaped by predominant infertile soils. Thus, the resilience and competitive status of sclerophyll vegetation is partly constrained, allowing a higher probability for rainforest to become more widespread. This may then allow more scope for variation in fire regimes to determine vegetation distributions (Jackson 1968; Bowman 2000; King 2004). Differences in habitat availability and fire weather under this climate result in different fire regime–vegetation interactions, compared with the mainland. This may represent the climate–fire margin of MTV dominance, where the inherent tendency of sclerophyll vegetation to dominate at broad scales (landscapes, regions and biomes) is diminished and unstable: that is, the zone of transition between alternative stable states dominated by these differing entities (Warman & Moles 2009).

The development of such a favorable coupling of habitats, functional types and emergent fire regimes can be invoked to explain the development and expansion of MTV through the Tertiary (Table 8.2). Such mechanisms can be inferred obliquely but not fully validated through pollen or macrofossils. Nonetheless, key elements of these mechanisms are described in contemporary research, such as: fire weather/ignition effects in determining boundaries of high-latitude rainforest (Kitzberger & Veblen 1999; Veblen *et al.* 1999, 2008); lack of resilience in Gondwanan rainforest lineages (Hill & Read 1984; Kirkpatrick & Dickinson 1984); and differential flammability (Kirkpatrick & Dickinson 1984).

Fire and climate are likely to have had an interactive role in the rise of major lineages that typify Australian MTV. Such a role may be extended into the Quaternary, after sclerophyll vegetation came to predominate, albeit with significant fluctuations. Quaternary shifts in the balance of rainforest and sclerophyll (and within-sclerophyll composition – casuarinas and *Callitris* vs. eucalypts) may therefore be seen as the outcome of coupled shifts in habitat availability and fire regimes, rather than simply the unilateral outcome of either climate or fire (e.g. Kershaw *et al.* 2002). A more refined appraisal of the relative contributions of fire weather and habitat availability as determinants of fire is required to understand the contemporary nature of MTV and the effect of humans and their changing populations in the late Pleistocene and Holocene.

Table 8.2 Mechanisms responsible for transitions from rainforest to sclerophyll vegetation during the Tertiary in Australia

Mechanism	Response
Changing fire weather, ignition and climate	Fire has a non-linear response to moisture/climate (Bond & Keeley 2005; Pausas & Bradstock 2007). Given the likely domain of Tertiary climate, increasing dryness caused an increase in fire activity.
Habitat transformation under a drying climate	Drying created new, more open habitats exploited by sclerophyll species with superior drought tolerance and competitive ability to exploit space. Fire reinforced the transition to sclerophyll vegetation through nutrient losses and additive effects of changing climate on fertility. An ongoing increase in amount of core and facultative habitat available for sclerophyll species resulted.
Influence of life history and resilience traits	Sclerophyll species have traits that enhance persistence and exploitation of large gaps created by fire, and rapid replenishment of regeneration capacity. Rainforest species lack these traits. If sclerophyll vegetation is more resilient to relatively frequent disturbance, then fire would play a role in maintaining the grip of sclerophyll vegetation on newly available habitat.
Influence of flammability	If sclerophyll vegetation is more flammable (ignitable and sustainable combustion) than rainforest, then under altered circumstances (see above), an increase in the amount of fire may result. This may not only reinforce acquisition of habitat (see above), but also result in further attrition of rainforest at its margins, promoting further sclerophyll expansion.

Given the antiquity of key families, MTV, with its high diversity of woody plants and pronounced sensitivity to variations in fire return interval, can be regarded as an *old fire world*. It displaced vegetation where fire may have been less prevalent and important as an ecological driver – a world largely of very limited fire. This notion implies domination by a suite of vegetation types and emergent fire regimes, which result via mechanisms driven by particular climatic settings. MTV dominates habitat provided by poor soils in interaction with moderate rainfall and decadal-scale fire recurrence, dictated by fire weather/ ignition syndromes characteristic of a temperate climate. Accordingly, differing settings of climate may alter these respective mechanisms and tilt the balance to suit dominance by other fire regimes, habitats and plant functional types, as shown by the contemporary interplay between MTV and rainforest in southern Australia.

MTV transcends a broad sweep of the non-linear relationship between fire regimes and available moisture. Given the position of MTV, both an increase and decrease in fire frequency may occur in response to declining moisture. Arguably, the potential for contemporary temperate rainforest, and its antecedents, to predominate lies at the extreme left of this relationship (Fig. 8.7).

The boundary with MTV (*old fire world*) is demarcated within the region where fire responds with a monotonic increase to declining moisture. What other combinations exist in continental Australia and how do these relate to MTV and its environments?

In marked contrast to MTV are the tropical grasslands and savannas dominated by grasses with the C_4 photosynthetic pathway. Although their distribution is strongly controlled by climate, the origins and rapid expansion of C_4 grasses in the late Tertiary in conjunction with development of the tropical monsoonal climates appear to have been strongly influenced by fire (Bond & Keeley 2005; Keeley & Rundel 2005). Indeed, their very maintenance today is dependent on regular fires to prevent encroachment by sclerophyll woodland (see Chapter 7). This model implicitly involves the interaction of climate, life history and flammability as discussed above. Specifically, the monsoonal climate, selecting for high-productivity warm-season grasses (C_4 plants) capable of producing heavy fuel loads of highly combustible herbaceous foliage, and unlimited lightning ignitions set the stage for a highly fire-prone landscape. The capacity of C_4 grasses to rapidly regrow leads to an almost annual fire frequency, a fire regime incompatible with most woody growth forms. All of this derives since the end of the Tertiary and thus represents a *new fire world* in contrast to the older origins of the MTV *old fire world* (Box 8.1).

> **Box 8.1** Old vs. New Fire Worlds
>
> C_4 grasses in Australia encompass both perennial (e.g. *Themeda*) and annual (e.g. *Sorghum*) life-history patterns in conventional tussock growth forms (Mott & Groves 1994). However, the endemic Australian genus *Triodia* exhibits a remarkable hemispherical growth habit derived from needle-like tillers (Rice *et al.* 1994). Hummock grasses form a prominent component of arid and semi-arid communities, dominated by shrubs or trees throughout arid Australia (Mott & Groves 1994), but also extending into the wet tropics (Russell-Smith & Stanton 2002). The hummock growth habit of interlocking tillers results in an ideal flammable arrangement of live and dead material (Bradstock & Gill 1993). Hummock grasses are sclerophyllous and adapted to the infertile habitats (Rice & Westoby 1999) that predominate across the continent. By contrast, C_4 tussock grasses predominate in deeper, heavier textured soils, with or without woody dominants (Mott & Groves 1994).
>
> C_4 grass-driven fire regimes are linked to available moisture, with high frequencies (1–5 yrs) in the monsoonal tropics and decadal frequency in arid landscapes where anomalous pluvial events periodically provide sufficient connectivity of biomass for major fires (Allan & Southgate 2002; Southgate & Carthew 2007). Relative to MTV, the composition of arid and tropical
>
> *Continued*

Box 8.1 (*cont.*)

communities containing dense C_4 grasses is insensitive to variations in fire return interval (e.g. Williams *et al.* 2002; Wright & Clarke 2007). This reflects the relatively high frequencies of fire driven by these grasses.

In southeastern Australia the MTV existing under a summer rain climate is suitable for warm-season C_4 grasses. Here the understories of sclerophyll woodlands on non-sandy substrates have a significant C_4 component (Mott & Groves 1994; Prober *et al.* 2002). This contrasts with MTC sclerophyll woodlands in southwest Australia, which comprise very different ecosystems in which C_4 grasses are poorly represented (Murphy & Bowman 2007). In the east edaphic factors come into play, leading to extensive temperate C_4-dominated grasslands on fertile, igneous substrates (Lunt & Morgan 2002).

The striking exceptions are the infertile habitats (rock and sand) within the *old fire world* of MTV which lack C_4 cover. The exception is mallee shrublands (semi-arid) that often contain *Triodia* on sandy substrates (Rice & Westoby 1999; Bradstock & Cohn 2002a). In these regions (*c.* 32 35° latitude), gradients between heath and mallee-heath (− *Triodia*) and mallee (+ *Triodia*) may be steep and related to overall rainfall (e.g. Pausas & Bradstock 2007). These represent an interface between "new" and "old" fire worlds, where the balance between space/competition, habitats and flammability could be finely poised. The nature of this balance will be critical in these habitats, given their diversity and sensitivity to changes in fire regimes. Would an alteration in seasonality and amount of rainfall have minor, incremental effects or could boundaries be poised in a manner that leads to a major realignment of these fire worlds (i.e. a grass fire cycle beyond the resilience of woody taxa, see Fig. B12.1.1)? Alternatively, will elevated levels of atmospheric carbon dioxide (CO_2) bolster the dominance of woody species in temperate vegetation, via enhanced water use efficiency?

The temperate, infertile MTV habitats represent the last frontier remaining unconquered by C_4 grasses in Australia. Candidate invaders include indigenous (e.g. *Triodia*) and exotic grasses. Predictions of twenty-first-century climatic change indicate increases in summer rainfall in southern MTV regions (CSIRO 2007) along with a tangible chance of overall decline. Such shifts may synergistically change ignitions, fire weather and habitat availability to favor these invaders in MTV regions. While this is speculative, attention is warranted given that diverse communities of MTV fringe the southern regions of the continent. A realignment of *fire worlds* in this way could leave the diverse enclaves of mainland MTV stranded in terms of available ecological space. Given evidence that the composition of biomes is relatively inflexible over evolutionary and ecological time (i.e. species migrations tend to be confined within similar biomes, Crisp *et al.* 2009) the implications of rapid global change are manifestly serious.

Thus, Australian *fire worlds* can be understood as the emergent outcome of opposing processes, each driven, in part, by climate and geology. Among these, MTV is arguably the one with the most at stake in the future, due to an internal composition that is sensitive to subtle shifts in fire regimes and a delicately poised continental position. More formal exploration of the implications of global change may come from application of appropriate landscape (King *et al.* 2006; 2008) and Dynamic Global Vegetation Models (Scholze *et al.* 2006; Lenihan *et al.* 2008) that represent the processes governing habitat availability, life history selection and fire regimes and their coupling to productivity and climate. Such insights, complemented with appropriate information from the field, are required to understand the fate of diverse MTV communities, which are an evolutionary legacy that is characteristically Australian.

Conclusions

Australia is unlike other MTC regions in that fire-prone MTV is extensively distributed across the southern part of the continent and transcends climatic boundaries with relatively subtle changes in community structure and composition. This MTV often occurs on very old and infertile soils, much like MTV in South Africa. Fire regimes tend to be crown fires and vary with fire return intervals from decadal to multidecadal. Fire-adapted vegetation appears to be ancient but the origin of fire-prone vegetation is a matter of some controversy, with some arguing that it is an emergent property of sclerophyll vegetation and others that fire promoted the emergence of sclerophylls. A substantial proportion of the MTV has focused reproduction on a single postfire pulse of recruitment and many of these taxa have abandoned the resprouting habit. These are clearly fire-dependent species, as are a significant number of short-lived postfire ephemerals. A variety of evidence supports the conclusion that fire functions as a nonequilibrium determinant of diversity in MTV communities through the disruption of competition. However, the system is sensitive to short and long fire intervals. Evidence of fire determining community distributions and boundaries via differential flammability within MTV is generally lacking. This may be because the influence of species-specific flammability traits on fire behaviour is often overwhelmed by the influence of severe fire weather.

Section III

Comparative Ecology, Evolution and Management

Here we utilize those points of convergence and divergence between mediterranean-type climate (MTC) ecosystems to develop a synthesis that reveals emergent properties not evident by study of any one region alone. Comparative study of plant traits, functional types and community responses to fire provides insight into selective factors driving the evolution and ecological assembly of fire-prone plant communities. Feedback processes are crucial to understanding evolution on such landscapes. Fire provides a challenge to understanding selective forces because, although inclusive fitness theory can explain fire-adaptive traits, such traits are dependent on community-level assembly that contributes to fire spread. MTC regions exhibit differences in climate and geology that have led to diverse fire environments, and account for many differences in trait evolution and community assembly. Humans have long been attracted to MTC regions but have not always adapted successfully to these fire-prone landscapes. Urban and peri-urban populations have been highly vulnerable to wildfires in some MTC regions, with differences in vulnerability between regions being due largely to innate differences in fuel loads of indigenous vegetation types and profound differences in population density.

9 Fire-adaptive Trait Evolution

Until relatively recently the importance of fire and the origin of fire-adaptive traits have received minimal attention from paleoecologists, and appreciation of this importance has varied across the different mediterranean-type climate (MTC) ecosystems. For example, Axelrod (1973) and Raven & Axelrod (1978) wrote extensive treatises on the origins of the California flora, and yet gave little or no mention to the issue of fire in the evolution of these taxa. Hopper (2009) suggests that fire has only been an incidental factor in the evolution of the Western Australian flora. These investigators have weighed climate and soils far above fire as an important evolutionary driver in these plant assemblages and have downplayed this component of community assembly (see Fig. 1.4).

Axelrod (1989) even went so far as to suggest fire was irrelevant to the evolution of California chaparral. Although he acknowledged that fire could have played a role in the spread of chaparral-like vegetation during the late Tertiary (2–10 Ma), he insisted that fire had played no significant role in the origin of "adaptive types." In his view, "Several lines of evidence suggest that the modern fire-adapted taxa may not reflect an evolutionary response to fire. The diverse adaptations to fire probably represent features that originated without the stimulus of fire. . ." Contrary to this belief, we suggest there is sufficient reason to accept a fire origin for many fire-adaptive traits in mediterranean-type vegetation (MTV), and that fire has been a potential ecosystem process on landscapes far longer than the late Tertiary (Bowman *et al.* 2009; Pausas & Keeley 2009).

Fire History from the Paleozoic

By the beginning of the Paleozoic Era (540 Ma; Fig. 9.1) the atmosphere had sufficient oxygen to carry fire, and ignition sources from lightning, but lacked fuel. In a sense the world was poised for fire and waiting for plants to emerge onto land, as the earliest Silurian land plants are associated with charred remains (Glasspool *et al.* 2004). These authors interpreted this early presence of charred litter and coprolites as evidence of a low-temperature surface fire. During the past decade there have been a number of Paleozoic studies reporting fire impacts on vegetation, and even a suggestion that this was a regular ecosystem process. Despite the presence of mesic tropical forest elements, these Paleozoic landscapes were

Era	Period	Epoch	Began (Ma)
Cenozoic	Quaternary	Holocene	.01
		Pleistocene	1.8
		Pliocene	5.3
		Miocene	23
		Oligocene	34
		Eocene	54
	Tertiary	Paleocene	65
Mesozoic	Cretaceous		145
	Jurassic		200
	Triassic		251
Paleozoic	Permian		299
	Carboniferous		359
	Devonian		416
	Silurian		443
	Ordivician		488
	Cambrian		542

Fig. 9.1 *Geological timescale based on the 2004 timescale endorsed by the International Commission on Stratigraphy.*

heterogeneous and fire-prone seasonally dry forests were also present (Falcon-Lang *et al.* 2009).

The fossil record also provides evidence that characteristics of contemporary fire regimes were already present in Paleozoic and Mesozoic ecosystems. For example, smouldering turf fires were interpreted for an Early Devonian site (Edwards & Axe 2004) not unlike ground fires in peat swamps today. Other Devonian fires include surface fires in pro-gymnosperm forests with an apparent fire regime of frequent understory fires that burned ferns and shrubs desiccated during the dry season (Cressler 2001; Collinson *et al.* 2007; McParland *et al.* 2007), not unlike present-day northern hemisphere conifer forest fire regimes. This fire regime has been observed under different climates. Falcon-Lang (2000) described laminated sediments with fires every 3–35 yrs in pro-gymnosperm forests under a tropical monsoon climate, indicating a frequent surface fire regime similar to the present-day ponderosa pine fire regime under a monsoon climate in the southwestern USA. Similar frequent understory burns in Jurassic gymnosperm forests (e.g. Fig. 9.2) occurred under an analogous summer-dry mediterranean-type climate (MTC) (Francis 1984). In contrast to these surface fire regimes, crown fires are evident in charred apices of wetland lepidodendron forests of the Carboniferous, and like contemporary crown fire regimes, fire intervals were very long, on the order of 100–1000+ yrs (Falcon-Lang 2000). Equivalent crown fires today are common in high-latitude conifer forests and temperate shrublands (Keeley *et al.* 2009a).

Fig. 9.2 Araucaria *is a southern hemisphere taxon that dates to the Jurassic and was an important component of the Jurassic gymnosperm forests that Francis (1984) reported had tree-ring patterns indicative of a mediterranean-type climate (MTC) and charred fossil evidence of light surface fires. (Photo of contemporary* Araucaria *forest in Chile by Thomas Veblen.)*

Thus, the Paleozoic record shows that fire has been a potential factor in plant evolution since the origin of land plants, and later Mesozoic fossils reveal a plethora of evidence for a widespread presence of global fire (e.g. Scott 2000, 2010; Jones *et al.* 2002; Marynowski & Simoneit 2009). These demonstrations are particularly remarkable because they represent a fossil record that is highly biased against deposition of materials from fire-prone environments (Kemp 1981). These fires were apparently ignited by lightning and there is substantial evidence that those ecosystems occupied climatically seasonal environments (Finkelstein *et al.* 2005), one of the key elements to creating fire-prone ecosystems (see Fig. 2.1). In addition, fires in these late Paleozoic forests led to postfire successional sequences of different plant functional types, indicating successional sequences of plant replacement and recolonization (Glasspool 2000; Calder *et al.* 2006). As is the case today, these fires were sometimes massive landscape-scale events covering thousands of square kilometers of forests and causing massive postfire sediment loss (Nichols & Jones 1992). The earliest Paleozoic ferns were associated with fire and have characteristics of a disturbance-dependent nature (DiMichele & Phillips 2002). By the Late Cretaceous there is evidence that some taxa were already specialized for fire-prone environments (Watson & Alvin 1996; Collinson *et al.* 1999). These observations suggest that early in land plant evolution, fire was an ecosystem process with the potential for selecting fire-adaptive traits and affecting community assembly.

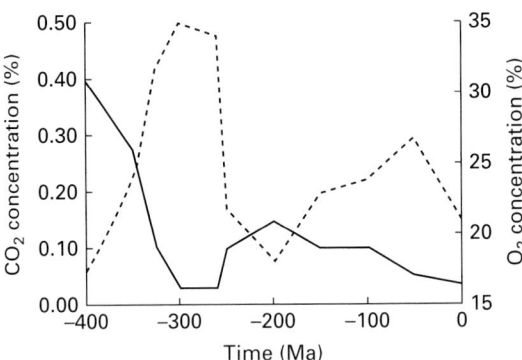

Fig. 9.3 Variations in atmospheric CO_2 (solid line) and O_2 (dashed line) over the past 400 million years predicted by the geochemical mass balance model (from Keeley & Rundel 2005 based on data by Beerling et al. 1998).These atmospheric changes likely had profound impacts on fire regimes as increased CO_2 availability would enhance fire activity by increasing fuel production and declines in O_2 would potentially restrict fire activity (see text).

Different Mesozoic fire regimes were actively selecting plant traits that in some instances are still evident in today's flora, for example in two California endemics, coast redwood (*Sequoia sempervirens*) and the interior giant sequoia (*Sequoiadendron giganteum*). Today redwoods occupy moist coastal forests with very long fire return intervals that generate high-intensity crown fires. As with many woody taxa in crown fire regimes, redwoods are vigorous resprouters from basal burls. The giant sequoia of interior mountains is associated with high lightning ignition landscapes that historically have burned in frequent surface fires, a fire regime that has not selected for resprouting capacity, but rather self-pruning and thick bark to withstand understory burning and reduce the potential for crown fires.

Earth history included periods of high and low fire activity, which Scott and Glasspool (2006) have suggested were tied to changes in atmospheric oxygen levels (Fig. 9.3). Oxygen, however, never dropped sufficiently to exclude fire, as evident by the paleofires described for the late Permian at both high (Glasspool 2000) and low latitudes (Uhl *et al.* 2007) and for the late Triassic (Jones *et al.* 2002). Presumably changes in climatic seasonality likewise could have affected the waxing and waning of fire activity, perhaps through changes in plant (fuel) structure (Belcher *et al.* 2010). The capacity to occupy drier seasonal habitats due to the origin of seeds may also have been a factor promoting the Paleozoic spread of fires (Rowe & Jones 2000; Uhl & Kerp 2003). Biotic changes in plant structure (e.g. Bond & Midgley 1995) likewise would have affected the incidence of paleofires.

Even prior to the Cenozoic radiation of angiosperms that led to our contemporary MTV, fire-adapted sclerophyll shrubland, analogous to modern MTV shrublands, dominated some Cretaceous landscapes (Batten 2002). For example, southern and eastern England during the Early Cretaceous had a moderate winter rain and summer drought climate analogous to contemporary MTC, complete with xerophytic plant adaptations and abundant evidence of periodic crown fires (Allen 1998). This vegetation dominated by dinosaurs was open low-growing "chaparral-like," comprising fire-prone conifers, cycads and ferns subject to periodic fires (Insole & Hutt 1994; Collinson et al. 1999). Charred remains point to numerous fossils that have been interpreted by Allen (1998) as fire adaptations, including serotinous cones, fire-stimulated hard seeds, and highly sclerotic fern indusia that were dropped when heated by fire, among others (see Scott 2000 for other examples of Mesozoic fire-prone environments). We hypothesize that such fire-prone environments have been present somewhere in the world throughout much of Earth's history. Even during periods of "equable" climates there have been gradients in moisture availability, with xeromorphic plants capable of carrying fire.

Early Cenozoic climates are widely considered to have been warm and humid with relatively little seasonal variability. Nonetheless, deposits from the Paleocene–Eocene boundary show evidence of widespread fires followed by seasonal rains that generated large charcoal deposits (Collinson et al. 2007). Much of this early history of fire is based on coal petrological analyses and such studies have not been widely conducted on Tertiary sediments; perhaps as a result we have far less evidence of fire in the Tertiary than in earlier periods (Scott 2000). Paleobotany has tended to emphasize taxonomy of macrofossils and been far less concerned about indicators of pyric conditions (Robinson 1989). Thus, while the Tertiary is represented by numerous fossil floras, nearly all have focused on the taxonomy of the floras and not on characteristics of the environments that would pertain to fire.

Perhaps the most extensive Tertiary evidence of fire is in Oligocene and Miocene coal beds from southeastern Australia and these attest to widespread fires in seasonally dry sclerophyllous shrublands (Martin 1996; Kershaw et al. 2002). Clear evidence of fire as driver of ecosystem change comes from the late Miocene, 5–10 Ma, where increased incidence of fires appears to have played a major role in establishing the dominance of Eucalyptus in Australia (Martin 1996; Bowman 2000), Also, Miocene coal deposits in Europe reveal a similar story of frequent fires in seasonally dry shrublands and swamps (Figueiral et al. 2002). The late Miocene rise in marine charcoal deposition (Herring 1985; Jia et al. 2003) is further evidence of Tertiary fire. However, there is no reason to interpret this as though Earth suddenly discovered fire in the late Tertiary; rather there appears to be a substantial change in the amount of fire-prone landscape at that time.

It has been hypothesized that the spread of C_4 grasses during the more seasonal climate of the late Tertiary was due to this increase in fire activity, which opened up woodlands and created environments favorable to C_4 grasslands (Bond et al. 2003; Keeley & Rundel 2005). The high productivity and flammability of C_4 grasses would have produced a feedback process that further increased fire activity, thus maintaining the grassland and savanna landscape. Few other ecosystem processes can account for the fossil record of C_4 grassland expansion at the expense of woodlands, particularly in light of the evidence that C_4 grasslands were expanding into more mesic habitats (Keeley & Rundel 2005). Further evidence for such a fire origin for these grasslands is that maintenance of these ecosystems is today dependent on frequent fires. Whether or not these represent qualitative changes in the earth system (cf. Bowman *et al.* 2009) or just an expansion of fire-prone environments is yet to be sorted out.

Origin of Fire-adaptive Traits

Many plant traits have been interpreted as adaptations to fire but that implies the trait was selected for that function, and this is often not known. It is considered important by some to distinguish adaptations from exaptations (Gould and Vrba 1982); the latter are traits that serve a particular function, but originated through selection for some other function. This is a theoretically valuable concept but linking traits to selective environments at the time of their origin is a huge challenge. Even if one could demonstrate a factor is responsible for the origin of a trait, it doesn't rule out other factors playing a role since evolutionary pressures do not necessarily act independently. Distinguishing between adaptations and exaptations may be shaded by one's perspective on what are believed to be the most important selective factors in the environment. For example, climate has long been considered the major driver in plant trait evolution; however, it is now clear fire has great antiquity and fire-prone environments have persisted through multiple climatic changes. The limitations to recognizing exaptations from adaptations are by no means unique to traits with adaptive value in fire-prone habitats, but apply to most traits and thus this historical definition of adaptations is fraught with numerous difficulties (Lauder *et al.* 1993; Keeley *et al.* 2011). In many cases it is doubtful one can clearly distinguish between adaptations and exaptations, and thus the term "apparent adaptation" may be appropriate.

Origin of Resprouting

Resprouting from vegetative structures after top-kill is a widespread regeneration mechanism following many types of disturbance, and although of apparent adaptive value in fire-prone landscapes, it is common in many vegetation types where fires are rare or uncommon such as tropical forests (Putz & Brokaw 1989; Kauffman 1991). In both tropical and temperate forest trees resprouting has adaptive value in recovering following major wind events such as hurricanes and tornados (Glitzenstein & Harcombe 1988; Bellingham *et al.* 1994; Paciorek *et al.* 2000), although fire cannot be ruled out as a factor since these wind events create fuel loads that induce widespread burning (Liu *et al.* 2008). Making a case for fire as the selective force in the evolution of resprouting is made difficult by the observation that resprouting has been around for a very long time, as suggested by its near universal

Fig. 9.4 *Tree ferns* (Dicksonia sp.) *with resprouting of apical meristem following high-intensity crown fire in* Eucalyptus *forest of Victoria, Australia. (Photo by Jon Keeley.)*

occurrence in woody dicots, and its presence in Mesozoic gymnosperm lineages such as *Ephedra*, *Ginkgo biloba*, *Sequoia sempervirens* and *Wollemia nobilis*.

Evidence for resprouting as a fire adaptation is often of the form of plant life histories consistent with a fire origin, although not mutually exclusive of all other factors. For example, ferns are a lineage with widespread resprouting in response to fire, and some lower Cretaceous fern floras appear to have consistently been associated with fire (Harris 1981; Scott 2000). The contemporary *Pteridium aquilinum* is a worldwide fire-resilient species that resprouts after burning under a wide range of tropical and temperate habitats (Gliessman 1978) and there is little reason to interpret this as an exaptation to some other environmental factor. *Gleichenia* is another taxon that has been associated with fire in Cretaceous (Herendeen & Skog 1998), Cenozoic (Collinson 2002), and Holocene environments (Black & Mooney 2006), and resprouts vigorously (Gillison 1969), as well as recruiting from spores following fires (Walker & Boneta 1995). Tree ferns (e.g. *Dicksonia* spp.) persist today in the understory of Australian *Eucalyptus* forests and survive high-intensity crown fires, with thick overlapping leaf bases protecting the trunk and allowing postfire resprouting from the apical meristem (Fig. 9.4). The lineage is deeply rooted in the Mesozoic and the widespread existence of fire evidence from these early forests in association with this and other fern taxa

(Collinson 2002; Van Konijnenburg-Van Cittert 2002) is consistent with a fire origin of resprouting in some fern clades.

A clearer example where fire has likely played a selective role in the woody plant evolution of resprouting is in northern hemisphere gymnosperms. Most lack any capacity for resprouting but the few that do (e.g. *Juniperus deppeana*, *J. oxycedrus*, *Pinus canariensis*, *P. serotina*, *P. echinata*, and *P. rigida*, *Pseudotsuga macrocarpa*, *Picea* spp., *Sequoia sempervirens*, *Taxus brevifolia* and *Torreya californica*) are all components of crown fire regimes and are in clades where resprouting is largely absent and of secondary origin (Stone & Stone 1954; Keeley 1981; Minore & Weatherly 1996; Keeley & Zedler 1998). A similar pattern is evident in southern hemisphere gymnosperms such as *Podocarpus* (Chalwell & Ladd 2005), *Cunninghamia* (Del Tredici 2001), and *Widdringtonia* (Keeley et al. 1999b). The latter genus is of interest because it comprises two relict species that are non-resprouters restricted to mesic sites with long-interval fire regimes (Manders 1986). The only other species is *Widdringtonia nodiflora*, a resprouter that has a widespread geographic range from Malawi to the Western Cape region; it occurs in fynbos and is tolerant of frequent fires (Keeley et al. 1999b).

Epicormic resprouting is one of the primary means of postfire survival in arboreal *Eucalyptus* in both MTC and non-MTC regions of Australia (Burrows 2002). In these trees epicormic resprouting arises from seemingly unique strips of meristematic cells well developed on the inner bark, providing a well-protected source of new shoots following even high-intensity fires. This structure is widespread in distantly related genera in the Myrtaceae, suggesting either an early origin for this fire adaptation or multiple origins within the family (Burrows 2010). Molecular phylogenies for the family suggest this trait arose very early in Myrtaceae and points toward fire having been an important evolutionary factor through most of the Tertiary (Crisp et al. 2011).

Another example where a case could be made for fire playing a selective role is in the postfire resprouting and flowering of the South African fynbos geophyte *Cyrtanthus ventricosus*. This species maintains preformed floral buds that remain dormant for years but are triggered by smoke (Keeley 1993b) so that they flower only within 1–2 weeks after a fire, regardless of the season (Le Maitre & Brown 1992). This illustrates one of the dicey problems of sorting out adaptation from exaptations. In herbaceous perennials resprouting *per se* is not likely an adaptation to fire (see Chapter 3) yet in *C. ventricosus* this trait has been fine tuned in ways that clearly reflect a selective role by fire. To a lesser degree this may be true as well of other geophytes from closed-canopy MTC ecosystems.

One thought experiment that purportedly showed resprouting in Californian shrublands as not an adaptation to fire (Lloret et al. 1999a) deserves closer examination. These authors demonstrated resprouting after an experimental fire in MTV matorral shrublands of mainland Mexico and concluded that because this region lacked a fire regime of frequent fires, as observed in contemporary MTC southern California chaparral, there was little reason to assume a relationship between fire and resprouting in either shrubland ecosystem. However, these authors failed to appreciate that the current California fire regime is anthropogenic and humans now account for well over 95% of all fires. There is no justification for the assumption that resprouting evolved under such a high fire frequency and the natural regime in southern California had fire frequencies of once or twice a century (see Chapter 5), which is the frequency to be expected in crown fire regimes (see Fig. 2.7). This is not outside the fire regime to be expected for their matorral shrublands in Mexico (Román-Cuesta et al. 2003; Torres-Rojo et al. 2007; Rodríguez-Trejo 2008; González-Tagle et al. 2008), and thus it is reasonable to conclude that fire exerts a similar selective role in both shrublands.

While we may never know the selective environment for postfire resprouting in any lineage, the fact that fire has been present in terrestrial ecosystems ever since plants began accumulating biomass, and that resprouting is a widespread trait in woody dicotyledonous plants, it is not an untenable suggestion that in some lineages resprouting is an adaptation to fire (Pausas & Keeley 2009). However, even in those clades where fire seems to likely have played a role in the evolution of resprouting, it is reasonable to imagine a scenario where fire has acted in concert with other disturbances, for example fire and herbivory (e.g. Johansson et al. 2010) and fire and wind (Liu et al. 2008).

Postfire Resprouting vs. Seeding

Resprouting is an ancient trait in land plants and in some lineages this likely has been maintained by fire. Although resprouting is a widespread response to a range of fire regimes, in many shrublands subjected to a predictable crown fire regime there also has been selection to capitalize on burned sites for reproduction. In most cases these postfire seeders have specialized their life cycle by delaying all recruitment to a single postfire pulse of seedlings (e.g. see Table 3.3). Although both obligate resprouters and facultative seeders are adapted to fire-prone environments, this evolutionary step toward postfire seeders was driven by changes in fire regime that created greater opportunities for seedling recruitment in recently burned environments (Fig. 9.5). Undoubtedly these changes were the result of both bottom-up and top-down effects and the importance of one or the other was likely not the same in different fire-prone landscapes. Also the timing of this evolutionary step may have been very different in the various MTC regions. For example, the widespread and very old infertile soils in southwestern Australia would have led to sclerophyllous foliage that contributed to flammable vegetation early in the Tertiary, whereas in other MTC regions aridity may have played a bigger role at a later point in time (see Chapter 10).

Once postfire seeding evolved, MTV was poised for a second major event. Taxa in a number of lineages focused all of their reproductive effort on postfire seedling recruitment to the extent that they abandoned the resprouting mode and depended entirely on the seemingly risky mode of only a single pulse of postfire seedling recruitment. These taxa are obligate seeders and represent a second important evolutionary transition (Fig. 9.5). Factors driving this transition could have been

Fire-adaptive Trait Evolution

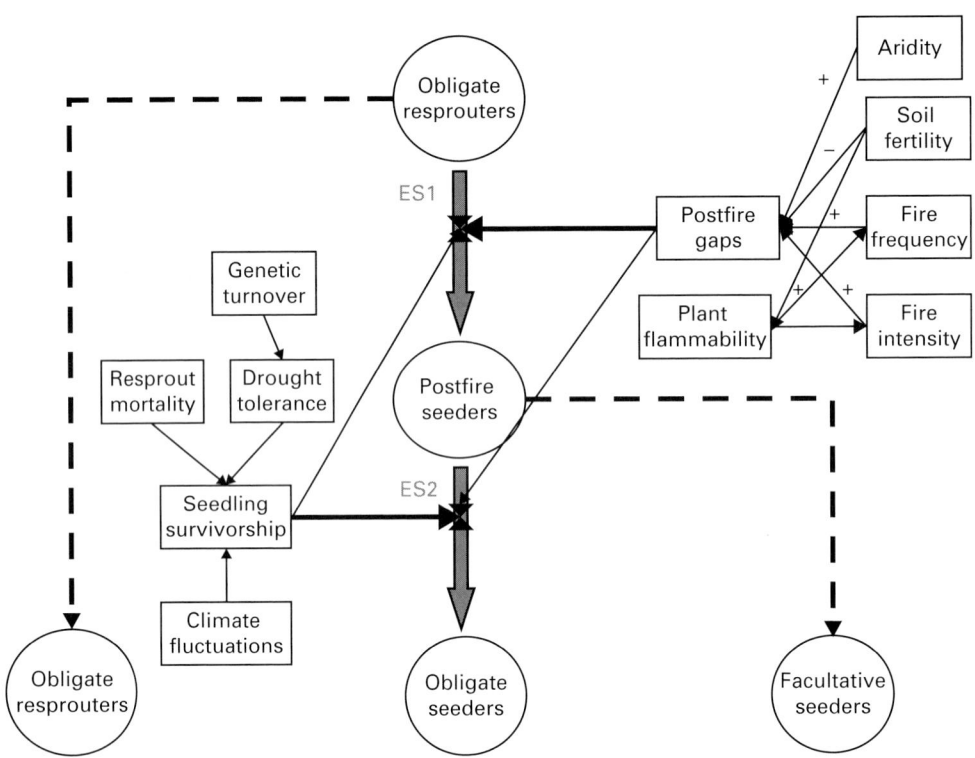

Fig. 9.5 *Model of factors driving the evolution of postfire seeders and the further specialization of obligate postfire seeders. Resprouting is hypothesized to be a trait contributing to persistence on a wide variety of fire-prone landscapes. With increases in the predictability of high-intensity fires capable of creating gaps for seedling recruitment, selection has delayed reproduction to a single postfire pulse of seedling establishment. Further changes in fire regimes that placed a greater premium on seedling recruitment and limited the value of resprouting have led to some lineages abandoning the resprouting habit and evolution of the obligate seeding life history. On rare occasions there apparently have been reversals in the ES2 arrow with obligate seeders giving rise to facultative resprouters.*

just an intensification of those responsible for the initial transition to postfire seeding. However, other factors that affected resprouting mortality and more rapid adaptation to changing climates by reduced generation time may have also played a role.

There has been much interest in trying to sort out the environmental correlates between resprouting and seeding. This question has been framed differently by various authors and there is need for some clarification. Often it is placed in the context of resprouting vs. seeding after fire. This has the potential for creating some confusion since most woody species that only resprout after fire (postfire obligate resprouters) also reproduce by seedlings during the inter-fire interval, so in effect they are seeders, just not postfire seeders. Thus, there are two evolutionary transitions to be addressed:

(1) What factors have led to selection for postfire seeding vs. recruitment independent of fire (evolutionary step 1 in Fig. 9.5)?
(2) What factors have led to selection against resprouting as a postfire regeneration mode (step 2 in Fig. 9.5)?

Fire-independent Recruitment

Postfire obligate resprouters, with fire-independent seedling recruitment, are characteristic of many dominant species in MTC shrublands of the Mediterranean Basin, California and Chile. It is of lesser importance in South Africa and minor importance in Australian shrublands (see Table 3.1). The lesser importance of this functional type in the latter two regions appears to be tied to low soil fertility as it does increase markedly in Australian forests on more fertile soils (French & Westoby 1992; Chalwell & Ladd 2005). This functional type represents a number of lineages with parallel evolution in different MTC regions. For example, obligate resprouters include many of the same genera in both northern hemisphere MTC regions and some of the same families, such as Anacardiaceae, in both hemispheres. Almost all have seedling recruitment that is independent of fire and a surprising number have similar bird-dispersed fruits (Fig. 9.6).

All of these obligate resprouters with fire-independent seedling recruitment appear to have originated in the early or middle Tertiary and have persisted seemingly unchanged to the present (see Chapter 10). They are a class of functional types that Herrera (1992) considered as relictual taxa that failed to adapt to contemporary fire-prone MTC conditions. In his view they represent evolutionary inertia that are present today merely by chance avoidance of random extinctions. Building on this idea, Valiente-Banuet *et al.* (2006) have suggested that their reproductive mode of fire-independent recruitment has persisted through seemingly impossible changes in climate by using "modern" arid-adapted taxa as nurse plants for recruitment.

These papers need serious scrutiny because of their assumptions about the lack of arid sites and thus lack of fire during much of the Tertiary, and their assumption that one can accurately determine time of origin with the fossil record. When dealing with arid land plants this is a very dangerous assumption (see Chapter 10). The idea that fire-independent recruitment is dependent on other taxa for facilitation is supported by a database that is based on expert opinion rather than hard data (Valiente-Banuet *et al.* 2006). Their study only demonstrates that these species can recruit into vegetation dominated by so-called Quaternary taxa, not that they are dependent on such sites for reproduction. In the California chaparral, these fire-independent recruiters not only are not dependent on other taxa for reproduction but they in fact reproduce best on sites dominated by these presumed relict species, and sites dominated by presumed modern species are far less favored (Keeley 1992c).

The perspective presented in both Herrera (1992) and Valiente-Banuet *et al.* (2006) is shaded by the mistaken view that seasonal climates, and fire-prone landscapes, are a new phenomenon associated with Quaternary climate changes.

(a)

(b)

Fig. 9.6 *Brightly colored bird-dispersed fruits typical of many obligate resprouting shrubs/trees with fire-independent seedling recruitment: (a)* Arbutus unido *from maquis in southern France, north-central Mediterranean Basin, (b)* Prunus ilicifolia *in Californian chaparral, (c)* Schinus polygamus *in Chile matorral, and (d)* Rhus *sp. in South African fynbos; this functional type is largely absent from Australian MTV shrublands. (Photos by Jon Keeley.)*

They also have a lack of appreciation for how heterogeneous landscapes could select for the assembly of diverse life history strategies in response to fire, and that this has likely been occurring for a very long time. The relative advantages of fire-independent and fire-dependent recruitment have undoubtedly changed over time but not necessarily along a single trajectory. Although broad portions of the landscape with more predictable fire have likely increased during certain Quaternary phases, the relatively rapid glacial–interglacial episodes would have altered these habitats, perhaps even favoring the vertebrate-dispersed

Fig. 9.6 (*cont.*)

obligate resprouters with better capacity for spatial dispersal that could more easily track landscape changes.

The persistence of obligate resprouters/fire-independent recruiters through the last part of the Tertiary and into the Quaternary illustrates remarkable stasis, a phenomenon not normally interpreted as a result of random extinctions that persist today only by chance as proposed by Herrera (1992). Stasis is hypothesized to result from taxa that occupy smaller, more specialized niche spaces (Ricklefs & Latham 1992), although it is commonly observed in widely distributed taxa (Eldredge *et al.* 2005). We hypothesize that these obligate resprouters have had a long evolutionary association with fire, but under a fire regime somewhat different from contemporary crown fires. These early Tertiary sclerophylls would have occupied pockets of edaphically stressful sites that resulted in less predictable and possibly less intense fires, factors sufficient to maintain selection for vigorous resprouting but not conducive to delaying reproduction to postfire conditions. On contemporary semi-arid landscapes these taxa often dominate in more specialized microsites such as on mesic slopes where rapid resprouting generates substantial competition for seedlings (see Chapters 4, 5 and 6), and this likely has always been a factor selecting for fire-independent reproduction.

Undoubtedly the proportion of landscape favoring obligate resprouters has shifted over geological timescales, but there is little about their current life history that suggests they are maladapted to contemporary landscapes or dependent on other taxa for facilitating recruitment. The strategy of not delaying reproduction to postfire conditions isn't surprising given that obligate resprouters are most abundant on sites less conducive to seedling recruitment. Considering that landscape heterogeneity leads to heterogeneous fire regimes, it is to be expected that evolution would select for, and communities assemble, diverse life history solutions to fire, including both fire-independent and fire-dependent recruitment strategies. Coexistence of evergreen fire-independent recruiters with evergreen fire-dependent recruiters in the same fire-prone environment should be thought of in the same light as winter annuals and summer annuals in summer-drought MTC environments. Neither of these annuals are better adapted than another, but rather both are subdividing resources in ways that promote coexistence (Chiariello 1989).

One way in which resources are subdivided between these two life history types is in different strategies of drought tolerance, which has implications for character syndromes associated with these recruitment strategies (Keeley 1998; Paula & Pausas 2006, 2011; Pratt *et al.* 2007, 2008; Saura-Mas & Lloret 2007). Fire-independent recruiters handle drought by avoiding it whereas fire-dependent recruiters have evolved greater tolerance to soil drought stress.

Physiologically, *drought avoiders* have anisohydric behavior in which daytime leaf water potential closely tracks soil water availability. These plants maintain suitable water potentials by accessing soil moisture sources with deep roots or by limiting growth to periods when shallow soil moisture is available. Evergreen species may become dormant but at the metabolic cost of leaf maintenance under conditions where photosynthetic carbon gain is not significant. Demographically, this makes seedling recruitment precarious on postfire sites due to the extended maturation period required to develop an adequate root system. Therefore, they have specialized reproduction on more mesic sites under the shrub canopy or appropriate gaps generated between fires. In other words, drought avoidance is not a viable strategy for seedlings on open burned sites because it is dependent on deep roots.

Drought tolerators have isohydric behavior and are able to maintain physiological activity and growth under conditions of low water availability by maintaining favorable gradients of water potential to their tissues through morphological, physiological and/or biochemical traits. Such tolerance has the advantage of allowing metabolic activity at levels of low water potential (high water stress) that would not be possible in drought tolerators. This contributes to their ability to establish well on severely drought-prone sites, allowing seedlings to capitalize on the availability of resources after fire. Where fires are highly predictable they have specialized reproduction to a single postfire pulse of recruitment.

Table 9.1 Character syndromes associated with postfire obligate resprouters and obligate seeders in MTC shrublands

Stresses or traits	Obligate resprouters (see Table 3.1)	Obligate seeders (see Table 3.4)
Water stress mode	avoiders (anisohydric)	tolerators (isohydric)
Mechanism	morphological (deep roots)	anatomical physiological
Potential drought-induced mortality		
Adults	very low	moderate
Seedlings	very high	moderate
Recruitment mode	disturbance free	disturbance dependent
Safe sites	under canopy	burned sites
Safe site availability		
In time	annually	10–100 yr intervals
In space	limited	extensive
Seed dormancy	weak/no	deep
Seedbank	transient (< 1 yr)	persistent (10–100 yrs)
Germination cues[a]	none	heat shock or chemicals from char or smoke
Dispersal strategy	spatial	temporal
Mode	vertebrates	passive or invertebrates
Shadow	wide	narrow
Season	autumn–winter	spring–summer
Seed size	large	small
Ecological niche width	wide	narrow
Biogeographical distribution	widespread	localized
Origin	early Tertiary	early Tertiary – Quaternary

[a] Many species from both life histories may have a cold stratification requirement.

Physiological, anatomical and morphological changes were thus likely an important part of evolutionary steps (ES) 1 and 2 in Fig. 9.5.

Coincident with these recruitment patterns is a suite of life history characteristics (Table 9.1), many of which are related to reproduction (García-Fayos & Verdú 1998; Keeley 1998). This is particularly evident in dispersal mechanisms. Fire-dependent reproduction has selected for seeds that fall near the parent plant and are deeply dormant in order to disperse through time, from one fire cycle to the next. Disturbance-independent recruitment requires fruits that are spatially dispersed in order to seek out appropriate safe sites, and thus they are wind or animal dispersed with limited seed dormancy. These dispersal strategies imply very different metapopulation dynamics between these life history modes. These transient seedbanks have the potential for recruiting annually between fires and are likely to be dependent on years of extraordinary rainfall for successful establishment. The observation that the seeds of this functional type are fire sensitive (e.g. French & Westoby 1996) should not be interpreted to mean such taxa are maladapted to fire-prone environments; rather they do not capitalize on fire for recruitment opportunities.

Thus, references to resprouters vs. seeders need to be qualified to emphasize that this is in reference to their postfire response, since both have seedling recruitment. In fire-prone ecosystems it is often assumed that there is a great advantage to postfire seedling recruitment over recruitment between fire cycles. However, recruitment by fire-independent recruiters has not been widely studied and the relative costs and benefits of continuous recruitment between fires vs. delaying reproduction to a single postfire pulse requires careful scrutiny (Gadgil & Bossert 1970). A single snapshot in time for older stands shows seedling (and sapling) to parent ratios within the range of values observed for the single pulse from postfire seeders (Table 9.2).

Although these obligate resprouters all appear to have origins in the early Tertiary, there is no reason to assume those landscapes were uniformly mesic and it is hypothesized these taxa originated on sites at the arid end of the soil-moisture gradient. The notion that these taxa were understory plants in mesic forests is untenable considering the physiology of present-day sclerophylls. As discussed in Chapter 10, those islands of suitable habitat were likely widely

Table 9.2 Seedling/parent ratios for fire cycles; fire-independent recruiter recruitment in long unburned stands and fire-dependent recruiters for the first postfire year

These should not be thought of as species-specific values as they will be shaded by factors such as stand age for fire-independent recruiters and stand age prior to the fire for fire-dependent recruiters.

Functional type	Species	MTC[a]	(# of sites)	Ratio seedling/ parent	Source[b]
Fire-Independent Recruiters (postfire obligate resprouters)					
(total cumulative seedlings and saplings present in stands > 50 yrs of age)					
	Heteromeles arbutifolia	Ca	(1)	77	9
	Prunus ilicifolia	Ca	(2)	33	1
	Quercus spp.	Ca	(5)	5	1
	Quercus suber	Me	(1)	5	16
	Rhamnus crocea	Ca	(4)	3	1
Fire-Dependent Recruiters					
(single postfire pulse of seedlings)					
Soil-stored seedbanks – Facultative seeders					
	Adenostoma fasciculatum	Ca	(25)	64	3
	Ceanothus spinosus	Ca	(10)	75	3
	Fremontodendron californicum	Ca	(62)	93	2
	Malachothamnus fasciculatus	Ca	(6)	546	3
	Malosma laurina	Ca	(5)	194	3
	Rhus ovata	Ca	(8)	80	3
	Ribes spp.	Ca	(8)	683	3
	Trevoa trinervis[c]	Ch	(2)	7	12

Table 9.2 (cont.)

Functional type	Species	MTC[a]	(# of sites)	Ratio seedling/parent	Source[b]
Soil-stored seedbanks – Obligate seeders					
	Arctostaphylos viscida	Ca	(18)	22	2
	Ceanothus crassifolius	Ca	(5)	110	3
	C. cuneatus	Ca	(60)	168	2
	C. greggii	Ca	(3)	208	3
	C. megacarpus	Ca	(9)	101	3
	Cistus ladanifer	Me	(1)	61	11
	C. monspeliensis	Me	(1)	22	10
	Erica umbellata	Me	(1)	80	11
	Rosmarinus officinalis	Me	(1)	27	11
Canopy-stored seedbanks – Facultative seeders					
	Banksia attenuata	Au	(2)	10	6
	B. candolleana	Au	(2)	2	6
	B. menziesii	Au	(2)	1	6
	B. ruscifolia	Au	(1)	19	15
	Widdringtonia nodiflora	SA	(19)	2	4
Canopy-stored seedbanks – Obligate seeders					
	Banksia baxteri	Au	(2)	2	15
	B. coccinea	Au	(2)	6	15
	B. hookeriana	Au	(2)	55	6
	B. leptophylla	Au	(2)	141	6
	B. pulchella	Au	(2)	3	15
	B. speciosa	Au	(2)	7	15
	Beaufortia elegans	Au	(5)	84	7
	Hesperocyparis (Cupressus) forbesii	Ca	(6)	2	17
	Hesperocyparis (Cupressus) sargentii	Ca	(10)	66	5
	Hakea obliqua	Au	(5)	5	7
	H. polyanthema	Au	(1)	8	15
	H. smilacifolia	Au	(1)	26	15
	Protea eximia	SA	(4)	2	14
	P. lorifolia	SA	(4)	8	13
	P. lorifolia	SA	(24)	6	14
	P. punctata	SA	(4)	3	14
	P. repens	SA	(8)	3	13
	P. repens	SA	(16)	2	14

[a] Au, Australia, Ca, California, Ch, Chile, Me, Mediterranean Basin, SA, South Africa.
[b] 1, Keeley 1992a; 2, Keeley et al. 2004; 3, Keeley et al. 2006b; 4, Keeley et al. 1999b; 5, Ne'eman et al. 1999; 6, Enright & Lamont 1989b; 7, Bell et al. 1987; 9, Pelton 1984; 10, Ladd et al. 2005; 11, Quintana et al. 2004; 12, Segura et al. 1998; 13, Bond et al. 1995; 14, Bond et al. 1984; 15, Lamont et al. 1999; 16, Pausas et al. 2006d; 17, Zedler 1981.
[c] Recruits after fire but is not "fire-dependent."

dispersed in space. With the origin of postfire seeders, characteristics associated with enhanced flammability (Fig. 9.5) were likely to be an important driver not only for the seeders but for providing suitable habitat for obligate resprouters as well. One of the striking differences between MTC regions is the disproportionately high representation of obligate resprouters in northern hemisphere systems and the depauperate representation in South Africa and Australia. This may derive from different origins for fire-prone sclerophyll shrublands in these regions and in particular the early origin of highly flammable sclerophylls over large expanses of nutrient-poor soils in these southern hemisphere MTC regions (see Chapter 10).

One of the potential genetic effects of the *persister* life history is that resprouting genets may be extremely old and the potential exists for greater accumulation of deleterious alleles from somatic mutations (Wiens et al. 1987). This high genetic load would potentially result in reduced seed set and enhanced selection for resprouting (Lamont & Wiens 2003). This hypothesis could explain the anomalously nearly non-existent seedling recruitment in some resprouters in Cape fynbos (*Retzia* spp.), *Banksia elegans* in Western Australian woodlands (Bond & Midgley 2003), and in the monotypic genus *Xylococcus* in California chaparral (J.E. Keeley personal observations).

Fire-dependent Recruitment

This describes postfire seeding that is restricted to a single pulse after fire, and is well developed in many woody genera in MTC shrublands, as well as some MTV crown fire shrublands distributed in other climatic regimes, such as Florida scrub and southeastern Australian shrublands. It has been widely considered to be a relatively recent Quaternary phenomenon (Raven & Axelrod 1978; Axelrod 1989; Herrera 1992; Ackerly 2004a; Valiente-Banuet et al. 2006); however, this designation is largely based on the absence of Tertiary fossils for these taxa. The dependence in science on negative evidence is always weak but especially so when it comes to the fossil record, which is highly biased against semi-arid fire-prone floras. As discussed in more detail in Chapter 10, despite the lack of a Tertiary record, many of these presumed "Quaternary species" originated much earlier in the Tertiary; for instance, molecular clock estimates place the origin of *Fumana* and *Helianthemum* much earlier than the fossil record dictates (Guzmán & Vargas 2009a). In North America the highly disjunct chaparral pockets far outside the MTC in Arizona and northeastern Mexico (see Fig. 5.1) containing postfire-seeding *Arctostaphylos pungens* and *Ceanothus greggii* suggest early origins outside California, prior to their widespread appearance in California Quaternary sediments. The near total absence of Tertiary fossil floras from the southwestern United States makes it difficult to test that idea, but it does illustrate the extraordinary bias involved in using fossil records for pinpointing origins.

Although there is reason to believe the timing of the evolution of the postfire seeding trait might have varied across different MTC regions, the primary factors may have been very similar. The most important was the reliable creation of gaps after fire, which both limited survivorship of resprouting shrubs and increased available resources for seedlings (Fig. 9.5). Increasing aridity and decreasing soil fertility would increase gap size and also affect plant structure in ways that increase flammability, which in turn would affect fire intensity and frequency, both of which would increase gap size and predictability.

Delaying reproduction to postfire conditions requires the accumulation of a seedbank during the inter-fire interval; two modes of seed storage include soil-stored and canopy-stored seedbanks. Soil-storage is widespread throughout most MTC crown-fire ecosystems whereas serotiny is largely of importance in South Africa and Australia (see Chapter 3).

Soil-stored seedbanks
Heat shock, chemicals in smoke and charred wood provide signals that trigger germination in seeds from fire-prone environments. These mechanisms enhance fitness by cueing germination to postfire conditions when light, water and nutrient resources are abundant. Thus traits enhancing postfire-seeding mechanisms are viable candidates for true fire adaptations.

Heat breaks dormancy of hard-seeded species with water-impermeable seed coats by several mechanisms: heat shock can trigger germination by rupturing the seed coat layer, by shortening after-ripening, and by desiccation (Brits et al. 1993). Species have different heat dose optima for germination, and in some cases there is a tendency for obligate seeders to have higher seed dormancy, higher heat-stimulation requirements and lower heat-induced seed mortality than facultative seeders (Paula & Pausas 2008; F. Moreira & J.G. Pausas unpublished). However, this is not the case in all floras (Keeley 1987).

Heat shock is not a specific cue to fire as soil heating by sun may also trigger germination on unburned open sites with bare mineral soil. Although soil temperatures experienced on sun-exposed sites are much lower than those generated by fires, the heat dose (i.e. temperature × time, e.g. Paula & Pausas 2008) may be similar due to the long exposition time on sun-exposed sites. However, the relationship between seed germination and the different combinations of temperature and exposure time are complex and differ between species (Keeley 1991).

Chemicals released by combustion and transferred to seeds by smoke or charred wood (here referred to as "smoke") appear to be a highly specific germination cue for fire and indicative of a rather specific fire adaptation. Smoke-stimulated germination in species from non-fire-prone ecosystems has been cited as contrary evidence (Pierce et al. 1995), but this conclusion does not consider the fact that smoke is a complex mixture of thousands of chemicals, many of which occur in diverse ecological settings and stimulate many plant processes (Keeley & Fotheringham 2000). Hopper (2003) raises questions about whether smoke-stimulated germination constitutes an adaptation because in the Australian flora it appears to have arisen numerous times in taxa with an origin as far back as the Cretaceous. A similar pattern is evident in the California flora as a number of smoke-stimulated species appear basal in lineages with a long Tertiary history

(Pausas & Keeley 2009). However, considering that fires have been well documented since the Paleozoic, we see little reason to not expect traits such as smoke-stimulated germination to have had a very early origin.

Smoke-stimulated seed germination of species that regularly recruit seedlings after fire is reported for diverse angiosperms and from MTC regions of South Africa, Australia, California and the Mediterranean Basin, and has been invoked as an example of convergent evolution that has arisen multiple times in different lineages (Keeley & Bond 1997). On the basis of other examples of physiological convergence such as C_4 photosynthesis with a variety of different biochemical pathways in different lineages (Roalson 2007), this model would predict that unrelated species would evolve mechanisms that are triggered by different components in smoke, and there is some support for that idea (Keeley & Fotheringham 2000). The mechanism of smoke overcoming dormancy does not appear to be the same in all smoke-stimulated taxa. In some taxa it works directly on physiological barriers to germination and in others it is associated with embryo development (Keeley & Fotheringham 1997, 1998) and in others it overcomes external environmental allelopathic inhibitors (Krock et al. 2002).

In contrast, Flematti et al. (2004) contend that smoke-stimulated germination in postfire-recruiting species represents a universal response to a single organic molecule present in smoke. This compound, karrikinolide, a type of butenolide, draws such a claim because it turns out to be an effective germination stimulant in not just species that recruit after fire but a vast array of plants where fire is not involved in their life history, for example many agricultural species (van Staden et al. 2000). It has been proposed that this is a universal seed germination trigger that can be produced by other forms of disturbance, thus has likely been part of plant evolution for much of the Tertiary (Chiwocha et al. 2009).

Although this is clearly an intriguing finding, the conclusion in papers by Flematti and Chiwocha that this is the universal trigger for postfire seed germination is premature. Smoke-stimulated fire ephemerals are known that fail to respond to this butenolide (Downes et al. 2010). Other chemicals in smoke are also known to trigger germination (Keeley & Fotheringham 1997). The primary question about karrikinolide and other butenolides is how can they cue germination to precisely the immediate postfire year when they appear to be widespread in nature and can trigger germination at extraordinarily low concentrations (10^{-7} M or lower)? We have no field studies to date showing whether even these levels are present in soils after fire, but more importantly that they disappear after the first couple of years when germination generally ceases. These butenolides are apparently produced by other disturbances in the environment, yet a great many postfire species will not germinate after disturbances other than fire. Additionally, smoke also produces a butenolide that inhibits the stimulatory effect of smoke and it is unknown what balance of different butenolides is needed to trigger germination (Light et al. 2010). Thus, there is the need for research on the physiological and ecological role of this and other compounds in smoke.

Canopy-stored seedbanks

Serotiny is a special case of bradyspory in which seeds are maintained in "cones" (here we take botanical liberties by using this term for both true gymnosperm cones and serotinous angiosperm fruits) on the parent plant and released suddenly in response to an environmental trigger. It is a trait that appears frequently in gymnosperms (Pinaceae and Cupressaceae) in most MTC regions, but in the southern hemisphere it is also found in about a dozen angiosperm families (see Chapter 3) and is widely distributed in the Proteaceae (see Box 7.1). Serotiny involves delayed seed release and often seeds may be stored within cones for decades and then rapidly dispersed following fires in canopy-consuming crown fire regimes. Postfire seedling establishment is very successful, in contrast to the occasional seeds dispersed during the inter-fire interval, which have very low seedling survivorship (Bond 1984). Although delayed seed release occurs in some arid-land herbaceous species as a mechanism for timing germination to rainfall events, including annuals (Ellner & Shmida 1981), whereas woody species that accumulate seeds from several annual crops appear almost exclusively in fire-prone environments.

Thus, it would seem inescapable that fire was the selective factor responsible for delaying reproduction to a single pulse of postfire recruitment. However, Axelrod (1980) contended that in the case of serotinous pines, fire was a recent anthropogenic factor and that serotiny evolved in response to drought. Not only did he fail to recognize the importance of fire throughout land plant evolution, but he also did not recognize that serotiny, which times cone opening to postfire conditions, would not be an effective means of avoiding drought; rather drought should select for a bet-hedging strategy with multiple seed dispersal events throughout the plant's lifetime. Despite the inescapable importance of fire as a selective force, fire need not be the only selective factor and sometimes it may be acting in concert with other stresses such as granivory (Lamont *et al.* 1991).

Northern hemisphere gymnosperms in crown fire regimes are mostly postfire seeders with serotiny, and none have soil-stored seedbanks, suggesting phylogenetic constraints on adapting recruitment to postfire conditions in crown fire regimes. This may be tied to the presence of cones in ancestral species that pre-adapted some taxa to canopy storage, or to the lack of appropriate seed coat layers that limited the possible evolution of seed dormancy in the soil. The latter may be a consequence of the apparent origin of gymnosperm serotiny from arboreal ancestors in surface fire regimes where long-term soil-stored seedbanks have never been selected (Keeley & Zedler 1998; Schwilk & Ackerly 2001). However, phylogeny and ecology may both be involved since fire return intervals are commonly much longer in these northern hemisphere MTC regions and, reflecting the arboreal ancestry, the serotinous conifers are all very long lived, which would be critical for serotiny under such fire regimes. In the southern hemisphere where crown fire regimes prevail, many angiosperm plant families have numerous serotinous taxa. Such taxa are in lineages that also have taxa with soil seed storage, suggesting phylogenetic constraint is an unlikely explanation for canopy seed storage.

One of the key differences between soil storage and canopy storage strategies is resilience to unpredictable fire cycles. When the fire interval exceeds the life span of the parent plant, soil storage is a viable strategy (e.g. Keeley et al. 2005b) but canopy storage is not (Lamont et al. 1991; Enright et al. 1998; Lamont & Enright 2000). Another important distinction is inherent differences in potential lifetime reproductive output. Canopy seed storage is limited by determinant growth patterns that restrict lifetime seed accumulation to the available canopy space, whereas seed output is potentially much greater in species that disperse seeds for soil storage. Seedbank studies for canopy-storage plants have shown that the magnitude of seedbank storage is between 10 and 10^2 (occasionally 10^3) seeds per plant (Gill & McMahon 1988; Cowling et al. 1987; Witkowski et al. 1991; Lamont et al. 1999; Bradstock & Cohn 2002b). Seedbank estimates are much more difficult to obtain for soil-borne seeds and few have provided seed/parent ratios, but in a single good year most are capable of producing between 10^3 and 10^4 seeds per plant (Keeley 1977; Luis-Calabuig et al. 2000). However, seed output and seed storage are difficult to compare because competitive outcomes are not determined solely by numbers but by seed size and seedling attributes. Nonetheless, postfire seedling recruitment is often much greater for soil-storage species than for canopy-storage species (Table 9.2).

The evolutionary forces selecting serotiny over soil storage may be diverse (Lamont et al. 1991), but some hypothesized factors seem more likely than others. For instance, it has been suggested that postfire serotinous seed release is a form of mast seeding and this synchronized seed release may satiate postdispersal predators (O'Dowd & Gill 1984). It seems likely that predation is an important driver behind the high incidence of serotiny in South African and Australian MTC ecosystems. Indeed, this pressure has been a factor selecting for the tough woody cones in diverse lineages. Despite this extra investment in protective structures, seed predators can destroy a large fraction of the canopy seedbank (Groom & Lamont 1997; Mezquida & Benkman 2004).

Lamont and Enright (2000) hypothesized that greater reliability of precipitation in Australian and South African MTC ecosystems was a factor selecting for serotiny because seed germination is restricted to the year of seed release. In their view less reliable rainfall as in California and the Mediterranean Basin would favor soil-stored seeds, which could hold over to subsequent years if the first postfire year were unfavorable. However, there is little empirical support for this model as seed carryover after the first postfire year is largely inconsequential in many soil-stored seedbanks. For example, in chaparral shrubs the first postfire year recruitment typically comprises 95–100% of the total seedling recruitment during early seral stages, even if rainfall is below average (see Table 3.3). In addition, if rainfall unpredictability were a critical selective factor, there is nothing about the serotinous habit that would prevent gradual release over multiple years (e.g. in the serotinous *Hesperocyparis* (*Cupressus*) *forbesii* second postfire year seedling recruitment comprises 5–10% of the total, J.E. Keeley unpublished data).

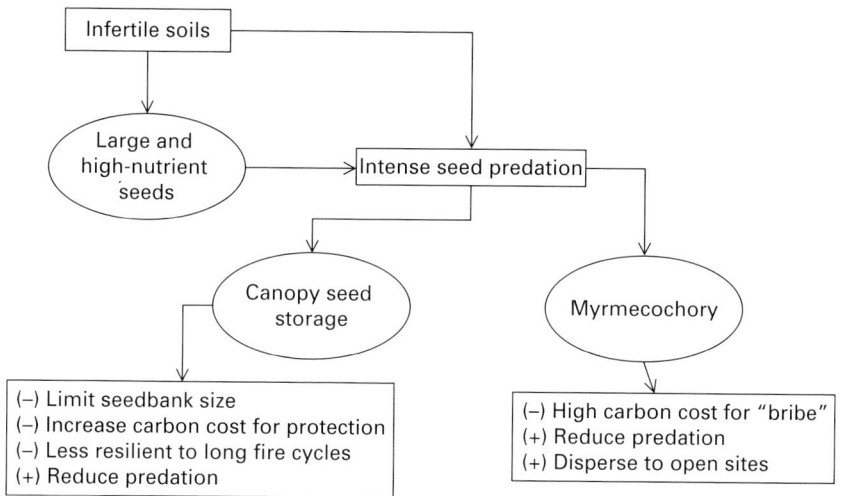

Fig. 9.7 *Hypothesized effects of nutrient-poor soils in fire-prone ecosystems on serotiny and myrmecochory in Australian and South African MTC ecosystems.*

Perhaps the greatest weakness of the rainfall reliability theory is that serotiny in some lineages appears to be Tertiary in origin with important speciation events occurring over the last 10 Ma (see Chapter 10). However, the subtle differences in rainfall predictability between the northern and southern hemisphere MTCs have likely fluctuated greatly over that period of time, in concert with Milankovitch cycles and other longer-term changes (Bennett 1990, 2004).

One factor that has persisted through much of the Tertiary and ties together centers of serotiny in Australian heathlands and South African fynbos is the widespread occurrence of highly weathered nutrient-poor soils (see Fig. 1.5), and thus nutrient-poor fire-prone vegetation (Fig. 9.7). There may be a link between soils and serotiny since the oligotrophic soils in heath and fynbos put a premium on nutrients for seeds (Stock *et al.* 1990; Vaughton & Ramsey 2001; Groom & Lamont 2010). Seedling establishment in these soils is favored in species with larger seeds with greater nutrient levels (Stock *et al.* 1990; Jurado & Westoby 1992; Bond *et al.* 1999) and this makes them a target for seed predators already stressed by the oligotrophic conditions in this ecosystem. This pressure is even greater considering that the high nutrient stores in seeds will limit seed number over that compared with less nutrient stressed soils. In general, these costs are hard to recoup over the higher production of smaller seeds (Moles *et al.* 2004) typical of species with soil-stored seedbanks, suggesting there are other rewards to canopy storage.

It is hypothesized that interactions between selection for large seeds and avoidance of seed predation have been important drivers in the evolution of serotiny in these oligotrophic environments (Fig. 9.7). Serotiny puts significant amounts of carbon, which is more readily available than other nutrients, into protecting the seeds against predispersal predation and since total seedbank size is generally lower than in soil-stored seedbank species, this predation presents a substantial threat to successful reproduction. In this context, the absence of this strategy in Chile and the limited role in northern hemisphere MTC regions may not be surprising since they do not face such nutrient-limited conditions, and much larger seedbanks are feasible for soil-stored seeds, as indicated by generally greater seedling recruitment density. Another potential factor is the much higher fuel loads and higher fire intensities experienced in MTC shrublands on more fertile soils (see Chapter 2), which would make aerial seed storage more vulnerable than in shrublands of less fertile soils. For example, it was estimated in the serotinous fynbos shrub *Widdringtonia nodiflora* that on the majority of sites > 80% of the cones were destroyed by high fire intensity (Keeley et al. 1999b).

Seed dispersal
Postfire seeders disperse seeds in time more than in space (Keeley 1992d) and thus most have relatively passive dispersal. One exception is myrmecochory (a dispersal syndrome characterized by specialized seeds that attract ant dispersers); the contrast between MTC regions in the importance of this form of dispersal is striking. On infertile soils in Cape region fynbos and Western Australian heathlands there are literally hundreds of species that disperse seeds by myrmecochory (Bond & Slingsby 1983). Although myrmecochory is found in the other MTC regions, the total number of species in all of the other three is about an order of magnitude lower than in either fynbos or heathland. The connection between myrmecochory and poor soils is inescapable, even within regions (Milewski & Bond 1982), and stands as one of the more dramatic cases of evolutionary convergence (Lengyel et al. 2010). One of the longstanding hypotheses has been that in order for seedlings to establish successfully they need to find pockets of high-nutrient soil and myrmecochory is a mechanism for targeting such sites (Westoby et al. 1991). However, tests of this hypothesis in both Australia and South Africa have failed to support it (Rice & Westoby 1986; Bond & Stock 1989; Schatral et al. 1994).

An alternative strategy to targeting dispersal to high-nutrient sites is to produce large seeds with high nutrient concentrations (Stock et al. 1991), but one consequence is that such seeds are ready targets of predation on the soil surface (Andersen 1982). Two means of reducing this impact would be serotiny and myrmecochory (Fig. 9.7). The advantages of serotiny for reducing predation are discussed above and the means by which myrmecochory reduces predation is as follows. Myrmecochorous ants remove seeds from exposed sites on the soil surface and thus reduce predation by other ground-dwelling granivores (Bond & Breytenbach 1985; Westoby et al. 1991). The lipid-rich elaisome is the reward these ants receive for not consuming the seed (Keeley 1992b). This elaisome reward directly reduces predation as well as indirectly affecting predation since the presence of these myrmecochorous species would be expected to displace other ground-dwelling invertebrates and vertebrates that do prey on seeds (e.g. Brown & Davidson 1977; Andersen & Yen 1985). Myrmecochorous ants further reduce predation by producing chemicals deceptive to other seed-predator ants (Pfeiffer et al. 2010).

Promoting Postfire Recruitment with Enhanced Flammability

We hypothesize that the evolution of postfire seeding has been coincident with the selection for traits that enhance flammability and thus increase sites for recruitment (Fig. 9.5). In the paleorecord there is evidence that changes in plant structure are associated with increased fire activity (Belcher *et al.* 2010), but this is interpreted as the result of climatic change altering plant structure and, as a consequence, changing fire regimes. The idea that plant structure might evolve to enhance flammability is something rather different. The idea was first proposed by Mutch (1970), and although he saw a connection between fire-prone communities and species with enhanced flammability, he didn't tie it to postfire seeding. Evolution of flammability has been criticized on a number of grounds (Snyder 1984; Troumbis & Trabaud 1989). As with many fire-adaptive traits, it is reasonably argued that there are other selective factors such as drought or herbivory that could account for flammability of leaf and stem characteristics. Another argument is that fire is a complex of biotic and abiotic factors and thus not affected by individual plant traits. Perhaps the harshest criticism has been that selection for flammability is group selectionist and unlikely to evolve in the context of inclusive fitness theory that forms the basis of modern evolutionary theory. This criticism may have been avoided had the evolution of flammability initially been tied to enhancing postfire sites for fire-dependent seedling recruitment.

This last issue was directly addressed by Bond and Midgley (1995) who placed flammability in the context of postfire reproduction and provided sufficient reasons to consider it a possibility that flammability characteristics could arise through selection on individual fitness. They reasoned that if plants with fire-dependent reproduction evolved characteristics that enhanced flammability or intensity, which in turn created greater gaps for postfire seedlings by negatively impacting obligate resprouters, then such traits could evolve within inclusive fitness theory. Modeling studies suggest several genetic mechanisms for evolving flammability and contend that it is a form of niche construction that alters environments to enhance fitness (Kerr *et al.* 1999; Schwilk & Kerr 2002). Some of the traits that contribute to greater flammability in postfire recruiter species include volatile compounds that promote ignition and fire spread (Rundel 1981b), retention of dead wood in the canopy that enhances combustion of associated live foliage (Zedler 1995a; Schwilk 2003), and structural differences in leaf and stem placement that likewise promote combustion (van Wilgen *et al.* 1990b).

Thorough tests of this hypothesis are lacking; however, in chaparral, some postfire seeder species such as *Adenostoma fasciculatum* and species of *Arctostaphylos* and *Ceanothus* fail to self-prune dead branches. *Ulex parviflorus* (see Fig. 4.5) is an outstanding example of an obligate seeder from the Mediterranean Basin that retains large amounts of dead biomass (Baeza *et al.* 2006, 2011). It is difficult to imagine other selective factors accounting for such retention. The selective value for retention is that it greatly increases fire ignition and intensity, over what would occur if those branches were dropped and remained as surface fuels (Schwilk 2003). Fire severity measures also show a positive relationship with obligate-seeding

Ceanothus seedling density in chaparral (Keeley *et al.* 2005b) and in two species of Proteaceae in fynbos (Bond *et al.* 1990), and increased growth rates of seedlings in Mediterranean serotinous *Pinus halepensis* (Ne'eman 2000; Pausas *et al.* 2003). In addition, associated non-fire-recruiting obligate seeders generally drop dead branches (W.J. Bond unpublished data). Similarly, tree species with flaky barks that readily carry fires into the crown typically have good postfire regeneration capacities, such as the stringy-bark group of *Eucalyptus* (Burrows 2001). However, in this complex genus the evolutionary nexus between postfire seeding and bark is complex as many bark types are associated with postfire seeding.

Keeley and Bond (1999) have hypothesized enhanced flammability was a selective force behind semelparous flowering cycles of many bamboos. Delayed reproduction and semelparity (monocarpy) generate massive amounts of fuel, and the gregarious clonal distribution produces a contiguous fuel load, all of which encourage the propagation of high-intensity fires. This effectively eliminates canopy trees and canopy tree recruitment. Mast flowering synchronizes seedling recruitment with the creation of safe sites in canopy gaps. Semelparity, in addition to setting the stage for disturbance, also concentrates reproductive allocation to the optimum time for recruitment.

Another selective basis for flammability is that by retaining dead branches and foliage in the canopy it reduces soil heating around soil-stored seedbanks and underground vegetative structures (Gagnon *et al.* 2010). This hypothesis has not been tested and although it may apply to flammability characteristics in some taxa, it seems unlikely that it applies very widely. For example, obligate seeding taxa may be split between those with soil-stored seeds and others with canopy seed storage. One would expect that this would select for very different flammability characteristics, but a casual inspection of MTC communities fails to support that expectation. One of the intriguing aspects of this theory is that it provides a mechanism for obligate resprouters to compete against seeders by not retaining dead fuels in the canopy and accumulating sufficient surface fuels to kill soil seedbanks. Since these obligate resprouters don't recruit after fire they may reduce seedling and sapling competition. This effect is most evident where surface fire regime forests are juxtaposed with crown fire shrublands. The prolific postfire regeneration in the latter contrasts with the relatively depauperate regeneration in the former, and the sharp boundary is controlled in part by the higher soil-surface fire intensity in the forest communities. Even within surface fire regimes pines that recruit in early postfire years generate surface fuels capable of achieving substantially higher surface temperatures than litter from species such as oaks that do not recruit after fire (Williamson & Black 1981).

Evolution of Non-resprouting Obligate Seeding Taxa

A second significant evolutionary innovation in postfire seeding was to eliminate resprouting altogether and depend entirely on postfire seedlings (Fig. 9.5). Wells (1969) first made the surprising observation that many woody shrubs in California

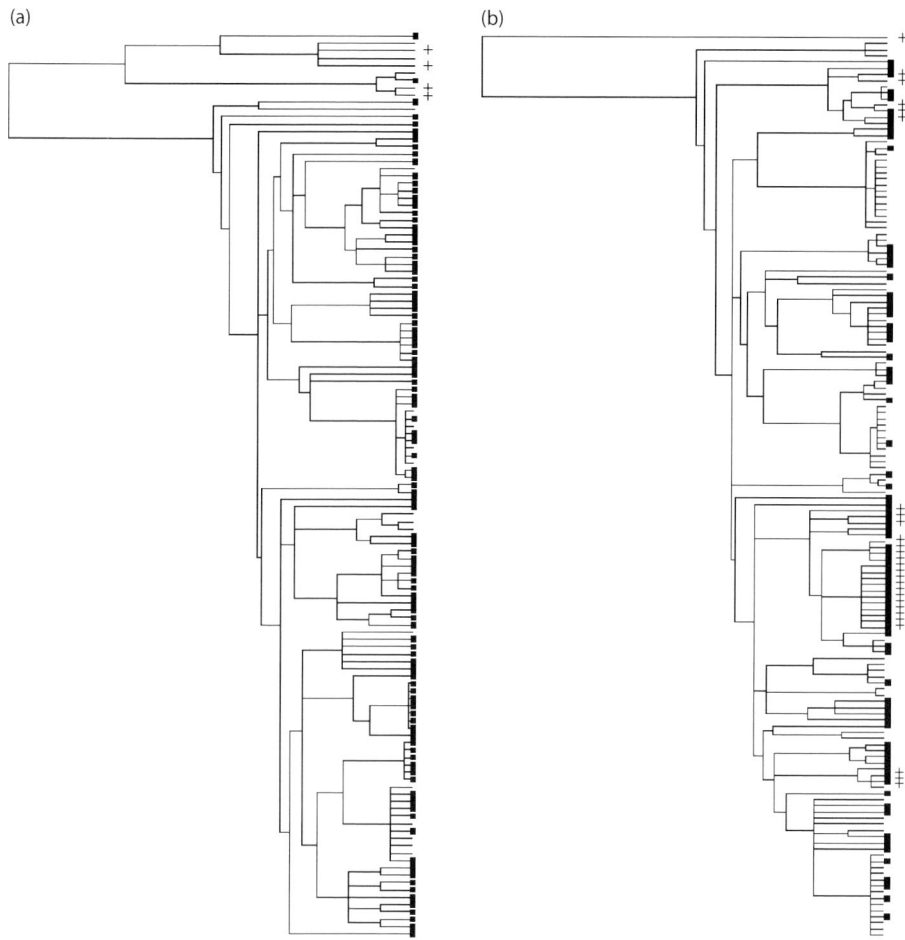

Fig. 9.8 *Distribution of resprouting (squares) and serotiny (crosses) in the phylogeny of (a) the main woody species in fire-prone ecosystems of California and eastern Spain (Mediterranean Basin), and (b) the main woody species in the mallee vegetation of southeast Australia. The phylogenetic signal (niche conservatism) of resprouting is highly significant for both trees (P < 0.01). (Elaborated from Pausas* et al. *2006b; Pausas & Bradstock 2007.)*

chaparral failed to resprout after fire, despite this being a nearly ubiquitous trait in woody dicots (e.g. Fig. 9.8). In this community there are roughly 15 genera of shrubs; all have resprouting taxa and most are obligate resprouters with fire-independent recruitment. Two of the genera, *Arctostaphylos* and *Ceanothus*, have postfire seedling recruitment; some taxa in these two genera are also resprouters but others are non-resprouters and are termed obligate seeders. Subsequent studies (Le Maitre & Midgley 1992; Pausas *et al.* 1999; Bell 2001) have reported obligate seeding species in woody genera from crown fire regimes in MTC parts of South Africa, Australia and the Mediterranean Basin (see Table 3.4). These shrubs recruit seedlings in the first postfire year and most are in genera that include species with both resprouting and postfire seedling recruitment. Most obligate seeders lack any capacity for resprouting; however, facultative seeder populations exhibit a range of resprouting levels and this is a function of species-specific characteristics and tolerances to fire, as well as to characteristics of the fire event, particularly fire intensity and degree of prefire and postfire drought.

Wells (1969) contended that resprouting was an ancestral trait and loss of the ability was derived. In a broad sense this is certainly true as nearly all woody angiosperms have the capacity to resprout and where obligate seeding has evolved, the closest sister genera are nearly always resprouters. Within the very large genus *Erica*, there is evidence that obligate seeding arose as a result of suppression of bud development responsible for lignotuber formation (Verdaguer & Ojeda 2002a). However, within some lineages there is evidence that resprouting has been derived from obligate seeding ancestors (Bond & Midgley 2003; Boatwright *et al.* 2008). This even appears to be the case at the subspecific level as illustrated by Californian *Arctostaphylos* spp., which have both resprouting and obligate seeding subspecies (J.E. Keeley unpublished data). Also, aberrant resprouting individuals of non-resprouting species have been observed in Lamiaceae in the Mediterranean Basin (J.G. Pausas personal observations), in Ericaceae and Fabaceae in the South African Cape region (Ojeda 1998; Schutte *et al.* 1995) and *Banksia* and *Hakea* in Australia (R.A. Bradstock personal observations).

Since the initial observations of highly diverse obligate seeding genera in MTC regions, at least five hypotheses have been put forth to explain the origin of this life history.

(1) Wells (1969) hypothesized that the obligate seeder mode had been selected for because it forced the population to turn over 100% of the individuals every fire-induced generation and was able to evolve more quickly, and Raven (1973) suggested such obligate seeders were thus better able to track changes in the environment from generation to generation.
(2) An ecologically based explanation was proposed by Keeley & Zedler (1978; see also Keeley 1986) who contended that obligate seeding populations would gain an advantage under conditions where it was more advantageous to regenerate after fire by seedlings than by resprouts (Fig. 9.5). On one hand the postfire strategy is one of *fire recruiter* and on the other hand one of a *fire persister* (Keeley 1991). Indeed, Bond & Midgley (2001) have proposed the concept of the *persistence niche*; thus, the balance between factors favoring recruiting over persisting will drive the evolution of reprouting and obligate seeders. What are those conditions? In the immediate postfire environment seedlings are at a distinct disadvantage in close competition with resprouts so conditions that favored large postfire gaps would favor seedlings and not resprouts. This model hypothesizes that obligate seeding is selected for when postfire environments produce safe sites for seedlings and unfavorable conditions for resprouts.

There are a multitude of factors that could work against postfire survivorship of resprouters, including weak resilience to high fire intensities and long fire-free periods, since resprouting is only adaptive when the fire cycle is shorter than the plant's life span, whereas soil-stored seedbanks can persist under very long fire cycles. Whether or not such factors favored obligate seeders over resprouters would be controlled strongly by the mode of seed storage. Long fire return intervals might favor soil-stored seedbanks but definitely not canopy storage as these taxa, like the resprouters, are at a disadvantage if they die before fire. This is one of the key reasons why models based solely on fire frequency are not likely to fully capture the conditions selecting for the non-resprouting mode (see #5 below).

Other situations that would produce gaps would be along arid margins where growth and survival of shrubs over many decades is precarious and postfire gaps are large, and this is generally consistent with the distribution of postfire seeders (Keeley 1977; Pausas 2001; Clarke & Knox 2002; Meentemeyer & Moody 2002). Some factors driving the obligate seeding mode are illustrated by several Californian *Arctostaphylos* species, which have both resprouting and obligate seeding subspecies: *A. parryana*, *A. peninsularis* and *A. manzanita* (J.E. Keeley unpublished data). Both morphotypes are found in ecosystems subjected to crown fires but the obligate seeding taxa are distributed in open woodland habitats where postfire gaps are large and the resprouting taxa are in dense chaparral where close competition after fire has likely selected for resprouting over seeding.

A more generalized version of this idea is the resource-based model proposed by Clarke that hypothesizes trade-offs between resource availability and resprouting (Clarke *et al.* 2005). Presumably as resource levels increase, resprout success increases due to unfavorable conditions for seedling recruitment in the face of resprout competition. In addition, such resource gradients affect fuel production and as fuels increase they potentially increase fire frequency (Pausas & Bradstock 2007). Thus, increased resprouting dominance at higher resource levels may be due to both reducing opportunities for seedling recruitment as well as creating higher fire frequency that extirpates non-resprouting populations prior to reproductive maturity.

Ecologically, it would appear that facultative seeders include the best traits of both postfire resprouting and seeding. Thus, the evolution of obligate seeding would appear to hinge on a selective advantage to losing the resprouting mode (Fig. 9.5). This would require that there be costs to diversion of resources to resprouting, and these costs would need to be sufficient to reduce seed production and thus reduce seedling recruitment, which ultimately would translate into reduced establishment of adults. In some facultative seeders there does not appear to be trade-offs between the extent of prefire flowering and postfire resprouting (Cruz & Moreno 2001b). One trade-off may be in earlier flowering of juvenile seeders, perhaps due to different patterns of resource allocation from resprouters (Bell & Pate 1996; Verboom *et al.* 2004; Schwilk & Ackerly 2005). In general, there is no consistent pattern of greater seed output by obligate seeders over facultative seeders (Keeley 1977; Enright et al. 2007).Thus, ecological models need to address the cost:benefit ratio between resprouting and seeding in order to find a solution to why selection has worked to generate the non-resprouting mode.

(3) In considering trade-offs, however, it has not been widely appreciated that the costs of resprouting may not lie entirely in diversion of resources at the stage of seed production. Diverting resources during seedling growth toward the production of adventitious buds and for starch storage to lignotubers could reduce seedling survivorship under very arid conditions and thus select for the non-resprouting mode with a greater capacity for deep roots and taller shoots. Indeed, greater seedling survivorship is commonly observed in obligate seeders over facultative seeders (Keeley & Zedler 1978; Thomas & Davis 1989; Lloret et al. 1999b). Comparisons of starch storage in roots consistently show much greater concentrations in juvenile resprouters than in seeders in Australian, South African and Californian shrubs (Pate et al. 1990; Bell & Ojeda 1999; Verdaguer & Ojeda 2002b; Knox & Clarke 2005; Schwilk & Ackerly 2005). However, it should be noted that such postulated differences in allocation patterns and growth rates are not universally observed in resprouter and non-resprouter species (Chew & Bonser 2009).

Trade-offs in resource allocation at the seedling stage have the potential for explaining the pattern of increasing obligate seeders in MTC summer-drought conditions. On fertile mesic sites under summer-rain conditions stiff competition with grasses greatly favors resprouting seedlings over obligate seeder seedlings (Clarke & Knox 2009). Level of starch resources diverted to lignotubers has a demonstrable effect on resprouting following top removal (Walters et al. 2005), and this allocation strategy could be very effective under conditions of intense seedling herbivory. These different patterns of internal resource allocation might be accentuated on very oligotrophic sites such as in South African fynbos, where Ojeda (1998) hypothesized that diversion of resources for lignotubers takes away from resources needed to develop mycorrhizal associations in seedlings. In his view, this cost is sufficient to favor non-resprouting seeders under summer-drought conditions, thus explaining the depauperate representation of resprouting in the MTC part of South Africa, and their much greater representation in summer-rain climates.

(4) Another hypothesized cost associated with resprouting is that it enforces a multistemmed growth form incapable of competing with single-stemmed non-resprouters in height (Midgley 1996; Kruger et al. 1997). Although multistemmed resprouting shrubs often do not attain the maximum heights of single-stemmed non-resprouting shrubs, there is nothing about the resprouting mode that prevents selection for a monopodial growth form if there was always a competitive height disadvantage to being multi-stemmed (e.g. Fig. 9.9). Also, the fact that the tallest tree in the world (*Sequoia sempervirens*) is a resprouter doesn't help that case. Additionally, the selective factors determining height are complex and involve trade-offs in photosynthetic rates and shade tolerance. In some cases obligate

Fig. 9.9 *Monopodal growth form of resprouting* Arctostaphylos rainbowensis *from California chaparral with 50-cm diameter lignotuber; meter stick for scale. (Photo by Jon Keeley.)*

seeding shrubs have much higher photosynthetic rates than resprouters, lower leaf area indices, higher photosynthetic compensation points and lower shade tolerance (Parker 1984; Verdú 2000; Ackerly 2004b). Resprouters and obligate seeders represent very different character syndromes and simple measurements of height alone are unlikely to explain competitive advantages associated with regeneration mode (Givnish 1988).

Although no doubt a non-resprouter height advantage has selective value in some environments, for example loss of resprouting has been implicated as a trade-off for height growth in *Eucalyptus regnans* by Waters *et al.* 2010, it is not the broad geographical explanation for the evolution of the non-resprouting mode these authors have suggested. If a competitive height advantage were the main driver behind evolution of non-resprouting, one would expect this to increase along a resource gradient where shrub density increases. However, in MTV shrublands under various climatic regimes, resprouters dominate as resources increase and non-resprouting seeders become more important on open semi-arid sites where competition for height is of much less importance (Keeley 1986; Clarke *et al.* 2005; Pausas & Bradstock 2007).

(5) Frequency of fire and other disturbances has been proposed to be a widespread factor controlling the selective balance between resprouters and non-resprouting seeders by Bellingham & Sparrow 2000. According to their model,

Fig. 9.10 *Transition from crown fire chaparral to surface fire regime of mixed conifer forest in the Sierra Nevada Range of California (Photo by Jon Keeley.)*

under a regime of severe disturbance resprouting would be selected against when disturbance frequency is low; however, as a general prediction, the nearly ubiquitous distribution of resprouting in woody dicots from most environments contradicts that conclusion. Resprouting is common in non-fire-prone or fire-resistant communities, although postfire seeding is not (e.g. Gill 1997; Campbell & Clarke 2006). For shrublands, the Bellingham & Sparrow model predicts that non-resprouting obligate seeders will dominate under very low disturbance frequency and gradually be replaced by resprouters as frequency increases. Indeed, under very short fire return intervals non-resprouting species will be eliminated whenever the fire return interval exceeds the time to reproductive maturity (Zedler *et al*. 1983; Nieuwenhuis 1987; Bradstock 1990; Pausas 2001). However, long intervals between fires (low fire frequency) will have very different selective values dependent on whether seed storage is in the soil or in the canopy. Obligate seeders with hard-seeded soil-stored seedbanks could be favored by long fire intervals (Keeley 1977) but such a regime would select against obligate seeders with canopy seed storage (Lamont *et al*. 1991).

The primary limitation to the Bellingham & Sparrow (2000) model is that it considers resprouting and seeding as continuous functions of fire frequency or fire severity, and it ignores other components of the fire regime, in particular fuel structure and its control on fuel consumption. Fire frequency has radically different impacts on resprouting success in shrublands under crown fire regimes than in forests with surface fire regimes and this problem is not solved by scaling against productivity as they have done. For example, the sharp boundary between a shrubland and a forest (Fig. 9.10) represents a tipping point in which the selective factors for life history characteristics change state from a crown fire selective regime to a surface fire selective regime. Despite radical differences in fire intensity and frequency, both systems may have equal levels of resprouting and non-resprouting seeding species, but for very different reasons that are tied to the different selective environments of a surface fire regime vs. a crown fire regime (see Chapter 2). Because of the effect of such spatial thresholds on fire behavior (e.g. Slocum *et al.* 2010), the most successful analyses of trait evolution have been ones that have focused on patterns within a single fire regime (e.g. Ackerly 2004a; Pausas *et al.* 2004b).

The ambition of developing a unifying model of life history that applies across a wide range of plant communities (Sparrow & Bellingham 2001) is not feasible without consideration of fire regime characteristics, particularly in the fuels consumed. Different growth forms exist in very different selective environments that are critical to understanding the selective value of resprouting. Complicating the picture even more is the fact that some growth forms such as trees change their resprouting capacity at different life history stages (see Chapter 3). We do agree with these authors that there is a strong interaction between disturbance history and site productivity, but maintain that the primary impact is on fire regime characteristics (see Fig. 2.7) and the subsequent impact this has on life history options (Keeley & Zedler 1998). As in Sewell Wright's (1932) adaptive landscape model, we might think of a similar metaphor with different fire regimes as representing adaptive peaks in which selection pressures vary from one fire regime to another.

A Model for the Evolution of Non-resprouting Shrubs

To summarize these arguments, it should be recognized that evolutionary and ecological models discussed above are not mutually exclusive and there is reason to believe they have worked in concert to promote the postfire obligate seeding life history. It may be useful to frame the evolution of this life history by analogy with arguments surrounding the evolution of semelparity and iteroparity (Keeley 1986). If we change our reference from annual cycles to fire cycles then facultative seeders represent an iteroparous and obligate seeders a semelparous life history. Charnov & Schaffer (1973) concluded that the evolution of iteroparity would be expected when the average clutch size of a semelparous organism was increased by P/C individuals, where P and C are adult and juvenile survivorship, respectively. Thus, obligate seeding would be expected when resprouting success is low and/or when seedling success is high. In environments where resprouting success is high, P/C would be high, and obligate seeding would evolve only when large increases in successful seedling recruitment could be achieved from diversion of resources from resprouting to seed production. In contrast, where resprouting success is low, obligate seeding could be adaptive with far less increase in seedling recruitment and survivorship.

Bond & van Wilgen (1996) investigated the trade-offs between resprouting and seeding with a simple logistic model and found that different components had different effects on the outcome. Although both resprouter mortality and level of

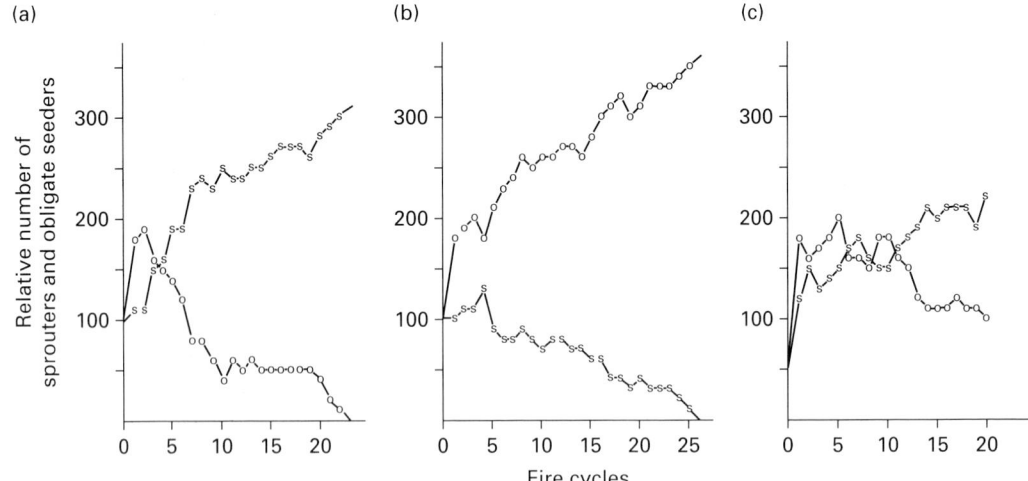

Fig. 9.11 Modeled competition between facultative seeder (s) and obligate seeder (o) life histories under different parameters of (a) equal seed production for both life histories and 20% postfire mortality of resprouts in "s," (b) three times greater seed production for "o" and 20% mortality of resprouts in "s" doubled every fifth fire cycle, and (c) two times greater seed production in "o" and 20% mortality of "s." (From Keeley 1986.)

obligate seeder seedling recruitment were important, the most sensitive component was the seedling survival of obligate seeders and the subsequent successful establishment as a mature reproductive adult. Thus, factors contributing to seedling success, such as adaptation to arid sites enhanced by greater numbers of sexual generations or allocation of resources to growth over storage (Fig. 9.5), may be critical to the evolution of obligate seeding.

Simulating competition between facultative and obligate seeding species predicted that an obligate seeder taxon could drive a facultative seeder to extinction with three times greater successful seedling recruitment for the former and 20% resprout mortality each fire cycle (doubled to 40% every fifth cycle) (Fig. 9.11). The model predicts that as resprouting success declines there is a proportional decrease in the seedling recruitment differential required to drive the evolution of obligate seeding. These predictions are consistent with empirical studies that suggest trade-offs in seedling recruitment differentials and resprouter mortality. In postfire California chaparral, relative to the parent population size, a *Ceanothus* obligate seeder had 1.8 times greater seedling recruitment than the facultative seeder and this resprouter experienced 67% mortality of resprouts, whereas different trade-offs were evident in a congeneric pair of *Arctostaphylos*; the resprouter had only 40% mortality but the obligate seeder had 9.8 times greater seedling recruitment (Keeley 1977).

Comparisons in other MTC regions have generally found similar patterns of seedling recruitment differentials between obligate seeders and facultative seeders (Thomas & Davis 1989; Hansen et al. 1991), but most have not reported resprouting success and this makes the comparison of more limited value. Complicating some comparisons is recognition that there is no reason to expect contemporary patterns are indicative of selective environments that occurred millions of years ago (see Chapter 10). The genetic consequences of greater selection for seedling over resprout regeneration would further enhance the advantages for recruitment in postfire environments. These fire recruiters as a group exhibit physiological and anatomical traits that favor recruitment in open sites under stressful soil-drought conditions (S.D. Davis et al. 1998; Paula & Pausas 2006; Pratt et al. 2007, 2008). Under summer-rain conditions such sites might be restricted to rocky outcrops as in eastern Australia (e.g. Clarke et al. 2005) but under MTC these conditions are much more widespread. The greater number of sexual generations leading to faster selection for novel recombinant types could enhance inclusive fitness of individuals and increase the capacity for recruiting in fire-induced gaps and feed back to increased selection for obligate seeding. This mirrors advantages of herbaceous taxa over woody taxa in climatic niche evolution (Smith & Beaulieu 2009).

Obligate seeders might also enhance their ability to track environmental changes through their greater mobility between generations. As environments change, such as through glacial and interglacial Pleistocene cycles, resprouting taxa would be at a decided disadvantage in tracking these environments, since dispersal is critical to finding new sites as the conditions deteriorate at present sites.

The Biography of Postfire Resprouting and Seeding

With increasing focus on global patterns there is a need to understand the relative importance of postfire responses in vastly different ecosystems representing a diversity of growth forms and fire regimes. Although one can detect local patterns of resprouting across different fire frequencies (e.g. Clarke & Dorji 2008), global predictions need to consider fire in combination with climate and geology (see Fig. 1.4). The level of resources based on climate and soils will have a deterministic role in setting the range of potential growth forms and these in turn will form the fuels for determining fire regime characteristics (Pausas & Bradstock 2007). Functional types will be dictated by resources, fire regime and growth form, with substantial shifts in type driven in some systems by herbivory and modified by phylogenetic constraints.

Scaling resprouting vs. seeding responses to site productivity is complicated by differences in fire regimes. In addition, growth forms with different rates of maturity will differ in fire sensitivity. The important parameter for scaling would be the ratio of time to maturity over fire return interval. Seeders would be at a decided disadvantage when that ratio drops below 1.

Potential growth forms vary along the resource gradient from trees where resources are high to herbaceous plants at the lowest end of the gradient and multiple strata of herbs and trees at intermediate levels. The tipping point between dense woody growth and herbaceous vegetation will differ relative to resources as a function of fire frequency. Functional responses of resprouting and seeding will vary depending on growth form and fire regime.

The most resource limited conditions will limit options and favor annual species that only seed and may not produce sufficient fuels to support fire. As resources increase, fuels will increase and then the potential for fire is dependent on ignitions. When ignitions are not limiting, the seasonal distribution of precipitation will play a key role in determining postfire response. Under MTC summer drought and high fire frequency, annual grasses and forbs will dominate with a mixture of resprouting geophytes that increase with decreasing drought stress. Under warm growing season conditions such as tropical C_4 grasslands there is increasing dominance by resprouting grasses with a minor component of annuals. As fire frequency declines in both systems there will be increasing dominance by resprouting woody species, with the exceptions of MTC obligate seeders discussed above. As resources are added, functional type options increase but the combination of temperature, precipitation and soil nutrients will lead to very different growth form options and productivity, all of which will provide the fuel base for driving different fire regimes (Pausas & Bradstock 2007). For a given fire regime one can make broad generalizations about resprouting and seeding responses (see Chapter 3) and one can make very useful generalizations across resource and disturbance gradients (Clarke & Knox 2002). Indeed, there is even the possibility of drawing parallels with other organisms; for example, the persistence strategy of marine annelids to catastrophic winter storms has been likened to the resprouting mode of fire adaptations (Barry 1989). However, it stands as a major challenge to produce broad geographical generalizations. Several reviews have provided additional perspectives on these patterns (Bond & Midgley 2001, 2003; Vesk & Westoby 2004).

Life History and Diversification

MTC ecosystems are among the most remarkable plant biodiversity hotspots on Earth with many species-rich genera (Cowling *et al*. 1996; Myers *et al*. 2000). The greatest concentration is in MTC South Africa and Australia, but with some large woody genera also in California. These are all fire recruiter genera and most are dominated by obligate seeding species; and it has been suggested that the evolution of this life history mode was key to their diversification (Wells 1969; Raven 1973; Cowling 1987).

Although both resprouters and seeder life histories are resilient to fire, life history confers somewhat different population dynamics (Keeley 1986; Bond & van Wilgen 1996; Pausas 1999) with potential impacts on rates of evolution (Segarra-Moragues & Ojeda 2010). Because fire kills the entire population of obligate seeders, this life history mode results in a shorter generation time than is the case with resprouters, thus potentially increasing their molecular evolutionary rates and, ultimately, diversification (Wells 1969; Wisheu *et al.* 2000). Furthermore, the resulting non-overlapping generations of seeders increase the probability of manifestations of genetic novelties associated with different conditions present in each generation (Cowling & Pressey 2001), and thus it has been theorized that seeders should speciate more than resprouters, and consequently diversify more as well.

Although there have been several attempts to test the differential diversification of seeders vs. resprouters (Bond and Midgley 2003; Lamont & Wiens 2003; Verdú *et al.* 2007; Boatwright *et al.* 2008), all have found limited support for a significant difference between these life history modes. These results suggest that seeders have not diversified more than resprouters, although it is to be expected that differences would be rather subtle since many of the resprouters studied were not obligate resprouters, and thus the comparisons are between taxa that replace the entire population each fire cycle with taxa in which a variable proportion of the population is replaced each fire cycle.

Complicating this comparison is the fact that there are also potential genetic advantages associated with resprouters as repositories of genetic innovations. Obligate seeding species subject their gene pool to intense selection because each fire results in a new generation. Thus, the characteristics of individuals that are selected by circumstances associated with one fire may not be subsequently selected if the environment changes. This is evident in that obligate seeder populations are generally much more phenotypically homogeneous than resprouter populations, which often exhibit substantial clonal variation (e.g. Keeley *et al.* 2007; Premoli & Steinke 2008). Considering that environmental change is not monotonic but cyclical at various scales, species capable of retaining suboptimal variants under certain conditions may ultimately be selected in the long run. This could have been of particular selective value during the Quaternary alternation between glacial and interglacial episodes.

As discussed more fully in Chapter 10, different fire histories are associated with different patterns of diversification in lineages and these appear to distinguish themselves between the northern and southern hemispheres. In the southern hemisphere the origins of genera such as *Banksia* are tied to nutrient-poor sites, which have selected for sclerophyllous fire-prone vegetation, and these occurred very early in the Tertiary or even earlier. As a consequence vast expanses of fire-prone landscape potentially would have provided the stage for fire-driven diversification in *Banksia* over much of the Tertiary, accounting for its spectacular speciation by the Eocene (Hill & Christophel 1988). In contrast, in the northern hemisphere early Tertiary fire-prone landscapes were likely marginal habitats on widely disjunct sites. Although these habitats selected for sclerophyllous fire-prone vegetation, the discontinuity of fuels would have led to a less predictable fire regime, one that selected for resprouting, but a regime lacking sufficient predictability and intensity to select for postfire seeding until climatic changes later in the Tertiary. Not surprisingly most of the northern hemisphere fire-prone genera with a clear linkage to early Tertiary origins are depauperate genera of obligate resprouters (e.g. see Table 3.1). Thus, fire-adaptive traits, climate and geology (see Fig. 1.4) all seem to have played a role in accounting for different patterns of diversification in resprouters and seeders.

This interaction between fire and geology can be invoked in explaining patterns of fire traits in many MTC genera. The South African grass *Ehrharta* comprises both resprouters and obligate seeders and much of its radiation appears to be tied to substrate heterogeneity (Verboom *et al.* 2004). In California large outcrops of serpentine substrate also have had selective effects on seeders vs. resprouters (Safford & Harrison 2004).

Orogenic changes such as Quaternary mountain building also have been considered a factor in the diversification of the MTC taxa such as the Californian genus *Arctostaphylos*, which is dominated by obligate seeding species (Raven & Axelrod 1978). Although intuitively appealing it needs to be recognized that there is little evidence that radiation was this recent. The genus comprises two lineages, both of which are almost entirely composed of postfire seeders, and the majority of species are obligate seeders. Many of the species are localized endemics and some are strictly confined to unusual substrates, e.g. *A. myrtifolia* found only on outcroppings of Eocene lateritic soils, and others are serpentine endemics (Parker *et al.* 2009). Many of the endemic species are in coastal southern and central California and some of these ranges were up to 2000 m in the Miocene (Stadum & Weigand 1999).

Pleistocene mountain building has also been used to explain the origin of many narrow endemics in the very large woody Californian genus *Ceanothus*, most of which are obligate seeders (Raven & Axelrod 1978). However, this ignores the substantial evidence that there was significant elevational variation throughout the Miocene (Wolfe *et al.* 1998) that could have provided habitat differentiation. In addition, latitudinal replacement of localized endemics within the state appears to be more important than altitudinal variation and rain shadows (Cody 1999). One could argue that only about 10% of the extant species are reported from the middle to late Miocene fossil record, and thus most are more recent. However, assigning these fossil impressions to contemporary species is problematical (Edwards 2004), which is not surprising since no specialist today could identify most species solely from leaf impressions. The difficulty in answering this question of timing of diversification with fossils is that much of the present diversity in *Ceanothus* comprises localized endemics. Even if they were extant during the Tertiary, macrofossil samples comprise only a tiny fraction of that landscape and thus sampling theory alone would predict that relatively little of the diversity would have been recovered from fossils. In support of recent speciation in *Ceanothus*, one could cite the fact that there is very little molecular separation evident between species within subgenera; however, rather than indicative of recent separation, this also could be explained by introgression, which is well known in the genus (Hardig *et al.* 2000).

Some diverse genera have radiated relatively recently, for example the Mediterranean Basin *Cistus* (Guzmán *et al.* 2009), possibly in response to intensification of the MTC summer drought since this is a drought-deciduous taxon. More recent Quaternary speciation is also supported by molecular studies for the largely herbaceous South African genera such as *Heliophila* and *Ehrharta* (Verboom *et al.* 2003; Linder 2005), also perhaps in response to intensification of the MTC summer drought. The herbaceous Australian Restionaceae in contrast appear to have had a constant rate of speciation since their early Tertiary origin (Linder *et al.* 2003). These include both facultative and obligate postfire seeders, suggesting a long history with fire and possibly a factor driving diversification.

Life History and Community Structure

At a localized scale the distribution of species reflects abiotic filters, such as climate, fire and soils, as well as biotic interactions such as competition and facilitation, and the resultant community composition is thought to be constrained in predictable ways. Community assembly rules are based on the concept that there are not only restrictions arising from species-specific responses to the environment but from the presence or abundance of other species as well (Diamond 1975; Weiher & Keddy 1995). The rules are modified by assembly histories and multiple stable equilibria are possible (Chase 2003). We are only beginning to unravel the rules for fire-prone ecosystems but it appears that life history is an important determinant.

Ultimately adaptive evolution is the mechanism by which organisms make the appropriate fit to a given environment and may be the primary factor accounting for presence or absence in a given community (e.g. Cody & Mooney 1978). Equally important are ecological sorting processes, whereby functional traits are influenced by the trait combinations present in other taxa to the extent that non-random combinations are common (Ackerly 2003). This is illustrated in the two lineages of the Californian genus *Ceanothus* with different postfire regeneration modes, as well as different morphologies and physiologies, which are more likely to co-occur than by chance alone (Fig. 9.12). The genus dates to the Eocene and the divergence in the two lineages dates to a basal split 18–39 Ma (Hardig *et al.* 2000). The *Cerastes* exhibits extraordinary niche conservatism in that all species are obligate seeders, whereas the other subgenus *Ceanothus* has both obligate and facultative seeders. Niche conservatism also is quite evident with respect to serotiny and resprouting in lineages from MTC ecosystems of both hemispheres (Fig. 9.8).

Other examples of non-random sorting are found in MTC crown fire shrublands. Because seeders are favored by predictable crown fires, the proportion of seeder species in a woody community is expected to increase with the frequency of fires (assuming frequency does not exceed time to reproductive maturity). One comparison in the Mediterranean Basin communities under frequent crown fires showed an overrepresentation of postfire seeders relative to what would be predicted by the regional species pool (Verdú & Pausas 2007). This is also evident in timing of postfire recruitment in the semi-deciduous sage scrub shrubland in

Fire-adaptive Trait Evolution

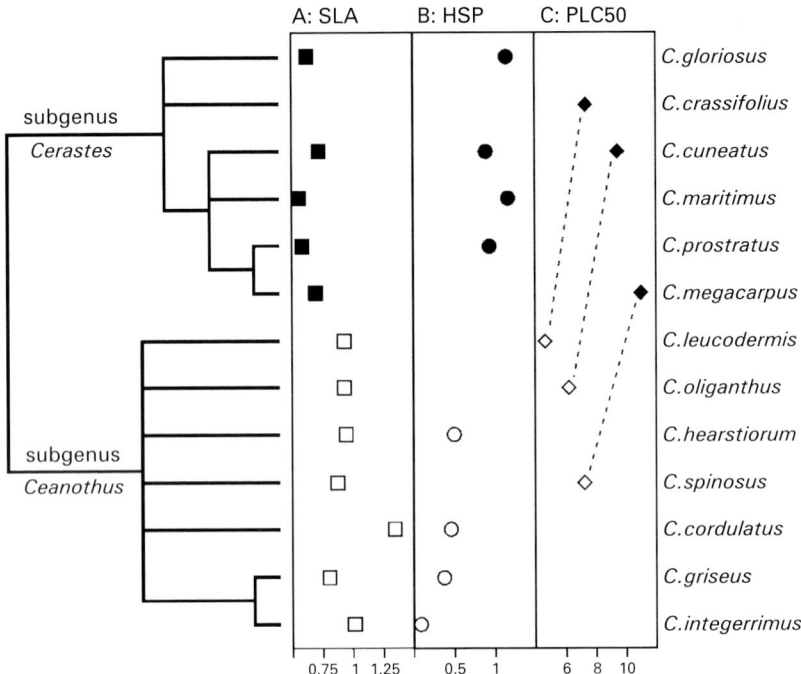

Fig. 9.12 *Partial phylogeny of* Ceanothus *species illustrating divergence in regeneration strategy and other functional traits. A: specific leaf area (log (mm² mg⁻¹)), B: heat shock proteins, C: drought tolerance ($\Psi_{50\% \ survival}$ MPa) and dashed lines connecting sympatric species pairs along an altitudinal gradient. (From Ackerly 2003.)*

California where different recruitment patterns combine within communities in non-random associations (Keeley & van Mantgem 2008).

Many of the postfire seeder genera are quite diverse (see Table 3.4) and this phenotypic clustering of seeders is considered an example of evolutionary conservatism when taxa have radiated into environments with very similar fire regimes. This of course is a function of scale, as at the biome level conservatism appears to be the norm in plant evolution (Crisp et al. 2009). This can be detected by a higher mean relatedness (lower mean phylogenetic distance) between all pairwise species of a fire-prone community than the one expected under a null model (Fig. 9.13), and also higher than in a community where such a strong filtering process as fire is not present (Verdú & Pausas 2007).

Niche conservatism is evident in the North American *Ceanothus* where the very diverse *Cerastes* lineage comprises only obligate seeding species that have radiated into different crown fire shrubland communities in California, Arizona and eastern Mexico (Keeley 2000). However, another chaparral genus, *Arctostaphylos*, illustrates a pattern more consistent with an adaptive evolution model in that 10 of the 16 resprouting species have subspecies that are obligate seeders, reflecting a close tracking of changing fire regimes: the resprouting populations are

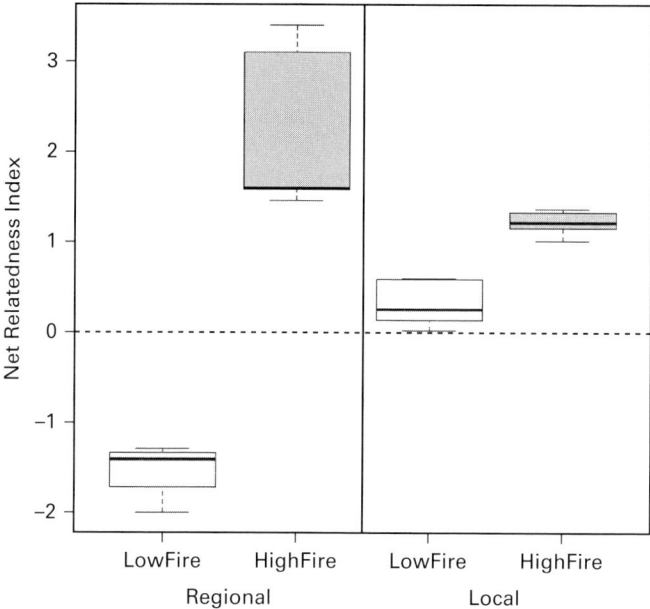

Fig. 9.13 *Net relatedness index (NRI, i.e. standardized form of the mean phylogenetic distance) of woody species coexisting in communities in contrasted crown fire regimes (LowFire vs. HighFire). Note that low mean phylogenetic distance implies high net relatedness. At regional scale, "LowFire" corresponds to mountain communities living in zones that rarely burned, and "HighFire" are warm and dry coastal communities subject to a high frequency of crown fires. At local scale (under the same climate), "LowFire" corresponds to communities growing in fertile soils while "HighFire" are communities growing on poor soils where flammability is higher. In comparison with community null models, HighFire communities show higher NRI than expected by chance (phylogenetic clustering), which indicates the importance of habitat filtering in shaping fire-prone communities. (Elaborated from Verdú & Pausas 2007 and Ojeda et al. 2010.)*

distributed in closed-canopy chaparral where postfire resprouting likely limits seedling recruitment opportunities and obligate seeding populations are distributed in more open communities (J.E. Keeley unpublished data). Similar environmental selection is suggested by the switching between lignotuberous resprouters and obligate seeders multiple times in the endemic clade of South African Cape Podalyrieae tribe of Fabaceae (Boatwright et al. 2008).

Conclusions

Fire-prone landscapes began with the origin of land plants, and fire has potentially been a factor whenever climate and soils combined to create sufficient biomass and seasonality to generate continuity of available fuels. Thus, trait evolution in fire-prone plant communities is closely linked to ecosystem processes that contribute to fire spread. Lightning ignitions vary in abundance and seasonal timing, but over most (but not all) land surfaces they have been a ready source of ignitions for fires. The most dependable and ubiquitous plant trait that provides resilience in fire-prone landscapes is resprouting ability. With life histories such as herbaceous perennials, which have an annual cycle of top dieback followed by growing season regrowth, some level of pre-adaptation was involved in adapting to fires. However, in woody plants there are many lineages where it seems likely that resprouting was selected for and has been continually maintained by fire. The Late Cretaceous and early Tertiary development of sclerophyllous MTV initiated a new era in plant–fire relations. Today this vegetation dominates MTCs but has its origins in climates and/or soils that shared primarily two characteristics with contemporary MTCs: (1) growing seasons conducive to production of sufficient biomass to form contiguous fuels and (2) seasonality that varied in timing and regularity but was sufficient to make the vegetation fire-prone and sufficient to limit growth so that fires were stand-consuming crown fires. Under extreme conditions there was selection for fire-dependent reproduction because the prefire closed canopy limited recruitment opportunities and the postfire conditions opened up gaps sufficient for seedling populations to exploit resources made available only by fire. The late Tertiary intensification of summer drought greatly expanded the amount of landscape available for regular high-intensity crown fires and associated with this was an expansion of MTV in MTC regions. However, even on the most fire-prone landscapes present today, fire regimes are not always conducive to generating gaps sufficiently predictable for seedling recruitment. Thus, MTC communities often comprise mixtures of obligate resprouters and obligate seeders, each exploiting different portions of available niche space.

10 Fire and the Origins of Mediterranean-type Vegetation

The mediterranean-type climate (MTC) is widely agreed to have been in place in all five MTC regions since at least the late Pliocene (see Fig. 9.1), ~2 Ma, with much of the contemporary mediterranean-type vegetation (MTV) present and contributing to a highly fire-prone environment. There is far less agreement on: (1) the timing of the origin of the MTC, (2) the timing of and factors responsible for the origins of MTV, and (3) the paleohistory of fire and extent to which it has played a role in the origins of MTV. Ample evidence exists to suggest a much earlier origin of MTC and MTV.

A widely held paradigm is that many of the woody sclerophylls that comprise MTV are much older than the Pliocene and thus have not adapted to contemporary fire-prone MTC conditions (Axelrod 1989; Herrera 1992; Verdú et al. 2003; Ackerly 2004a). Most of these have origins in the Tertiary Period of the early Cenozoic and are viewed as relictual taxa that represent evolutionary inertia and are present today merely by chance avoidance of random extinctions.

This needs further examination for several reasons. The MTC appears to be older than commonly accepted and it developed gradually over a period of 10–20 million years and thus its date of origin is a matter of how one defines MTC. Also, the critical environmental features that have played a selective role in MTV species may not have changed since their origin in the Tertiary; for example, soil drought *per se* may be the key selective factor rather than summer soil drought, or predictability of winter rains may be a key selective factor as it has shaped numerous life history characteristics. Both soil drought and winter rains preceded the MTC yet persist under a MTC. Under this model a species could exist through multiple climatic changes and be tracking the same proximal environmental conditions. Lastly, regardless of the seasonal distribution of rainfall, MTV likely has had a long association with stressful substrates experiencing seasonal soil drought, and thus has been fire-prone throughout much of its evolutionary history. Under this model the primary change to the environmental template brought about by the MTC was an expansion of drought-prone landscape (Fig. 10.1) and the potential this had for fire spread. This scenario though does not preclude more recent innovation of some fire-adaptive traits or species radiation in some groups in response to the expansion of drought-prone landscapes in MTC regions.

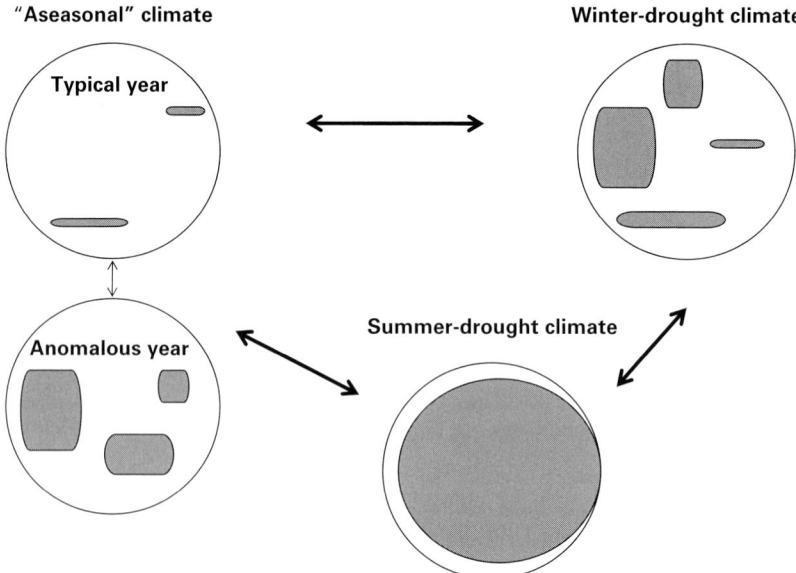

Fig. 10.1 *Schematic model of changes in drought-prone landscape under different climatic regimes. Aseasonal climates are those that, on average, have precipitation in all seasons of the year and droughts are not annual events but may occur multiple times a decade, whereas the other two climates have regular annual droughts. Depending on structural characteristics of the vegetation and seasonal distribution of lightning ignitions, these drought-prone sites are also likely to be fire-prone sites.*

Origin of the Mediterranean-type Climate

Climatic seasonality and soil nutrient stress are key to understanding the origin of fire-prone MTV, and these certainly precede the MTC in all regions. Such climatic characteristics have waxed and waned with movement of the continents, changes in ocean circulation, orogeny, earth orbital patterns and associated factors (Parrish 1998). The MTC, by having an annual drought in the hottest months of the year, has greatly expanded fire-prone landscapes (Fig. 10.1).

In prior discussions on the timing of MTC origins, there has been some level of disagreement in the scientific literature and part of this is semantic in that authors have focused on different questions. For example, in discussing origins of the MTC Axelrod (e.g. 1973) considered the timing of the MTC in coastal California regions, whereas other authors have taken a more global perspective on the paleohistory of this climatic pattern, such as Beard (1977) who considered the question of a MTC in Australia, which appears to have arisen in regions adjacent to the contemporary MTC of southwestern Australia.

The most straightforward approach to uncovering paleoclimates is from physical models that consider changes in earth system processes such as Antarctica's glaciation and subsequent effects on ocean currents. The advantage of this approach is that one can theoretically predict climates for any given time and location. However, this approach is fraught with difficulties due to the complexity involved, as climate is a function of numerous factors: orogeny and its effect on atmospheric circulation and rain shadows, plate tectonics and movement of plates into new climatic zones, the thermohaline cell, ENSO events and Milankovitch variations, to name just some of the parameters to be considered (Graham 1999).

An indirect approach is to use fossil floras and deduce past climates based on corresponding climates of related contemporary species (Wolfe 1995). Although widely used, this approach has a number of limitations, not the least of which is that many areas of the world lack fossil floras for time periods of interest. Another major problem is that these macrofossil floras are not an unbiased sample of the landscape; rather, except for fossil sites generated by volcanic ash deposition, they largely sample wetland vegetation and some unknown portion of the surrounding landscape. This is a particular problem if one is concerned about fire-prone landscapes since they tend to be on the arid end of the moisture gradient. Landscapes with mosaics of mesic and arid patches will provide limited information on the arid fire-prone portion of the flora until it reaches some unknown level of landscape dominance. Those arid environments where fire has likely been an important factor may not even have had suitable wetlands for fossil deposits and thus lack a record entirely.

Since the goal of this discussion is to better understand the origins of fire-prone sclerophyllous-dominated communities of MTV, which are widespread in contemporary MTC regions, we will begin with the history of climatic factors that play a role in fire-prone systems. Climatic seasonality (at some scale; i.e. annual, decadal or longer) is a necessary condition for any fire regime (see Chapter 2) and such patterns have been present off and on since the mid-Paleozoic (see Fig. 9.1) rise of land plants (Scott 2000, 2010; Pausas & Keeley 2009). Throughout the Mesozoic, large annual seasonal variations in rainfall were associated with mid-latitude continental margins due to the alternation of high and low pressure centers over continents and adjacent oceans (Parrish *et al.* 1982). Despite the widespread characterization of global climates in the Mesozoic as "equable," there is evidence that some mid-latitude regions in the northern hemisphere experienced seasonally dry conditions, even climates analogous with contemporary MTCs, comprising wet winters and hot dry summers (Francis 1984; Allen 1998).

Early Cenozoic Climates

Most lineages contributing to our contemporary sclerophyllous MTV originated sometime in the Late Cretaceous or early Cenozoic, and were shaped by climates that fluctuated markedly during different epochs. Paleoclimatic inferences point to the earliest Tertiary epoch, the Paleocene, as showing marked seasonality over the more equable climate of the Late Cretaceous and with a steep north–south temperature gradient (Davies-Vollum 1997). This was followed by the Eocene, which globally was a time of high temperatures and limited seasonal variation (Raven & Axelrod 1972). However, conclusions about equable Eocene climates are in part an artifact of the general coarsening of temporal resolution with increasing distance back in time, as climatic oscillations on timescales of 10–100 kyr are permanent features of earth history (Bennett 2004).

The Eocene world does not appear to have been highly conducive to frequent widespread burning; however, it was not completely lacking in seasonality conducive to periodic fires. Decadal-scale oscillations can create droughts conducive to fires, even in otherwise aseasonal climatic regimes. Also, the model of universal equable conditions is disputed by modeling studies that show substantial annual seasonality effects in continental interior landscapes (Sloan & Barrón 1990; but cf. Wing & Greenwood 1993). For example, during this epoch seasonal aridity was evident in central Australia (Greenwood 1994) and the southwestern USA (Peterson & Abbott 1979). Consistent with regional variation in seasonality is the middle Eocene appearance of shrubs and grasses in parts of southwestern North America, interpreted as signs of decreased precipitation and increasing severity of the dry season (Frederiksen 1991). The region is thought to have had a winter drought and summer rain climate and a very long dry season that was able to sustain high-intensity forest fires. In coastal southern California late Eocene climates appear to have had an annual rainfall of 500 to 600 mm that was distinctly seasonal (Abbott 1981; Frederiksen 1991). With this level of precipitation there could have been sufficient fuels to carry fire, and the seasonality would have created burning conditions on a regular basis. Thus, there is reason to expect that early in the Tertiary, fires were an ecosystem factor on some landscapes, despite more equable conditions in other parts of the landscape. These fire-prone pockets potentially spread fire to more mesic landscapes during longer-interval cyclical droughts, expanding fire-prone vegetation, as observed today in tropical forest types (Cochrane 2003).

Oligocene

Cenozoic plate tectonics moved the continents closer to their current configuration (Milne 2006), contributing to global circulation patterns homologous to contemporary patterns. Middle Oligocene glaciation of Antarctica, coupled with the formation of Drakes Passage resulting from movement of South America away from Antarctica, led to cold water flow into northern latitudes (Axelrod *et al.* 1991). This strengthened high-pressure systems so that drier climates spread more widely than previously seen in the Cenozoic. These changes contributed to marked climate change during the transition from Eocene to Oligocene and seasonality increased over much broader portions of the globe (Eldrett *et al.* 2009). During the Oligocene, global temperatures dropped by more than 10 °C, climates became much more seasonal and arid landscapes expanded (Wolfe 1994). Potential fire-prone environments undoubtedly expanded during this time.

Several investigations have suggested late Oligocene as the origin of MTC in some regions. On the basis of proposed circulation patterns, Beard (1977), contended that a large area north of the contemporary MTC in southwest Australia was MTC at 25 Ma or earlier (Hopper & Gioia 2004). However, aseasonal tropical climates were the norm over much of southern Australia at this time (Christophel & Greenwood 1989). Using sea surface temperatures and Antarctica glaciation, Linder (2005) suggested a MTC for the Cape region of South Africa during the early Oligocene followed by more mesic conditions in the early Miocene.

In North America, paleosol characteristics indicate the presence of a summer-drought climate in southern Colorado during the late Oligocene (Wolfe and Schorn 1989). This seasonality potentially existed throughout the southwestern portion of interior North America and these lower latitudes were possibly more arid, which would be consistent with the very limited Oligocene fossil floras for the region. The Pacific slope of California was much moister and definitely not MTC at that time, as revealed by the substantial subtropical macrofossil record (Axelrod 1973).

Miocene

The Miocene began with moderate temperatures and increasingly became more seasonal in both temperature and precipitation. These were global patterns driven by changes in sea surface temperatures and Antarctica glaciation as well as orogenic uplift.

Western North America has a rather extensive Tertiary macrofossil record, making climate reconstruction more feasible than in some other regions. Miocene temperature and rainfall heterogeneity was enhanced by the Sierra Nevada rain shadow that had developed by the mid-Miocene (Crowley *et al.* 2008). However, such topographic influences appear to have been present much longer, as recent studies point to an earlier Late Cretaceous uplift for the northern Sierra Nevada (Busby & Putirka 2009; Cassel *et al.* 2009), where the present-day range represented the western edge of the Nevadaplano that extended east across the present Great Basin (Henry 2009). Today the northern range reflects a reduction in mean elevation from early Tertiary levels but greater relief (Hren *et al.* 2010).

Using an abundant macrofossil record, Axelrod (1989) concluded that the MTC in California was a Quaternary phenomenon and that Tertiary climates were dominated by summer rains. He took great stock in the observation that mid and late Miocene floras had taxa that were allied to contemporary species now confined to summer-rain climates (e.g. Eastern Deciduous Forest taxa: *Carya, Diospyros, Nyssa, Robinia, Ulmus,* and *Zelkova,* and subtropical taxa: *Persea, Magnolia and Cinnamomum,* Raven & Axelrod 1978). Even for fossil floras with many contemporary chaparral and sclerophyllous woodland taxa, he considered any representation of summer-rain-allied taxa as indicative of a summer-rain climate.

One example is the 7-Ma Piru Gorge flora in which 5 of the 22 species were considered summer-rain indicator species; based on this, Axelrod (1982) concluded that summer rains were still persistent in the region. Condit (1938) offered an alternative interpretation and pointed out it is to be expected that as summer rains declined during the Miocene one would not see the immediate loss of summer-rain-dependent species as they would persist for a considerable time in riparian zones or other wetlands. Consistent with this interpretation, Piru Gorge, like many macrofossil floras, was a wetland site; the majority of the 22 species were wetland taxa including *Typha*, *Equisetum*, three species of *Salix*, one of *Platanus*, and six species of *Populus* (Axelrod 1982). Thus, the persistence of a few summer-rain-allied species may not be an indicator of summer rains. Condit's (1938) insight is consistent with contemporary patterns in that some taxa widely distributed in summer-rain regions, such as *Amorpha*, still persist throughout MTC California largely within riparian zones. In an analogous manner, MTC chaparral sclerophylls such as *Malosma laurina* and *Hesperocyparis* (*Cupressus*) *forbesii* persist much further south into very arid parts of Baja California by restricting their distributions along riparian watercourses. *Washingtonia* palms currently restricted to riparian sites in MTC California may stand as another example. In southern Australia, the restriction of rainforest species to riparian corridors similarly has been interpreted as a response to declining summer rains (Kemp 1981).

Although the MTC appears to have established in coastal California somewhat later, several lines of evidence point toward an early Miocene origin for the MTC in more interior portions of western North America. At 17 Ma, macrofossil floras in southwestern Nevada were already lacking subtropical and summer-rain taxa, indicative of marked climate change and increasing summer aridity (Wolfe 1964). Axelrod and Schorn (1994) also reported a major floristic change at 15 Ma in western Nevada, marked by the rapid disappearance of deciduous hardwoods allied to those now in summer-rain southeastern USA, and this was attributed to increasing summer drought. Based on the minimum summer rain currently tolerated by contemporary summer-rain taxa, it was estimated that summer precipitation was about a third of the annual total at 15 Ma. However, this estimate assumed that the few remaining summer-rain taxa were indicative of the regional climate and were not relicts persisting along riparian corridors or other wetland refugia. By late Miocene (7 Ma) a distinctly MTC was in place in southern California (Fig. 10.2), although Axelrod (1982) contended there was still about 15% summer rain, a value he considered necessary to account for relictual summer-rain taxa.

Consistent with an early origin for the MTC in North America is the Miocene origin of *Aesculus californica* (Hippocrastinaceae). This shrubby tree is endemic to the MTC region of California. It exhibits an undeniable adaptation to summer drought in its summer deciduous habit (Fig. 10.3a), which is quite distinct from the mostly winter deciduous genus. Molecular data indicate an Eocene or earlier origin for *Aesculus* and the split of *A. californica* sometime in the Miocene (Xiang et al. 1998), which suggests significant summer drought present at that time. One could argue that this taxon originated under summer-rain conditions and the summer-deciduous habit developed later; however, this seems unlikely because

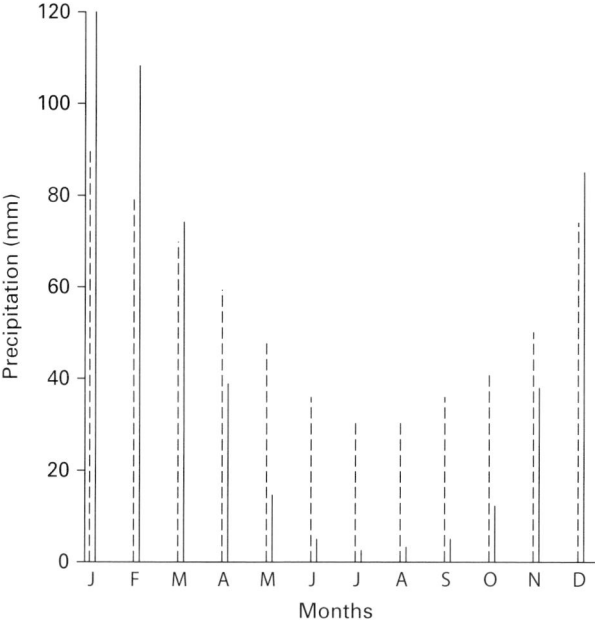

Fig. 10.2 Axelrod's (1982) inferred seasonal precipitation for late Miocene (7 Ma) southern California. Paleoprecipitation (dashed bars) was estimated to have been 650 mm annually and present precipitation (solid bars) is 500 mm).

Fig. 10.3 Aesculus californica, *a Miocene-derived taxon exhibiting traits clearly evolved in response to the MTC. (a) The summer deciduous habit, (b) fleshy stems that store carbohydrates for later transport to developing fruits during the leafless summer period, and the silvery stems that increase albedo are all interpreted as adaptations to the summer drought environment. (Photos by Jon Keeley.)*

evolution of the summer-deciduous habit required co-ordinated evolution with other traits. In *A. californica* and the northern Baja California endemic *A. parryi* there is a close linkage between summer deciduousness and the complex life history involving spring carbohydrate storage in fleshy stems, followed by the transport of these stores during the summer leafless period to fill its massive fruits (Fig. 10.3b; Mooney & Hays 1973; Mooney & Bartholomew 1974).

Timing of the MTC in the Mediterranean Basin also appears to be earlier than the late Pliocene or Quaternary age assumed by some (e.g. Axelrod 1975; Herrera 1992). On the basis of paleosol data and paleoclimate modeling results, Tzedakis (2007) concluded that the MTC appeared intermittently in the early Tertiary and was firmly established between mid and late Miocene, in line with other evidence (Thompson 2005; Brachert *et al.* 2006). Although the contemporary climate of severe summer drought is generally thought to have been continuous since the late Tertiary, this drought was ameliorated during Pleistocene glacial episodes (Tzedakis 2007).

Australian paleoclimates were markedly affected by continental movement and the southward migration contributed to more rapid Miocene increases in aridity than observed in other parts of the globe (Hill & Brodribb 2001). Although seasonal dryness was widespread by late Miocene, it likely established much earlier in the interior, but fossil records are absent from the interior as well as from many other parts of the continent (Greenwood 1994). In southwest Australia, it appears that a mildly seasonal MTC was present from very early Miocene (20 Ma) with increasing intensification to the present (Hopper & Gioia 2004; Martin 2006).

Although the contemporary MTC in the Cape region of South Africa is a more recent phenomenon, seasonally arid conditions conducive to fire-prone landscapes were present in the region since at least mid-Miocene (Deacon *et al.* 1983; Scher & Martin 2006; Cowling *et al.* 2009). The lack of Tertiary fossil floras for South Africa makes climate reconstruction problematical, although based on phylogenetic considerations it appears that seasonally dry environments were present during the mid to late Miocene (Linder 2005). On the basis of changes in the Benguela Current it has been suggested that intensification of the MTC in the Cape region began 5 Ma, although other estimates place it in the late Miocene at 8.9 Ma (Hopper *et al.* 2009). Pliocene uplift in the Cape region undoubtedly contributed to enhanced climatic heterogeneity.

In central Chile, early Eocene tropical paleofloras were gradually replaced due to a subtropical seasonal climate with biennial precipitation (Hinojosa 2005). Due to the very slight rain shadow of the Andes in the Miocene, the region is thought to have experienced a highly seasonal (and thus fire-prone) environment with summer monsoonal rains coming from the east (Hinojosa *et al.* 2006). Beginning about 15 Ma the Andean uplift raised the mountains to about half their present height by 10 Ma and eventually to elevations of 3000–6000 m by the end of the Miocene (Reynolds *et al.* 1990; Gregory-Wodzicki 2000). This uplift effectively cut off the flow of moist air that had previously moved across the South American continent in summer. Chile appears to have slowly developed a MTC that began about the same time as the initiation of the Andean uplift and continued until about 8 Ma when modern atmospheric circulation patterns were largely in place (Armesto *et al.* 2007). A major effect of the Andean uplift was the drastic reduction in lightning activity (Fig. 10.4) and thus, from the late Miocene until humans arrived, the western side of the Andes in northern and central Chile lacked a predictable source of ignitions.

In summary, Tertiary climates have largely varied in the amount of landscape that was potentially fire-prone. However, even during the most equable climatic epoch of the Eocene, seasonally arid climates conducive to fire have existed in some regions. On those landscapes that on average were aseasonal, longer decadal-scale droughts could have contributed to fire-prone conditions. Seasonal distribution of drought appears to have varied spatially and temporally. In some regions middle Tertiary aseasonal climates gave rise to winter-drought and summer-rain regimes, whereas in other regions it is apparent that tendencies toward a MTC of winter rain and summer drought appeared relatively early in the Miocene. Some reports suggest alternating conditions of winter rain and summer rain at different times since the Oligocene. From the middle Miocene on, it appears that the severity of the summer drought component of the MTC intensified and there is little evidence that it developed from a dramatic climatic shift at the end of the Tertiary. The coupling of drought with high summer temperatures did not create qualitatively different drought conditions, requiring an entirely new mode of adaptation. Rather the MTC greatly expanded the portion of landscape subject to drought stress. When combined with sufficient winter rainfall to produce a contiguous cover of sclerophyll vegetation, fire regimes became more predictable and this represented an important selective force. Late Tertiary increases in fire-prone landscape were not restricted to MTC regions but were evident in other ecosystems such as subtropical savannas under monsoonal climates (e.g. Keeley & Rundel 2005).

Origins of Mediterranean-type Vegetation

Our understanding of MTV origins is based on macrofossils of leaves, microfossils of pollen and molecular clock estimates from genetic studies of contemporary taxa. The quality and depth of record for these metrics vary greatly between regions. Some parts of southern Australia and western North America have extensive Tertiary vegetation records, whereas South Africa and Chile are not well represented. The macrofossil record is the most extensive; however, it largely pertains to woody species with annual leaf turnover, and herbaceous species are not well represented. Palynology (study of fossil pollen) has provided some insights into herbaceous origins, but pollen often cannot be recognized below the generic or sometimes family level.

What is apparent from the fossil record is that some sclerophyllous MTV taxa appear to predate the MTC, although the extent to which this is true is dependent

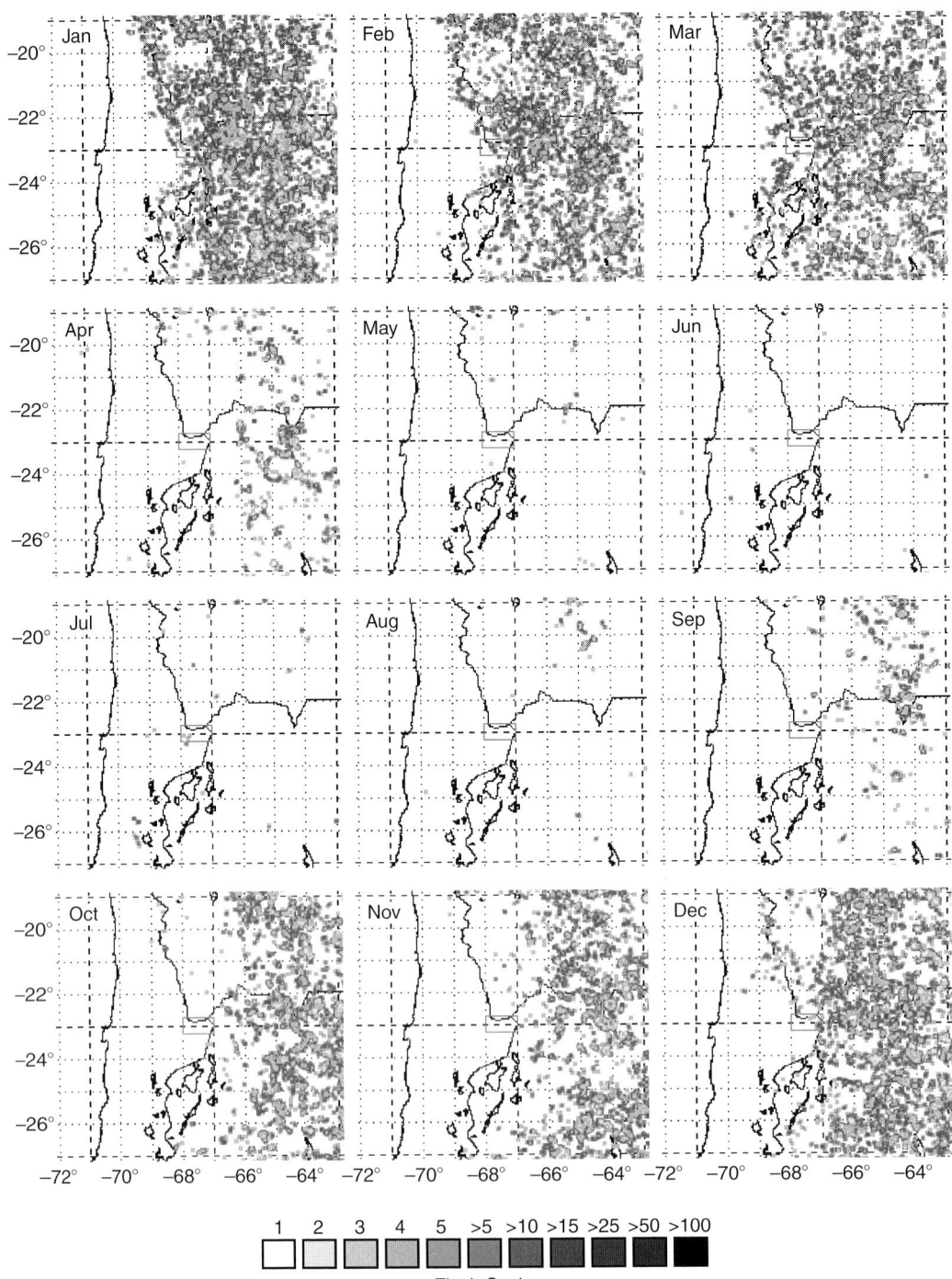

Fig. 10.4 *Monthly spatial lightning distributions recorded by the Lightning Image sensor in northern Chile, illustrating the blocking influence of the Andes on summer storms reaching the western side of the range. (From Sakamoto & Radford 2004.)*

on the interpretation of how climates developed within each of the regions considered. Seasonality and summer aridity apparently developed gradually over a period of 10–20 million years and the contemporary MTC is an extreme expression of a climate that has been a long time in the making.

Not only are there big unknowns in the timing of the MTC origin, the climatic progression is unclear. Was the predecessor a bimodal rainfall regime with progressive loss of summer rains or a winter drought/summer rainfall regime that switched the timing of both drought and rainfall. A potential model in southwestern North America is that the bimodal rainfall regime of Arizona chaparral was the early Tertiary climate and this gave rise to the winter-drought vegetation in northeastern Mexico and the summer-drought vegetation of California. If true this raises serious questions about the interpretation of phenology comparisons made between winter-drought and winter-rain regions (Verdú et al. 2002).

However, complicating the role of climate is the observation that on some landscapes substrate appears to have played a strong role in selecting for sclerophylly (Loveless 1961). This makes it problematical to categorize the selective environment responsible for the evolution of sclerophyllous MTV taxa. We are rather limited in our ability to draw firm conclusions about the exact selective environment driving plant traits. For example, the conclusion that sclerophylly in southwest Australian heathlands was strictly an adaptation to nutrient stress and neither drought nor fire played a role (Hopper 2003) is unwarranted. Despite the early origins of sclerophylly on poor soils, under a regional climate capable of supporting mesic forest this would not rule out a role for soil drought. Indeed, these coarse-textured soils are capable of generating significantly greater drought stress than other soils (Sperry & Hacke 2002). Annual droughts typical of MTC regions are not required to provide a selective advantage to sclerophyllous leaves as contemporary sclerophyll vegetation exists under aseasonal environments with longer-interval droughts such as in southeastern Australia. Also, although this latter region lacks a predictable annual fire threat, longer-interval cyclical fires are commonplace and there is little reason to assume this was not the case during the early evolution of sclerophyllous heathlands. Indeed, it has been argued that the absence of forests on nutrient-stressed sites cannot be accounted for without consideration of fire (Bond 2010), and thus fire has potentially been a factor since the origins of heathlands in South Africa and Australia. In short, interactions between climate, fire and geology need to be considered in understanding the selective environment of plant traits.

Selective Environment for MTV

Drawing conclusions about adaptations vs. exaptations in contemporary MTV requires consideration of the scale at which proximal selective factors operate. The niche relationships of the MTC paleofloras typically have been interpreted within a narrow niche space that includes mean temperature, total precipitation and its seasonal distribution (Raven & Axelrod 1978; Ackerly 2004a) or substrate

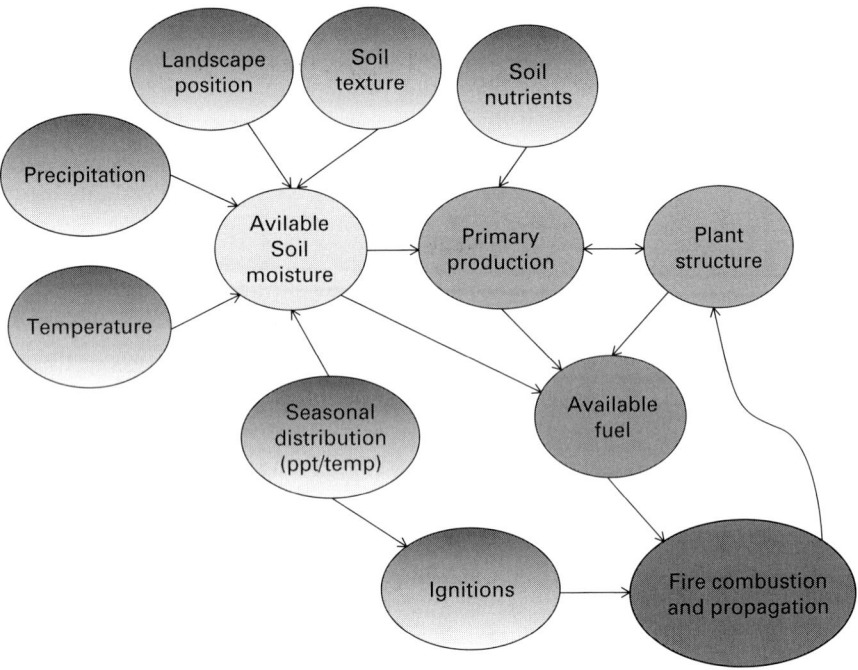

Fig. 10.5 *A conceptual model relating direct and indirect interactions between fire activity and available fuels, including the influence of plant structure, primary production and available soil moisture. Equally important in determining soil moisture are proximal site factors as well as regional climate.*

age (Hopper 2009). However, on fire-prone landscapes (see Fig. 1.3), climate and geology interact to affect soil moisture stress and this has direct effects on plant flammability as well as indirect feedback effects of fire on the evolution of plant traits.

In some ways the selective factors driving the origin of taxa, as well as specific traits that affect fire response, are a function of the scale at which the problem is viewed. It is common to think of seasonal climatic parameters as the most critical, but of course from the plant's perspective it is much more proximal factors (Fig. 10.5) that are most important. For example, seasonal patterns of primary production are directly affected by the water deficit or difference between potential evapotranspiration and available water (Stephenson 1990). Water deficit, or available soil moisture, is controlled by both precipitation inputs and by evaporative demand of the atmosphere. In MTCs drought occurs in the summer when high temperatures exacerbate soil-drought deficits by increasing evaporative demand. However, soil texture, soil depth and topographic drainage patterns are other factors, beyond the seasonal timing of drought, that contribute to soil moisture deficits. As an illustration, southwestern North American chaparral under bimodal rainfall conditions in Arizona may have summer water deficits

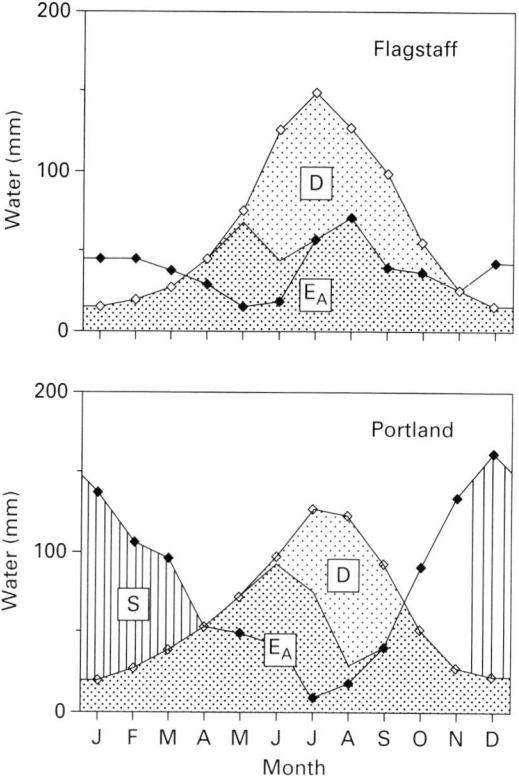

Fig. 10.6 *Water balance for MTC site and for bimodal winter rain and summer rain climate in Arizona. Actual evapotranspiration (E_A, dense stippling), precipitation (solid diamonds) and potential evapotranspiration (open diamonds) contribute to water deficit (D). S, winter surplus. (From Stephenson 1990.)*

equivalent to that in the MTC of Portland, Oregon (Fig. 10.6). Under summer-rain conditions, when evapotranspiration rates are substantial, plants on severe sites may actually be exposed to summer-drought conditions and unable to fully utilize this summer moisture input (Vankat 1989). A similar phenomenon is observed in the eastern Cape region of South Africa where MTV similar to the Cape fynbos is distributed under a climatic regime of even rainfall throughout the year, but due to high summer evapotranspiration it exists under essentially a winter-rainfall climate (van Wilgen 1984).

On any given landscape soil moisture stress can vary markedly dependent on microsite characteristics. Plants distributed on exposed rocky ridgelines or equator-facing slopes with shallow soils may experience soil drought stress when other parts of the landscape are less stressed, and plants sort out along such gradients. Thus, landscapes comprise a mosaic of plants tolerant of varying degrees of drought stress and with different strategies for coping with water stress (e.g. Mooney 1989; Mahall *et al.* 2010). Climatic seasonality plays a key role in

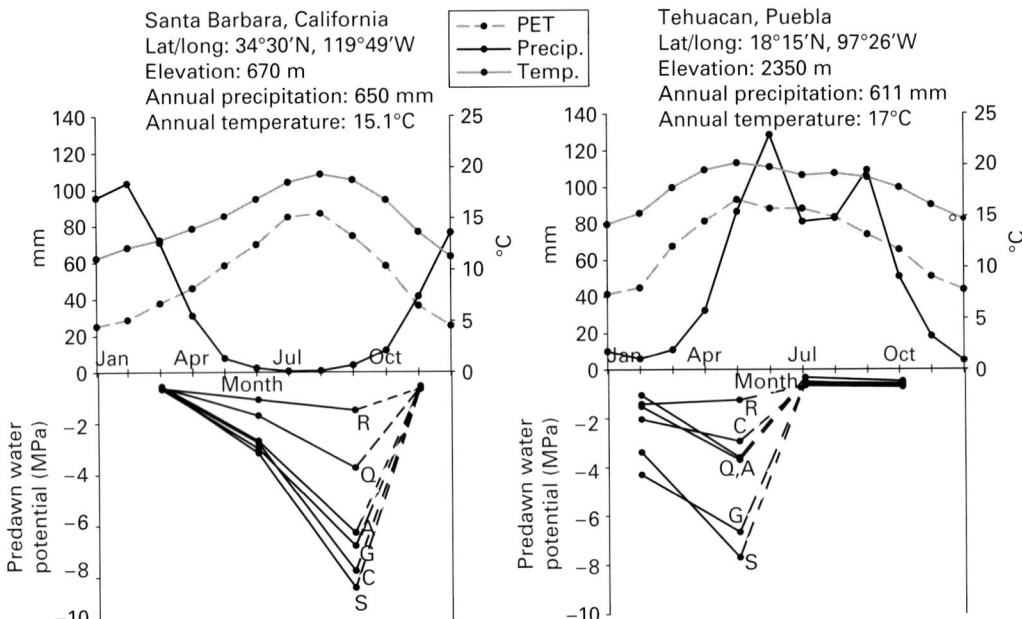

Fig. 10.7 *Comparison of climate diagrams for (upper left) a MTC California site with (upper right) a winter drought/summer rain climate in northeastern Mexico. Physiological responses of analogous functional types of the dominant MTV at both sites show the pattern of seasonal pre-dawn water potentials in (lower left) California and (lower right) Mexico. (From Bhaskar et al. 2007.) R, Rhus spp.; Q, Quercus spp.; C, Ceanothus spp.; A, Arctostaphylos in California and Comarostaphylos in Mexico; G, Garrya spp.; S, Salvia spp.; PET, potential evapotranspiration.*

dictating the portion of the environment experiencing potentially stressful soil drought conditions. However, within similar plant functional types, those under summer drought conditions appear to possess many of the same physiological drought responses as those in winter drought conditions (Bhaskar et al. 2007). This is supported by comparisons of water relationships between present-day MTC California chaparral and winter drought and summer rain Mexican chaparral where intra-site variations between functional types are greater than inter-site variations within the same functional type (Fig. 10.7). In other words different functional types have adapted to specific microsite characteristics of soil moisture availability more so than to synoptic weather patterns. It seems likely that trait evolution in semi-arid landscapes may follow very similar trajectories regardless of the seasonality of rainfall. This is not to suggest that the timing of rain is irrelevant since it can affect the amount of landscape exposed to drought stress (Fig. 10.1).

Thus, it may be instructive to think of MTV as being adapted to seasonal drought stress and the accompanying effects this has on plant traits and in turn their effect on fuel structure, which is a major determinant of the fire regime. It may well be that the major distinction between mid-Tertiary landscapes and Quaternary landscapes in most MTC regions is that in the former case seasonal drought-stressed sites occupied a much smaller proportion of the landscape and thus were subject to a less predictable fire regime.

The model that species track proximal soil moisture conditions more than average annual climatic patterns is perhaps not surprising when one considers that the life span of most species is between 1 and 30 Ma, and thus due to Milankovitch and other climatic cycles they are almost guaranteed to experience multiple climates (Bennett 1990). It is commonly believed that species simply migrate to stay matched with the precise climate under which they evolved. Alternatively though, species that have persisted through multiple climates may have adapted to features of their environment that are common across multiple climates. In light of this, Bennett (2004) has questioned the validity of trying to distinguish adaptations from exaptations (see Chapter 9). Over time as climates change there will always be reshuffling and ecological sorting of species into new assemblages. Despite this sorting process species may still be within their potential climatic niche and thus one need not necessarily view adaptive evolution and ecological sorting as mutually exclusive options.

The demonstration that one cannot adequately predict contemporary global vegetation patterns without including fire (Bond *et al.* 2003, 2005) might very well be applied to paleohistorical vegetation patterns as well. Most MTV taxa have traits that are reasonably interpreted as being derived in response to fire. In addition, plant traits appear to have the potential for significant effects on flammability and fire spread (Bond & Midgley 1995; Schwilk 2003; Bond & Keeley 2005). These considerations suggest that plants can control fire in ways that determine the ultimate assembly of MTC plant communities and thus climate may be only a subtle indirect effect.

Lastly, perhaps drought has been overemphasized as a selective factor in MTC taxa. What about the selective role of winter rains under mild temperatures? A substantial proportion of taxa have their life histories closely linked to the winter growing season, from germination to flowering. In this respect comparison of species in common between the California MTC chaparral and the Arizona bimodal rainfall chaparral is instructive. These regions share many species of shrubs and postfire herbs that are winter active (Fotheringham 2009). For example, flowering phenology shows little difference in two of the dominant shrub species that presently occur in both California and Arizona chaparral as they track winter rains and not summer rains (J. Vankat unpublished in Keeley & Keeley 1988). If we assume the Arizona chaparral represents an earlier climatic condition under which the MTC chaparral evolved (e.g. Axelrod 1989) then it is apparent that the selectively important features of the MTC appeared before a true MTC developed.

Considerations in Evaluating MTV Origins

The dominant species in MTV are sclerophyllous-leaved shrubs and trees but the communities in MTC regions comprise a diversity of other growth forms. Tracing their evolutionary history requires recognition of several factors:

(1) Although macrofossils provide the most extensive record of past landscapes, it is not an unbiased sampling of past environments; rather it can be highly biased against arid land plants and thus against fire-prone vegetation. Other than fossil sites resulting from volcanic ash, deposition of fossils usually occurs in wetland sites, and plants growing far from water are poorly represented (MacGinitie 1969; Spicer & Wolfe 1987; Robinson 1989). What portion of the upland vegetation is represented is not known, although it has been suggested that plants farther than 1000 m from a lowland deposition site may not be represented (Rich 1989). Illustrative of this is the observation that most North American Miocene or later fossil floras from western North America have one or more species of the water-dependent *Salix*, and commonly a half dozen or more other wetland taxa. Thus, poor representation of arid land plants in a fossil flora limits our ability to determine the origin of these taxa and history of arid land vegetation and ecosystem processes such as fire. As arid land vegetation expands there is a greater chance that lowland deposition sites will record these taxa. Bias is strongest with macrofossils, as pollen fossils may sample over a much wider part of the landscape. The macrofossil record is highly biased against herbaceous taxa, except for wetland taxa such as *Typha*, *Carex* and *Equisetum*. The pollen fossil record is less biased against growth form but heavily biased in favor of wind-pollinated taxa as well as being a less specific taxonomic indicator, usually only to genus and sometimes only to family. An additional bias is that arid land ecosystems may lack suitable deposition sites and thus not leave any fossil records (e.g. a possible example may be the lack of Miocene floras from southwestern North America).

(2) In trying to understand the origins of the MTC plant communities, the focus should not be based strictly on a reconstruction of events within the present-day MTC regions (e.g. Axelrod 1973, 1989). Climate considerations above suggest that semi-arid MTC may have developed in other regions and moved during the late Tertiary into the present MTC regions (e.g. Kemp 1981). It is to be expected that the origins of many MTV taxa were outside the present MTC region and those taxa migrated into present MTC regions in the late Tertiary or more recently.

(3) Macrofossils are often used to reconstruct the paleohistory of plant communities but since they fail to include most of the herbaceous component it could lead to erroneous conclusions about ecosystem structure. For example, contemporary California and Arizona chaparral are dominated by many of the same evergreen taxa, but Arizona differs markedly from California in the presence of C_4 bunchgrasses and duel winter and summer rain postfire floras (Fotheringham 2009). These two chaparral systems represent very different plant communities, but if the only information available were macrofossil assemblages one might conclude these were the same ecological assemblages, or if viewed over time they might be considered representative of ecological stasis arising from similar environments (e.g. DiMichele *et al.* 2004).

Northern Hemisphere Origins (Laurasia)

The evolution of contemporary sclerophyllous vegetation is largely a Tertiary phenomenon, a period that opened the Cenozoic Era 65 Ma. Modern dicotyledonous angiosperm floras became widespread in the early Tertiary, a time when the present-day continents of North America, Europe and Asia formed the massive continent of Laurasia. The continents separated during the Eocene (56–35 Ma), although plant migration routes between continents persisted further into the Tertiary (Raven & Axelrod 1974; Denk *et al.* 2010). At this time Laurasia vegetation was dominated by mixed deciduous hardwood, evergreen coniferous forests, mesic temperate woodlands and subtropical forests. We know almost nothing about azonal communities of more restricted distribution.

As the continents separated, a number of Eocene genera (or ones of earlier origin) were left in both western North America and Eurasia and these account for the present-day occurrence of woody sclerophyll taxa such as *Arbutus*, *Cercis*, *Cupressus* (*sensu lato*), *Pinus*, *Prunus*, *Quercus*, *Rhus* and *Rhamnus*, and non-sclerophylls *Aesculus*, *Juglans*, *Staphylea* and *Styrax*, in both regions (Axelrod 1975; Wolfe 1975; Fjellstrom & Parfitt 1995; Milne 2006). Although lacking a macrofossil record, the distribution of a number of herbaceous taxa (Raven & Axelrod 1978) also would appear to have been part of this flora as today they are represented in both North America and Eurasia; for example, *Antirrhinum*, *Datisca*, *Erodium*, *Galium*, *Helianthemum*, *Lotus*, *Salvia*, *Trifolium* and *Triodanis*. The conifer genus *Tetraclinis* extended widely across western North America in the Tertiary despite its restriction today to scattered parts of North Africa and the western Mediterranean Basin (Kvaček *et al.* 2000). Other Eocene or older chaparral taxa such as *Ceanothus* (Richardson *et al.* 2000b), *Heteromeles* (Phipps 1992), and *Malosma* (Miller *et al.* 2001) remained restricted to North America. *Nerium*, *Olea*, *Punica* and *Pyracantha* are Eocene taxa that apparently remained restricted to Eurasia (Palamarev 1989).

There is a tendency to see contemporary Mediterranean Basin floras as products of recent Quaternary events (Suc 1984), in part because of our greater familiarity with the climatic oscillations of that period. However, there is genetic evidence indicating that the modern geographical structure of Mediterranean plant populations may be traced back to the Tertiary history of taxa. Cork oak (*Quercus suber*), an emblematic MTV sclerophyll tree, exhibits patterns of genetic drift consistent with the Oligocene and Miocene breakup events of the continental margin and lack of detectable chloroplast DNA changes for the last 15 Ma (Magri *et al.* 2007). Thus, climates conducive to MTV appear to have been present for a significant portion of the Tertiary and may have led to much earlier origins than commonly assumed.

Although Eocene climates were considered to have been warm, mesic and equable, parts of southwestern North America were experiencing semi-arid climates (Peterson & Abbott 1979) conducive to sclerophyllous shrublands. In the middle Eocene in southern California, and elsewhere in North America, pollen

Fig. 10.8 *Islands of xeric chaparral (arrowed) within a matrix of more mesic woodlands in northern California as a model of Miocene chaparral distribution. (Photo by Jon Keeley.)*

profiles showed a decline in species diversity associated with decreasing precipitation and an increasingly severe dry season (Frederiksen 1991). Despite this lower community diversity, one might expect the landscape mosaic of mesic and arid assemblages would have contributed to higher beta diversity. The Eocene-Oligocene (35 Ma) transition resulted in continental-scale cooling and drying and set the stage for further expansion of sclerophylls on most continents in the northern and southern hemispheres (Raven & Axelrod 1972). By the mid-Oligocene drier and colder climates had spread sufficiently to eliminate many tropical lineages from temperate latitudes (Axelrod *et al.* 1991), although little consideration has been given to the interaction between climate and fire in driving these changes.

In North America, sclerophyllous chaparral vegetation came into its own in the Oligocene (Wing 1987), although unlike contemporary landscapes dominated by pure chaparral, this early chaparral was restricted to rocky well-drained slopes, rain shadows or coarse-grained low-fertility substrates, and formed a mosaic with more mesic woodlands (e.g. Fig. 10.8). The Creede flora (27.2 Ma) in southern Colorado is one example (Wolfe & Schorn 1989). The climate was dominated by winter rain or snow and summer rainfall was low. This chaparral occupied the drier slopes around a mixed woodland and comprised largely two species of *Cercocarpus*, plus a few other chaparral taxa such as *Mahonia* and *Ribes*. Wolfe and Schorn (1989) disagreed with Axelrod's (1987) contention that these chaparral elements were understory species in woodlands and posited that they likely were present on more arid ridgelines and other surrounding sites. In western North America and as far south as southeastern Mexico, *Cercocarpus* was common in the fossil record through the Oligocene and early Miocene (de León & Cevallos-Ferriz 2000). Contributing to its success were its light plumose wind-dispersed fruits (only taxon in contemporary chaparral with well-developed wind dispersal), which would have had significant advantages in finding suitable habitat when such sites were dispersed as arid islands within a landscape of more mesic woodlands.

On early Miocene sites in the interior (e.g. present-day Nevada) most chaparral genera were represented, often by taxa indistinguishable from modern species. These included species of *Arctostaphylos*, *Ceanothus*, *Cercocarpus*, *Fremontodendron*, *Garrya*, *Heteromeles*, *Mahonia*, *Malosma*, *Prunus*, *Quercus*, *Rhamnus*, *Rhus* and *Ribes*. Through the Miocene, as rainfall seasonality increased, chaparral patches expanded, and increased patch size almost certainly contributed to increased fire activity in these seasonally dry habitats with highly flammable sclerophyll foliage.

Throughout the early to middle Miocene (~17 Ma), associations with affinities to contemporary chaparral were best developed away from the coast in more arid and less equable interior parts of North America, such as the western edge of the Great Basin (Axelrod 1939, 1973, 1985; Wolfe 1964). Middle Miocene deposits from northern parts of the Great Basin reveal that *Artemisia* shrub and grassland dominated these landscapes much like they do today (Davis & Ellis 2010). Chaparral elements likely developed further south, but we have little information since there is a lack of Miocene fossil floras for most of southwestern North America. The present-day chaparral patches that persist in bimodal-rainfall Arizona or the winter drought, summer rain climate of northeastern mainland Mexico (see Fig. 5.1) may be relics of a broad Tertiary distribution of chaparral taxa in southwestern North America, even though chaparral was not dominant in coastal California at that time.

Arctostaphylos (Ericaceae) and *Ceanothus* (Rhamnaceae) are two of the largest woody genera in North America and their center of radiation is in California chaparral; species are also present in Arizona and northeastern Mexican chaparral. The latter genus is known from the early Tertiary fossil record at high latitudes; however, *Arctostaphylos* is not recorded until the Miocene, when chaparral elements had already moved into southern California (Axelrod 1939). The lack of an earlier history is something of an anomaly. The nearest relative appears to be *Arbutus*, but not the present North American species, rather the Mediterranean Basin taxa (Hileman et al. 2001). This is consistent with an earlier origin for *Arctostaphylos*, perhaps prior to the Eocene splitting up of the continent of Laurasia, which included the present-day continents of North America, Europe and Asia. Alternatively, allies of the Mediterranean Basin *Arbutus* could have persisted in North America after the breakup of Laurasia; however, the several *Arbutus* taxa known from Tertiary deposits have all been allied with North American species (e.g. Axelrod 1939; Wolfe 1964). We hypothesize that *Arctostaphylos* originated early in the Tertiary and persisted for much of the Tertiary in arid shrublands of southwestern North America, but due to the lack of fossil assemblages from these arid lands, its history is not fully recorded. This would be part of the Southwestern Chaparral element Axelrod (1950) recognized, and presumed remnants of this persist in the summer rainfall chaparral of Arizona and northeastern Mexico.

By the late Miocene (7 Ma), chaparral had migrated into southern California (Axelrod 1982). At this time the seasonal distribution of rainfall had changed substantially and climates were progressively looking more like a MTC (Fig. 10.2). As drought shifted toward summer, the combination of low rainfall and high temperatures increased the portion of landscape favoring fire-prone sclerophyllous vegetation. Also, as the inland sea in California receded and the coastline moved westward to its present location, more extreme winter temperatures and lower rainfall likely combined to reduce chaparral in the interior parts of the southwest, leaving it largely a Californian vegetation.

One feature of the fossil record is that the many arid-adapted woody taxa are unknown prior to the formation of the MTC in the late Pliocene. For example, in the Mediterranean Basin the very widespread Cistaceae genera, *Cistus*, *Fumana* and *Helianthemum*, and Fabaceae *Cytisus* and *Genista* are unknown from these fossil assemblages and likewise for the Californian taxa that occupy the most arid end of the moisture gradient, such as *Salvia* spp. (Lamiaceae), *Hesperoyucca* (*Yucca*) *whipplei* (Liliaceae) and *Adenostoma fasciculatum* (Rosaceae). This observation has been used to infer a relatively recent origin for these taxa in response to the MTC (Axelrod 1973; Herrera 1992; Valiente-Banuet *et al.* 2006). However, not only are these arid fire-prone sites unlikely places for fossil deposition, but other evidence calls this generalization into question. In the Mediterranean Basin, some fossil records and molecular clock estimates place the origin of *Cistus*, *Fumana* and *Helianthemum* in the Miocene or earlier (Guzmán & Vargas 2005, 2009a).

Illustrative of the inadequacy of the fossil record to address questions of origin for semi-arid fire-prone vegetation is *Adenostoma fasciculatum*. It is the most ubiquitous shrub in California chaparral and it not only has no Tertiary history it has almost no Quaternary fossil record; it is only known from the late Pleistocene, in a packrat midden (Wells 2000), pollen in marine sediments (Heusser 1978) and in the La Brea tarpits (Warter 1976). The genus is morphologically quite distinct from its nearest relative, *Chamaebatiaria*, a desert shrub from the Great Basin and recorded (or a close relative) from Oligocene fossils (Wolfe & Schorn 1989). Morphologically the only other *Adenostoma*, *A. sparsifolium*, appears more closely aligned with *Chamaebatiaria* than *A. fasciculatum*. Both *Chamaebatiaria* and *A. sparsifolium* are obligate resprouters and *A. fasciculatum* is a facultative seeder. When the switch from obligate resprouter to facultative seeder occurred is unknown, but a Tertiary origin for *A. fasciculatum* cannot be ruled out. Conceivably it was part of the interior Tertiary Southwestern Chaparral (Axelrod 1950) and is absent from Arizona and northeastern Mexican chaparral today due to its apparent intolerance of low winter temperatures (Keeley & Davis 2007).

Other than a few wetland species, herbaceous plants are absent from most macrofossil records, which is unfortunate because postfire floras in northern hemisphere MTC shrublands are dominated by a diverse assemblage of annuals and we know relatively little about their origins. Raven and Axelrod (1978) considered much of the diversification of herbaceous groups in California to be concentrated in the Pleistocene because this was the time of greatest intensification of the summer drought climate. Although this may be true, it says little about the origin of this postfire flora, which is clearly not dependent on summer drought as there is a diverse postfire winter annual flora in the bimodal winter and summer rainfall climate of Arizona chaparral, including many of the same taxa as in California chaparral (Fotheringham 2009). This postfire ephemeral flora appears to be highly dependent on winter rainfall and not on the duration or timing of drought, although California's MTC may have played a role in speciation of these taxa. Based on what is known about the Tertiary history of winter rain there is every possibility many of these lineages had very early origins.

The Mediterranean Basin history is broadly similar to western North America. Being part of Laurasia, the basin shares a lot of early history with North America and the early Tertiary sclerophyll forest vegetation was quite similar to that in western North America (Axelrod 1975). The two regions have a number of genera in common due to vicariance of taxa originating in the Eocene or earlier. As in North America, sclerophyllous vegetation arose in the Tertiary on edaphically generated drought-prone sites on a landscape dominated by more mesic broadleaf forests, and then expanded in the mid to late Miocene (Palamarev 1989; Barrón *et al.* 2010).

One landscape difference between these two regions is the broad latitudinal gradient over which plants could shift their distribution in response to repeated climatic changes in North America and more limited opportunities in the Mediterranean. Plant migration may have been inhibited at times due to the Mediterranean Sea, although it most definitely was not a complete barrier to long-distance dispersal (Guzmán & Vargas 2009b).

In both Europe and North America, *Pinus* is in some respects the *Eucalyptus* of the northern hemisphere in that it is a fire-adapted sclerophyllous-leaved tree that dominates MTC ecosystems. Also, like the eucalypts, the pines predate the MTC and today are widely distributed far outside the MTC of both Europe and North America, as well as in Central America and Asia (Richardson 1998). The origin of *Pinus* dates to the Mesozoic (Willyard *et al.* 2007), but its radiation is a Cenozoic phenomenon (Millar 1998). Sometime in the Cretaceous the Diploxylon subgenus separated from the Haploxylon subgenus. Many of the latter taxa have radiated into climatically stressful habitats such as deserts or subalpine environments, whereas most of the Diploxylon taxa are tied to fire-prone habitats (Keeley & Zedler 1998; Schwilk & Ackerly 2001). Faced with intense competition from the rapidly radiating angiosperms, pines and other gymnosperms were at a decided disadvantage in warm and mesic conditions (Bond 1989). As a consequence, climates through much of the Eocene limited pine distribution to either high latitudes or high elevations, in both Europe and North America (Millar 1998).

It is thought that an important time of radiation in *Pinus* was the Oligocene as the marked cooling and drying conditions would have greatly expanded potential habitats, in both Europe and North America (Klaus 1989; Millar 1998). At this time in North America, pines were successful along the southern part of the Rocky Mountain Cordillera. *Pinus ponderosa* or allied taxa expanded throughout this range down into southern Mexico. *Pinus ponderosa* moved into California in early Miocene, before an intense summer drought climate had developed, and even today its widespread distribution indicates no specific relationship to summer drought climates (see Chapter 5). Based on molecular studies it appears that early to middle Miocene was the time of greatest radiation in *Pinus* (Willyard *et al.* 2007).

Southern Hemisphere Origins (Gondwana)

Sclerophyllous-leaved MTV, characteristic of dominants in MTC climates today, has been present on Australian landscapes since at least the Eocene, and some lineages have persisted to the present (Hill & Brodribb 2001). Throughout the warm and humid conditions of the Eocene, much of Australia was dominated by rainforest trees, podocarp gymnosperms and pteridophytes. At this time the region of contemporary MTC, southwestern Australia, had *Nothofagus* (Fagaceae) dominants, with Proteaceae and Casuarinaceae subdominants. However, even at this very early stage of the Tertiary, sclerophyllous taxa such as *Banksia* and *Dryandra* (Proteaceae) were present (Beard 1977) and by late Eocene in central Australia, rainforest was displaced by sclerophyll vegetation on slopes and ridges (Martin 2006). Sclerophylls may also have played a successional role following dry periods that are indicated by sedimentological evidence in the form of silcretes (Kemp 1981). Prominent sclerophylls such as *Banksia* were already quite diverse (Hill & Christophel 1988). They were pinnately lobed taxa that are today mostly restricted to the southwest.

In southwestern Australia vast expanses of landscape have nutrient-deficient soils that have been continually weathered since the end of the Cretaceous and after greatly accelerated lateritization in the Oligocene and Miocene (Johnstone *et al.* 1973). These oligotrophic soils favored the spread of large contiguous populations of xeromorphic heathland and other sclerophylls (Hopper 1979). The present-day sclerophyllous flora is thought to have evolved from old lineages that originated during the Late Cretaceous or early Tertiary in isolated pockets of highly weathered nutrient-deficient soils on landscapes dominated by rainforests. Fossil records show that half the families and genera in kwongan heathland were present in the Eocene (Lamont 1982).

This sclerophyll vegetation developed in response to nutrient-poor lateritic soils and has persisted through a remarkable range of climatic changes (Hopper 1979, 2009), although it has been argued that fire played a more important role in determining this competitive balance (Bond 2010) and this will be discussed more fully in the next subsection. Hill (1994) hypothesized that sclerophylly originated in response to nutrient-deficient soils and this preadapted those taxa to xeric conditions. One test of this hypothesis has been comparison of evolutionary transitions that include xeromorphic traits. One such trait in Proteaceae is the stomatal crypt, where stomata are deeply sunken inside surface indentations of the leaf. In Australian Proteaceae stomatal crypts have evolved at least 11 times, always associated with lineages in arid environments (Jordan et al. 2008). Although sunken stomata are associated with a number of selective factors (Gibson 1983), in Proteaceae their origin has been interpreted as an indicator of aridity and evidence that early origins of sclerophylly in response to nutrient-poor soils pre-adapted these taxa to aridity. This idea has been supported by phylogenetic reconstructions of lineages that have evolved stomatal crypts (Mast & Givnish 2002).

In southern Australia Hill (1994; Paull & Hill 2010) suggests that Casuarinaceae taxa such as *Casuarina* and *Allocasuarina* were present in the early Tertiary rainforests but not found in the fossil record because they were restricted to dry microsites such as ridgetops, and they expanded rapidly when the climate dried later in the Tertiary. Other mid-Tertiary radiations include the rainforest progenitors of the Elaeocarpaceae that radiated widely during the Oligocene and Miocene, producing a diversity of arid-adapted shrubs (Crayn et al. 2006). In the late Oligocene to early Miocene there was a shift away from *Nothofagus*-dominated forests toward forests dominated by sclerophyllous-leaved Myrtaceae and Casuarinaceae, more characteristic of the contemporary MTV found throughout southern Australia (Hill et al. 1999).

Eucalyptus (Myrtaceae) is not only the most diverse but also the most dominant component of Australian landscapes. These sclerophylls thrive in nutrient-poor arid sites and although fossils date the group to at least the Eocene, usually in association with mesophyllous rainforest, they likely occupied more arid or nutrient-poor sites within those communities (Holmes et al. 1982). As with the northern hemisphere *Pinus*, Miocene was the time for *Eucalyptus* expansion, and given its exceptional fire-surviving and even fire-promoting adaptations, it probably evolved in environments with a high fire frequency. Grasslands also developed during this period, likely promoted by fire (Bowman 2000; Bond et al. 2003; Keeley & Rundel 2005) and it is interesting to speculate what sort of coevolutionary relationship they might have had with the eucalypts.

Origins of South African MTC fynbos and related shrublands are poorly recorded in the macrofossil record but some palynological records exist. One characteristic that appears similar to Australia is the widespread presence of sclerophyllous-leaved Proteaceae since the Eocene (Coetzee et al. 1983). Sclerophyllous taxa originated in arid pockets during the Oligocene (Goldblatt 1978) and many fynbos elements were present in small patches of "skeletal soils" in the mountains, where soils were too shallow to support forest (Cowling et al. 2009).

As was the case in California, the equable conditions along the coast during the early Miocene supported temperate and tropical forest elements and few representatives of fynbos sclerophylls (Coetzee & Muller 1984). Pollen evidence shows that by middle Miocene these coastal riverine sites were gallery forests of mixed tropical forest but with fynbos well established in the area, despite the presumed winter drought, summer rain climate (Coetzee & Rogers 1982).We lack much of a Miocene record from interior, and potentially more arid and climatically seasonal sites thought to be sites of early fynbos evolution (Axelrod & Raven 1978). An early origin of fynbos taxa is necessary to support many of the patterns of speciation evident from molecular studies.

Although it has long been the belief that speciation in both the Cape of South Africa and southwestern Australian sclerophyll shrublands is largely a recent phenomenon, this is not true for many lineages (Edwards & Hawkins 2007; Hopper et al. 2009; Sauquet et al. 2009a). For example, it appears that much of the present diversity in the Cape MTV is the result of recruiting diverse lineages over the entire Tertiary, such as several clades of Iridaceae, *Pelargonium* (Geraniaceae), and *Indigofera* (Fabaceae) (Linder 2005), *Muraltia* (Polygalaceae) (Forest et al. 2007), *Protea* (Valente et al. 2009), and the legume tribe Podalyrieae (Boatwright et al. 2008). It is estimated that 60% of the Cape flora originated in the Tertiary and the remainder in the Quaternary (Linder & Hardy 2004), although the vast majority have apparent origins within the last 10 Ma (Verboom et al. 2009). Other prominent South African taxa such as the Ericaceae were apparently absent in the region until late Miocene and so a more recent radiation is inferred (Deacon et al. 1983). Molecular clock estimates for the Poaceae *Ehrharta* indicate most speciation was confined to the Miocene and Pliocene (Verboom et al. 2003) and Quaternary radiations are evident in the Brassicaceae *Heliophila* (Linder 2005; Mummenhoff et al. 2005).

Within lineages dominant in both Australia and South Africa such as the Restionaceae, the radiation appears to have started substantially earlier in Australia. Rates of lineage accumulation have been constant since the early Tertiary in Australia whereas there was a Miocene acceleration in the Cape, differences perhaps due to different geomorphological histories (Linder et al. 2003). Eocene to early Oligocene radiations are also evident in Australian *Eucalyptus*, Casuarinaceae and *Banksia*, suggesting Australian radiations were earlier than South African radiations (Linder 2005). In the Haemodoraceae both regions are characterized by pulses of speciation since middle Miocene (Hopper 2009). Although the southwest Australian flora is more phylogenetically diverse, it includes numerous lineages with low diversification rates (Sauquet et al. 2009b).

Thus, the present-day MTV Gondwana landscape comprises a mix of lineages with Tertiary radiations and with Quaternary radiations, as well as Tertiary relicts that failed to radiate (Hopper & Gioia 2004; Linder 2005; Hopper et al. 2009). Gradual increases in seasonally dry conditions during the Miocene and lack of any single trigger seem to characterize radiation of these southern hemisphere MTC floras (Linder 2005). While this may be true overall, another way to express this is that lineages were stimulated by apparently different triggers at different times in the Miocene, Pliocene and Pleistocene. In the relatively short time since the beginning of the Pliocene a substantial amount of diversification has occurred as the MTC intensified and presumably the amount of fire-prone landscape increased.

In the third southern hemisphere MTC region in central Chile, temperate woodland communities extended across the continent in the Oligocene, and left biogeographic evidence of their past presence in the disjunctions of numerous woody plant genera between central Chile and southeastern Brazil. Examples of such disjunctions include *Araucaria, Bomarea, Crinodendron, Lithraea, Mutisia, Myrceugenia, Perezia, Persea, Podocarpus, Quillaja, Viviana,* and *Weinmannia* (Landrum 1981). These Tertiary forests gave rise to a mixed sclerophyllous forest by early Miocene, remnants of which survive today in relict patches throughout the coastal ranges (Hinojosa *et al.* 2006). Taxa with affinities to contemporary MTC *Kageneckia* (Rosaceae) were present in a Paleocene fossil flora of southern Chile (Axelrod *et al.* 1991). Miocene matorral bordered more mesic forest and occurred in small, isolated pockets, but relatively little is known of the late Tertiary vegetation changes that led to its spread. Sclerophyll forests of the Coast Ranges of central and southcentral Chile today show strong legacies of the pre-Pleistocene structure and composition of sclerophyll forest paleofloras. These forests and woodlands have been biogeographically isolated for several million years, but had relatively limited impact from glacial events in the Pleistocene. The similarity between these contemporary forests and late pre-Pleistocene paleofloras indicates evolutionary stability of surviving lineages under relatively moderate climate cycles (Villagrán and Armesto 2005; Hinojosa *et al.* 2006).

In summary, as in the northern hemisphere, sclerophyllous taxa that gave rise to MTV lineages developed gradually during the mid- to late Tertiary and were widely distributed by the mid-Miocene. Sclerophyll shrubs persisted in pockets of stressful substrates with limited nutrient and/or water resources. However, the widespread persistence of highly weathered low-fertility soils in southwestern Australia potentially gave rise to an earlier dominance of sclerophylls (Hopper 2009). It is widely believed that the Quaternary Period was a time of rapid speciation in all MTC regions, presumably due to rapid climate changes, alternating glacial and interglacial episodes, increasing aridity and mountain building, and fires. However, the uniqueness of these changes has been questioned (Bennett 2004; Sauquet *et al.* 2009a) and molecular studies fail to support a model of sudden Quaternary radiations in most southern hemisphere MTC lineages. More broadly, Willis and Niklas (2004) tested whether Quaternary speciation rates were substantively higher than for previous periods and found no evidence to support this assertion. The evidence for MTV origins in both the northern and the southern hemispheres shows that species radiations varied in their timing between different lineages. Some groups radiated throughout the Tertiary and others appear to have radiated more recently in the Quaternary.

Fire and Paleohistory of MTV

Northern Hemisphere Fire Responses

Throughout the Cenozoic, landscapes in the Mediterranean Basin and western North America have had sufficient seasonality and fuels to generate fire-prone conditions in marginal habitats with flammable sclerophyllous vegetation. Although there are reports of Tertiary fires from both regions (e.g. Frederiksen 1991; Cichocki 1998; Figueiral et al. 1999, 2002; Kolcon & Sachsenhofer 1999), how widespread fire was is unknown. Miocene and Pliocene macrofossil floras from western North America are relatively silent on the issue of fire, and it is not clear if this is because fires were rare or because of a lack of adequate searching, or because the macrofossil record has an inherent bias against recording semi-arid fire-prone landscapes. Considering the widespread presence of fire-adapted forest and shrubland taxa in the region today, many of which appear to be little changed since their Tertiary origin, it seems unlikely fire was not a feature of some ecosystems.

Evidence for Tertiary fire is substantially lower than that for Paleozoic and Mesozoic fire (see Chapter 9). Part of this may be methodological in that Mesozoic fire evidence is derived from microfossil charcoal (Scott 2000) and Tertiary studies in western North America are largely based on macrofossil leaf impressions (e.g. Axelrod 1982; Wolfe et al. 1998). Also, the Mesozoic fire evidence is largely from mire peatlands and these habitats are relatively uncommon in western North America. Nonetheless, a comparison of peatlands through time suggests a marked decline in peatland fire activity from the Mesozoic to the Tertiary (Diessel 2010). Factors responsible for this fire pattern are unclear and it is unknown whether or not it is a useful signal for fire activity on the broader landscape. Additionally, it is unknown whether or not this change in fire activity is tied to the Tertiary shift from gymnosperm and pteridophyte dominated to angiosperm dominated landscapes, although this has been proposed (Bond & Scott 2010). The potential for Tertiary expansion into marginal semi-arid habitats that the angiosperms encouraged could have resulted in altered global fire regimes and a change in landscape distribution of fire so that peatlands no longer comprised the dominant fire-prone landscape.

One of the few northern hemisphere Tertiary studies of fire is based on estimates of charcoal deposition in Pacific Basin marine sediments (Herring 1985). This study supports the idea that fire has been present through the last half of the Tertiary but the amount of highly flammable landscape increased 10–100 fold in the late Tertiary (Fig. 10.9). Although there are good reasons for implicating late Tertiary climatic changes in this increased fire activity, equally important were structural changes in plant fuels at both an individual as well as the landscape scale. This late Tertiary increase in fire activity has been linked to substantial ecosystem shifts in subtropical environments from woodlands to C_4-dominated grasslands (Keeley & Rundel 2005). It is apparent that in

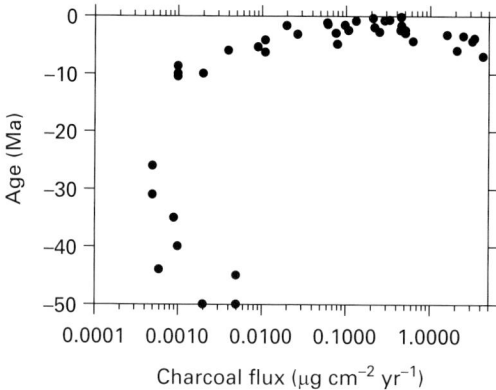

Fig. 10.9 *Estimated charcoal flux from aeolian transport to a Pacific Ocean deep drilling core. (Data from Herring 1985.)*

subtropical latitudes pockets of these C_4 grasses persisted through much of the Oligocene and Miocene but expanded rapidly at the end of the Miocene to form extensive grasslands, and this appears to be fire driven. This is a useful model of fire-adapted MTV, which likely persisted in isolated fire-prone pockets for a significant part of the Tertiary and then greatly expanded at the end of the Tertiary.

The Herring (1985) data cannot reflect much about fire activity in the Mediterranean Basin but samples from the eastern Pacific may reflect fire activity along the Pacific Coast of North America. The timing of increase in fire activity (Fig. 10.9) was coeval with the late Tertiary expansion of chaparral in coastal California. In this case early Miocene islands of chaparral on drought-prone substrates, embedded in a landscape of less flammable woodlands, expanded as the intensification of the MTC extended the drought-prone landscape. As the sclerophyllous MTV expanded its range it would have increased the predictability of fire activity, which in turn would have fed back into an even greater expansion of chaparral. An empirical model of how this process might have worked is illustrated by the demonstration on oceanic islands of increasing fire activity with increasing island size due to the greater probability of lightning-ignited fires on larger islands (Wardle *et al.* 1997). Additionally, in the case of chaparral, feedback processes would potentially have affected selection for flammability and other fire traits. As chaparral flammability increased it would have further eroded mesic woodlands already losing ground to climate changes, as outlined by Jackson (1968) and Bowman (2000) for the replacement of tropical forests by Australian sclerophyll forests. The paleoclimate data indicate an increase in summer drought in the late Tertiary, which further contributed to greater expanses of fire-prone landscape (e.g. Fig. 10.1).

The marine charcoal deposition record (Fig. 10.9) undoubtedly represents a localized record of fire activity on the west coast of North America. Due to the prevailing westerly winds these records would not provide an adequate picture of fire activity in the more interior regions. This is unfortunate because macrofossil records suggest an early Miocene rise of chaparral elements in the southwestern portion of the Great Basin. On the basis of climate models it seems likely that semi-arid chaparral-like vegetation was more widely distributed at this time in the southwestern portion of North America, but there is a lack of Miocene fossil sites from the region to confirm this. Depending on the extent of fire-prone landscape, fire activity could have started earlier in this southwestern chaparral than suggested by Herring's (1985) data. Conceivably, much of the early evolutionary history of MTV and fire may have occurred outside California and some fire-adaptive traits may have been present at the time of late Miocene expansion of chaparral into California.

Can we place the origin of fire-adaptive traits, such as resprouting, lignotubers, flammability, fire-dependent recruitment and obligate seeding (see Figure 9.5) within the context of the geological history for chaparral? The majority of genera are known from Eocene fossils; all are vigorous postfire resprouters and some have lignotubers (see Chapter 9), suggesting an early origin for resprouting and perhaps lignotubers as well. Whether or not these traits arose in response to fire is not known; however, sclerophyllous shrubs likely inhabited patches of fire-prone landscape since their origin and thus there is little reason to rule out fire as a potential selective factor.

Perhaps the trait most clearly associated with fire is postfire seeding, as the delay in reproduction to a single postfire pulse of recruitment would require, among other things, a highly predictable fire regime (see Figure 9.5). Several genera with postfire seedling recruitment, and roots back to the Eocene, include *Ceanothus*, *Garrya*, *Fremontodendron* and *Malosma*. This is a highly conserved trait as most if not all species in these genera have similar recruitment patterns, and this trait may have an early origin as well.

One of the traits with the clearest dependence on a predictable fire regime is the risky strategy of coupling loss of resprouting ability with postfire seeding; the obligate seeding mode. This event likely took place early in some lineages. For example, in the species-rich Californian shrub genus *Ceanothus*, the split between *Ceanothus* and *Cerastes* subgenera is placed in the Oligocene (Jeong et al. 1997), and in the latter group obligate seeding is highly conserved as all species in this subgenus are of this functional type. The other speciose genus in chaparral with obligate seeding species is *Arctostaphylos* and they first show up in the fossil record in the early Miocene (Brown 1935; LaMotte 1952), although for reasons discussed above they potentially originated earlier in the southwestern USA. An early origin for obligate seeding in this genus is suggested by the presence of this strategy in species endemic to Tertiary substrates (e.g. *A. myrtifolia*, Gankin & Major 1964) and the presence in species that occur far outside California, including both winter drought, summer rain chaparral in northeastern Mexico and bimodal rainfall sites in Arizona. Unlike the *Cerastes* lineage of *Ceanothus*, obligate seeding is not highly conserved and associated with habitat shifts in *Arctostaphylos*, and there have been switches between this and facultative seeding in several subspecific *Arctostaphylos* taxa (see Chapter 9).

Postfire seeding is also widespread in the Mediterranean Basin and includes numerous shrubs: *Erica* spp. (Ericaceae), most Cistaceae, and many Fabaceae and Lamiaceae, the majority of which recruit from dormant soil-stored seedbanks (see Chapter 4). In the Cistaceae, *Cistus*, *Fumana* and *Helianthemum*, among others, are mostly postfire seeders and all have origins in the Miocene or earlier (Guzmán & Vargas 2005, 2009a). *Cistus* is of interest because it is one of the few truly woody obligate seeding genera in the Mediterranean Basin, and this trait is highly conserved (see Table 3.4). *Cistus* appears to have had an early origin in the Oligocene outside its present range but radiated sometime after late Miocene within the Mediterranean Basin. Noteworthy is the fact that the *Helianthemum* has one disjunct species that is a postfire obligate seeder in California chaparral, although whether this represents a vicariance event from an early origin or a recent long-distance dispersal event is unknown.

Postfire seeding from serotinous cones characterizes many gymnosperms associated with crown fire shrublands in the Mediterranean Basin and California: *Pinus* spp. and Cupressaceae taxa (*Cupressus*, *Hesperocyparis* (New World *Cupressus*) and *Tetraclinis*). There are numerous life history traits that are correlated with serotiny that enhance postfire recruitment success (Keeley & Zedler 1998) and patterns of trait evolution support the conclusion that fire has played a major role in the evolution of serotiny (Schwilk & Ackerly 2001). California serotinous pines in subsection Attenuatae (*Pinus attenuata*, *P. muricata* and *P. radiata*) originated in Mexico or Central America during the middle Tertiary (Millar 1998), suggesting an early association with predictable fire regimes. Axelrod (1988; Axelrod & Deméré 1984) contended that by late Miocene they dominated a substantial portion of the southern California landscape, but with the Quaternary intensification of the summer drought were restricted to islands within a chaparral-dominated landscape. Alternatively, Millar (1999) has suggested the current disjunct populations are not the result of contraction of a broader distribution but a long-standing metapopulation pattern due to periodically fluctuating climates.

Very little is known about the timing of fire-adaptive traits in herbaceous taxa. The precise timing of recruitment to postfire conditions in the highly diverse California chaparral herbaceous flora indicates a highly evolved association with fire. This association is so extraordinarily linked with fire these taxa are often referred to as pyro-endemics (see Chapter 5). Soil-stored seedbanks can survive more than a century and dormancy is broken by heat shock in some species but in the majority of lineages it is broken by chemicals from biomass combustion. Families with this "smoke"-stimulated germination include Hydrophyllaceae, Scrophulariaceae, Boraginaceae and Asteraceae to name just a few. The origin of smoke-stimulated germination is unknown but is potentially very ancient (Pausas & Keeley 2009). The postfire annuals in California chaparral are very diverse but little is known about the timing of diversification. Raven & Axelrod (1978) implied that this was a recent Quaternary phenomenon; however, there is little evidence of this and earlier origins cannot be ruled out. It is widely believed that speciation is relatively rapid in herbaceous taxa and there is empirical evidence for this (Smith & Donoghue 2008). The primary factor appears to be the shorter life span in herbaceous plants that increases the number of generations relative to woody plants. However, in the case of postfire annuals, this is not true since they do not complete a life cycle annually but rather only after fire, much like woody postfire seeders (C.J. Fotheringham personal communication).

In the Mediterranean Basin, diversity of annuals and other herbaceous species increases after fire and some have germination that can be stimulated by heat and smoke (see Chapter 4). However, there is no specific postfire ephemeral flora as in California. Essentially all postfire species are widely distributed in other disturbed habitats and few if any have deep seed dormancy that is only triggered by fire cues. There are several possible hypotheses to explain this difference: (1) fire has played a longer and more selective role in California chaparral (Pausas *et al.* 2006b); (2) a longer human presence in the Mediterranean Basin (see Chapter 13) has disrupted natural processes, causing the extinction of postfire endemic taxa; or (3) there was a former pyro-endemic flora that has undergone selection during the Holocene for occupying other anthropogenic disturbances.

Not all northern hemisphere sclerophyllous shrubs that comprise an important part of MTC shrublands have embraced fire by modifying their reproductive cycle to delay recruitment to postfire conditions. Many taxa persist in the face of repeated fires by vigorous resprouting that rapidly recoups prefire canopies. All have Tertiary origins and some authors have suggested these are just the lucky survivors of random extinctions (Herrera 1992; Valiente-Banuet *et al.* 2006). Alternatively, a strategy of not concentrating reproduction to a single postfire pulse could be very adaptive for these obligate resprouters as they invariably occur on more mesic fertile sites where rapid resprouting limits gaps for postfire seeding (see Chapter 9).

Southern Hemisphere Fire Responses

Charcoal levels indicate fire was present but infrequent in the Eocene of southern Australia, it increased in the Oligocene, and was substantially higher in the Miocene (Fig. 10.9) (Martin 1996; Kershaw *et al.* 2002). A similar middle Miocene spike in fire activity has also been reported for New Zealand (Pole 2003). These patterns suggest changes in fire regimes but they can be interpreted multiple ways. Eocene fires could have been generally less frequent, as this record implies, or they could have been much more frequent but occurring as localized fires that did not frequently contribute evidence of fire to deposition sites (Martin 1996). In other words fires could have been concentrated on localized sites exposed to annual cycles of soil aridity or to broader areas subjected to periodic droughts at a decadal or longer scale. Due to the bias inherent in the fossil record, such sites would be poorly represented or missed entirely. Evidence that this likely is the case comes from molecular phylogeny studies of the Myrtaceae. Distantly related genera exhibit a seemingly unique capacity to regenerate after fire from an anatomical anomaly comprising strips of meristematic tissue buried beneath the full thickness of the bark (Burrows 2010). This trait appears to be rather ancient in the group and suggests an important presence of fire in the environment throughout much of the Tertiary (Crisp et al. 2011).

Kemp (1981) suggested that the Tertiary trend of replacement of rainforest by open sclerophyll vegetation may have been either a response to climate change or to fire activity opening up closed-canopy forests and drying out understory fuels. As suggested in Fig. 1.4, this may be an artificial distinction as climates and geology interact with fire in driving plant assembly. Kemp (1981; see also Jackson 1968; Bowman 2000) suggested that for fire to produce such landscape changes it would not require frequent fires, just a regime with fire cycles shorter than the life span of the tree. Support for this model is the fact that there was widespread rainforest disappearance during the Miocene from inland regions in association with regular seasonal burning (Martin 1990).

In southeastern Australia, as sclerophyllous vegetation expanded, charcoal levels increased in the late Oligocene and remained high into the Miocene (Kershaw et al. 2002; Holdgate et al. 2007). A key question is which is cause and which is effect. Fire would favor sclerophyllous woodlands and shrublands, but these taxa would also promote fire spread. Tertiary fossil floras show a clear correlation between vegetation type and amount of pyrofusinite (fossil charcoal) in a successional sequence where the driest vegetation was also the most fire-prone, according to Blackburn & Sluiter (1994). These authors contended that burning occurred after a change to drier conditions and development of a more sclerophyllous vegetation; that is, fire was a response to, rather than a cause of, vegetation changes. However, there is a rich body of literature that would argue the opposite (e.g. Bond & Midgley 1995; Schwilk 2003; see also Chapter 3). Hill (1990) contends that the fossil flora provides excellent evidence for a Tertiary increase in abundance of *Eucalyptus* associated with increased charcoal levels; when one considers the range of apparent fire adaptations in this taxon it is no surprise (e.g. Burrows 2002; see also Figure 3.3c,d).

Phylogenetic studies reveal that the mallee habit with lignotubers has arisen independently several times in *Eucalyptus* in several geographic regions, suggesting widespread influence of fire in the Tertiary (Hill 1990). In this regard it is important to recognize that resprouting and lignotubers are two distinct traits; the former is widespread in woody dicots and the later is restricted to mostly MTV (see Chapter 3). Evidence for a fire origin of lignotubers in *Eucalyptus* is implied in the apparent loss of the lignotuber in taxa that have radiated into more mesic fire-free sites.

In southwestern Australia Tertiary fire has potentially played a rather extensive role in the evolution of this unique flora. Vast expanses of nutrient-deficient soils have persisted since the Cretaceous and this is hypothesized to have played such an intensive selective role that these landscapes were buffered from Tertiary climatic fluctuations (Hopper 2009). These sclerophyllous shrublands have apparently dominated these landscapes continuously throughout the Tertiary. As is evident today, as long as there is some seasonality, these sclerophyll growth forms are highly fire-prone and these landscapes have potentially had a predictable fire regime throughout the Tertiary. This is supported by the extensive development of resprouting, lignotubers and postfire seeding in many lineages with early Tertiary origins (Hopper 1979, 2009; Orians & Milewski 2007).

Many fire-prone Cape fynbos lineages date from the early Tertiary and the existence of seasonal climates makes it likely that fire has been a factor somewhere on this landscape throughout the latter half of the Tertiary. In one of the more ecologically diverse Restionaceae genera it appears that ancestral resprouting taxa were restricted to shallow rocky sites and evolution involved radiation of seeding on a wider range of substrates, as well as reversals (Hardy & Linder 2005).

Based on molecular studies discussed above it is apparent that many fynbos elements were in existence in the Oligocene as pockets of MTV on stressful substrates or more broadly in interior parts of South Africa. Cowling *et al.* (2009) hypothesize that lightning-ignited fires in these flammable pockets of vegetation would have carried fire sufficiently to erode the extent of more mesic thicket and forest vegetation, thus creating new habitats as a sort of niche construction process. Contemporary examples of this phenomenon are well described for the southern Cape region (Watson & Cameron 2002). The widespread occurrence of nutrient-poor soils would have inhibited thicket and forest and favored fire-prone sclerophyllous vegetation even at this early date. Perhaps a reflection of this early role of fire in driving fynbos diversity is by comparison with the relatively non-flammable succulent karoo vegetation. Verboom *et al.* (2009) found that the vast majority of lineages within karoo were young and apparently closely tracked climate changes in the late Tertiary. In contrast, fynbos endemic lineages exhibit a much broader age distribution, with some originating in the Oligocene and earlier. Fire is likely one critical environmental factor that has contributed to the decoupling of fynbos radiation from climate.

By middle Miocene summer drought conditions would have added to the fire-prone nature of the sclerophyll shrublands and in concert with fire regime shifts acted to further spread the MTV. From middle to late Miocene black debris in cores taken in the Cape Basin (Udeze & Oboh-Ikuenobe 2005) suggests a long presence of fire in the region. During this time there were several important fynbos clades that began intensive speciation (Linder 2003), perhaps driven by an increase in fire-prone landscapes and increased topographic heterogeneity (Cowling *et al.* 2009).

Bruniaceae is an interesting Cape family to look at in terms of fire adaptations. It has an origin that dates back to the Cretaceous but most of the radiation appears to be Tertiary and dated at 18–3 Ma (Quint & Claßen-Bockhoff 2008). The family has many postfire seeding species and exhibits both serotiny and soil-stored seeds with smoke-stimulated germination, and thus provides a possible family for looking at the timing of different fire responses.

Several Cape genera experienced rapid speciation toward the end of the Tertiary. These include shrubs such as *Phylica* (Rhamnaceae) and herbaceous perennials such as *Moraea* (Iridaceae). Such speciation is attributed to mountain building (Cowling *et al.* 2009), although the increased spread of MTV due to increasingly severe summer drought, coupled with increasing fire activity (Fig. 10.9) would also explain these patterns.

One interesting difference between the northern hemisphere MTC regions and the southern hemisphere MTC regions in Australia and South Africa is the much more limited presence of obligate resprouters with fire-independent seedling recruitment and mostly fleshy-fruited vertebrate-dispersed seeds. This functional type is poorly represented in the two southern MTC ecosystems of fynbos and heath. South African Cape fynbos does have occasional *Rhus*, *Olea*, *Diospyros* and *Heeria*, but this functional type is largely absent from southwestern Australian shrublands. In an earlier subsection it was hypothesized that, in the northern hemisphere in the early Tertiary, this functional type would have been adaptive in an environment where sclerophyllous shrubs were restricted to islands of suitable stressful sites within a matrix of woodland. Under those conditions long-distance vertebrate dispersal would be of immense selective value and fire-independent recruitment an advantage on pockets of vegetation with a less predictable fire regime.

A long history of extensive, highly weathered infertile substrates in southwest Australia (Hopper 2009) would have created a highly fire-prone vegetation and the continuity of landscape fuels increased fire predictability. These habitats would have selected for postfire seeders with localized dispersal and not provided an impetus for vertebrate dispersal to target isolated islands of sclerophyll shrublands, as hypothesized for northern hemisphere shrublands. An explanation for the limited representation of obligate resprouters in these two southern hemisphere MTC regions could be tied to nutrient-deficient soils. In the northern hemisphere MTC ecosystems it is hypothesized that this mode is adaptive on mesic fertile sites where vigorous resprouts limit gaps for seedling recruitment. Infertile substrates in southern hemisphere MTC landscapes may limit the size of postfire resprouts so that there is always a sizeable gap resource for seedling recruitment after fires, thus providing less incentive for recruitment between fires.

Central Chile is particularly interesting in that today it seems to have little if any natural source of ignitions due to blockage of summer thunderstorms by the Andes (Fig. 10.4). The sclerophyllous matorral shrublands are highly resilient to anthropogenic fires as almost all species resprout, and many do so from preformed buds in lignotubers, a structure largely associated with fire-prone vegetation. Also a few hardseeded species in the Fabaceae and Rhamnaceae do recruit some seedlings after fire from weakly dormant seedbanks. These characteristics suggest an early association with fire that perhaps is on the decline since completion of the late Miocene Andean uplift eliminated sources of ignition only a few million years ago. Examples of earlier matorral vegetation may be represented by the fire response of shrubs and herbs in southern Chile and just over the border in western Argentina, which is an area that still regularly has lightning-ignited fires and weak MTC (see Chapter 6). In addition to well-developed lignotubers there is evidence of a somewhat diverse postfire annual flora (see Fig. 6.7).

Conclusions

MTC began forming at least as early as mid-Miocene and strengthened through the later Tertiary, often in regions outside of contemporary MTC regions. This pattern of winter rains and summer drought was well developed in all regions by the end of the Tertiary, and since then its intensity has waxed and waned with various climatic cycles. Associated with the MTC are shrublands, woodlands and forests dominated by sclerophyllous-leaved taxa. Much of this MTV has its origins prior to the fully developed MTC and the factors primarily responsible for its development are seasonal droughts, low or moderate fertility soils and fire. Sites of limited resources have selected for the sclerophyll leaf functional type, and coupled with periodic droughts these conditions have led to fire-prone vegetation, and feedback processes on trait evolution have likely contributed to enhanced flammability.

Many recent studies have addressed issues of origin of MTC species, and almost all have related these patterns to changing climates and soil and have largely ignored fire as an important evolutionary driver. However, since at least the middle Tertiary fire has been a potential ecosystem process on landscapes associated with contemporary MTC regions. Fire regimes varied in predictability and frequency in association with the extent of sclerophyllous vegetation and scale of seasonal drought and these have been influenced by both climate and geology. Seasonality is a prerequisite for fires and has been present at annual, decadal and longer cycles throughout much of the evolutionary history of MTV. The seasonal timing of drought is not a determinant of fire-prone landscapes but it potentially affects the amount of landscape conducive to fire. The primary effect of the MTC was to couple drought with high temperatures, which did not qualitatively affect niche characteristics as much as it quantitatively expanded the extent of fire-prone landscape. Synergism associated with this expansion of fire-prone landscape affected the predictability component of the fire regime and had profound impacts on success of fire-dependent species. This model explains the widespread distribution of highly fire-prone sclerophyllous vegetation in MTC regions and more localized occurrences of MTV in non-MTC regions such as in southeastern Australia and North America.

All five MTC regions are dominated by sclerophyllous shrublands and there are three separate stories of fire, plant evolution and community assembly. In the two northern hemisphere systems the origins of most sclerophyll shrubs began in the early Tertiary amidst a landscape of mesic forests and woodlands. Their origins are hypothesized to have been on marginal sites more susceptible to drought stress, such as ridgelines and course-grained substrates that would have inhibited woodland and forest development. The size of these islands of suitable habitat likely varied along north–south gradients in temperature as well as along gradients from coast to the interior, and varied in relationship to orogenic uplift. Seasonal conditions conducive to fires were possibly not annual and the probability of lightning-ignited fires would have been a function of habitat island size. These conditions selected for the obligate resprouting mode, suitable for persisting through periodic fires, but not modifying their reproductive biology to concentrate reproduction to postfire conditions. This strategy is tied to fire regime characteristics, as well as to a life history syndrome that includes strategies for avoiding drought stress with deep roots and dispersal strategies for finding other suitable island habitats amidst a sea of more mesic woodlands. Deep roots promote rapid resprouting, which on moister sites limits the window of opportunity for seedling recruitment and has been an additional factor selecting against postfire seeding in many of these taxa. With increasing aridity postfire gaps became larger and more predictable and were critical to the evolution of postfire seeding. As the predictability of gap formation increased, some lineages took the extra hazardous step of eliminating resprouting and becoming obligate seeders. Postfire seeding, including obligate seeding, originated at different times in different lineages and includes apparent origins in the Oligocene, Miocene and Pliocene.

Southern hemisphere MTC regions of South Africa and southern Australia are distinct in the expanse of highly leached and infertile soils and in some cases these have persisted since the Cretaceous. Sclerophyllous MTV developed early in the Tertiary and evidence of fires extends back to the Eocene, with increasing fire activity into the Miocene. The greater expanse of flammable vegetation would have increased the predictability of fire and this may have selected for postfire seeders earlier than in northern hemisphere systems. This would also have selected against the obligate resprouter mode with vertebrate seed dispersal since the early evolution of sclerophylls may not have been restricted to small islands of habitat. Additionally, low-fertility soils would have reduced the capacity for resprouters to dominate and thus increase gaps for postfire seedling recruitment, which would have selected against fire-independent recruitment as in obligate resprouters.

Higher-fertility soils on Chilean landscapes represent a very different southern hemisphere MTC story. Although lacking much of a fossil record we surmise that early Tertiary evolution of sclerophylls was similar to the history described for the northern hemisphere. As in these northern systems, fire-adaptive traits such as resprouting, lignotubers and postfire seeding would have evolved on marginal sites under various climatic scenarios of seasonality. However, during the late Miocene, the Andean uplift almost totally blocked summer storms from bringing lightning ignitions into central and northern Chile. This barrier has persisted to this day and explains the presence of certain fire type traits on a landscape with few natural sources of ignition. Chilean matorral represents an ecosystem in transition from fire-prone to a non-fire-type system, and such transitions have likely occurred in other biomes throughout evolutionary history.

11 Plant Diversity and Fire

Mediterranean-type climate (MTC) regions are some of the most botanically diverse landscapes in the world (Table 11.1). They are among the 25 global hotspots of diversity in both richness of species and endemics (Myers *et al.* 2000). Occupying a bit more than 2% of the Earth's surface these landscapes hold 15–20% of the world's total vascular plants (Cowling *et al.* 1996; Rundel 2004). Between the five regions there is extraordinary variation in temporal and spatial patterns of vascular plant diversity and the relationship between fire and diversity is quite different across the five MTC ecosystems.

Differences between MTC regions are evident at many scales but one of the frequently noted differences is the regional species density or number of species per unit area. To put this in perspective we need to recognize that one of the commonly held generalizations about species diversity is that it increases with area (Fig. 11.1a,b). This species–area relationship is understandable since there are constraints on the number of individuals that can sustainably occupy a given area. Thus, as area increases, the probability of encountering more species increases. However, despite the observation that the number of species increases with increasing area is one of the few "laws" in ecology (Lomolino 2001), there are exceptions. Dissimilar environments often have very different species richness. Thus, this species–area relationship only approaches the status of a "law" when describing patterns in nested samples (Dunn & Loehl 1988); that is, samples of different size taken from within the boundaries of larger samples so that species from the smallest sample unit share environmental features with larger sample units (Box 11.1). There is no clearer demonstration of this than the species–area relationship observed for total regional diversity between the five MTC regions (Fig. 11.1c). The glaring lack of fit to an idealized species–area relationship (Fig. 11.1a) points up some of the important differences in diversity between these MTC regions. These patterns are the result of complex responses to subtle variations in climate, not so subtle variations in geology, and to their interaction with fire, as well as to phylogenetic and biogeographic histories.

Fire and Community Diversity

One of the key differences in diversity between MTC plant communities is tied to differences in vegetation structure, which reflect differences in soil fertility, and to

Table 11.1 Mediterranean-type climate regions and biogeographic characteristics

	Mediterranean Basin	California Floristic Province	Central Chile	Cape Floral Region	Southwest Australia
Original primary vegetation (km²)	2 362 000	324 000	300 000	74 000	309 850
Remaining vegetation (%)	4.7	24.7	30.0	24.3	10.8
Plant species	25 000	4426	3429	8200	5469
Endemics (%)	52	48	47	69	79
Dominant shrublands	maquis/garrigue	chaparral	matorral	fynbos	heath
Fire-dependent ephemeral flora	none	dominant	none	minor	minor
Geophytes	moderate	moderate	moderate	very abundant	very abundant
Annuals	abundant	abundant	abundant	minor	minor
Colonization from outside burn	abundant	rare	none?	none	none
Shrubs resprouting	most species	~ half of species	most species	~ half of species	~ half of species
Shrubs with fire-stimulated seedling recruitment	many species in a few genera	many species in a few genera	1 or a few species	many species, many genera	many species, many genera
Prefire diversity	low in closed canopy; high in disturbed sites	low in closed-canopy shrublands; higher in open sites	low in closed-canopy shrublands higher in open sites	generally high	generally high

the relative change in vegetation structure after fire. Northern hemisphere MTC ecosystems largely are dominated by moderately fertile soils and, in the absence of recent disturbance, often develop closed-canopy shrublands and woodlands (see Chapters 4 and 5). The dominant sclerophyllous-leaved species typically form a monolayer of tall (2–5 m) evergreen shrubs that shade out most understory species. Consequently community diversity tends to be rather low, comprising mostly shrubs and a few vines and lianas. In the understory there may be sparse populations of a few annual and herbaceous perennial species but these populations fluctuate with annual precipitation and often disappear under drought conditions. In the soil, however, there is a rich diversity of long-lived dormant seeds, bulbs and corms that are triggered to germinate and resprout in response to fire. As a consequence these communities exhibit dramatic increases in diversity after fire (Table 11.2).

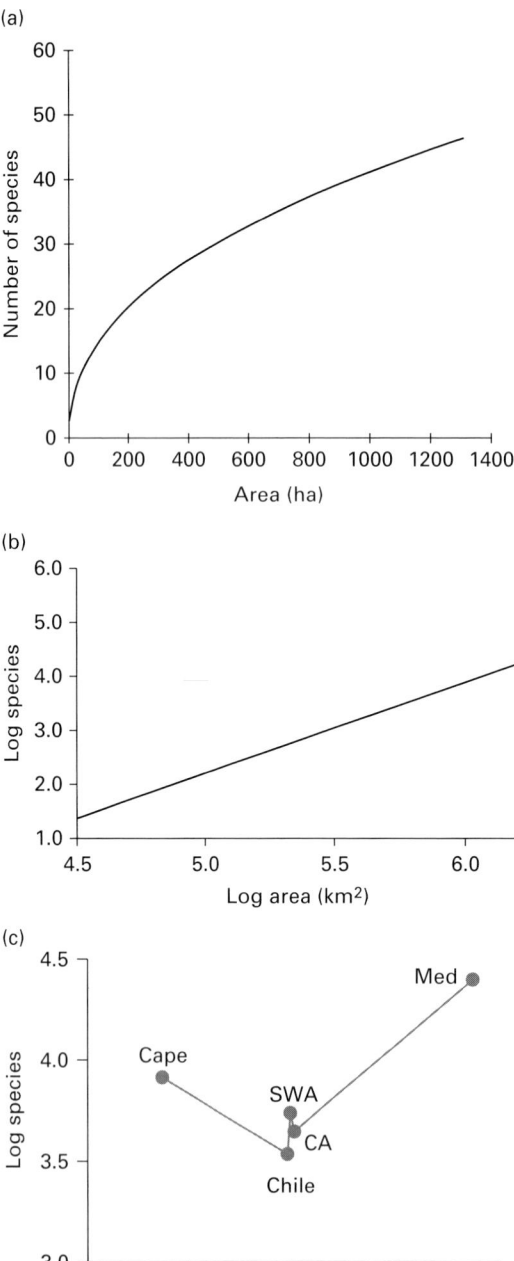

Fig. 11.1 *Idealized species–area relationship (a) on an arithmetic scale, (b) log-log scale and (c) observed species–area relationship for the five MTC regions. Med, Mediterranean Basin; CA, California; Chile, central Chile; Cape, Cape region of South Africa; SWA, southwest Australia.*

Box 11.1 Effect of Sampling Design on Diversity Measures

Nested sample designs are the only way to ensure the species–area curve increases with increasing area. All of the California data presented in Table 11.2 were recorded from nested plots with the sample design in Fig. B11.1.1. Nested within this 0.1-ha plot are ten 100-m^2 subplots with two 1-m^2 quadrats nested in the subplots. One concern about the use of nested designs in species–area studies is with the statistical analysis, as least squares regression analysis assumes that samples at different scales are independent. Nested designs potentially can result in dependence between estimates at different scales, although this need not always be the case. In a study of 90 sites in California chaparral this was tested by randomly subsampling without replacement so that each scale comprised a different subset of 30 sites. Species–area regression equations, Standard Error and r^2 were nearly identical when calculated for these independent unnested plots or when calculated from the nested plots for all 90 sites (Keeley & Fotheringham 2005).

Data outside California reported in Table 11.2 have been collected with other designs and it has been proposed that plot shape and orientation can greatly alter community diversity estimates (Harte et al. 1999; Stohlgren et al. 1995). However, field comparisons fail to show such differences (Keeley & Fotheringham 2005). Plot size can have an effect and as a general rule the bulk of the aboveground and belowground plant should be contained within the plots where it is recorded, otherwise one is sampling diversity over a larger niche space than assumed by the plot size. In postfire shrublands, 1-m^2 plots are large enough to contain both a good sampling of herbaceous plants as well as seedlings and resprouts of woody species. Where sampling designs may run into trouble is when the plot size is substantially smaller than the growth forms being sampled. For example, Condit et al. (1996) reported that highest diversity for tropical forests was obtained in very long 2-m bands, very likely because the roots and canopies of these tropical trees were occupying niche space outside the sample plot and in effect they were "sampling" from a much larger area.

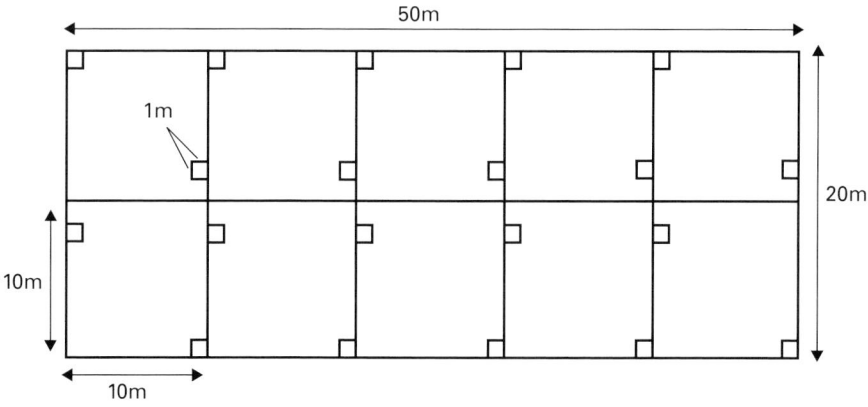

Fig. B11.1.1 *Nested 0.1-ha sample plot used in Keeley & Fotheringham (2005).*

Continued

Box 11.1 (*cont.*)

Slope orientation potentially has an effect on recorded richness. It has been suggested that community diversity is maximized when the long axes of rectangular plots are oriented perpendicular to the elevational contours (Stohlgren *et al.* 1998). Although elevational changes in species composition are well documented, there is no evidence of such an effect at the local 0.1-ha scale. Indeed, it has been suggested (Keeley & Fotheringham 2005) that on steep slopes in semi-arid MTC regions, plots parallel to the contour might actually increase diversity as species tend to follow drainage patterns up and down the slope, with greater species turnover along the contour as populations of different species encounter different microscale drainage patterns.

Interpreting differences in species–area curves at the regional level is complicated by one factor not usually encountered at the local level. Community-scale species–area curves generally measure area along a plane parallel to the ground surface, whereas larger-scale area measurements use map projections, which on level terrain can be equated with surface area, but on steep terrain, map projections underestimate the actual ground surface area. Thus, when one considers the role of landscape heterogeneity, any increases in diversity in mountainous regions must be partially the result of the additional ground surface area hidden in such area estimates. Some forestry studies of community diversity have used slope-corrected measurements, which creates the same sampling artifact, and thus they are sampling diversity in areas larger than the actual area parallel to the ground surface.

Two of the southern hemisphere MTC communities differ structurally from those in the north. Many of the prominent South African Cape and southwest Australian ecosystems exist on highly infertile substrates (see Fig. 1.5) that do not lend themselves to tall-stature closed-canopy shrublands. Shrubs of different stature contribute to multiple strata interspersed with a greater diversity of growth forms, often including various Restionaceae (see Fig. 7.7) and other herbaceous perennials (Naveh & Whittaker 1980). Community diversity is relatively high and there are only modest increases in species richness after fire (Table 11.2). Chilean shrublands occur on more fertile substrates and, without disturbance, will form closed-canopy shrublands and woodlands with little understory. However, due to the absence of fire-stimulated species, they do not exhibit a postfire increase in diversity (S. Keeley & Johnson 1977; Armesto & Gutierrez 1978) as observed in other MTC communities.

The role of vegetation structure in controlling diversity is evident even within regions. In eastern Mediterranean Basin shrublands, open-canopy disturbed sites have more than three times greater alpha diversity than closed-canopy associations (Table 11.2). Although South African fynbos and Australian heathlands

Table 11.2 Species richness reported at 1-m² and 1000-m² scales for the MTC and related regions

	Time since fire (yrs)	Species richness 1 m² $\bar{x}$ (range for sites)	Species richness 1000 m² $\bar{x}$ (range for sites)	Sample # sites	Data source
California chaparral	2	9.5 (2.9–22.1)	58.0 (27–106)	250	1
	2	12.2 (4.8–20.6)	53.0 (33–85)	28	2
	4	6.6 (3.0–11.4)	35.0 (10–57)	28	2
	mature	5.5	28.0 (16–34)	10	3
	mature	1.4	8.8	2	4
Arizona chaparral	2 (yearly total)	7.8 (3.1–12.0)	76.5 (27–102)	40	5
	2 (spring)	4.0 (2.0–8.1)	46.6 (16–72)	40	5
	2 (autumn)	5.3 (2.4–10.9)	55.0 (22–73)	40	5
Mediterranean					
Western garrigue	1	8	29	1?	6
Western garrigue	2	6.7 (5.6–7.8)	48.8 (42–56)	2	7
Western garrigue	5	8	28	1?	6
Western garrigue	(closed canopy)	2–4	21–33	4	13
Eastern maquis	(closed canopy)	–	39.0 (21–57)	2	4
Eastern maquis	(open canopy)	–	125.0 (84–179)	4	4
Eastern maquis	(open canopy)	–	154.0 (147–162)	1(5yrs)	8
Eastern *Pinus halepensis*/maquis					
Pole-facing slope	2	–	86 (80–92)	2	9
Pole-facing slope	10	–	94 (91–97)	2	9
Pole-facing slope	mature	–	82 (78–86)	2	9
Equator-facing slope	2	–	84 (80–88)	2	9
Equator-facing slope	10	–	124 (123–125)	2	9
Equator-facing slope	mature	–	131 (127–135)	2	9
Chilean matorral	20–25	7.7	–[a]	3	6
South African fynbos	6	9.5	86.7	3	10
	8	11.0	64.7	3	10
	8	16.6	80.0	3	10
	mature	15.2 (9.8–26.6)	63.8 (41–141)	20	3
	mature	16.1 (12.8–24.5)	66.4 (41–93)	9	11
	mature	13.7 (3.8–24.1)	68.8 (26–143)	17	12
Western Australia					
Heathland	mature	12.9	69.4	30	6
Heathland	mature	13.3	65.0	7	4
Banksia woodland	mature	15.5 (11–19)	69.3 (59–81)	9	14
Mallee	immature?	7.8	48.9	13	6
Mallee	mature?	6.1	49.0	6	4

1, J.E. Keeley, unpublished; 2, Keeley & Fotheringham (2005); 3, Bond (1983); 4, Naveh & Whittaker (1979); Keeley & Fotheringham (2003b); 5, Fotheringham (2009); 6, Westman (1988); 7, Pausas et. al. (1999); 8, Aronson & Shmida (1992); 9, Kutiel (1997); 10, Schwilk et al. (1997); 11, Cowling (1983b); 12, Cowling (1990); 13, Chiarucci et al. (2001); 14, Bridgewater & Backshall (1981).
[a] The very high 0.1-ha values reported by Westman (1988) are not valid measures as they were based on adding species lists from smaller plots taken over an area much larger than 0.1 ha (G. Montenegro personal communication, 3 July 2003).

generally comprise open-canopy vegetation due to stressful substrates, reduction in understory species richness often occurs on sites that support taller stature shrubs (Cowling & Gxaba 1990; Keith & Bradstock 1994; Specht & Specht 1999). After fire the change in diversity is substantially higher on those sites with greater prefire canopy coverage (Specht 1981; Bond & Ladd 2001). This effect of oligotrophic soils is evident in other shrubland comparisons. For example, comparing sclerophyllous shrublands on sandy substrates in non-MTC Florida with MTC California chaparral reveals that the more open Florida scrub community has higher diversity prior to fire and much less increase in diversity after fire (Carrington & Keeley 1999). A similar situation is evident in nutrient-stressed serpentine chaparral with higher diversity prefire and a less striking postfire diversity increase compared with chaparral on non-serpentine substrates (Safford & Harrison 2004) and a similar fire response is evident in serpentine and non-serpentine grasslands (Harrison et al. 2003).

As a result of these structural differences between MTC communities, and the manner in which diversity interacts with fire, comparisons of diversity between mature communities (Cowling et al. 1996) have not fully captured diversity patterns between MTC regions. In general, MTC communities have remarkably similar diversities (Table 11.2) but peak diversity is closely tied to fire in some but not others. When recently burned sites are compared, the diversity patterns are remarkably similar. At the 1-m^2 scale, all regions typically have between 9 and 15 species but all regions show high variability with anywhere from 2 to 25 species. At the 0.1-ha scale MTC communities typically have between 50 and 80 species, although site to site variability is extraordinary, ranging from 10 to 180. Differences between MTC ecosystems emerge as they recover from fire. In the northern hemisphere, MTC shrubland diversity declines markedly with canopy closure, but in the Cape of South Africa and southwest Australia diversity remains high in their more open fynbos and heathland communities, respectively.

Fire does not affect diversity solely by niche differentiation but rather there are stochastic processes related to postfire recruitment that can play roles in determining community diversity (Laurie & Cowling 1994; Lamont & Witkowski 1995). For example, lottery type recruitment of postfire seeders has the potential for altering community diversity patterns. Population fluctuations in different fire cycles cause species to change their contributions to postfire diversity and this is potentially controlled by stochastic effects between, during and after fire.

Regional Variation in Factors Driving Fire–Diversity Interactions

Despite the apparent convergence in dominance of fire-prone sclerophyll shrublands across all five MTC regions, and remarkable similarities in postfire regeneration (Fig. 11.2), there are marked differences in the relationship between diversity and fire (more details in Chapters 4–8).

Fig. 11.2 *Postfire communities in MTC ecosystems illustrating (a) early successional* Cistus *recruitment in maquis from southern France, (b)* Pinus halepensis *seedlings in recently burned woodland in Israel, (c) postfire annuals in California*

(d)

(e)

(f)

Caption for Fig. 11.2 *(cont.)*
(d) annual Loasa *beneath a shrub skeleton after fire in central Chile, (e)* Pillansia, Watsonia *and other geophytes in recently burned fynbos from the Cape region of South Africa, and (f) postfire heathland in southwestern Australia. (Photos by Jon Keeley.)*

Mediterranean Basin

Shrubland diversity increases markedly after fire due to resprouts and seedlings from species present prior to fire, coupled with a large ephemeral herbaceous flora arising from dormant bulbs/corms and seeds. Herbaceous perennials comprise a large portion of this diversity on many sites. On rare occasions annual plants disperse into burned sites from surrounding disturbed sites (Bonnet & Tatoni 2004). In garrigue and maquis, diversity is typically low during the first year but increases during the first decade as more species colonize from outside the burned area and persist until canopy closure (Trabaud & Lepart 1980). During early succession diversity is sensitive to precipitation (Kutiel 1997). As the canopy closes in, this ephemeral flora diminishes but persists on open sites between shrubs or in grasslands. These shrubs are core species widely abundant within sites and widely distributed across landscapes (Pärtel et al. 2001), as is evident in California shrublands (Keeley et al. 2005c).

Intensive land use, particularly grazing, has replaced fire in many of these shrublands (Zohary 1962; Naveh & Whittaker 1979), and resulted in a mosaic of grasslands and shrublands often with very different species composition from undisturbed shrublands (Alados et al. 2004). These open unburned stands have substantially higher diversity than burned sites (Verdú et al. 2000; Keeley & Fotheringham 2003b; de Bello et al. 2007). At the more arid eastern end of the basin, on sites with a history of intensive livestock grazing and fire, open shrublands produce some of the highest plant diversities observed in any MTC region (Table 11.2). These sites often average 125–150 species per 0.1-ha scale with or without recent fire, which is nearly double the species richness observed in other MTC ecosystems (Table 11.2), although potentially comparable to the extraordinary diversity observed in mediterranean-type vegetation (MTV) dominated by longleaf pine in the southeastern USA (Platt et al. 2006). In the eastern Mediterranean Basin these high levels of diversity persist even in years of subnormal rainfall (Aronson & Shmida 1992). It has been hypothesized that this may arise from the much greater regional species pool to draw from in this region at the confluence of floras from Europe, Asia and Africa (Keeley & Fotheringham 2003b). For example, Israel has nearly nine times greater density of species than Spain (Danin 2001).

California

California shrublands follow a pattern of diversity changes in response to fire that is similar to the Mediterranean Basin. Following fire in closed-canopy shrublands there is a large postfire flush of annual species from dormant seedbanks. However, postfire herb succession in California differs in several respects. Most prominent is the fact that a significant proportion of the annuals in the postfire flora are strict *pyro-endemics*, which are generally not found on other disturbed sites as in the Mediterranean Basin. These fire endemics are largely from families best developed in western North America, in particular the Hydrophyllaceae and Polemoniaceae (Keeley et al. 2006b).

This postfire endemic annual flora is fire dependent with strict smoke-stimulated germination (Keeley & Fotheringham 2000), but despite its disturbance-dependent life history, persistence of these taxa is sensitive to invasion by non-native grasses and forbs. In the absence of human interference this is not normally a threat because canopy closure generally excludes non-native aliens from the site. However, under repeated fires and other disturbances non-native annuals invade and contribute to increased fire frequency that feeds back into enhanced alien invasion (see Chapter 12). As this occurs, the postfire endemic annual floras and their seedbank are lost from the site, apparently from competitive exclusion with the more aggressive aliens. As this invasion process proceeds there is reason for concern over regional loss of these postfire endemics. These observations in California suggest the hypothesis that the absence of such a postfire endemic flora in the Mediterranean Basin may in part be an artifact of human disturbance history, which is 50 times longer than in California and has resulted in only 5% of the original vegetation still persisting (Table 11.1).

During early postfire succession in California chaparral there are interactions between precipitation, diversity and time since fire. Peak diversity is typically in the first year or two and largely comprises annual species but it is also sensitive to annual variations in precipitation. Diversity declines during early succession, but even as late as 5 yrs postfire, high precipitation events can again trigger rapid increases in annual species diversity, almost to levels in the immediate postfire year (Keeley et al. 2005a, 2006b). However, this later seral community is very different from the immediate postfire community as it does not include the postfire endemic annuals, and 30–50% of the species may be ones not present on the site in the immediate postfire years. These "colonizers" are mostly not ruderal species that have invaded from outside the burned perimeter, but rather species that were present within the burned area immediately after fire and have expanded their populations in subsequent postfire years. This mass effect (Shmida & Wilson 1985) results in populations of new species spilling over into sample plots where they were previously absent, although they typically comprise a very small proportion of the total cover. Thus, if we consider the total postfire flora over the first 5 yrs, community-scale diversity would be substantially richer than reported in Table 11.2. Such temporal niche separation of species has been reported in Australian shrublands as well (Fox 1995). Eventually, as the shrub canopy returns over 5–10 yrs, species diversity plummets regardless of rainfall.

Community-scale plant diversity in California shrublands is a multifaceted problem (Keeley et al. 2005c) and involves: (1) Life history specialization to the temporal heterogeneity in resources created by a predictable fire cycle, (2) species-specific responses to characteristics of disturbance events, in particular fire severity, (3) niche specialization on different features of the landscape that vary between growth forms, (4) growth forms that further subdivide resources along other axes such as soil characteristics, phenology, etc., (5) species-specific differences in response to annual fluctuations in resource availability tied to the critical role of precipitation in these semi-arid landscapes, and (6) mass effects due to metapopulation dynamics whereby fluctuations in disturbances and resource availability result in occasional localized expansion of highly successful core species' populations. During early succession the importance of these factors may change rapidly such that similar diversity levels in the immediate postfire year, and in subsequent years, may be due to different factors. Although diversity appears to be driven by disequilibrial processes, the fact that peak diversity occurs postfire from residual species present prior to the fire suggests a long-term equilibrium whereby community composition has stabilized according to species-specific niche specialization.

Arizona chaparral under a non-MTC presents an interesting contrast. The dominant sclerophyll shrubs in this community represent MTV under a non-MTC bimodal rainfall regime. The postfire flora arising in the first spring after fire is somewhat lower than observed in California. However, the Arizona chaparral communities produce a second flora in the autumn and the total flora for the year is markedly higher than in California chaparral (Table 11.2). The bimodal rainfall regime allows for greater niche separation of postfire annuals, with the spring flora taxa having affinities to the California chaparral flora, and the autumn flora taxa to subtropical floras (Fotheringham 2009).

Each of the five MTC regions has a forest component to greater or lesser degrees, and these communities exhibit a variety of fire regimes. Many conifer forests in California have a surface fire regime and the understory regeneration is typically shrubs and herbaceous perennials from resprouts and seed germination (Keeley *et al.* 2003; Knapp *et al.* 2007). Diversity at scales of 1–1000 m^2 are roughly half that observed in shrubland communities (Keeley & Fotheringham 2003b). There is a low level of postfire annuals in mixed conifer forests (Knapp *et al.* 2007). Some of these ephemerals apparently arise from fire-stimulated germination of dormant soil-stored seedbanks; however, almost all are found on other disturbed sites and none appear to be strict fire-following species. Unlike crown fire shrublands where diversity is resilient to a wide range of fire severities (Keeley *et al.* 2008), in forests high fire severity punches holes in the canopy and increases diversity when understory species colonize these gaps (Keeley *et al.* 2003). In contrast, diversity in these forests is much more resilient to frequent fires than diversity of crown fire shrublands.

Chile

Chilean matorral diversity typically declines after fire because the community lacks a pool of dormant soil-stored seeds and vegetative structures that can colonize after fire. This is apparently due to the lack of a predictable source of natural fire ignitions in central Chile: the absence of summer lightning storms is because they are blocked by the Andean Cordillera (see Chapters 6 and 10). A potential exception to this pattern is matorral shrublands in the southeastern end of the MTC region where the Andes are low enough to allow summer lightning storms, and thus a predictable fire regime. Here there is some evidence of a postfire annual flora comprising species of Hydrophyllaceae, Boraginaceae and Portulacaceae (see Fig. 6.7).

Open Chilean matorral characteristically has its highest diversity concentrated in the understory of the mature shrubs (S. Keeley & Johnson 1977) and this pattern remains after fire (Fig. 11.2d). Diversity increases slowly as species colonize from outside the burned area (Armesto & Gutierrez 1978; Gómez-González & Cavieres 2009).

Cape Region of South Africa

The species diversity of South African fynbos is relatively high in the absence of fire and shows only modest increases with fire (Table 11.2). Increased diversity after fire arises largely from geophytes present as dormant rhizomes, bulbs or corms in the soil that are triggered to resprout by fire. Geophytes are an important component of all MTC floras, and most are stimulated to flower by fire. Such stimulation can be seen dramatically in large expanses of *Watsonia* that can arise from rhizomes after fynbos fires. The Cape *Cyrtanthus ventricosus* is the only clearly fire-dependent geophyte known from MTC regions. This "fire-lily" shows an extraordinary relationship to fire, with bulbs possessing preformed flower buds that remain dormant until smoke triggers flowering, usually within days of a fire, regardless of season (Keeley 1993b).

The absence of a dramatic postfire increase in diversity is in part due to the fact that there are many geophytes present in mature fynbos, although their presence becomes more obvious due to increased flowering after fire (Le Maitre & Brown 1992). Often overlooked are a small number of diminutive annuals in families such as the Scrophulariaceae and Campanulaceae that have smoke-stimulated germination and are largely restricted to burned sites (C.J. Fotheringham unpublished data).

Western and South Australia

The postfire changes in species diversity of southwest Western Australia heathlands share many features with the Cape region shrublands, with relatively high prefire diversity and modest increases after fire (Russell & Parsons 1978; Bell & Loneragan 1985). Although resprouting of geophytes and other herbaceous perennials comprises much of the increased postfire diversity, there are some *fire ephemerals* (Bell et al. 1984; Pate et al. 1985). As in the Cape postfire flora, these annuals comprise a relatively minor proportion of postfire biomass with most being diminutive plants, usually less than a gram dry weight per plant. The low-fertility soils likely play an important role in limiting the importance of annuals due to the difficulty that fast-growing annuals have in sequestering sufficient nutrients to reach reproductive maturity on infertile soils.

Eucalyptus-dominated woodlands and forests comprise a mixed collection of forest types with diverse fire regimes; however most are highly resilient to understory surface fires. Compared with California conifer forests, the relatively open canopies of eucalypt woodlands and forests allow the development of a rich understory flora that persists either from resprouting or seed germination after fires (Bell & Koch 1980; Wardell-Johnson et al. 2007). In Western Australian forests as many as 5–10% of the species are postfire ephemerals (Burrows & Wardell-Johnson 2003).

Insights from Community Species–Area Relationships

Regional species–area curves (e.g. Fig. 11.1a,b) often differ between landscapes and statistical coefficients of these curves are thought to reveal important

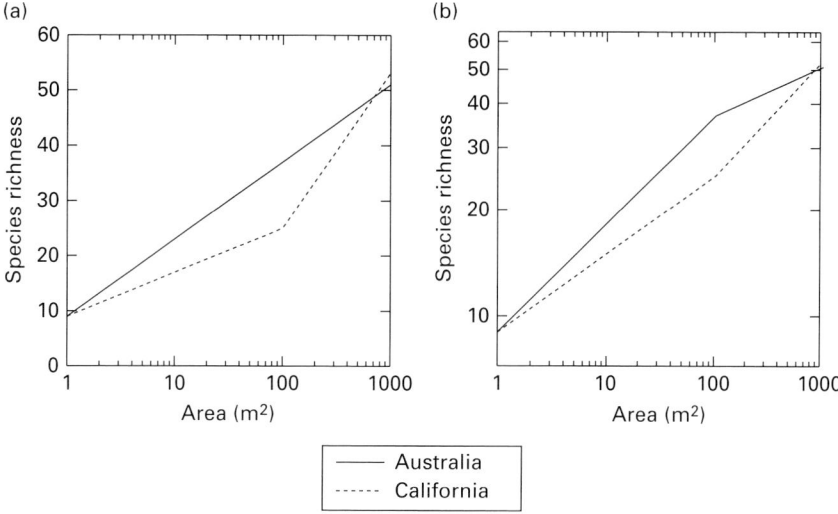

Fig. 11.3 *Species–area curves for California chaparral and southwest Australian heathlands with the (a) exponential model plotted on a semi-log scale and (b) power model plotted on a log-log scale.*

characteristics about global scale diversity patterns (Rosenzweig 1995). Far less attention, however, has been focused on comparative studies of this species–area relationship at the scale of individual plant communities and the potential insights it might give into factors driving community assembly (Keeley & Fotheringham 2003b).

MTC plant communities differ markedly in their community-scale species–area relationships and there is reason to believe the differences are driven by different functional types selected in response to the climate, fire, geology filter (see Fig. 1.4). For example, following fire, species–area curves are very different between California chaparral and Western Australian heathland communities. Across the scales from 1 to 1000 m^2, chaparral fits a power model and heathland exhibits a better fit to an exponential model (Fig. 11.3).

It is hypothesized that these different model fits are the result of different dominance–diversity relationships tied to life history differences (Keeley & Fotheringham 2003b) and this is evident in very different species–abundance curves (Keeley 2003).

Chaparral dominance–diversity patterns fit a geometric model (Fig. 11.4a,b) and species–abundance curves are very broad with many uncommon species indicating diversity is controlled by the strong dominance of a few species with many subordinate species. This arises because most of the postfire flora comprises obligate and facultative seeders and a small number of vigorous resprouters, which often dominate postfire cover. Although postfire endemics are usually abundant, the majority of annuals occur in relatively small numbers and much of this annual diversity is relatively transient. The vast majority are *satellite*

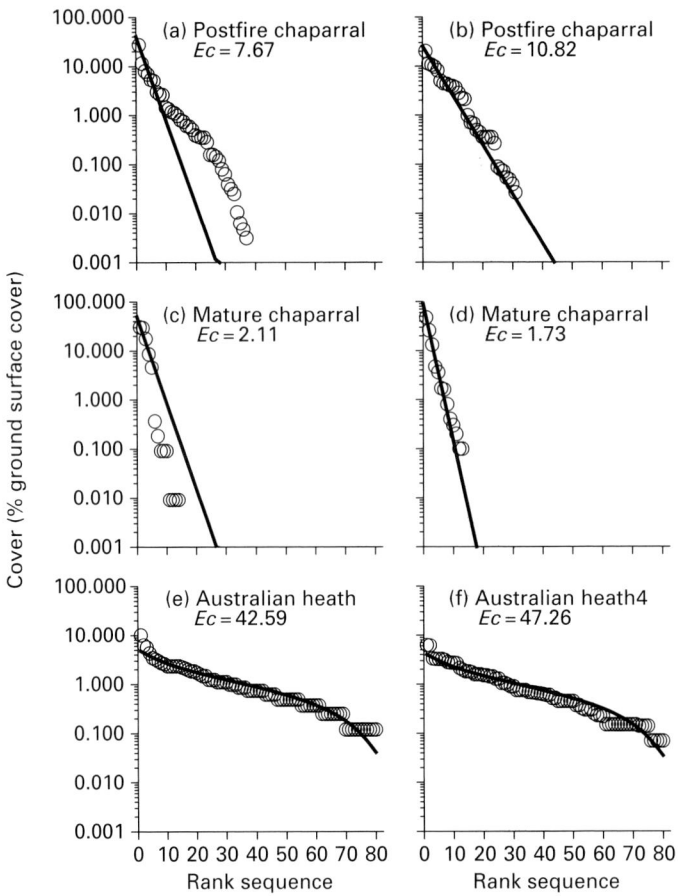

Fig. 11.4 Dominance–diversity curves for (a,b) Californian and (c,d,e,f) Australian shrublands based on cover (from Keeley & Fotheringham 2003b). Californian sites fit a geometric model and Australian sites showed a significant fit to MacArthur's broken stick model. Ec, Whittaker's equitability index.

species, being found in a very small fraction of quadrats within a site as well as within a small fraction of sites (Keeley et al. 2005c). Based on dispersal capacity this does not appear to be driven by metapopulation dynamics but rather microhabitat specialization.

In contrast, Australian heathland species are more equitably distributed, as illustrated by their narrow species–abundance distribution and dominance–diversity relationship that fits MacArthur's broken stick model (Fig. 11.4c,d). The implication of this model is that populations reach a stable equilibrium without the development of dominance by any one species (Whittaker 1972). This leads to an exponential species–area curve because most species in a community are common and occur within a small subsample of the area, leading to a sharp increase in diversity between 1 and 100 m^2 evident by a linear fit to a semi-log plot

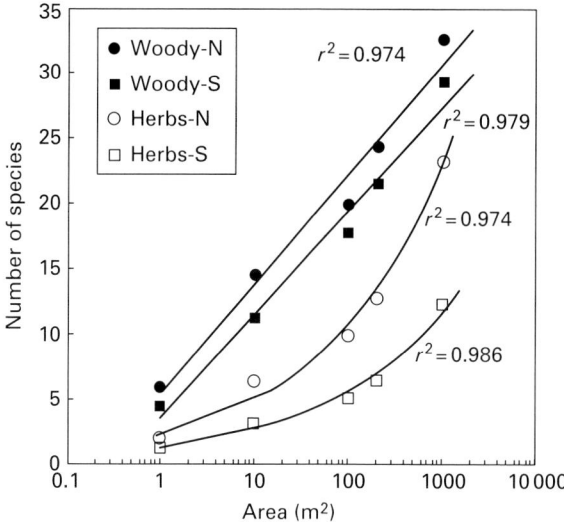

Fig. 11.5 *Species–area relationships for different growth forms (N and S refer to slope aspect; from Pausas et al. 1999) and similar patterns have been observed in California chaparral (Keeley & Fotheringham 2003b).*

(Fig. 11.3). An important contributor to these patterns is the fact that these heathlands are dominated by perennials, and more than three quarters are resprouters (George *et al.* 1979). Thus, following fire there are fewer opportunities for community reassembly and less chance of dominance by one or a few species.

Thus, comparing the species–area curves between communities provides some insight into the compositional differences in functional types. Those types determining the exponential vs. power model are evident even when the species–area relationship is observed within communities. For example, woody plants typically follow a different species–area relationship from herbaceous species (Fig. 11.5).

The climate, fire, geology filter controlling different fire-response functional types in these MTC ecosystems appears to be strongly influenced by soils. The more fertile Mediterranean Basin shrublands match Californian communities closely in their species–area power law fit (Pausas *et al.* 1999) and dominance–diversity relationships (Basanta *et al.* 1989). In contrast, the low-fertility South African fynbos resembles Australian heathlands in its fit to an exponential model (Bond 1983; Cowling 1983b; but cf. Schwilk *et al.* 1997).

Landscape Effects on Community Diversity Patterns

Landscape variation in resources is an important environmental filter in community diversity patterns (Keeley *et al.* 2005c) and is likely a primary factor responsible for the wide disparity in 0.1-ha scale species richness, ranging from 10 to 180 in the MTC communities surveyed in Table 11.2. Fire increases light and nutrients more so in some communities than in others and species with fire-adaptive traits such as lignotubers and dormant seedbanks that delay reproduction to a single pulse of recruitment can capitalize on these fire-induced changes in resources. Landscape variation in substrates can also alter community structure, which affects the extent to which closed-canopy shrublands develop and thus affects fire-driven diversity patterns (Cowling *et al.* 1989; Carrington & Keeley 1999; Safford & Harrison 2004).

In addition to these deterministic factors, there are also stochastic effects that control diversity in fire-prone communities. Event-dependent factors such as fire intensity or the coincidence of high rainfall after fires can play key roles in postfire diversity patterns, as demonstrated in both southern and northern hemisphere MTC regions (Yeaton & Bond 1991; Richardson *et al.* 1995; Keeley *et al.* 2005c). Elevated fire intensity works through diminishing seedbanks and resprouter survivorship, both of which contribute to postfire cover and species richness. Drought years after fire reduce soil moisture and thus diversity, and the impact of postfire drought years may vary with landscape location. Long-term studies of diversity in fire-prone Cape fynbos suggest that other stochastic factors are important determinants of community diversity (Thuiller *et al.* 2007).

A model illustrating how landscape variation in plant communities and fire history affect community diversity patterns in California shrublands is illustrated in Fig. 11.6. In this model distance from the coast influences growth-form dominance, which in turn affects community heterogeneity and site-specific abiotic conditions. Both of these contribute to local richness. Fires in this region vary in frequency relative to distance from the coast and create different landscape mosaics of stand age. Older stands have a greater amount of dead fuels and alter local fire intensity, which in turn affects postfire cover and richness.

Fire and Regional Diversity

Different processes appear to determine plant diversity at different spatial scales (Crawley & Harral 2001). Fire, however, is one of those ecosystem processes that can affect diversity at most spatial scales, including communities, landscapes and region. Common metrics include *beta*, *gamma* and *delta* diversities, terms that have been applied inconsistently by numerous authors (Tuomisto 2010).

Landscape mosaics of different fuel structure will contribute to heterogeneous fire regimes and provide diverse habitats that affect both floral and faunal diversity at the landscape scale (Romme 1982). Even for landscapes with the same fire regime, mosaics in fuels can lead to a patchwork of different stand ages that contribute to landscape diversity (Clark *et al.* 2002; Uys *et al.* 2004). Such patterning is less likely in shrubland-dominated crown fire regimes where fire spread is dependent on sufficient canopy fuels or strong winds, and these factors contribute to large landscape-scale fires. In forest types where fires are spread by surface fuels, patchiness in fuels or different seasons of burning can create a patchwork of

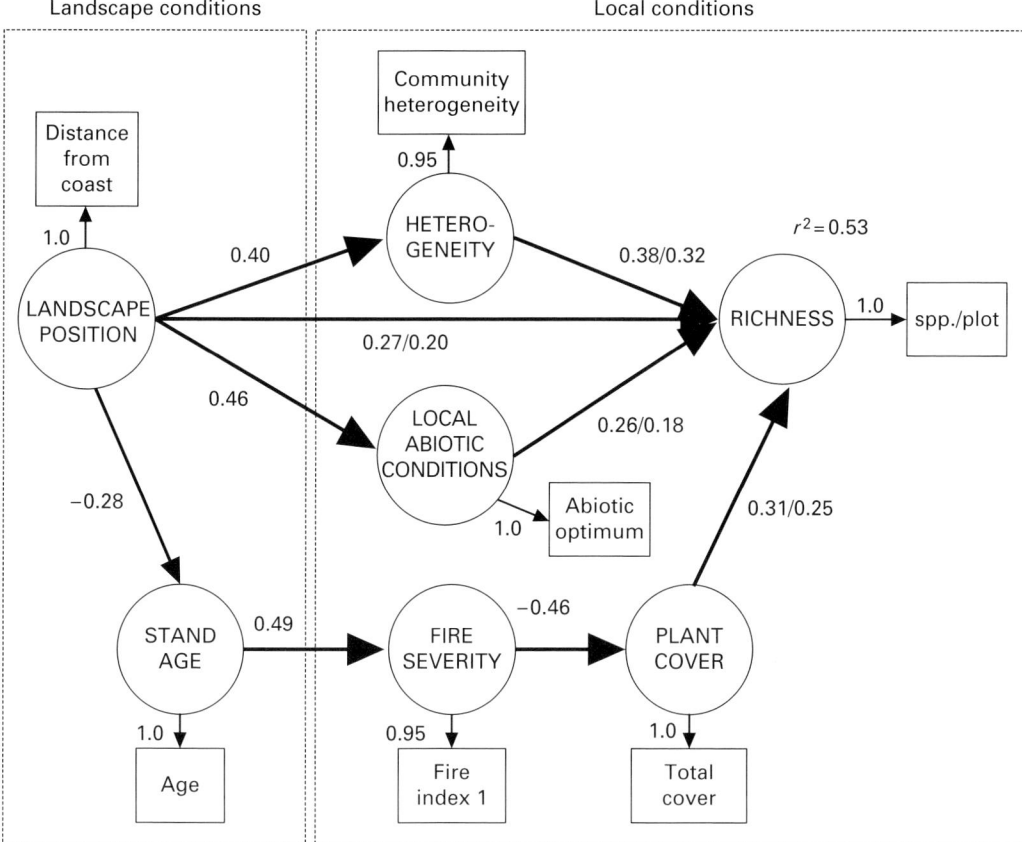

Fig. 11.6 Structural equation model of direct and indirect effects on postfire diversity in chaparral 5 years after fire (Grace & Keeley 2006). Circles signify latent variables used to represent theoretical concepts of interest, while rectangles represent observed variables that serve as indicators of latent variables. Numbers associated with paths from latent to observed variables represent loadings, while numbers associated with paths between latent variables represent path coefficients.

different fire histories and this can have profound impacts on landscape patterns of plant and animal diversity (Fox 1983; Ferrenberg et al. 2006; Knapp et al. 2007).

Regional diversity of MTC landscapes is among the highest recorded globally (Cowling et al. 1996; Myers et al. 2000), and since these are also among the most fire-prone systems in the world, fire should be given some consideration for its contribution to regional diversity. This would seem to be borne out in the fact that many taxa within these regions are not just resilient to fire but also dependent on fire and owe their origin to fire-selected traits.

Fire can also affect regional species diversity through localized selection of endemics that are restricted by substrate and/or climate. Where this is most apparent is in woody obligate seeding taxa, a functional type where postfire recovery is entirely dependent on seedling recruitment. This risky strategy makes such taxa vulnerable to fires during the juvenile phase, and localized extirpation follows fires at short intervals. Cowling (1987) hypothesized that localized extinction of obligate seeding fynbos populations has increased the isolation of taxa and contributed to their divergence. Landscape diversity is thus promoted because of subtle differences in substrate to which these fragmented populations adapted. Where less subtle substrate variation exists, such as with mosaics of ultramafic soils in California, landscape-scale diversity between serpentine chaparral patches is high and obligate seeding endemics contribute significantly to these patterns (Harrison & Inouye 2002; Safford *et al.* 2005). Perhaps a similar mechanism accounts for obligate seeding chaparral genera following a predictable pattern of species turnover along climatic gradients, both latitudinally and from the coast to the interior (Cody 1986).

These landscape effects also add to regional diversity and differences in fire regimes may explain some regional differences in diversity between MTC communities. One example may be the extraordinarily high floristic diversity observed in the Cape region of South Africa, with species richness markedly higher than other MTC regions, including the very diverse southwest Australian flora (Fig. 11.1c). Extreme landscape heterogeneity is often cited as a primary driver in the high Cape floristic diversity (Cowling & Lombard 2002), and is thought to result from mountainous terrain contributing to habitat heterogeneity. However, such terrain also has huge impacts on fire regimes, potentially increasing fire frequency due to the effect of steep slopes in driving fires when dead fuels and/or winds are insufficient to carry active crown fires. More frequent predictable fires could potentially impact regional species pools by speeding up the rate of evolution in fire-dependent obligate seeding taxa (see Chapter 9). Metapopulations have been hypothesized to be a key factor in the evolutionary differentiation of such populations (Harrison 1998), and fire would be one of the more likely mechanisms behind the hypothesized metapopulation effects.

Conclusions

Fire affects diversity differently in each of the five MTC regions. On nutrient-stressed soils in the Cape region of South Africa and in Western Australia the vast majority of the diversity is from perennials and their presence is not strictly tied to fire. In these Gondwana heathlands, shrub spacing is sufficient to allow a diversity of growth forms to persist between fires and postfire diversity is only modestly greater than prefire diversity. The more fertile MTC communities of the northern hemisphere tend to have closed-canopy conditions with low understory diversity. In these landscapes fire creates a marked change in resource availability and associated with this is a huge increase in diversity after fires, from both annual and perennial herbaceous species. In a comparison of postfire communities, no single MTC community stands out as being particularly higher in species diversity than another at small scales. Communities exhibit some significant differences with disturbed sites in the eastern Mediterranean Basin having substantially higher diversity than other MTC communities. Factors driving community diversity are very different between MTC regions apparently because communities are structured very differently. This is reflected in marked differences between northern and southern hemisphere plant communities in species–area relationships, dominance–diversity patterns and species–abundance curves.

12   Alien Species and Fire

A large diversity of alien plants is found in most mediterranean-type climate (MTC) regions and fire is sometimes closely linked to their ability to invade natural ecosystems. This is a concern because aliens often upset natural ecosystem processes, and thus are a major management concern. These five regions not only differ in their contributions of non-native plant species to other regions, but also vary in their susceptibility to invasion by alien species, something often referred to as a community's invasibility.

Fire is a key factor behind plant invasions into natural plant communities and particularly critical is the timing of propagule availability and characteristics of the fire regime. Fire also interacts with geology in dictating functional types that become pernicious invasive problems. For example, on coarse-textured low-fertility soils in two of the southern hemisphere MTC regions, shrubs and trees are among the most aggressive invasives, and are capable of invading seemingly undisturbed intact shrublands. However, on more fertile soils such as in California and Chile, grasses and other herbaceous species are bigger threats, but invasion typically requires disturbance and under some circumstances fire can effect *type conversion* from woody vegetation to alien-dominated grasslands.

One of the important characteristics of many invasions in MTC regions is the fire-promoting capacity of the invading species (Brooks *et al.* 2004). These species are favored by fire and have vegetative traits that promote further fire in the system. Such fire-promoting feedback processes characterize a *grass fire cycle* where initial invasion promotes further success of fire-promoting grasses (D'Antonio & Vitousek 1992). However, there are actually two very different fire-promoting grass invasion processes that are functionally quite different, one being more typical of subtropical ecosystems and the other of MTC ecosystems (Box 12.1). In some cases aliens may change fuel structure sufficiently to alter fire behavior from surface fires to crown fires (e.g. Fig. 12.1).

As with other interactions between fire and community characteristics in MTC ecosystems, different patterns are evident with respect to alien plant invasion (Fox 1990); South Africa and southwest Australia show similar patterns due to infertile soils, whereas the moderately fertile soils of California and Chile promote quite different patterns, and the Mediterranean Basin stands alone in that it is a major source of invasive plants for other MTC regions and

Fig. 12.1 *Italian thistle (*Carduus pycnocephalus*) preferentially establishes under savanna oaks and outgrows competing native forbs and grasses, often reaching the canopy and potentially forming ladder fuels that can convert savanna surface fires into crown fires. (Photo by Jon Keeley.)*

not highly susceptible to invasions itself. Hopper (2009) has also suggested invasive plants follow very different patterns in these regions based on geology, although, as discussed below, the many generalizations that are required to fit with his old vs. new soil syndromes generally do not fit well for fire-prone MTC regions.

Mediterranean Basin

Particularly noteworthy is the fact that this region (including adjacent areas of Eurasia) is the source for many species invasive in other MTC regions. About two thirds of the non-native species in California, more than that in Chile and slightly less than that in South Africa and South Australia, are of Mediterranean Basin origin (Groves 1991). The most widely distributed of such invasive species are those associated with disturbance and it is widely thought that their origins in the Mediterranean Basin are tied to a long history of anthropogenic disturbance from fire, agriculture and domestic livestock grazing (Naveh 1974; Rundel *et al.* 1998). Since our major livestock animals such as cows, goats, sheep and horses have their origin in Eurasia (Diamond 1997), it seems likely that some of the weeds favored by lifestock husbandry had their origins long before domestication. This association likely selected for traits associated with some of the most aggressive weeds, such as rapid growth in high light environments and animal-facilitated dispersal (Malo & Suarez 1997).

Box 12.1 Two Grass Fire Cycles

A very valuable contribution to understanding invasive plant mechanisms is the notion that grasses can alter fire regimes in ways that promote their further invasion (D'Antonio & Vitousek 1992; D'Antonio 2000). This grass fire cycle is not a single model but we can distinguish two very different grass fire invasion processes that alter fire regimes in profoundly different ways. One is the perennial grass invasion of fire-sensitive woodlands and forests and another is the annual grass invasion of fire-adapted shrublands (Fig. B12.1.1).

The grass invasion of fire-sensitive woodlands is dependent on elevating fire intensity and killing woody plants that regenerate poorly after fire. This usually involves perennial grasses because the higher biomass of these aggressive grasses is required to generate sufficient fire intensity to promote canopy gaps, and the underground vegetative structures of perennial grasses protect them during high-intensity fires, as well as providing a bud source for regeneration. An example of this invasion process is the fire-driven C_4 grass invasion of tropical woodlands (Mack & D'Antonio 1998). In temperate regions the invasion of the C_3 bamboo-like grass *Arundo donax* also fits this model (Coffman et al. 2010).

This strategy is ineffective in fire-adapted shrublands because these shrubs are resilient to high fire intensity; nonetheless, some shrublands are

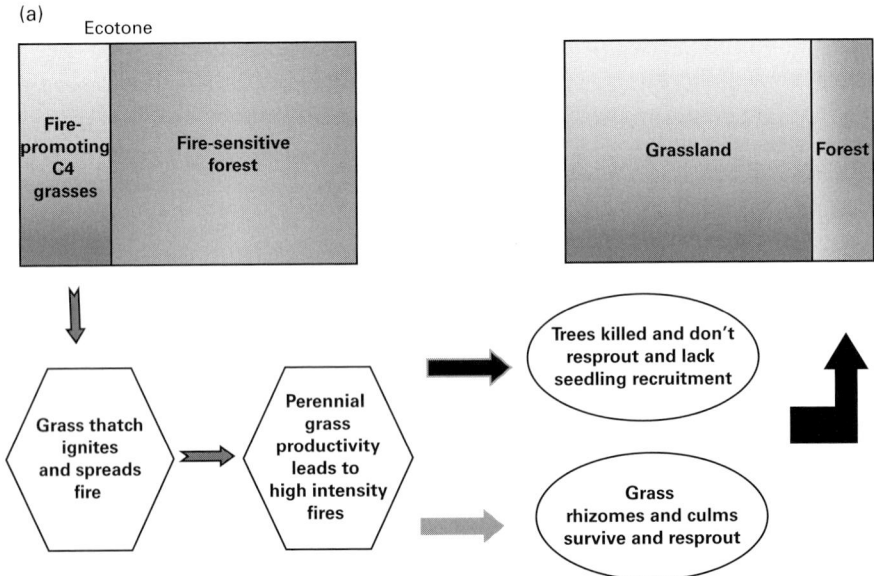

Fig. B12.1.1 *Grass fire cycles: (a) invasion of fire-sensitive forests by increasing fire intensity and (b) invasion of fire-adapted shrublands by increasing frequency and decreasing fire intensity.*

Continued

Box 12.1 (cont.)

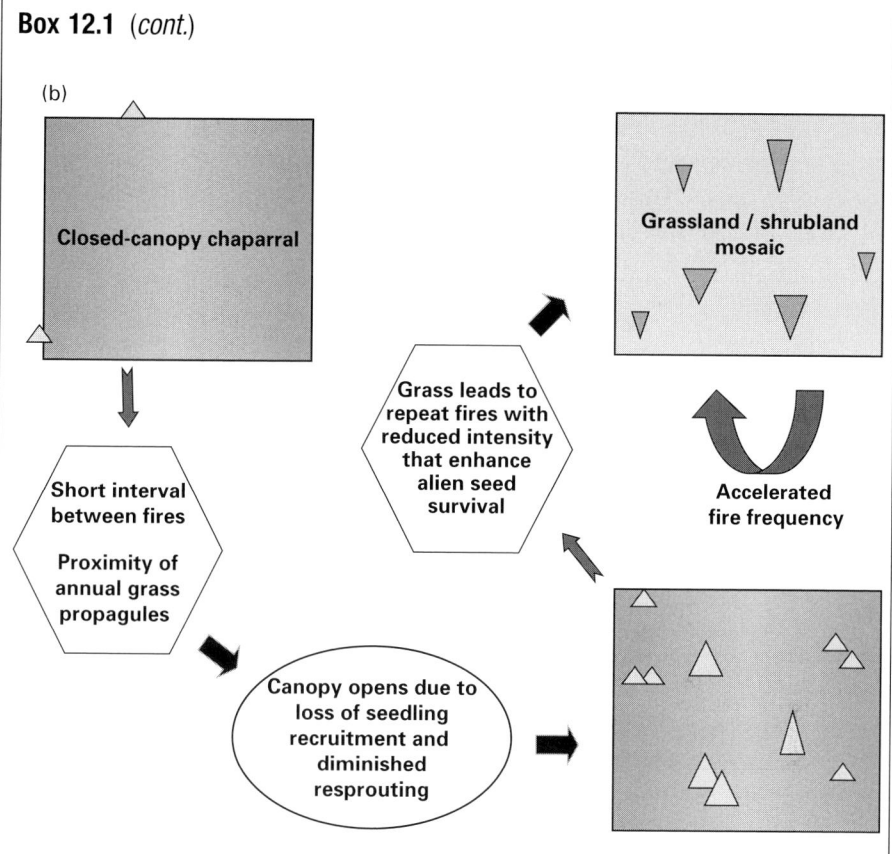

Fig. B12.1.1 (cont.)

vulnerable to grass invasion that works on a very different principle. MTC shrublands are sensitive to short fire intervals as many postfire seeder species are dependent on a sufficient number of years (5–15 yrs) between fires in order to reach maturity and replenish seedbanks (Zedler et al. 1983; Jacobsen et al. 2004). Under a frequent fire regime, annual grass invasion is promoted because these grasses can invade rapidly and produce flashy fuels that promote reburning. If a repeat fire occurs before shrub canopy closure, the lower shrub biomass leads to lower fire intensity, which promotes survivorship of annual grass seeds that are typically exposed on the soil surface (Keeley et al. 2005d). Type conversion of shrublands to grasslands in turn decreases water-holding capacity of the soils (Williamson et al. 2004), further adding to persistence of annuals. With each succeeding short-interval fire the shrub canopy becomes sparser and annual grasses and forbs spread and further increase probability of repeat fires.

Human activities conducive to the origin and spread of invasive species have been present in the region throughout the Holocene. As a consequence, movement of invasive species from one area to another has been going on for a very long time and has prompted the distinction of archaeophytes and neophytes to describe, respectively, old (before 1500) and new alien introductions (Pyšek 1998). Neophytes often are found in similar habitats as those favoring archaeophytes (Chytrý *et al.* 2008).

Despite being the source for many MTC invasive species, the Mediterranean Basin has a number of naturalized alien species as well. However, considering both the European and North African sides of the basin, these non-natives represent only about 5% of the flora compared with, for example, about 20% in California (di Castri 1989; Le Floc'h *et al.* 1990). For Spain, Italy, Greece and Cyprus Arianoutsou *et al.* (2010) reported 782 neophytes, but only 30 were widespread across all four countries. Most are roadside weeds or agricultural pests and not aggressive invaders of natural ecosystems, although roughly half of all neophytes do colonize fire-prone habitats including maquis, garrigue, phrygana and grasslands.

The Mediterranean Basin represents a particularly interesting area to investigate the relationship between invasive species and fire as it is the home to most of the invasive annual grasses that produce prominent grass fire cycles (Box 12.1) in other MTC regions. Adaptive modes of response to fire in woody species of Mediterranean shrublands exhibit parallel patterns to those seen in the other four MTC regions, and natural fire frequencies are similar to those of California (see Chapter 4). Many of the same genera of annual grasses and forbs that are highly invasive in some other MTC regions are also present and become established after fire. What makes the Mediterranean Basin dramatically different from the other regions, however, is the remarkable resilience of shrub species to frequent disturbance and seeming resistance to alien invasion.

Trabaud (1990) reviewed the resilience of a wide variety of Mediterranean Basin shrublands to repeated fire and general resistance to alien plant invasion. Although a number of naturalized alien species may briefly colonize recent burns, they seldom persist even under regimes of frequent fire. Several factors appear to be at work here. One is the limited number of aggressive alien species from other regions of the world. Another is that many shrublands are dominated by vigorous resprouting shrubs that quickly recapture space after fire. Relative to most other MTC regions, the Mediterranean Basin has many fewer non-resprouting obligate seeding shrubs (see Chapter 3). Thus, canopy cover increases rapidly, providing fewer opportunities for herbaceous aliens to establish. Even on those sites with obligate seeding shrubs, most of which are sub-ligneous, the vegetation exhibits high resilience to short fire return intervals. Many such shrubs reach sexual maturity in the third year after a fire, and some of them can sometimes produce seeds at 1–2 yrs (Carreira *et al.* 1992; Roy & Sonié 1992). Similarly, pine forests in the Mediterranean Basin have been found to be generally resilient to short fire return intervals (Thanos & Doussi 2000; Eugenio & Lloret 2006).

Although examples of grass fire cycles are few in the Mediterranean Basin (Naveh 1974), a positive feedback has been described between the expansion of the native tussock grass *Ampelodesmos mauritanica* and fire hazard in the Iberian Peninsula and its relation to tree cover decline (Vilà *et al.* 2001). With shorter fire return intervals (Box 12.1) the invasion success and contribution to community biomass of *A. mauritanica* increases. Modeling simulations have shown that landscapes can abruptly switch from regimes of small localized fires to extensive fires as a result of the spread of *A. mauritanica* (Grigulis *et al.* 2005).

Despite what appears to be a higher level of resilience to disturbance in Mediterranean Basin shrubs compared with the other MTC regions, herbaceous dominance and gradual alternation of stand dominance certainly occurs under land-use regimes with frequent disturbance from fire and/or grazing (Delitti *et al.* 2005). It is interesting to note that the herbaceous community formed in cleared and grazed fuel breaks in the eastern Mediterranean is equally diverse and rich in species of herbs as undisturbed garrigue but quite different in composition, favoring an early flowering, geophyte-rich assemblage of native species (Hadar *et al.* 1999). This is quite different from the dominance of alien annual grasses and forbs that colonize fuel breaks in other regions. Nevertheless, studies suggest that the Mediterranean Basin can expect to experience increasing invasion and establishment of alien species in the future under increasing disturbance and global change (Lloret *et al.* 2004; Gritti *et al.* 2006; Lambdon *et al.* 2008). In addition, some invasive shrubs are increasing after fire and becoming problematic, at least locally. This is especially relevant in areas with relatively high precipitation and calcium-free (acidic) soils, as in western Iberia. Some Australian shrubs find such conditions suitable and are becoming invasive. Examples are *Acacia* species (mainly *A. dealbata* and *A. longifolia*) and some *Hakea* species (*H. sericea*); they have recruitment stimulated by fire and are invading some ecosystems in Portugal. *Eucalyptus* (mainly *E. globulus*) trees are extensively planted in several places in the Mediterranean Basin, and although they are not currently considered invasive, species with fire-stimulated germination may become invasive in the near future.

California

This MTC landscape has proven to be particularly vulnerable to alien plant invasions. The extreme topographic heterogeneity with the highest and lowest elevations in the continental USA has led to significant variation from coasts to mountains and deserts.

Alien Plant Invasion and Type Conversion of Chaparral and Sage Scrub

MTC shrublands in the foothills and coastal ranges of California exhibit a marked sensitivity to short fire return intervals (Zedler *et al.* 1983; Jacobsen *et al.* 2004; Syphard *et al.* 2006). Natural fire frequencies absent any human influence in chaparral and sage scrub would likely have been in the range of every 30–100 yrs or more (see Chapter 5). Today the region is dominated by an anthropogenic fire regime and on most of the landscape fires are much more frequent.

Fires in undisturbed, relatively alien-free shrublands have limited alien presence after fire due the lack of an alien seedbank. Even when aliens are present in gaps in dense chaparral, fire intensities from these high fuel volume systems are capable of killing alien seedbanks (Keeley 2001). Those aliens that do colonize burned sites are typically shaded out as the shrub canopy closes in the first decade or two after fire. However, the native postfire annuals and other short-lived species in the early seral communities form dense contiguous surface fuels when they die back during the summer and autumn drought. With humans providing a ready source of ignitions such communities are susceptible to reburning before the shrubs have fully recovered and replenished their seedbanks. Given a source of alien grass and forb propagules, these reburned sites are readily invaded, initiating an annual grass fire cycle (Box 12.1) that can lead to degraded shrublands or complete type conversion to grasslands (Zedler *et al.* 1983; Haidinger & Keeley 1993; Jacobsen *et al.* 2004; Syphard *et al.* 2006; Cox & Allen 2008; Fleming *et al.* 2009). Multivariate modeling supports the hypothesis that the primary drivers of invasion are the ready source of alien propagules at the time of fire and the speed of shrub canopy closure (Keeley *et al.* 2005d).

Frequency of burning required to effect this type conversion varies with vegetation and site conditions. In California sage scrub, fires at less than 5-yr intervals are often required for type conversion but longer-interval fires can also displace sage scrub under more arid conditions (Keeley *et al.* 2005d). In chaparral, fire-return intervals of 5–15 yrs often initiate invasion and set the community on a trajectory toward type conversion. Over much of southern California there are extensive areas of type conversion from sage scrub and chaparral to alien-dominated annual grass and forb land. Sage scrub has suffered the greatest extent of type conversion, primarily because it is distributed in the lower elevations and subject to the greatest concentration of anthropogenic fires (Keeley 2006a). Jacobsen *et al.* (2009) suggest that water-use patterns of sage scrub species makes this community more vulnerable to alien invasion.

Type conversion of chaparral communities to grasslands or herblands almost certainly began in California with the activities of Native Americans (Keeley 2002b). The vegetation cover of the Coast Ranges today comprises a mosaic of chaparral, woodland and grassland. About a quarter of the landscape is covered by annual grasslands that are largely dominated by alien grasses and forbs from the Mediterranean Basin, which lack any obvious climatic or edaphic factors to account for their distribution (Huenneke 1989). Natural lightning ignitions in this region would not have occurred frequently enough to allow the establishment and maintenance of this open community, and thus they are undoubtedly an artifact of an anthropogenic increase in fire frequency (Wells 1962). We know that Native Americans in California actively used fire as a management tool to open up chaparral stands (Anderson 2006), although prior to European introductions

Fig. 12.2 *Type conversion recorded for Malibu Canyon, Santa Monica Mountains, California: left, natural chaparral landscape and right, landscape dominated by alien annual grasses after three fires in 12 years. (Based on Jacobsen et al. 2004; photos by Anna Jacobsen and Steve Davis.)*

these would have been dominated by native annuals and other herbaceous plants. This frequent burning would have generated a quasi-disequilibrium that perhaps made it easy for European weeds to take hold and accounts for their very rapid invasion during the early nineteenth century.

The dominance of annual grasses in California from the Mediterranean Basin certainly dates only to the Euro-American period, and management practices over the past two centuries have greatly expanded and maintained the area of the landscape that has been type converted from native shrublands to grasslands (Keeley 2002b). The extent of conversion of chaparral and sage scrub stands to alien-dominated annual grasslands across California continues to increase at a rapid rate today (Rundel 2007). On a broad scale, this landscape conversion can be seen in the extensive coastal areas and foothills of southern California that are dominated today by annual grassland (Hamilton 1997). With the exception of small areas of special edaphic conditions, most of these grasslands are degraded areas that once supported sage scrub or chaparral (Freudenberger *et al.* 1987; Keeley 1990a; Minnich & Dizzani 1998). On a local scale, this phenomenon of type conversion is widespread along major transportation corridors where human fire ignitions are common in the flash fuels of grasses along highway verges. Many areas with short fire return intervals lose their less resilient woody shrub dominants and are converted to annual grasslands, often with a few highly resilient perennials surviving (Fig. 12.2). This is a matter of some concern for several reasons: (1) loss of native biodiversity, (2) increase in alien plant spread, (3) change in hydrological processes due to switch in functional type from deep-rooted shrubs to shallow-rooted grasses, (4) increase in fine fuels and thus lengthening the fire season, and (5) change in carbon storage capacity.

Although it appears that the majority of these grasslands arose from type conversion of shrublands, on some sites they apparently displaced native perennial grasslands (Hamilton 1997) and on other sites annual forb-dominated *herblands* (Hoover 1936). In all cases the origin of these alien annual grasslands was dependent on increased frequency of fire and other disturbances such as intensive animal browsing (Zedler *et al.* 1983; O'Leary & Westman 1988; Haidinger & Keeley 1993; Cushman *et al.* 2004; Jacobsen *et al.* 2004). Air pollution has been suggested as an alternative factor but there is no evidence that this alone is sufficient; however, it is possible that type conversion has been enhanced by ozone (Westman 1979) and nitrogen pollution (Padgett & Allen 1999; Allen *et al.* 2000). One study reporting for some sites a stronger relationship between alien grasses and pollution than between alien grasses and fire (Talluto & Suding 2008) requires further study as the researchers used a database of only larger fire events; thus in the highly polluted portions of southern California they only captured 5–10% of all fires.

Once established, annual grasslands can be very persistent and the return of native shrublands is a rather slow process (Keeley 1990a; Minnich & Dezzani 1998; Stylinski & Allen 1999; Rundel 2007; Cox & Allen 2008). On sites formerly dominated by sage scrub, recolonization may occur within a decade or so if disturbance is removed and seed sources are in close proximity (Freudenberger *et al.* 1987). This is promoted by the highly dispersed propagules in most sage scrub species (Wells 1962). When chaparral is type converted, natural restoration is slow because shrubs with good dispersal establish poorly on open sites and those capable of recruiting on open sites have poorly dispersed seeds (see Chapter 5). The process where shrubs are observed to invade annual grasslands is often referred to as *shrub encroachment*, but this is a misnomer since they are actually recolonizing and restoring the natural communities (Keeley 2005). Encroachment is appropriate to describe the spread of alien shrubs into grasslands.

Some of the management practices that have long been used to reduce fire hazard in California shrublands have now been shown to have a mixed benefit as they also promote invasions of non-native grasses into native shrublands (Keeley 2006b). This is particularly evident with programs to reduce hazardous fuels along the urban–wildland interface. These programs reduce the amount of flammable biomass in such areas and thus potential fire intensity but at the same time the type conversion promotes the invasion of alien annual grasses and forbs into these open habitats. Without careful management of these grassy areas, flammability is locally enhanced because of the nature of these flashy fuels that ignite easily and can carry fire through much of the year. These problems are also evident in other fire-related management activities such as fuel breaks (Merriam *et al.* 2006; Potts & Stephens 2009). A related theme that has been studied relates to the management implications of fuel breaks. Fuel breaks along ridgelines or across shrubland slopes tend to promote alien plant invasion and play a role as a seed source for the

Fig. 12.3 *Fuel break on USFS lands in southern California illustrating the inevitable type conversion from native shrublands to alien-dominated annual grasses and forbs. (Photo by Richard W. Halsey.)*

establishment of alien grasses and forbs in adjacent wildland areas (Fig. 12.3). These fuel breaks may act as corridors that bring alien plants into wildland areas. Other corridors such as railroads are likewise implicated as sources of alien plant invasions (Bangert & Huntly 2010).

Despite the key role that fire plays in the spread of alien species in California, there are management agencies that advocate the use of prescription burning as a means of controlling some alien species. Although dependence on this or any other eradication method alone is doubtful as a means of sustainable control, it may have some short-term value (Box 12.1).

Alien Plant Invasion in Montane Conifer Forests

Alien plant threats are greatest in the foothills and coastal plain; they decline markedly with elevation (Mooney *et al.* 1986; Schwartz *et al.* 1996) and are generally relatively minor in undisturbed closed-canopy conifer forests (Keeley *et al.* 2003). In closed-canopy forests invasives are likely to increase following high-intensity fires (Keeley *et al.* 2003) but in more open forests invasives increase with a decrease in severity (Franklin 2010a). Both historical and niche-based factors may be responsible for this pattern. Mountain habitats have been occupied by humans for a relatively short time in California; thus, disturbances conducive to invasion have not been available until relatively recently and so time may be a factor. Less intensive land use in mountain habitats of the Old World also may be responsible for fewer potential invasive species in high-elevation sites, although this has not been demonstrated. A related historical explanation is that the higher-elevation sites have not had sufficient time for adequate propagule dispersal (Rejmánek 1989). Alternatively, character syndromes of most successful invasive species in California include rapid growth rates in high light environments. The low light environment on the forest floor of conifer forests may not be compatible with the life history of most potential invaders.

Disturbance cycles are frequently tied to alien plant invasions and so some of the basis for limited alien presence in montane conifer forests is tied to a century of management policy that has excluded fire. This fire suppression policy in most coniferous forests in the western USA has until relatively recently been highly successful in excluding fire, extending the fire return interval to far longer periods than under pre-Euro-American conditions (see Chapters 5 and 13). This policy, while promoting a number of problems of dense stands of understory saplings and accumulated large biomasses of downed litter, nevertheless restricted the invasion of alien grass species by maintaining a dense canopy cover. Active management today to reinstate more natural fire return intervals through prescribed burns has restored more natural ecosystem processes but has also enhanced forest vulnerability to alien invasions. This has been a notable problem in prescribed burns in ponderosa pine forests of the Sierra Nevada where the open conditions created by burning, although similar to natural conditions, has greatly promoted invasion of cheatgrass (*Bromus tectorum*) into these communities (Keeley & McGinnis 2007). This condition has led to a management challenge in choosing between restoring natural fire regimes or altering those fire regimes to favor communities of native species, with the latter having the potential for long-term impacts on forest structure, increased vulnerability to crown fires and potential cascading effects on new invasions of alien species (Keeley 2006b).

Fire and Alien Species in Mediterranean-type Climate Arid Lands

Prescription burning in sagebrush ecosystems at the desert margins of the MTC of California has been widely used to increase rangeland resources for livestock grazing (Keeley 2006b). *Artemisia tridentata*, the dominant shrub in these ecosystems, is intolerant of fire and is quickly replaced by more palatable herbaceous plants. Natural recovery of these stands is very slow and requires decades (Harnis & Murray 1973). Massive invasions of cheatgrass into these sagebrush ecosystems degraded by fire have created another classic grass fire feedback cycle which promotes maintenance of cheatgrass dominance by shortening fire return intervals (Mack 1981).

The widespread invasion of *Bromus madritensis* into the western Mojave and Sonoran Deserts in California began in the twentieth century and has continued, producing a widespread cover of flammable grass (Brown & Minnich 1986; Rogers & Vint 1987; Brooks 1999). Areas that formerly had no continuity of fuels to carry a fire now experience large fires thousands of hectares in size. Many native shrub species and succulents have little or no resilience to fire and are strongly and negatively impacted by the introduction of a disturbance regime that includes fire (Brooks & Minnich 2006).

Chile

Central Chile stands out among all MTC areas because natural fires are extremely rare due to a general absence of lightning and other natural ignitions (see Chapter 6), although this appears to be a relatively recent phenomenon beginning in the late Tertiary (see Chapter 10). Thus, fire dynamics in central Chile provides an interesting case study for evaluating the importance of fire on plant invasions (Pauchard *et al.* 2004, 2008). Relatively few data are available on the interactions of alien species and fire in central Chile, but it appears that frequent fire in matorral does favor alien over native herbaceous species (Ávila *et al.* 1981; Sax 2002). Matorral and chaparral share a similar community architecture and are largely invaded by the same species of annual grasses and forbs from the Mediterranean Basin (Gulmon 1977; Sax 2002; Pauchard *et al.* 2004; Jimenez *et al.* 2008).

The Chilean matorral is invaded by alien plant species from other mediterranean zones of the world where fires have been a recurrent component of disturbance regimes throughout the Cenozoic. It has been hypothesized that anthropogenic fires in central Chile may promote the invasion of alien plants with adaptive traits not present in the native flora (Muñoz & Fuentes 1989; Segura *et al.* 1998; Holmgren *et al.* 2000a, 2000b; Gómez-González *et al.* 2008; Figueroa *et al.* 2009). Low-intensity fires have been shown to not significantly affect the emergence of native herbs but do lead to increased alien species richness. High-intensity fires beneath the canopy of closed and open matorral shrublands negatively affect the seedling emergence of both native and alien species, but more strongly in native species (Gómez-González & Cavieres 2009).

As in California, extensive areas of evergreen matorral in central Chile have been type converted to alien-dominated annual grasslands (Holmgren *et al.* 2000a), with high frequency of anthropogenic fires (Kunst *et al.* 2003). Populations of invasive annual plants rebound quickly after fires whereas native perennials recover slowly, much as in California chaparral, and thus the annual grass fire cycle (see Figure B12.1.1b) fits well in the Chilean matorral. Floristically the invasive floras are remarkably similar between California and Chile (Bustamante *et al.* 2005). One difference between these two systems is that the chaparral has obligate seeder species whereas the Chilean matorral lacks this functional type, and so the postfire regeneration of open spaces is more limited in the matorral, thus making the system more susceptible to invasion. In addition, most shrubs of the Chilean matorral need shade for recruitment (Fuentes *et al.* 1984) and thus open spaces remain available to alien annuals for long periods of time. However, there remain disagreements among authors on the importance of fires in the invasibility of plant communities in central Chile (Figueroa *et al.* 2004).

Invasive woody legumes colonizing cut-over forest lands in south-central Chile have almost certainly had a major impact on ecosystem processes and fire dynamics but these remain poorly studied. *Ulex europaeus* (gorse), for example, has become established over thousands of hectares and increases its relative dominance with fire. Positive feedbacks of invasive woody species and fire frequency have been shown for *Teline* (*Genista*) *monspessulana* (French broom) in areas dominated by pine plantations in south-central Chile, with fire favoring the presence and dominance of this species (Pauchard *et al.* 2008).

South Africa

The fynbos shrublands of the Cape region of South Africa have been strongly influenced over broad areas by invasive woody species. Alien species of *Hakea* from Australia and *Pinus* from California and the Mediterranean Basin have become widely established in fynbos, as have Australian *Acacia* species in riparian habitats of the Cape region. These invasions have had major ecosystem impacts including the promotion of fires with higher intensity, with potentially negative impacts on native seedbanks (Holmes 2002).

Once alien trees have established, they typically grow faster and taller than indigenous species, and after one or two fire cycles form closed stands with reduced light penetration and altered nutrient cycling patterns, litterfall, and fuel properties (Richardson *et al.* 2000a). Such stands may replace fynbos vegetation and their impacts intensify with time elapsed since invasion (Holmes & Cowling 1997b). Invasions by shrubs and trees have been the subject of extensive studies because of the ecological impacts that these invaders have had on competition with native species, hydrologic flow, nutrient enrichment and increased fire hazards (Holmes & Cowling 1997a, 1997b; Le Maitre *et al.* 2002; Richardson *et al.* 2004; Yelenik *et al.* 2004).

Although fynbos vegetation is well adapted to moderately frequent fire return intervals (van Wilgen *et al.* 1992b), natural fires promote the spread of invasive shrubs and trees. One impact is that fires in areas heavily infested with dense stands of alien woody species are less easily ignited and spread more slowly than in pristine fynbos, where there is an abundance of fine material in the herbaceous layers (van Wilgen & Richardson 1985; van Wilgen 2009). However, under extreme weather conditions, the high fuel loads of invaded stands promote fires that burn with very high intensity. Such high fire intensities are difficult to contain and are potentially more damaging to ecosystem structure and processes than natural fires in fynbos vegetation. While fire may exacerbate invasions, there is some evidence that it may provide a valuable control on these invasives as well (Box 12.2).

Compared with studies of the ecosystem impacts and flammability issues related to invasive woody species in the Cape region of South Africa, there has been relatively little concern placed on environmental problems posed by the presence of alien grasses in the Cape region (Milton 2004; Musil *et al.* 2005). Much of the apparent inability of annual grasses from the Mediterranean Basin to become widely established and ecologically significant in the region relates to the extreme oligotrophic conditions presented by soils formed from the widespread Table Mountain sandstone. Nevertheless, there are reports that alien grasses have increased in abundance, especially in low-lying areas (Vlok 1988). This increase in

Box 12.2 Control of Invasive Plants with Prescription Burning

Although fire promotes alien plant invasions in MTC ecosystems, it is also advocated as a management tool to be used to control some particularly noxious invasive plants (DiTomaso *et al.* 2006). There are two very important caveats about use of fire to control alien species. One is that although short-term reductions in some alien species can be accomplished, long-term sustainable control is far less likely. Secondly, while fire may control a particular target alien species, unless burning is accompanied by active native plant restoration, this target will often be replaced by other alien species rather than by more desirable native species (e.g. Fig. B12.2.1). Similar issues can be raised with regard to other presumed control methods such as goat grazing (e.g. Thomsen *et al.* 1993).

Where sustainable control of target alien species with fire has been most clearly demonstrated is with woody aliens. For example, the Mediterranean Basin postfire seeding shrub *Cytisus scoparius* and related "brooms" are invasive shrubs in degraded grasslands in California. They are weak resprouters but maintain dormant seedbanks with fire-stimulated germination. A single fire

Fig. B12.2.1 *Example of unanticipated outcomes from prescription burns planned to target certain species. Prior to burning, this landscape in the Santa Monica Mountains National Recreation Area of Ventura County, California, was dominated by the non-native annual ripgut grass* Bromus diandrus, *which was the target of the burn that occurred to the left of the road. Ripgut grass persists in the unburned area to the right of the road but was replaced by an even more noxious alien, black mustard or* Brassica nigra, *to the left of the road. (Photo by Dr. Marti Witter, National Park Service.)*

Continued

Box 12.2 (*cont.*)

generally promotes the invasion process but close-interval repeat fires are often sufficient to reduce resprouting capacity and eliminate the seedbank, thus locally extirpating this shrub (Swezy & Odion 1998; Odion & Haubensak 2002; Alexander & D'Antonio 2003). This treatment appears to be sufficient for sustainable control of this target species although such close-interval fires inevitably favor alien grasses and forbs (Jacobsen *et al.* 2004; Keeley 2006a) and are not an appropriate management strategy for shrubland ecosystems.

South African fynbos is heavily invaded by postfire seeding shrubs from Australia such as *Hakea sericea* and fire alone will only act to spread this invasive. Effective control has been achieved by first cutting, drying and burning; however, this management results in much greater fire intensities and reduced native regeneration (Richardson & van Wilgen 1986). For this and other woody species, fire alone does not seem to be sufficient to eliminate invasion of fynbos (Holmes *et al.* 2000). There has been considerable study focused on restoration of invaded lands by the removal of infestations of invasive trees. A variety of approaches have been used including burning standing vegetation, felling and burning the dead fuels, and felling followed by removal of most dead fuels to reduce fire intensity, as well as biocontrol agents (Holmes *et al.* 2000; Pretorius *et al.* 2008; Richardson & Kluge 2008; Roura-Pascual *et al.* 2009). Burning standing vegetation is often not successful because of seed released from woody cones and subsequent postfire seedling establishment. While felling and burning kills seedlings and thus reduces the density of invasive plants, the high fire intensities involved may have detrimental effects on native species. Felling and brush removal followed by fire has shown good success but is expensive. A clear benefit of removing invasive shrubs and trees, however, has been shown in increased water supply in natural watersheds (Le Maitre *et al.* 2002; Gorgens & van Wilgen 2004).

Fire is frequently advocated as a control method for many herbaceous alien species. One example of the pitfalls in using prescription burning is in the application of spring burning to control yellow starthistle (*Centaurea solstitialis* L.). This European pest is established in parts of western North America and particularly noxious because it greatly alters livestock range conditions. Short-term reductions in this species can be achieved with repeated burning. However, this thistle has a relatively long-lived seedbank, and longer-term study shows that it re-establishes once burning is halted (Fig. B12.2.2). Clearly, prescribed burning provides only temporary reduction and does not lead to sustainable control of this alien; and may exacerbate the alien situation (Keeley 2006a).

This outcome should not be surprising since most alien herbs are opportunistic species that capitalize on disturbance. Prescription burning is a non-specific eradication method that affects both the target species and potential

Continued

Box 12.2 (cont.)

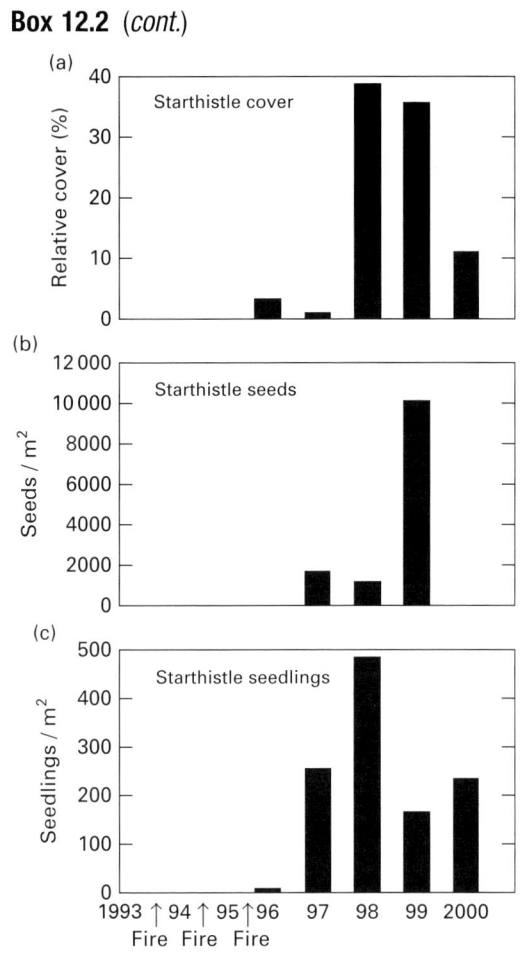

Fig. B12.2.2 *Yellow starthistle* (Centaurea solstitialis) *cover and seed and seedling production following three consecutive annual burns applied to extremely dense populations of this noxious alien weed. Immediate postfire results were very promising (DiTomaso et al. 1999), but follow-up studies indicate that burning destabilized these grasslands and allowed subsequent reinvasion once burning was stopped (Kyser & DiTomaso 2002).*

competitors in the community. The temporary disequilibrium creates opportunities for the most aggressive colonizing species and often this will include the target species. In most cases there are better methods than fire for specifically targeting a species of concern. Sustainable control of most aggressive weeds is likely going to occur only when natural, intact ecosystems are restored. When restoration of natural ecosystems is desired, more often than not there will be the need for active restoration of native species along with targeted eradication of specific alien species.

ecological presence of alien grasses has been attributed to rangeland deterioration caused by land use changes associated with plowing, landscape clearing and burning, and soil nutrient enrichment resulting from fertilizer run-off and nitrogen-fixing leguminous species (Milton 2004).

Concern has been raised that while alien grasses may not represent a major issue in landscape flammability today in the Cape region, this problem of grass invasion may increase in the future with global climate change. A recent study addressing this issue used climate modeling to predict the future distributions of 29 species of C_3 annual grasses from the Mediterranean Basin (Parker-Allie et al. 2009). The authors concluded that future climate warming will broadly hinder the spread of mediterranean annual grasses in South Africa, with all but one species predicted to have a contraction in their current range of distribution. However, the predicted shifts in distributions of these grasses into pristine higher elevations with climate warming in the Cape region may alter existing fire regimes and thus pose a threat to the natural vegetation. The authors also pointed out that rising atmospheric CO_2 levels could mitigate the negative impacts of climate warming and thus allow alien C_3 grasses to persist in more arid lowlands.

Southwestern Australia

Much like the situation in South Africa where oligotrophic soils have restricted the extent and relative dominance of establishment of annual grasses from the Mediterranean Basin, the extensive areas of heathland, mallee, and eucalypt woodland growing on poor soils in the MTC region of southwestern Australia have been relatively impervious to the invasion of alien annual grasses. However, one exception to this pattern lies with local areas of eucalypt woodlands on richer soils which are readily invaded by mediterranean alien grasses (Abensperg-Traun et al. 1998).

Banksia woodlands of southwestern Australia have been impacted by many changes since European settlement 150 yrs ago. These changes included fire frequency and intensity, nutrient inputs, and the introduction of alien species (Bridgewater & Backshall 1981) followed by an increase in exotic propagules (Fisher et al. 2009a). While *Banksia* woodlands are dominated by native species that are fire adapted (Hobbs & Atkins 1990), when fires become too frequent (Hopkins & Griffin 1989) or there are impacts from human activities that enrich soil nutrient pools, the resulting increase in resource availability makes the community vulnerable to invasion (Milberg & Lamont 1995).

As a result of impacts associated with frequent fire and nutrient enrichment, many areas of *Banksia* woodland have become altered from a species-rich, sclerophyllous, tree and shrub dominated community to one with an understory dominated by alien annuals and/or herbaceous perennials (Bridgewater & Backshall 1981; Fisher et al. 2009b). The invasion of these species has been described as an example of a classic grass fire feedback cycle, with the light grass fuels promoting more frequent fires because of their high flammability (Baird 1977). A similar pattern of sensitivity of native resprouting Proteaceae to frequent fires has been documented in mediterranean-type vegetation, including both shrublands and woodlands, of New South Wales (Bradstock & Myerscough 1988).

The ability of invasive perennial grasses and herbs to alter the structure and function of *Banksia* woodlands has been well documented in a series of studies focusing on South African *Ehrharta calycina* and *Pelargonium capitatum* as examples of invaders (Fisher et al. 2009a, 2009b). These two plants represent growth forms unlike those typically present in the understory of *Banksia* woodlands. Once established from propagules in sites with high fire frequency these resprouting species change the understory dominance from native reseeding species to alien resprouting species. The extensive root system of *E. calycina* provides this species with advantages over native species in competing for limited water and nutrient resources. Studies have shown that *E. calycina* is very competitive postfire but not a strong competitor in intact bushland (Baird 1977). Rapidly growing herbaceous species, with *Pelargonium capitatum* as a well-studied example, have been shown to interfere with the success of slow-growing native seedlings (Hobbs & Atkins 1991). Once established, flammable grasses such as *Ehrharta* provide a positive feedback on increasing fire frequency, which results, with short fire intervals, in eliminating native species that do not have adequate fire-free intervals to reproduce and replenish canopy- and soil-stored seedbanks, and sustain underground storage organs.

Studies in fragmented mallee shrublands have shown that alien grass invasion is limited by both the lack of propagules and low nutrient levels, and that fire and/or fragmentation disrupts these limits (C. Gosper et al. 2010 unpublished manuscript). Along the perimeter edges of mallee stands, the biomass of invasive grasses is increased with propagule availability and elevated nutrient levels, with fire having a neutral to negative effect. Nutrient-enriched edges are susceptible to invasion with or without fire. Away from perimeter edges, neither fire nor fragmentation by interior roads enhanced invasive plant abundance or biomass. Overall, fire does not promote alien grass invasion in mallee and is thus available as a viable disturbance strategy for biodiversity conservation in large native vegetation stands.

Conclusions

Invasive species problems are of far less importance in the Mediterranean Basin than in other MTC regions. The Mediterranean Basin, however, has been a rich source of invasive species that have spread widely around the world. In California fire has played a major role in driving annual grass and forb invasion of many shrub communities, the dynamics of which are captured in a variation of the classic grass fire cycle. Widespread loss of native shrublands and type conversion to alien annual grasslands are due to the lack of resilience among most woody species to the anthropogenic-based short fire return intervals, whereas most alien herbs are favored by this fire regime. Similar changes likely occurred in Chile but due to European colonization hundreds of years earlier than in California, we have little documented evidence of type conversion. The extent of impacts from invasive herbaceous species is far less widespread in the Cape region and southwestern Australia where oligotrophic soil conditions restrict the establishment of invasive annual grasses and forbs from the Mediterranean Basin. Invasion and establishment of woody acacia and pines in the Cape region have impacted fire dynamics by increasing fire intensities.

Although shrubland and woodland landscapes in the Mediterranean Basin have been greatly impacted by humans, both resprouting and obligate reseeding woody species in the region appear to exhibit higher levels of resilience to disturbance, including short fire return intervals, than those generally exhibited by woody species in the other four MTC regions. Nevertheless, an inherent resistance to alien species establishment in this region is threatened by changes in land use and global climate change.

All five MTC regions face expected increases in fire frequency under climate change scenarios. Invasive species represent threats not only through their direct impacts on community structure and composition, but also by promoting flammable herbaceous invaders that have the potential to induce powerful feedforward processes and thereby fundamental changes to ecosystem function.

13 Fire Management of Mediterranean Landscapes

The hazardous mediterranean climate, highly flammable vegetation, and rugged terrain, all important elements of fire behavior, become problems only in the presence of people. People recreate and build homes in the mediterranean wildlands because of the delightful climate and will continue to do so as long as space is available. People start most fires, and their mere presence tends to warp fire suppression strategies because fire agencies must protect lives and property threatened by fires rather than "back off" and build fire lines around fire perimeters.
Carl C. Wilson (1979a), Chief of Division of Forest and Fire Research,
USFS/Pacific Southwest Forest and Range Experiment Station

Human presence in mediterranean-type climate (MTC) regions has differed markedly in the length of human occupation; however, there are remarkable similarities in how early inhabitants altered fire regimes and how modern societies deal with the fire hazard. Here we draw on the history of human impacts outlined in the regional reviews (see Chapters 4–8), the problems created by nineteenth and twentieth century management practices, and conclude with twenty-first century problems and future options. As discussed throughout this book, MTC ecosystems are highly fire adapted but, as illustrated here, contemporary societies have not fully adapted to balancing fire hazard risk and resource needs on these landscapes.

Early Human Fire Use and Impacts

Fire has been a widely utilized management tool throughout the history of humankind (Pyne 1995). Early hunter–gatherers utilized fire to manage for plant and animal resources. Fire also played an important role in early domestication of crops as clearing off woody or other perennial vegetation would have required fire on many landscapes. With domestication of livestock it was an important tool for increasing forage.

Early human history has been very different in the five MTC regions. In the Cape region of South Africa and the Mediterranean Basin there has been a presence of modern humans for more than 100 000 yrs (Deacon 1983, 1992; Carbonell *et al.* 1995). Throughout this period humans had the capacity to utilize fire and, considering the limited tools early humans had for manipulating their environments, it seems inescapable that they would have actively used fire when appropriate. For example, frequent burning of fynbos enhanced availability of geophytes on which early inhabitants depended (Le Maitre & Brown 1992). In the Mediterranean Basin fire has been utilized for converting woody vegetation to herbaceous crops and rangeland, which intensified during the Holocene (Thirgood 1981; Blondel 2006).

Humans reached Australia more than 45 000 yrs BP, and some contend this was an accidental discovery from people on rafts caught in storms, because the continent would not have been visible from the next nearest land mass, New Guinea. Bowman (2003), however, suggested that fire may have played a role informing early peoples of a distant land mass since large smoke plumes from bushfires in northern Australia would have been visible from New Guinea. Late Pleistocene human colonization of North America also may have depended on fire while crossing the Bering Strait in order to survive the arctic conditions (Higuera *et al.* 2008). Aboriginal people in Australia and Native Americans in California and Chile are thought to have used fire regularly to facilitate resource acquisition (Pyne 1995; Anderson & Moratto 1996), in part because of very limited development of agriculture by the earliest inhabitants of these and other MTC regions (Diamond 1997). This type of resource acquisition has been described as "firestick" cultivation (Pyne 1995).

There is a tendency to view these original peoples as more in touch with their environments and sensitive to the proper use of fire in managing for sustainable landscapes: the "ecological Indian" model (Krech 1999). This model has been invoked to inform fire management as to appropriate management practices for maintaining biodiversity and natural ecosystem functioning (e.g. Pyne 2000; Abbott 2003). An alternative view is that early peoples used fire to maximize their fitness much like contemporary societies' impact on the environment was determined by the balance between resource availability and population density.

There is some commonality in human responses to their environments in widely disjunct MTC regions. Various modern human behaviors such as symbolic and technological complexity increased at the about the same time in the Holocene in both the Mediterranean Basin and South Africa, and it is hypothesized that they were the result of similar demographic factors that impacted lifestyles (Powell *et al.* 2009). The types of impacts that Holocene populations had on the environment involved resource exploitation, and sometimes the use of burning resulted in type-conversion of shrublands and woodlands to grasslands and other herbaceous-dominated associations. Landscape features played a role in determining impacts. For example, in California, remote areas in rugged terrain were left untouched whereas vegetation patterns in other parts of the landscape were transformed through repeated burning (Keeley 2002b). In the Cape region of South Africa geology played a role in that early pastoral peoples found fynbos on low-fertility soils to be of little grazing value, but on better soils did extensive fire-induced type conversion of renosterveld shrublands to grasslands (Krug *et al.* 2004). In the Mediterranean Basin forests and woodlands have been eliminated from low-lying flat landscapes by human activities (Vallejo *et al.* 2006).

The impact of 45 000+ yrs of Australian Aboriginal occupation on fire regimes is well described (Jones 1969), although the impact of this burning on vegetation changes is a matter of some debate. Charcoal deposition records show an increase in fire activity and a shift to more sclerophyllous vegetation during the early Aboriginal period for a number of sites (Singh et al. 1981; Luly 2001). However, vegetation shifts were not universally associated with occupied sites, shifts were not always sustained, and climate has not been ruled out as a factor in both increased fires and vegetation change (Lynch et al. 2007).

Humans also potentially influenced fire regimes indirectly through activities that affected landscape fuel patterns. For example, in both Australia and North America the Pleistocene colonization by early people contributed to the mass extinction of giant herbivores (Martin 1984; Owen-Smith 1987; Roberts et al. 2001; Ripple & van Valkenburgh 2010). As is illustrated by elephants today, these large mammals had the potential for greatly altering vegetation patterns and subsequent fire activity and thus it is to be expected that as landscape fuel patterns changed, fire regimes followed suit. Loss of much of the large mammal fauna in the Cape region of South Africa was accompanied by introduction of livestock and thus the impacts on fuel structure may not have been as marked (Hendey 1983).

Global studies of Holocene fire activity, as reflected in charcoal deposition records, find a strong link between climate and fire activity (Power et al. 2008; Marlon et al. 2009; Daniau et al. 2010). Some of these reports have used this correlation as evidence that humans played a minimal role in Holocene fires. In some regions, though, this fire record is biased against detecting human impacts; for example in California, charcoal deposition records come from high-elevation lakes in landscapes more or less saturated with natural lightning-ignited fires, whereas Holocene populations were distributed in the foothills and coastal plains where natural ignition sources were highly limited (Keeley 2002b).

Additionally, although many late Quaternary fire records are consistent with climate change, in most records there are inconsistencies that are best interpreted as evidence of early human use within a framework of changing climates evident on several continents (Gavin et al. 2007; Markgraf et al. 2007; Black et al. 2008; Gil-Romera et al. 2009). In Australia it is thought that the arrival of humans may have partially decoupled fire activity from climate (Kershaw et al. 2002; Lynch et al. 2007), and this effect intensified with increased population growth in the late Holocene (Hassell & Dodson 2003; Enright & Thomas 2008). Early Holocene (8000 yrs BP) mediterranean people increased fire incidence to levels typical of contemporary landscapes and this was associated with opening of closed-canopy woodlands, although there is evidence that changes were exacerbated during wetter periods (Colombaroli et al. 2008). In short, there is no reason to think of climate and humans as representing mutually exclusive factors determining fire activity and there is no clearer demonstration of this than recognition that although contemporary burning patterns are markedly influenced by humans, climatic signals are still evident in annual changes in fire activity.

Arguably the most extensive anthropogenic transformations have occurred in the Mediterranean Basin and in the last millenium humans have been responsible for major transformations of woodlands (Marsh 1864). Fire was the primary tool used to transform wooded landscapes to herbaceous associations (Trabaud et al. 1993; Henderson et al. 2005). Additionally, these land-use practices were widely transferred to all other MTC regions from the sixteenth to the eighteenth centuries (Pyne 1982, 1991, 1995), although sometimes new settlers assimilated native burning practices (Deacon 1992). Timing of European colonization played a role in landscape impacts. In, Chile, Europeans arrived relatively early and these landscapes have had intensive rural land use since the sixteenth century. Today there are very few parts of central Chile that have not been affected by goat grazing and wood collecting for charcoal production (Bahre 1979). These activities greatly reduced fuel continuity and undoubtedly affected burning patterns. Europeans arrived much later in California, only about 100 yrs prior to the Industrial Revolution of the 1880s that resulted in massive rural to suburban migration, and which minimized the duration of intensive land use. European colonization of both South Africa and Western Australia had variable impacts dependent on geology as ecosystems such as fynbos and kwongan on nutrient-poor soils were of limited value, but vegetation types on other soils were converted to agricultural and pastoral uses.

European colonists regularly utilized fire in MTC regions primarily for increasing pasturage (Pyne 1982; Rundel 1998). Colonists, in addition to adding new ignitions to the landscape, also reduced other sources of ignitions, either through the decimation of native populations or through regulations imposed against continuation of native burning practices. Sometimes these laws were rather harsh, as in the Dutch death penalty against second offenders in South Africa (Bands 1977). In addition, other land management practices altered fuels through grazing and timber harvesting, which in turn affected fire activity (Griffiths 2002). The net effect on the fire regime is often debatable and seldom does one find a consensus on such issues. For example, in Australia it is widely thought that burning increased with the arrival of Europeans (Singh et al. 1981; Banks 1989; Kershaw et al. 2002; Mooney et al. 2001). An alternative model is that Aboriginal burning was frequent and in harmony with the environment, and that this was immediately suppressed by European colonizers (Kimber 1983; Burrows et al. 1995). Fire-scar dendrochronology records from southwestern Australian forests reveal a marked increase in fire scars following European colonization (Burrows et al. 1995), which would normally be interpreted to mean an increase in fire use. However, proponents of Aboriginal fire use have argued that increased fire scarring following colonization was in fact due to a decrease in fire frequency because Aboriginal burning was so frequent it kept fuel levels below the threshold required to scar trees. In their view Europeans reduced the frequency of burning and this resulted in fewer, but higher intensity fires. Independent tests of this interpretation of fire-scar data failed to support it (Richards 2000) and other studies have found fire scars are a consistent indicator of fire frequency in Western Australian *Callitris* trees (O'Donnell et al. 2010), supporting an increase in fire activity with European colonization.

Contemporary MTC Fire Regimes

MTC plant communities are among the most fire-prone woody vegetation types in the world (Chuvieco et al. 2008). This is the result of wet winters with moderate temperatures, which are conducive to biomass production sufficient to carry fires, and an annual summer drought, which converts this biomass to available fuels. The low to moderate productivity of these communities generates large expanses of shrubland or woodland fuels that are conducive to high-intensity crown fires. MTC alone does not make this vegetation fire-prone, as demonstrated by the distribution of fire-prone mediterranean-type vegetation (MTV) outside MTC regions; pines in North America and *Eucalyptus* in Australia are good examples. The primary difference with fire regimes in sclerophyll vegetation in other climates is that in MTC regions there is an annual fire season, making these landscapes vulnerable to fires every year, whereas under other climates, this vegetation type may experience high fire danger at longer intervals, for example every 2 yrs (Hasson et al. 2009) or longer (Vines 1974) in southeastern Australia.

Large, high-intensity fires have long been associated with MTC vegetation (see the discussion of paleohistory in Chapter 10), and it is the colonization of these landscapes by people that creates a fire problem. MTC regions are highly attractive to people and all regions have high-density metropolitan areas closely juxtaposed to flammable landscapes. All MTC communities have roots to European societies and this has contributed to substantial transfer of culture and land management philosophy, strategies and tactics between regions. In all cases there is a clear recognition of the fire danger inherent in these landscapes and the need to reduce community vulnerability to fires, but at the same time protect natural resources. These ideas, however, have emerged at different times and in response to different events in each MTC landscape.

Human Impacts on Fire Regimes

Humans have had a multitude of impacts on fire regimes that include changes in frequency and timing of ignitions and changes in fuel load and landscape patterns of fuel distribution (Keeley et al. 2009a). In MTC regions where natural ignitions are frequently limiting, increased population growth has generally been associated with increased fire frequency. Humans have the ability not just to change numbers of ignitions but also to shift the natural season of burning to periods of severe fire weather with high winds and high temperatures. Fuel structure has been altered in diverse ways. When a policy of suppressing fires has been successful in excluding fires over broad regions for extended periods of time, fuels have accumulated beyond their historical levels. Fuels are also altered under intensive livestock grazing regimes in diverse ways. On rangelands this has the potential for reducing fine fuels and thus reducing fire events. In savannas this may be coupled with increased woody fuels, although this is a function of recruitment strategies of local woody species and grazing habits of livestock and of native grazers and browsers. Timber harvesting also produces changes in fuels that vary with extraction methods. In some forest types this can lead to much smaller, more flammable fuels and often increased surface fuels, greatly exacerbating fire hazard.

In the Mediterranean Basin repeated burning of shrublands to expand rangelands and agricultural lands has been replaced in the late twentieth century by the abandonment of traditional rural lifestyles, as well as by legislative changes that outlawed such burning (Henderson et al. 2005; Moreira & Russo 2007). The resulting recolonization of previously cleared landscapes by woody species has resulted in increased fuels that have contributed to a greater frequency of large high-intensity fires (Loepfe et al. 2010), and perhaps slower response time to fires due to fewer eyes to spot fires (Wrathall 1985) and fewer volunteers to fight fires (Bassi et al. 2008). Due to such fire events, as well as a cultural framework that prefers dense forests, land management has focused on reducing fires with intensive fire suppression and increasing forest growth (Seijo 2009). This has the potential for producing other undesired changes such as loss of native diversity (Hédl et al. 2010). Momentum is increasing for the replacement of this cultural-based management with a paradigm focused on sustainability of natural ecosystems within the context of natural fire regimes (Pausas & Vallejo 2008).

Although the vast majority of fires are anthropogenic, only a small number of these are intentional. In fact, prescribed burning is only used locally or sporadically in France, Portugal and Spain, and not allowed in Greece, Turkey and most of Italy. The technique is well established only in France, where it is used for fuel reduction of intact woodlands, and for reducing shrubland recolonization of former pastures (Rigolot et al. 1998).

In California the trend since the mid-twentieth century has been toward expansion of metropolitan population centers into wildland areas. This urban sprawl has placed more and more structures and people at risk and created a highly vulnerable wildland–urban interface that totals more than 5 million homes across 28 000 km^2 (Radeloff et al. 2005; Hammer et al. 2007). This is a severe problem in the southern half of the state where the population density is far higher than in any other MTC region (Table 13.1). In response to this growth pattern most of the last century has been managed under a policy of rapid deployment of fire suppression forces against all unplanned fires. Despite this policy, the total area burned in coastal and foothill landscapes has either remained unchanged or increased over the last 100 yrs (Keeley et al. 1999a; see Fig. 5.10). The success of this policy is measured by the fact that although the numbers of ignitions have increased rapidly over this period, in concert with population growth (Sapsis 1999; Syphard et al. 2007), fire suppression has been successful at keeping much of the landscape within the historical range of fire frequency and so far saved the majority of native shrublands from type conversion to alien grasslands (see Chapters 5 and 12).

In more remote mountain communities embedded in conifer forests throughout the western USA, the fire suppression policy has had radically different impacts. Under a fire regime of understory surface fires, fire suppression forces have been

Table 13.1 MTC region population density and big fire events in recent history

	Land area (km^2)	Population (million)	Population density (km^{-2})	Big fire events				Ignition sources	Notes
				Year	Size (ha)	Losses Buildings	Lives		
Mediterranean Basin:									
Portugal	92 400	10.6	114	2003	430 000[a]	200	21	Lightning, accidental	Heat wave and high wind gusts
Greece	131 900	10.7	87	2007	271 000[a]	2100	84	Arson, accident	Meltimi wind and heat wave Substantial losses of forest and orchards
California:									
northern	257 600	14.3	56	1991	620	3791	25	Accidental	Tunnel Fire, Diablo winds
southern	146 400	22.4	153	2007	109 500	2400	15	Power line	Witch Fire, Santa Ana winds following 17 months drought
Central Chile	756 900	16.0	21	—					
South Africa:									
Western Cape	129 400	5.4	41	2000	8000	8	0	Accidental	High winds, anomalously high alien plant fuels
Australia:									
Western Australia (southwestern MTC region)	356 700	1.8	5.0	1961	40 000	160	0	Lightning	Dwellingup fires, cyclonic winds
Victoria	227 400	5.2	23	2009	330 000	2029	173	Power lines	Black Saturday fires with extreme heat wave, high winds and record drought

[a] Annual total area burned.
Data from government websites, population estimates from 2006 to 2009, fire events from newspaper and web reports.

highly successful at nearly eliminating fires, and thus fire suppression can be equated with fire exclusion for much of the twentieth century. However, as the North American TV newsman Eric Sevareid was fond of saying, "Most problems begin as solutions" and this well describes the success story with fire suppression in western North American forests. After a century of fire exclusion, fuels have accumulated sufficiently to shift the fire regime from a surface fire to a crown fire regime. A similar situation appears to be true of montane conifer forests in the Mediterranean Basin (Fulé et al. 2008).

Prescription burning was routinely used during the first 70 yrs of the twentieth century in California as a means of "range improvement," which consisted of repeated burning for the purpose of converting shrublands to alien-dominated rangelands (Keeley & Fotheringham 2003a). Subsequently, prescription burning was widely advocated as a means of reducing hazardous fuels, but the close juxtaposition of wildlands with urban populations has proven too dangerous and it has all but been abandoned in coastal and foothill shrubland landscapes. However, it is widely utilized in montane conifer forests due to the remoteness of the localities and greater ease of controlling understory surface fires. Long-term repeated applications appear to have been highly successful at restoring historical fire regimes, at least in localized areas (e.g. Keifer 1998).

In central Chile landscapes surrounding major metropolitan areas have had a long history of intensive land management that has reduced fuel loads and fuel continuity, making catastrophic wildfires less threatening. In central Chile essentially all fires are anthropogenic although lightning is a source of ignitions further south. With increasing production of pine plantations in the southern part of the MTC region has come the potential for high-intensity crown fires that threaten timber resources. National efforts at fire prevention and fire suppression are focused on reducing destruction of resources such as these.

Fire suppression policy in the MTC Western Cape of South Africa was begun in the first half of the twentieth century and continues to the present. This is justified because of the growing populations and increasing anthropogenic ignitions, coupled with evidence that fires are more ignition dependent than fuel dependent in this region (van Wilgen et al. 2010). Increased human ignitions in mountains surrounded by urban areas have burned through younger fynbos stands more often than historically was the case, changing the fire return interval from 38 to 13 yrs and threatening slower maturing plants (Forsyth & van Wilgen 2008). Nonetheless, in the 1970s prescribed burning programs were initiated in mountain catchment areas, for purposes of protecting forests, watershed cover and reducing alien plant invasions. Since the mid 1980s, the areas burned in prescribed fires have declined, linked to changes in state agencies managing the land and a decline in financial and human resources for prescribed fires. Prescribed burning has also become more difficult, especially near the urban–wildland interface, because of issues of legal and financial liability.

In more remote areas of the Cape region, prescribed burning has been replaced by natural fire zones in which attempts are made to suppress anthropogenic fires while allowing naturally ignited *veld* fires to burn to their full extent (Seydack *et al.* 2007). Over the past 30 yrs under this program, the area burned by anthropogenic fires has declined as well as the frequency of fires. A similar program was initiated in the 1970s in remote wilderness areas of Sequoia National Park in California (Kilgore & Briggs 1972). In these largely conifer-dominated forests this program over the past 50 yrs has successfully allowed a return of fire where the fire suppression policy had largely excluded it during most of the twentieth century (Colllins & Stephens 2007).

Highly fire-prone MTV spans much of the southern edge of Australia, with strong MTC in the western half and gradually shifting to aseasonal rainfall conditions in the southeast. As a consequence, fire regimes vary widely according to local climate (Bradstock 2010). Impacts of European land use on MTV fire regimes are multiple. Vast areas of shrubland and woodland have been cleared for cereal cropping or utilized for rangeland grazing in drier regions and this has resulted in a decline in fire, among other changes to ecosystem functions (Hobbs 2002). Large areas of intact dry shrublands remain in sparsely inhabited coastal regions of southwestern Western Australia and South Australia (Table 13.1), where lightning ignites occasional large bushfires (McCaw & Hanstrum 2003). Suppression capacity and other management activities in these areas are often minimal.

Australian mosaics of open forests, woodlands and shrub-dominated communities are often juxtaposed with densely populated urban centers and intensive rural industries, and substantial resources go toward fire suppression efforts and prefire fuel treatments. This is particularly true for the southeast that includes the weakly-MTC Victoria and aseasonal climate region of New South Wales. Considerable areas of eucalypt forest are managed for timber production or as conservation reserves and these resources are vulnerable to frequent anthropogenic and lightning-ignited fires necessitating aggressive fire suppression (Bradstock & Gill 2001; McCaw *et al.* 2003).

One common pattern observed across all MTC landscapes is the role of humans in increasing fire frequency and the non-linear relationship between population density and fires (Syphard *et al.* 2007, 2009). As people move into wildland areas, fire frequency increases with increasing population density. However, there is a population threshold where further population increases result in decreased fire frequency, due to reduction in patches of fire-prone vegetation as well as increased infrastructure leading to rapid suppression of fire starts.

Fire Impacts on Human Populations

MTC regions differ in community vulnerability to catastrophic wildfires. These landscapes have been prone to such events for tens of millions of years (see Chapter 10), and historical records show that massive, high-intensity fires were a common feature in Australia and California at the time of European colonization (Edgell 1973; Bradstock 2008; Keeley & Zedler 2009). However, during the

(a)

(b)

Fig. 13.1 *Wildland–urban interface (WUI) in (a) southern France and (b) southern California. (Photos by Jon Keeley.)*

twentieth century, as population centers grew and expanded outward into these wildland areas (Fig. 13.1), these fires became increasingly destructive to property and lives (Martin & Sapsis 1995). Examples of recent such events are in Table 13.1. Although such fires typically last for days or even weeks, the period of destruction is often less than a day during bouts of extreme fire weather (e.g. Cheney 1979; Keeley *et al.* 2009b).

Some regions have repeatedly suffered devastating losses of property and lives; other regions, despite a regime of frequent fires, have been spared. Certainly a critical factor is population density, which determines the potential number of people at risk and varies by more than an order of magnitude between MTC regions (Table 13.1). However, other inherent factors that vary between MTC regions include: (1) vegetation type, which affects fuel loads (see Table 2.1): (2) terrain, which affects fire behavior; (3) climate, which controls fuel production and fuel moisture, and (4) fire weather, in particular severe winds that can generate "firestorms" (see Box 1.3).

One of the more extreme comparisons would be the impact of fires on the communities in southern California vs. Western Australia. Since the late 1800s southern California has experienced major chaparral fires encroaching on the urban environment and destroying homes and lives; and in the last 60 yrs these have occurred about every decade (Keeley & Zedler 2009). The frequency of such events has accelerated since 2000, with major events in 2003 (Keeley *et al.* 2004) and a repeat in 2007 (Keeley *et al.* 2009b). In contrast, Western Australia has not had such catastrophic fires (Table 13.1), although the region commonly experiences large bushfires (McCaw & Hanstrum 2003). On the surface it might seem as though fire management practices play a role since, as discussed below, fuel treatments using prescription burning are more widely applied in Western Australia than in southern California. However, fuel load is not a major determinant of whether or not a fire becomes catastrophic in southern California (Keeley *et al.* 1999a; Moritz *et al.* 2004).

More important factors include vast differences in population density, which result in many times more people at risk in southern California (Table 13.1). In addition, fuels differ radically between these two regions as chaparral stands inherently have substantially higher fuel loads than Western Australian heathlands and *Banksia* woodlands (see Table 2.1), which typically border urban areas. When fires ignite in southern California, control is always compromised by rugged terrain, which accelerates fire spread and reduces accessibility, and many homes are placed in these dangerous watersheds. In contrast, much of Western Australia is an ancient flat landscape with a few isolated mountain ranges. The worst fires in southern California are driven by extreme offshore winds known as Santa Anas that occur 15–30 days per year, but in Western Australia, severe winds are less predictable and in southwestern forest areas there is on average only one extreme fire weather day a year (McCaw & Hanstrum 2003). In short, much more catastrophic fires in southern Californian than in Western Australia are the result of differences in human demography and fire regime. Central Chile also seems not to have the same problem with catastrophic fires as southern California. The intensive land use that surrounds most metropolitan areas may be a factor in keeping fuel loads low, although this situation may be changing as rural populations are declining with increased migration to the major cities.

Historically wildfires have not been a major threat to urban environments in the Western Cape of South Africa. Some of this is because not only do the native fynbos shrublands have relatively low fuel volumes (see Table 2.1), but they are maintained at a low level by a fire frequency of roughly every 15 yrs. In contrast,

California chaparral inherently has higher fuel loads and fire return intervals of 15 yrs are at the lower threshold of tolerance; historically fire return intervals were double or triple that interval. Other factors are likely at work as well. The population density in the Western Cape is relatively high, although it doesn't approach the level of southern California. However, the urban growth pattern of the Western Cape appears to have avoided the extensive wildland–urban interface problem of southern California with developments and consequently has suffered less damage from wildfires. The closest the Cape has come to destructive fires are two that occurred in January 2000 on the urban fringe, burning a total of 8000 ha. Homeowners were evacuated and there was general panic, but there were very few structures lost and no one died (Table 13.1).

Similar to the situation in southern California is the major bushfire threat in the Australian state of Victoria, and adjacent states of South Australia and New South Wales. In contrast to Western Australia, fire losses have historically been very high, including lives, structures and livestock (Table 13.1). The reasons are similar to those in California: higher population density and substantially higher fuel loads comprising dense *Eucalyptus* woodlands and forests (see Table 2.1). It is of immense interest to our focus on MTC ecosystems because Victoria has a weak MTC, yet its neighbor South Australia to the west, with a stronger MTC, and its neighbor New South Wales to the east, with an aseasonal rainfall regime, both have equally severe bushfire problems (see Chapter 8).

For similar reasons (i.e. high population density and high fuel loads) the Mediterranean Basin has a history of catastrophic wildfires, from westernmost Portugal to the eastern end of the basin in Greece (Table 13.1). Commonly these fires have had extreme impacts on rural populations, and under severe wind conditions such as the Mistral or the Meltemia winds (see Box 1.3) there are often extensive losses in rural areas of lives, structures and agricultural crops.

Climate Impacts on Major Fire Events

Although annual summer droughts contribute to large fire events in all MTC regions, there are regionally unique factors. For example, the intensity of annual summer drought is greater in California than in some other MTC regions (see Box 1.1), which can affect summer fuel moisture. Also, timing of winter rains is different: for example, there is high probability of rains beginning in the autumn in the Mediterranean Basin, whereas this is rare in California. Very different synoptic weather conditions lead to major fires in different regions. For example, in Western Australia they are associated with the eastward passage of strong anticyclones, whereas in southeastern Australia dangerous fire weather is associated with warm, dry air masses generated by southern ocean cold fronts and/or tropical depressions (Luke & McArthur 1978; McCaw & Hanstrum 2003; Hennessy *et al.* 2006; Hasson *et al.* 2009).

Two factors closely associated with major fire events are extreme winds and anomalously long droughts. High-intensity wind storms associated with large fires in MTC regions differ in timing with potential effects on predictability of extreme fires (see Box 1.3). The mistral winds in the Mediterranean Basin are distributed in summer and autumn and thus they often occur at times of moderate fuel moisture content, reducing their likelihood of leading to major fires. Likewise, Berg winds in the Cape region of South Africa are autumn and winter winds and thus do not always coincide with low fuel moisture conditions. In contrast, the Santa Ana winds of southern California are absent in the summer and concentrated in late autumn when fuel moistures are at their lowest levels.

Unusually long droughts contribute to major fires by leading to very low live fuel moisture content at the time of fire, and by increasing the dieback of vegetation, thus producing rapid changes in dead fuel loads. In California major fires have been preceded by 1–2-yr droughts, and substantial droughts in the first decade of the twenty-first century correlate with a doubling of large fire events with catastrophic impacts on local human communities (Keeley & Zedler 2009). Droughts are likewise implicated in major destructive fires in Greece (Viegas 2004) and Australia (Hennessy et al. 2006; Bradstock 2008). In California most major fires occur in the autumn when live fuel moisture is at its lowest and this does not change with extended drought. Thus, it has been hypothesized that the association between major droughts and major fires is due to the short-term increase in dead fuels, which increases fire spread directly, and indirectly through enhanced spotting behavior (Keeley & Zedler 2009). Fire danger indices fail to capture these short-term changes in dead fuels and this may explain the limited explanatory value of such indices in southern California (Schoenberg et al. 2007). It is unknown to what extent recent severe droughts are tied to anthropogenically induced climate change. However, temperatures have been steadily increasing and do exacerbate drought impacts, and climate changes can lead to more extreme conditions in general (e.g. Hasson et al. 2009; Bradstock 2010).

Fire Management Strategies

Contemporary fire management goals are focused on minimizing impacts of fire hazard to human population centers and sustaining natural ecosystems. Regions and jurisdictions within regions vary greatly in the importance of these two goals. For environments where both human vulnerability and ecosystem sustainability are important, balancing these two goals is a challenge. Meredith (1996) put it rather succinctly; "The placing of houses in flammable bushland increases the risk factor associated with fire to such an extent that the need for trade-offs between fire protection and ecological management is greatly increased." Thus, an issue of potential conflict is how to find the appropriate balance between fire hazard reduction and resource protection.

Perhaps the biggest conflict between resource management and fire hazard management lies in the application of prefire fuel treatments such as prescription burning. A common management philosophy is that hazard-reduction burning will prevent wildfires, or make them easier to control, and will also protect native fauna and flora from the devastating effects of high-intensity fires. Whelan (2002) questioned the assertions that (1) high-intensity wildfires cause unnaturally high mortality rates of native flora and fauna, (2) an effective hazard reduction burning program would have no long-term detrimental effects on biodiversity and ecosystem functioning, and (3) broad-scale hazard reduction burning would ensure suppression of wildfires. Inevitably, the correctness of these assertions is a function of the natural fire regime. For example, they may be valid for a surface fire regime but not for a crown fire regime. Each of these assertions needs to be addressed at the regional level where consideration can be given to the individual ecosystems being managed.

Even when detrimental effects of hazard reduction treatments are recognized, fire managers historically have placed fire hazard issues ahead of resource issues, particularly on high population density landscapes. Oftentimes if a fire management treatment had any potential benefit in reducing fire hazard it was considered a legitimate strategy. However, even on populated landscapes, agencies increasingly are faced with having to resolve fire hazard reduction and natural resource issues in ways that minimize ecosystem impacts. Balancing these issues necessitates a cost-benefit analysis that goes much further than conventional (financial) cost-effectiveness metrics to include impacts on natural resources such as diversity loss and alien plant invasion (e.g. Meredith 1996; Omi *et al.* 1999). This is well captured by the president of the National Parks Association of New South Wales, Australia: "While the protection of life and property are paramount, it needs to be provided in a manner that also protects the flora and fauna and prevents other unintended side effects, such as soil erosion, pollution of catchments and air pollution" (Potter 2002). Alien plant invasion should also be added to that list.

Ultimately the balance between fire hazard reduction and resource protection will depend on management goals. On some landscapes intensive fuel management detrimental to resource sustainability may be appropriate. Although sacrificing patches of resources for protection of human populations is certainly justifiable, serious attention needs to be given to whether there are true benefits to the treatment that justify the resource loss. The most critical need in this regard is obtaining data adequate for evaluating the effectiveness of management options in achieving asset protection and balancing that level of protection against resource needs (Driscoll *et al.* 2010).

Natural Resources and Ecosystem Sustainability

The flora and fauna that comprise fire-prone communities are commonly described as fire-adapted species; however, many of these may be extirpated by particular fire regimes. Thus, it is more appropriate to think of these species as adapted to a particular fire regime, and if the system deviates from that fire regime then some species may be at risk. Deviations that are known to be risk factors include changes in fire frequency, fire season and fire intensity. In this sense fire *per se* should not be thought of as a disturbance, but rather ecosystems are disturbed by deviations from the natural fire regime.

Plants and animals differ markedly in how they respond to fires and to deviations in fire regime. In many fire-prone MTV communities plants have endogenous means of reproduction from stored seedbanks or vegetative buds, and colonizing species that come in from outside the burned area are relatively limited (see Chapter 3). Many animals in contrast are often extirpated from burned sites and must recolonize, and thus the metapopulation dynamics are critical to sustaining these ecosystem components. As a consequence, the fire regime characteristics that are critical for sustainability may differ between plants and animals; for example, fire size or patchiness often has little impact on plant recovery but huge effects on animal recovery (Bradstock *et al.* 2005; Parr & Andersen 2006).

Humans have disturbed fire regimes on many landscapes by altering fire frequency. In many instances they have increased ignition sources and thus decreased fire intervals. In other cases they have suppressed natural ignitions and increased fire intervals. The impact of altering fire frequencies (as well as other aspects of the fire regime) may be best predicted by examining thresholds of tolerance exhibited by life histories (Keeley *et al.* 1989; Gill & McCarthy 1998; Bradstock & Kenny 2003; Pausas *et al.* 2004b). In general, species face two threats from alterations in fire regime: immaturity risk, where fire intervals are shorter than the tolerable level, and senescence risk, where fire intervals are longer than tolerable (Zedler 1995b). These risks lead resource managers to be watchful for "thresholds of potential concern" when the fire regime is altered by too frequent or too infrequent fires (van Wilgen & Scott 2001).

Immaturity risk is a major concern in all MTC landscapes due to the high numbers of unplanned anthropogenic ignitions (Syphard *et al.* 2009). It also is a potential threat on landscapes where planned ignitions for fuel treatments create short fire intervals that inhibit recovery of some plant and animal populations (see Chapter 8). Risk increases when the interval between fires drops below a critical threshold of tolerance for a species and on moderately fertile soils can lead to type conversion of shrublands and woodlands to grasslands (Keeley 1995b; Jacobson *et al.* 2004; Syphard *et al.* 2006). This risk is greatest for those species dependent entirely on dormant seedbanks for postfire recovery (obligate seeders) and is measured by the time after fire required to replenish the seedbank. With animal populations it is measured by time to recolonize burned sites and interacts strongly with landscape patterns of burning and metapopulation dynamics.

Senescence risk is a potential threat to some populations that are short lived and have short-lived seedbanks. Documentation of such risks is largely lacking and one attempt to look for such risks in California chaparral shrublands unburned for more than a century failed to detect any evidence for senescence impacts on postfire recovery (Keeley *et al.* 2005b). In crown fire shrublands it is primarily a risk for serotinous species with aerial seedbanks (Bond 1980; Lamont *et al.* 2004a, 2007). In forests with historical fire regimes of frequent low-intensity surface fire, where fire suppression policy has been effective at excluding fire, some species populations may die out. As fuel loads increase with time since fire, they can lead to a fire regime shift toward higher intensity crown fire that could extirpate some species.

Fire hazard reduction treatments are sometimes incompatible with resource issues. One of the biggest threats to many plants and animals is the fact that prescriptions often require frequent rotations of burning in order to maintain fuel levels within specified limits (Morrison et al. 1996). Forest types with historical fire regimes of frequent surface fire may be amenable to this treatment as most species in these systems are adapted to such a regime (Keeley et al. 2009a). In such forest types fuel treatments may even enhance the habitat for some animal species (Craig et al. 2010).

However, crown fire regimes in general are more vulnerable to frequent fires. Nonetheless, even on these landscapes there are examples of prescription burning programs that have played an important role in protecting natural resources (Conroy 1996). In short, in some surface fire regimes, fire hazard reduction objectives may be compatible with resource objectives, whereas in crown fire regimes the best rotation for hazard reduction oftentimes is not compatible with resource sustainability. Factors other than frequency also may have impacts on resources. Sometimes safe prescriptions call for out-of-season burns in winter or early spring. This treatment may result in poor recruitment of postfire seeders (Bond 1984; Parker 1987; Brown et al. 1991) and increased mortality of resprouting species (Rundel et al. 1987). Poor out-of-season recruitment has been tied to heating effects on moist seeds (Rogers et al. 1989) and more intensive seed predation (Heeleman et al. 2008), but the truncated growing season prior to summer drought has to be a critical factor as well.

Where there is a necessity for applying prescription burning, managing sensitive ecosystems can take different courses. An approach widely applied to conifer forests of the western USA is restoration of fire regimes present prior to the nineteenth-century colonization by Europeans (Covington & Moore 1992). This is justified on the basis that these fire regimes were appropriate for maintaining sustainable populations of native flora and fauna (Millar 1996), and there is a vast database on historical fire regimes inferred from fire-scar dendrochronology for forests with surface fire regimes (Swetnam et al. 1999).

A related approach possible in wilderness areas where risks are lower is to allow naturally occurring fires to burn unchecked (Kilgore & Briggs 1972; van Wilgen et al. 1990a; Seydack et al. 2007). Since this approximates the "natural" fire regime it is presumed that these fires are conducive to long-term sustainable biodiversity. This seems to be a reasonable strategy in forests with surface fire regimes, or remote fynbos-dominated mountain ranges, but advocating it for southern California chaparral (e.g. Childers & Piirto 1989; Minnich 2001) seems dubious. Not only would this be an extreme fire safety hazard, but it is not justifiable based on the fact that the bulk of this landscape has experienced more burning during the twentieth century than historically was the case (Safford & Schmidt 2010).

However, some chaparral wilderness areas have been cut off from the heavy load of anthropogenic ignitions and there may be some resource benefit to adding fire. One way this could be accomplished is illustrated by the 2007 Zaca Fire, which burned over 94 000 ha (Keeley et al. 2009b), with about 38 000 ha due to a single backing fire ignited by fire crews (Vives & Boxall 2009). Although there are perhaps some resource benefits to this action, this amount of prescription burning represents a significant resource challenge. Fires in this vegetation are largely ignition limited and unless there is a substantial reduction in human ignitions, there is a high probability that a significant portion of this landscape will reburn within the next two decades, thus posing a major immaturity risk for some taxa.

Other regions have not embraced the notion of restoring historical fire regimes but rather designed management strategies using direct evidence of fire regimes required to support target species. One is to determine prescription burn intervals based on time to flowering maturity (Burrows & Friend 1998) or time to produce a viable seedbank (Keith 1996; Menges 2007). Another approach may focus on maximizing the density of target populations or on minimizing the risk of extirpation of species in the system (Bradstock et al. 1995, 1998a, 1998c; Gill & Bradstock 1995; Keith 1996). Still another is to use prescription burn intervals that are compatible with the widest range of flora and fauna thresholds of tolerance, with the expectation that this will lead to maximum biodiversity (Gill & McCarthy 1998). Sometimes this will require including a range of fire regime parameters beyond just fire frequency (van Wilgen et al. 1994). Critical to these approaches is the recognition of a need for regional approaches since interactions between fire and environmental heterogeneity mean that prescriptions for some sites may not be suitable for others (Williams et al. 1994; Bradstock & Kenny 2003; Burrows 2008). There is some evidence that regional and spatial variability in application of different prescription regimes will maximize regional diversity.

Biodiversity is just one of a number of resource costs that can be associated with fuel treatments such as prescription burning. For example, in kwongan heathlands of southwestern Australia, the conflict between reducing fire hazard and maintaining sustainable resources has a somewhat different character. Here protection of agricultural lands from fire spread from wildlands is thought to be enhanced by frequent burning of surrounding heathlands. However, these wildlands are a rich nectar source for the apicultural industry and it is believed that frequent burning is detrimental to optimum honey production (Bell et al. 1984).

Other ecosystem processes that may be affected by frequent fires include soil nutrient depletion. Fire increases losses both through volatilization as well as greater postfire soil leaching due to reduced plant cover (DeBano & Rice 1971). Such treatments may diminish ecosystem nutrients and alter the competitive balance in ways that affect community assembly. Frequent burning may also affect forest structure. For example, in jarrah forests of Western Australia 25 years of frequent prescription burning reduced soil nutrients, but the thinning of understory trees apparently reduced competition for water and increased tree diameter growth (Burrows et al. 2010).

Prescription burning is also utilized in the Cape region of South Africa to enhance water resources (van Wilgen et al. 1990a; Le Maitre et al. 1996). Clearing of invasive alien plants is another potential use of prescription burning in this MTC region. Its effectiveness is a function of the type of alien threat. In California the alien threat is largely from annual grasses and forbs and in general there is little evidence that prescription burning can provide any sustainable protection from such invasions (Keeley 2006b). However, the woody aliens appear to be controlled by fire in several MTC regions (see Chapter 12).

Fire Hazard and Community Vulnerability

Destruction of property and lives is a major management concern in both MTC and non-MTC regions with fire-prone vegetation. Risk of destruction, D, is considered to be a function of multiple factors impinging on the wildland–urban interface (Bradstock & Gill 2001) with the damage threat summarized as:

$$D = f(I, S, E, G, H) \quad (13.1)$$

where:

$I =$ the chance of a fire starting in the landscape
$S =$ the chance of the fire reaching the urban environment
$E =$ chance of it encroaching into the urban environment
$G =$ chance of fire propagating within the built environment
$H =$ chance of a fire, once within the built environment, resulting in the destruction of a building.

In fire-prone environments reducing D to zero is nearly impossible. Thus, the question to be addressed is which of these factors has the potential for producing the greatest reduction in D, and is cost effective. These factors are potentially addressed through actions focused on wildland management and on urban management.

Wildland Fire Management

Fire prevention, fire suppression and prefire fuel treatments are the primary means by which fire managers minimize I, the probability of fire starting, and once ignited, S, the probability of fire reaching the urban environment.

On most MTC landscapes anthropogenic ignitions vastly outnumber natural lightning ignitions and fire prevention has long been the primary focus for minimizing fire starts. Restrictions on fire use and advertising campaigns have historically been the primary management actions but there is room for new innovative approaches (Box 13.1).

Fire suppression strategies are well described in Pyne et al. (1996) and beyond the scope of this book. Critical to successful fire suppression is the ability to forecast severe fire conditions and different regions have developed indices unique for their specific needs. These fire danger indices assess fire potential, which includes factors associated with the ease of ignition, rate of fire spread, difficulty

Box 13.1 Reducing Fire Losses

MTC regions are subject to catastrophic wildfires that result in substantial property losses (Table 13.1). It has long been believed that by managing landscape fuels and aggressive fire suppression, governments could prevent fires from encroaching upon urban environments. However, with populations expanding ever closer to watersheds of dangerous fuels it has become apparent that not all fires can be stopped. Increasingly it is recognized that emphasis needs to shift toward other approaches to reducing fire losses, such as (1) reducing wildland fire ignitions, (3) altering urban growth patterns, and (3) altering fuel properties within the urban environment.

Fire prevention is the first step in reducing fire losses. In most regions there has been a long tradition of personal responsibility in not starting fires. Accidental fires are prominent and there are potential avenues for reducing these fires. For example, a great many fires ignite from power tools used in wildland areas so restrictions on use of such tools could reduce fire starts. In MTC regions most fires are anthropogenic and occur along roadways (Wilson 1979b; Keeley & Fotheringham 2003a; Wittenberg & Malkinson 2009; Curt & Delcros 2010), so potentially the placement of concrete barriers along roadways in high population density areas or landscaping with fire-resistant plants might reduce fire starts. Power line failure is responsible for a number of catastrophic fires (Box 13.2) and these fires could be reduced through placement of power lines underground.

Arson is another leading cause of catastrophic fires as ignitions are timed to the most extreme fire weather conditions. Although often the result of malcontents, on some landscapes this may exist as a form of political dissent (Seijo 2009), or as a means of appropriating unregistered forest lands (Briassoulis 1992). Increased patrols and remote cameras during severe fire weather are important steps to reducing arson fires. Success in tracking down arsonists is critical to discouraging future arsons. Equally important are severe penalties and these have been increased in both California and Australia. In a 2006 southern California fire that tragically resulted in the deaths of five fire fighters, homicide charges were brought against the accused arsonist.

One area with a large potential for reducing losses is urban planning that is more sensitive to fire issues. There is no question that some locations are inherently more dangerous than others in terms of wildfire losses. One might expect that fear of losing your home would induce homeowners to select carefully and this would be reinforced by insurance companies that refuse coverage for homes in dangerous watersheds. However, one example where these free market forces have not worked is in North America. After wildfire losses in the early 1960s made it difficult to purchase fire insurance, the California legislature created the California Fair Access to Insurance Plan, so that all people could get insurance, regardless of where they built their homes.

Continued

> **Box 13.1** (*cont.*)
>
> In response an industry-sponsored syndicate, the California Fair Plan serves as insurer of last resort for those deemed too high risk for conventional fire insurance, potentially increasing risky building decisions.
>
> Most fires that encroach into urban environments do so as embers that land on flammable structures. Due to mandatory evacuation orders, most homes are vacant at the time embers land on or adjacent to homes. Homeowners that are on site can go a long way in reducing losses due to ember-generated fires. Staying with your home in the path of a wildfire is dangerous and as a result most societies do not encourage homeowners to do so and often homeowners are required to evacuate. Australia has had a long history of trying to develop guidelines for the safe conduct of home protection. Early on it was described as *shelter in place* and later the *leave early or stay and defend* policy (www.rfs.nsw.gov.au/dsp_content.cfm?cat_id=1214). This is a voluntary policy and emphasizes the need for preparing homes in advance of fire threats so that they are fire safe, including changing home structures as well as clearing vegetation around the home. It has been suggested that losses in California could be reduced with such a policy (Stephens *et al.* 2009). However, after the disastrous 2009 fires in Victoria, Australia, this policy has come under some level of scrutiny (Cart 2009; Haynes *et al.* 2010).
>
> Reducing urban fuels is an avenue that has not been explored. In some landscapes the landscaping surrounding homes is highly flammable and this is particularly so when it is not well maintained (e.g. Fig. 13.5b). It appears the biggest culprit is branches overhanging homes and dropping litter that accumulates on the roof. Ignition is very likely when embers land on dry roof litter.

of control and fire impact. Fire danger rating or burning index is an integration of weather, fuels and other fire-related factors and is commonly used to direct fire management activities.

Indices are tailored to meet the needs of different regions. For example, the McArthur Forest Fire Danger Index, and modifications of it, is commonly used in Australia, or in Canada the Fire Weather Index or in the United States the National Fire Danger Rating System. These regional assessments use broad estimates of fuels based on published fuel models of live and dead fuels for different vegetation types. In many regions substantial effort has gone into producing regional fuel maps used in fire danger rating (Woodall *et al.* 2005). These indices are sensitive to short-term changes in weather parameters and include longer-term impacts of drought on fuel moisture (Andrews *et al.* 2003). They, however, do a poor job evaluating short-term changes in dead fuels due to drought-induced dieback of vegetation, which has been hypothesized to be a critical factor in driving large fire events in California (Keeley & Zedler 2009).

Although these indices are useful planning tools, they vary in importance to fire management. On non-MTC landscapes with aseasonal rainfall, fire danger indices play a key role alerting managers of severe fire weather conditions, often with long lead times (Roads et al. 2010). In MTC regions where the fire season is more predictable, they potentially play a less critical role, although large fire events do generally occur during periods of high burning index (e.g. van Wilgen & Scott 2001; McCaw & Hanstrum 2003; Trouet et al. 2009). Oftentimes, though, the critical fire parameter is severe winds that are seldom predicted with much lead time but dominate fire outcomes (Schoenberg et al. 2007).

Prefire fuel treatments have long been considered a key step in affecting S, the probability of fire spread in the wildland environment. Wildland fuel treatments create patches of landscape with reduced fuel loads. There are multiple objectives. One is to reduce the probability of fire spread, another is to reduce flame lengths and thus increase defensible space for fire fighters, and a third is to reduce loss of resources such as trees. The value of such treatments is a function of fire weather, as even very recently burned or otherwise treated patches have a low likelihood of stopping unplanned fires during severe fire weather (Keeley et al. 2009b; Price & Bradstock 2010).

Prescription burning in forest management is focused on reducing fire intensity and thus preventing understory fires from becoming crown fires and destroying trees. One of the longest and more successful programs is in the understory of giant sequoia (*Sequoiadendron giganteum*) forests in California (Fig. 13.2).

Fig. 13.2 *Prescription burn in* Sequoiadendron giganteum *forest of Kings Canyon National Park in response to more than a century of fuel accumulation due to highly effective fire suppression management. (Photo by National Park Service.)*

Fire-scar dendrochronology studies show that these forests historically burned at short intervals of ~10 yrs until government-sponsored fire suppression began in the early 1900s, after which fires were successfully excluded (Swetnam 1993). The largest groves in national parks have been routinely maintained for the last several decades with understory burning done under prescriptions that are sufficient to consume substantial quantities of dead surface fuels but not escape into the crowns of these majestic trees (Parsons 1995). Success is largely measured by the reduction of surface fuels and tree density to presumed pre-management levels (Keifer 1998). In similar conifer forests in California other metrics of success have modeled fire behavior and tree mortality (Vaillant et al. 2009). However, despite the apparent success of 50 yrs of prescription burning in this and other parks, there are socio-economic problems that limit its effectiveness. Due to air drainage patterns into populated foothill communities, windows of burning opportunity are limited by air quality restrictions and it is doubtful that the bulk of the landscape can be treated under current air quality and budget restrictions.

Australian *Eucalyptus* have an equally long record of prescription burning but on a much grander scale that allows one to investigate impacts on fire behavior. In southwestern Australian open forests, prescription burning has been widely practiced since the mid 1900s; a total of *c.* 114 000 ha in the 1950s to 314 000 ha in the 1970s (Gill 1986). In one of the longest running studies in southwestern Australia, over 50 yrs of prescription burning has changed the frequency and intensity of unplanned fires (Boer et al. 2009). Indeed, in most years the number of prescription burns outnumbered the unplanned ignitions. However, the impact of these planned fires on incidence and size of unplanned fires was short lived and barely evident after 6 yrs. While this program seems to be an important part of fire management in Western Australia, there are many reasons for questioning whether it is something readily implemented on other landscapes. The low population density in the region, coupled with flat landscapes and an extensive road network, plus relatively long periods of stable weather in spring and autumn may make this policy ideal for that region. All of these effects contribute to magnifying the effect of these treatments on unplanned fire incidence and size, relative to regions of high population density, rugged terrain and less stable weather. The fact that it is relatively ineffective after 6 yrs would be a concern if there are flora and fauna that require longer intervals between fires in order to maintain sustainable populations.

Prescription burning has been variably successful in other MTC regions. Two decades of prescription burning in South African mountain fynbos has not produced any obvious changes in wildfire regimes (Brown et al. 1991). Modeling results suggest that prescription burning would have little impact on the total area burned in Mediterranean shrublands (Piñol et al. 2007). In California chaparral the massive reburning of burned scars by fires only 4 yrs apart strongly suggests that prescription burning designed to create young fuels is not likely to be effective under severe fire weather on these landscapes (Keeley et al. 2004, 2009b). Fernandes & Botelho (2003) provide a comprehensive review of prescription burning results although some of the success stories are based on short-term results. Reports of successful changes in fire behavior need to be considered in the context of sustainability of treatments. It is to be expected that areas treated within 1–2 yrs will be effective but unless sites are to be continuously treated at this interval then it is not possible to conclude that in the long run this fuel treatment is effective.

Most who advocate the use of prescription burning are realistic in the expectations that these treatments will alter fire behavior by reducing fire intensity and providing safer access to fire fighters (Conard & Weise 1998). However, in California for the last few decades of the twentieth century the U.S. Forest Service (USFS) managed their chaparral landscapes with a strategy of prescription burning designed to create patch mosaics of different-aged vegetation that would eliminate large fires (Rogers 1982), as advocated by Minnich (1983; Minnich & Franco-Vizcaino 1999). The basis of the mosaic notion is that in the absence of modern fire suppression on chaparral landscapes, mosaic fuel patterns developed spontaneously from high frequency of lightning ignitions, and thus large fire events were non-existent, and the theory further stipulated that modern catastrophic fires are an artifact of fire management (Minnich & Chou 1997). However, there is a substantial body of evidence that disputes this model (Moritz 1997; Conard & Weise 1998; Keeley *et al*. 1999a). In short, it is clear that massive high-intensity wildfires are a natural feature of these landscapes and are not the result of fire suppression (Keeley & Zedler 2009). In 2005 the mosaic fuel strategy was largely abandoned by the USFS for southern California chaparral landscapes (USFS 2005). Mosaic burning, however, has been proposed for other MTC landscapes such as southwestern Australia (Burrows 2008), where it could very well be a viable strategy due to landscape and fire regime characteristics of that region.

The mosaic model fails on landscapes prone to severe fire weather conditions, where fires readily spread through young stands of reduced fuels or jump such stands by spot fires. In California chaparral the early seral stages of the dense highly flammable postfire ephemeral flora are incapable of stopping fires, oftentimes even under moderate fire weather (see Chapter 5). In many Australian *Eucalyptus* forests fuel accumulates so rapidly that fires readily spread within 5–6 yrs of a fire (see Chapter 8). Even where there are clear discontinuities in fuels due to stand age or topography, firebrands or embers of burning biomass are responsible for generating spot fires that enable a fire to overcome such obstacles (Cheney 1996). Spot fires are common near the fire front and they decrease with distance. Understanding spotting distance is important for predicting fire spread as well as threats to downwind communities. Spotting distance is a complex function of fuel structure and condition, fuel height, topography and winds (Albini 1979; Sardoy *et al*. 2007). Sometimes firebrands may be lofted high up in plume-dominated fires with the potential of long-distance transport. The probability of whether or not these firebrands initiate a spot fire will depend on size and material of firebrands, which determines how long it persists before extinguishing, and the fuels in which it lands. Typically in forested environments these spot fires ignite fine surface fuels, but in shrublands with dense canopies of fine dead fuels it is possible to ignite spot fires in the canopy.

Another prefire treatment designed to reduce fuels is fuel breaks, which are typically elongated areas of 50–1000 m in width in which woody fuels are cleared or at least greatly reduced by prescription burning, mechanical treatment or herbicides (see Fig. 12.3). Unlike prescription burns, which sometimes utilize fire return intervals compatible with sustaining native flora and fauna, fuel breaks oftentimes are designed to permanently alter fuel structure, and in shrublands they usually result in permanent loss of native flora and fauna (Keeley 2002a). In forests with surface fire regimes, fuel breaks do not require complete removal of overstory trees; rather forest thinning and removal of understory fuels produce what is referred to as a *shaded fuel break* (Agee *et al*. 2000).

A potential cost to fuel breaks is that they are readily invaded by non-native species and due to their typical configuration of following ridgelines are capable of moving alien species deep into wildland areas (Merriam *et al*. 2006). On moderately fertile soils grass and forb invasion results in flashy fuels that are more easily ignited than woody fuels. In forests, however, shaded fuel breaks typically leave substantial overstory tree cover and this may inhibit alien plant invasion (Agee *et al*. 2000).

Benefits of fuel breaks depend a lot on severity of fire weather and location. Sometimes under moderate weather conditions these fuel breaks may constrain fire spread by causing fires to die out due to lack of sufficient fuel to carry fire through the fuel break. However, under most weather conditions sufficient to create destructive wildfires, such breaks are highly ineffective as barriers (Cheney 1996). Under these conditions their primary value is as access areas of defensible space that can be used for backing fires or related fire-fighting activities. Studies of fuel break effectiveness in southern California have found that fuel breaks are relatively ineffective as barriers to fire spread and primarily of value for providing access to fire-fighting activities (Fig. 13.3).

Understanding costs and benefits of fuel treatments may be improved by economic analysis. A simple model of cost-benefit analysis (Fig. 13.4) would show that as the amount of management effort increases, implementation costs likewise rise but the damage from fires decreases. Rideout *et al*. (1999) predict that the total cost, that is, the sum of prefire implementation plus postfire damage, would not be a monotonic function of management effort but rather there will be some optimum point where the total costs are minimized. This is a welcome approach although several considerations are in order. One is that total management effort *per se* may not be the important metric but rather the strategic pattern of effort; that is, not *total area treated* but where they are located may be more critical. Ultimately there needs to be a clear articulation of resource costs associated with fires as well as with fire management practices.

On many landscapes fuel treatments have the greatest impact on fire outcomes when they provide conditions that reduce fire threats around values at risk, particularly at the wildland–urban interface (WUI) or areas compatible with

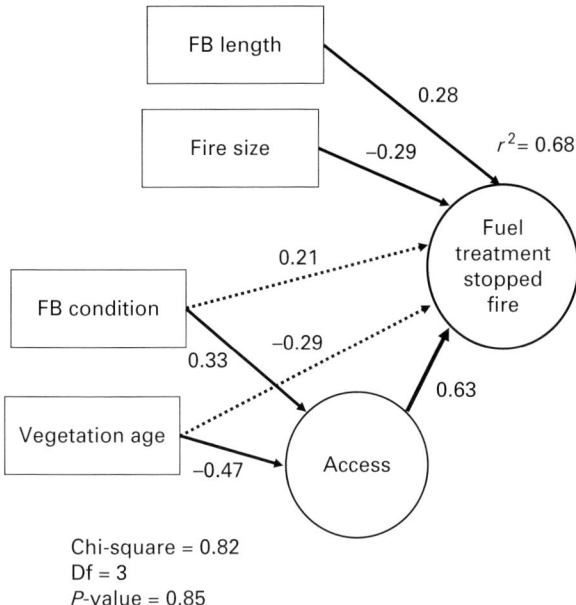

Fig. 13.3 Structural equation model of factors that directly and indirectly explain why fires stopped at fuel breaks (FB) in the Los Padres National Forest, California, USA. Solid arrows represent direct effects and dashed arrows represent indirect effects. Coefficients shown along arrows are standardized values. Circles represent endogenous (or dependent) variables in the model. (From A.D. Syphard, J.E. Keeley & T.J. Brennan unpublished.)

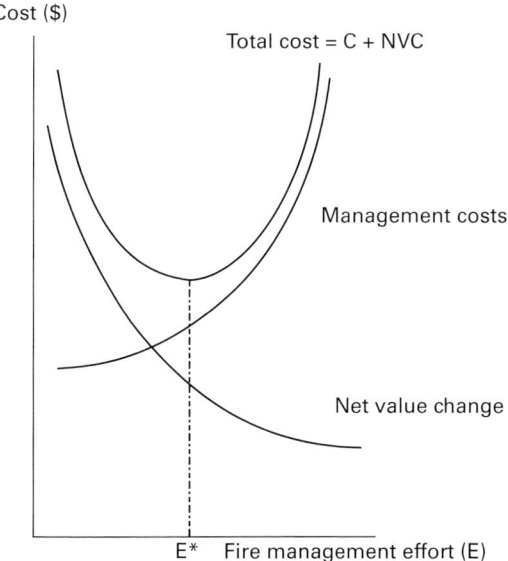

Fig. 13.4 Model of how management costs (C) and net value change (NVC) vary in response to fire management effort in fuel treatments. (From Rideout et al. 1999.)

fire-fighting activities such as backfires. As a consequence it is critical that strategic planning be a critical part of the process. Historically area treated has been the measuring stick for level of success but the sheer volume of treated landscape may not be the appropriate metric. Planning needs to consider where defensible space is most likely to be available under scenarios representative of the extreme fire conditions that generate the worse fire outcomes. One example of this approach is in the fire management plan for the southern California Santa Monica Mountains National Recreation Area (SMNRA 2004). Their planning process consisted of a serious of decisions about what landscape features comprised a suitable site for defensive fire activities and coupled with other data layers to generate the most strategic sites for fuel treatments. In general the plan pointed toward treatments along boundaries of the park with the urban fringes, a management approach shared with some Australian parks (Whelan 2002). Fuel treatment is inherently expensive and rates of treatment need to be high to have major effects. This boils down to the fact that we can only ever afford to intensively treat smaller rather than larger areas. Control over fire is therefore only ever local rather than global except where vegetation is permanently removed. Thus, it is about optimizing what we can realistically afford and sometimes the costs can be extraordinarily high for effective risk reduction (Bradstock et al. 1998b).

Wildland–Urban Interface Management

Historically there has been an inordinate amount of effort directed at solving the fire problem by focusing on the wildland fuels; that is, efforts at affecting factors I and S in Equation (13.1). Indeed, Cohen (2008) contends that a legacy of "an organizational mindset that persistently frames the wildland–urban interface fire problem in terms of fire suppression and control" has hindered work toward more effective alternatives in the urban environment. The history of fire control efforts in MTC regions, particularly in Australia and California, make it clear that wildland prefire fuel treatments and fire suppression cannot stop all fires from reaching the wildland–urban interface. It is becoming increasingly evident that greater effort needs to be focused on factor E, namely the probability of fire encroaching into the urban environment. Fires encroach in two ways: the radiant heat of the fire front ignites urban infrastructure (homes, fences, landscaping, etc.) or wind-driven embers (also known as firebrands) land and ignite components of the urban infrastructure.

Under severe fire weather conditions it is apparent that reducing fire encroachment into the urban environment is little affected by fuel treatments broadly distributed across the wildland landscape; rather perimeter-based strategies may be far more effective (Franklin 1987; Bradstock & Gill 2001). Even where forest fuel treatments significantly altered fire behavior, there are examples where this was not sufficient to prevent sizable losses of homes (Cohen & Stratton 2008; Safford et al. 2009). Oftentimes the problem is not as much with the forest fuels but with the urban fuels.

A significant amount of local preparedness for fires comes in the form of fuel alterations to create a buffer zone around homes, what is sometimes referred to as

(a)

(b)

Fig. 13.5 *Wildland and urban fuels. (a) Some homes lack sufficient clearance to prevent direct home ignition from wildland fuels, as in this postfire scene, 11 months after the 2009 fires in Victoria, Australia, which shows the speed of postfire resprouting in indigenous eucalypts. (b) Southern California urban landscaping presents substantial urban fuels that far surpass the wildland fuels evident in the background, and provide a bigger fire hazard when ignited by windblown embers. (Photos by Jon Keeley.)*

the *home ignition zone* in the USA (Cohen 2008) or the *asset protection zone* in Australia (Gill & Stephens 2009). Much of the effort has been focused on fuel treatments in a buffer zone around homes and other values at risk, both to reduce direct ignition of homes and landscaping. Failure to provide defensible space can lead to unfortunate outcomes (Fig. 13.5a). However, even adequate defensible

Fig. 13.6 *Prefire aerial image of urban housing adjacent to wildlands affected by the 2008 Freeway Fire in southern California. Flames on the seven houses on the upper perimeter indicate they were destroyed by the fire; houses at the WUI on the left have substantial clearance between homes and wildlands and houses on the right had a well-watered greenbelt. (From C.J. Fotheringham & J.E. Keeley unpublished data.)*

space sufficient to eliminate direct home ignition from the flaming front is not enough to ensure homes will not be lost. Commonly firebrands travel substantial distances and can result in housing losses even when significant fuel reduction buffer zones are in place (Fig. 13.6). Once fire ignites structures, fire spread within the urban environment can be both from direct heat flux from the burning structure as well as from embers transferred to other buildings.

Thus, E in Equation (13.1) is heavily dependent on the fuel level in the interface zone plus the firebrand load impinging on the urban environment, as well as weather and fire fighting resources. In both Australia and California it is apparent that firebrands are the primary cause of house loss (Ramsay *et al.* 1995; IBHS 2009). Fuel age, wind speed and vegetation structure all seem to be critical factors in determining ember spread to homes (Ellis 2003), and can be exacerbated by prior drought (Keeley & Zedler 2009). Wider buffer zones are sometimes justified on the basis that they will reduce firebrand load on urban environments. However, the source of destructive embers is probably one of the most critical knowledge gaps in fire management. We have little information on the spatial arrangement of fuel treatments for effective reduction in firebrand delivery to urban environments. This is particularly important to determine since evidence suggests firebrand transport can be many kilometers beyond a fire front, and perhaps much further in rugged terrain or under extreme wind conditions (Luke & McArthur 1978; Albini 1979). Embers are capable of destroying homes well over a half kilometer into urban environments (Chen & McAneney 2004).

Although the prescription for the size of the fuel reduction buffer zone varies with the site, generally 30 m is thought to provide adequate protection (Cohen 1999), except on steep slopes where it may need to be much wider (Butler 2009). Some local and state governments mandate a 30-m or more *clearance zone* around homes, but that term is misleading as clearing all vegetation is both unnecessary and undesirable. There is evidence that reducing the volume of dead fuels and thinning the density of live fuels is sufficient. Complete clearance not only increases the chances of erosion but also has negative impacts on biodiversity and aesthetics. In addition, complete clearance may lose some of the benefits of vegetation; for example, there is anecdotal evidence that trees with moderate fuel moisture may act as barriers to embers blowing into the urban environment.

Typically this buffer zone is created by modifying the natural vegetation through mechanical clearing, prescription burning or other treatments. However, often this is impractical or extraordinarily expensive (Bradstock *et al.* 1998b). An alternative approach is that of creating a "green belt" in which less flammable and irrigated vegetation is maintained within the buffer zone. Sometimes this may be in the form of agricultural lands such as orchards or grassy parklands or golf courses. However, creating such greenbelts requires coordination of local and regional planning efforts, and in most regions this problem remains unsolved.

Certainly one of the important determinants of factor E is due directly to land planning decisions. "There are members in the community who believe that they hold an inalienable right to live wherever, and however, they choose. It is not an inescapable fact that whenever homes and lives are threatened by fire, there is a community expectation that the firefighters will turn up and provide protection" (Koperberg 2003). Poor land-planning decisions put homes at risk, and road infrastructure increases numbers and distribution of ignitions as well as increasing access and effectiveness of fire suppression. Gill and Moore (1997) found that fire incidence in jarrah forests of southwestern Australia had remained constant over the past 40 yrs despite changes in roads, fire-fighting resources and fuel management during that time. Syphard *et al.* (2007) found that fires increase with increasing population density but as housing density increases and the wildland–urban intermix changes to an interface situation fires drop. Watersheds of hazardous fuels need special zoning that restricts residential use, something that has long been noted by fire managers (e.g. Arnold *et al.* 1951; Leisz & Wilson 1980).

Certainly part of the solution to reducing community vulnerability is more strategic land planning that considers avoidance of high fire hazard areas. Numerous factors will likely need to be considered in altering planning decisions (Moritz & Stephens 2008). Land zoning based on fire history is one obvious means of effecting change (Bovio & Camia 1997). While this is important for the future there are huge legacies of risk due to past planning decisions and the desire to rebuild and remain undefeated by nature following disastrous fires.

Urban Management

Factor G in Equation (13.1), chance of fire propagating, is a function of urban fuels that include home construction characteristics, landscaping and home attachments. Often these are a more significant threat to urban environments than wildland fuels (Spyratos *et al.* 2007). Predicting the impact of these factors on fire risk is complicated by housing density, socio-economic status and age of the development (Syphard *et al.* 2007; Luck *et al.* 2009). We know relatively little about the potential trade-offs of different landscape planting palettes on fire spread in the urban environment. Combustion studies have provided a basis for selecting landscaping plantings (Weise *et al.* 2003; White & Zipperer 2010). However, the widespread use of palm trees and pine trees is of some concern as these trees within the urban environment can act as ember catchers and relay embers. Exacerbating the situation is the observation that landscape changes in vegetation fuels over the past half century have in some regions been greater on the urban side of the wildland–urban interface (Fig. 13.5b). Although some plants appear to be more fire resistant than others, perhaps the primary factor determining impact on home ignition is the proximity of landscaping to structures (Wilson & Ferguson 1986) and the degree of litter accumulation on roofs (C.J. Fotheringham & J.E. Keeley unpublished data).

Home construction characteristics should make a difference as to whether or not it survives a wildfire, although most of the evidence for this is circumstantial, or based on small-scale tests (Dietenberger 2010). Shutters on windows and special attic vents to prevent ember entry into homes are some of the recommended structural changes that could lead to reduced building losses (Ramsay *et al.* 1995). In California, maps of fire hazard severity are constantly updated for the purpose of informing decision makers on the construction codes necessary to reduce losses from wildfires. It is interesting that an Australian study of bushfire losses found that over the past century the annual probability of building destruction has remained almost constant (McAneney *et al.* 2009). These authors suggested that despite improvements on several fronts, housing losses were the result of severe fire weather, and homes in the path of these firestorms stood little chance of surviving. This suggests that changes in construction materials may not provide the necessary level of protection and perhaps more drastic changes in urban growth patterns designed to avoid dangerous landscapes may hold the most promise for reducing wildfire losses.

Social and Political Constraints

Political and social factors (Box 13.2) are critical to minimizing D, risk of destruction (Equation 13.1). Preventing wildfires from reaching the urban environment is generally the responsibility of state and federal governments, and community planning is often predicated on the belief that these fire management agencies can eliminate wildfire threats. However, there is abundant evidence that D will never be reduced to zero by wildland management activities alone. Despite the scientific evidence for this, differing agendas by stakeholders often diminish the

Box 13.2 Socio-political Aspects of Catastrophic Fires

Wildfires that encroach on the urban environment and result in loss of property and lives have many ramifications. In high-density landscapes such as southern California losses from wildfires are often recovered through litigation of responsible parties. Recent examples illustrate widespread distribution of blame for such events. Following the 2003 fires in San Diego County, California, which included the 110 000-ha Cedar Fire, hundreds of homeowners filed a lawsuit claiming wrongdoing by one of the major insurance companies for underinsuring their properties (Marshall 2007). This insurance company in turn brought a lawsuit against the California Department of Forestry and city and county of San Diego to recover damages it incurred because of purported negligence on the part of these agencies in controlling the fire (Steele 2004). Due in part to expectations that the U.S. Forest Service should be able to completely control all wildfires, a homeowners' class action lawsuit was brought against the federal government (Scarcella 2008). Just 4 yrs later a significant portion of the area that burned in the Cedar Fire was reburned by the Witch Fire (Table 13.1; Keeley *et al.* 2009b). This fire putatively originated from a power line and soon afterwards the state government, homeowners and insurance companies filed suits totaling hundreds of millions of dollars against the regional utility company (Perry 2009; Soto 2010). In response the utility company threatened countersuit lawsuits against fire victims (Naiman 2009).

This power company acknowledged that power lines started several other large fires in 2007 and a total of 167 fires during a 5-yr period in San Diego County, California. Because these mostly occurred during extreme Santa Ana wind events it was proposed that this important source of catastrophic fires could be reduced by placing power lines underground in known corridors of high winds (Keeley *et al.* 2009b). The power company proposed an alternative plan that involved switching off power to all customers in high-risk areas, which comprised much of the eastern portion of the county, during Santa Ana wind events (Jones 2008). Although this blackout area would affect 60 000 customers for several weeks each year, and despite the likelihood that during blackouts many of them would rely on gas-powered generators (known to ignite fires from sparks), there was local political support for the plan (Lescure 2009). This issue is likely to continue to gain in importance as power lines have been implicated in 20% of the major fire events on state responsibility lands in California (Cal Fire data, unpublished). In this respect California is not unique as power lines have been implicated as the ignition source for a number of fires during the Black Saturday bushfires in Australia (Table 13.1).

value of scientific analysis and greatly complicate solutions (Goldstein 2007). This is specially problematic in the Mediterranean Basin where similar landscapes are managed by different agencies from different countries with very different socio-economic frameworks (e.g. North Africa vs. southern Europe).

There are marked differences between MTC regions in the politics of fire management with potential ramifications for solutions. In the USA fire management has become increasingly centralized in the federal government, initially as a cost savings effort to make fire management more efficient. One downside of this approach is that there has been for many decades an attempt to apply one model of fire management to most ecosystems. In particular the prescribed understory burning of conifer forests has been very successful on some landscapes but the long-standing effort to apply it to California shrubland ecosystems has been largely unsuccessful and has delayed consideration of other approaches. In contrast, Australian fire is managed, even on federal or Crown lands, by each state. One consequence is there is seemingly greater acceptance of diverse management approaches, and probably a greater capacity or potential capacity to tailor solutions to suit local environments and context.

In California one resolution would be greater regional and state planning in the location of urban developments, and less community control over planning decisions, which seems to be the root of other environmental problems as well (e.g. Pincetl 1999). In this region there is also a serious concern about the financial responsibilities for fires that cross local, state and federal jurisdictions. Too often the federal government pays the bulk of the expenses associated with protecting private property from wildfires. The existing framework insulates non-federal entities from the cost of protecting private property in urban environments (GAO 2009). Proposals that the federal government seek ways to recoup costs of fire suppression from communities (OIG 2006) have the potential of effecting greater responsibility in regional planning decisions. Such a financial burden in the future could contribute to better planning decisions as they pertain to placement of future developments.

In Australia steps are being taken in this direction. Planning approval for new developments on public and private land in New South Wales require approval by the Rural Fire Service – the government department responsible for co-ordination of fire prevention and suppression in the state. Developments are assessed on the basis of appropriate setbacks from vegetation, using physical criteria for heat impacts on structures that take into account worse case scenarios of weather and fuel, along with topographic factors specific to each site (NSWRFS 2006).

Postfire Management

Burn Severity Assessment

Fires in MTV often are high-intensity crown fires and may have severe impacts. Sorting out the factors of fire intensity, fire severity and ecosystem responses (see Fig. 2.4) are critical to the appropriate management response. Although progress is being made at remote sensing detection of fire intensity patterns during the fire, complete coverage of a fire is often hindered by smoke so that the post hoc fire severity (sometimes known as burn severity) measure of biomass loss and associated soil impacts is the common metric available for assessing subsequent ecosystem responses (Keeley 2009). On most landscapes fire or burn severity is mapped with remote sensing indices calculated from prefire and postfire images. This often correlates with on-the-ground measures of fire severity based on biomass loss (Rogan & Franklin 2001; Chafer et al. 2004; Hammill & Bradstock 2006; Keeley et al. 2008). Federal agencies in the USA have redefined fire or burn severity much more broadly to include many other parameters and termed this the Composite Burn Index (CBI) (Key & Benson 2006). Some studies report correlations between this field measurement and remote sensing indices of fire severity; however, it appears to be a poor field measure of fire severity in mediterranean shrublands (Sikes et al. 2006) and some forested ecosystems (Murphy et al. 2008). Keeley (2009) contends that many of the components of CBI are important measures of burn severity but raises serious objections to use of the composite index because it combines fire severity variables with ecosystem response variables such as postfire resprouting, and thus using it as a predictor of ecosystem response is circular.

Remote imaging techniques show great promise for detecting different patterns of fire severity and are increasingly important on large fires (see Chapter 2). The differenced Normalized Burn Ratio (dNBR), which is based on the Landsat TM sensor, is widely applied with variable success across different vegetation types (Hammill and Bradstock 2006). In shrubland ecosystems it is strongly correlated with field measurement of fire severity; however, due to the extraordinary resilience of these ecosystems neither fire severity nor dNBR predicts ecosystem response variables such as postfire cover (Keeley et al. 2008). In forests interpreting dNBR is more complicated since it often cannot detect understory fire impacts and thus is a better measure of the impact on tree canopies. Apparently a more useful metric is the relative dNBR (dNBR/prefire NBR) (Miller & Thode 2007). The absolute dNBR is a measure of the actual biomass loss from fire and thus a good surrogate for fire intensity in crown fire ecosystems but less so in forests with understory burning. However, the relative dNBR is a measure of change after fire and in forested ecosystems this is tied to tree mortality. In shrublands, and in other crown fire ecosystems where 100% aboveground mortality is to be expected, the relative dNBR is of limited value and unrelated to field measures of fire severity (Keeley et al. 2008).

Currently dNBR indices are widely used as predictors of hydrologic stability; however, the extent to which different degrees of fire severity are correlated with postfire hydrologic changes remains unclear. Postfire increases in soil water repellency due to hydrophobic soil layers is tied, albeit sometimes weakly, to fire severity (Lewis et al. 2006), although in some ecosystems soil hydrophobicity is unrelated to fire severity (Doerr et al. 2006). Although fire per se does affect hydrological functioning, there is little direct evidence that fire severity is a reliable predictor of hydrologic changes (Robichaud *et al.* 2000; González-Pelayo *et al.* 2006); indeed, Doerr *et al.* (2006) suggest that fire severity classifications are unsuitable for predicting fire impacts on soil hydrological responses. The primary reason is that ecological responses such as erosion, overland water flow and debris flows are affected as much by topography, soil type, rates of weathering, fire-free interval, and precipitation as they are by fire or burn severity (Moody and Martin 2001). Factors responsible for hydrologic responses to fire are multifactorial and we need better mechanistic models across a diversity of landscapes before we infer that fire or burn severity metrics are predictable measures of ecosystem responses.

Further research is required to fully understand the extent to which dNBR informs managers of potential ecosystem responses. In addition to the very different information in relative vs. absolute Landsat TM indices, there are many other remote sensors that potentially could provide better indicators of postfire management needs (Roldán-Zamarrón *et al.* 2006; Chuvieco *et al.* 2007; Holden *et al.* 2010).

Postfire Restoration

Emergency responses to postfire conditions are largely focused on protecting property and other human infrastructure such as roads from flooding and debris flows. Two approaches are utilized: measures directed at limiting the extent of erosion and landslides at the source, such as seeding or mulching, and measures directed at stopping these flows somewhere between the source and the values at risk, usually with barriers of some kind.

Postfire aerial seeding as a management practice has its roots in southern California as a flood control measure. This was partly due to an incomplete understanding of the natural capacity for rapid natural recovery in chaparral ecosystems and a perceived need to enhance winter herbaceous growth with roots that could assist in soil stability (Corbett & Green 1965). This was once a widely popular postfire management technique in California (Beyers 2004) and the Mediterranean Basin (Pinaya *et al.* 2000). Effectiveness of seeding depends on the reliability of early autumn rainfall patterns. For example, it appears to be effective in the Mediterranean Basin: rains come early and establish seeded grasses and forbs before heavy winter downpours, particularly on sites dominated by non-resprouting pines burned in crown fires (Vallejo *et al.* 2006). However, in southern California rains are later and in more intense storms making postfire seeding ineffective in most years (Keeley *et al.* 2006a). Throughout the western USA evidence is increasing that seeding is of limited effectiveness in reducing soil erosion (Peppin *et al.* 2010). Due to the fact that intense winter storms typically initiate the rainy season, often the ungerminated seeds are washed off slopes prior to germinating and in the majority of projects there is relatively little successful recruitment and when it does come the worst of the winter storms may be past. Terrain also affects the success of seeding. On steep slopes a significant amount of dry ravel comes off prior to the rainy season and this soil loss is unaffected by

Fig. 13.7 *Postfire slope stabilization by felling and placing* in situ *tree skeletons from 2007 fire on Mt. Olympus, Greece. (Photo by Jon Keeley.)*

seeding. In general, where dominated by resprouting shrubs or having a native postfire herbaceous flora, this native regeneration may be much greater than the seeded cover (Keeley 1996; Beyers 2004; Vallejo *et al.* 2006). An important consideration is that often there are better methods than seeding for reducing slope erosion, such as log barriers (Fig. 13.7), mulch or hay bales, which have proven to be more effective and far more predictable than seeding (Robichaud *et al.* 2000).

In terms of conserving naturally functioning chaparral ecosystems, seeding also has the potential for negative impacts. Where non-native species are seeded and the rains co-operate to establish these species, they have the potential for outcompeting the native species and altering the natural ecological balance. On some landscapes the short-lived non-native seeded grasses dry earlier than native species and increase the length of the fire season, promoting repeat fires (Zedler *et al.* 1983). In addition, seeded species sometimes escape and become aggressive invasives, such as black mustard (*Brassica nigra*), which was the favored exotic species used to seed after fires in southern California during the first half of the twentieth century. Today this invasive is a widespread pest throughout the region (e.g. see Fig. B12.2.1).

One approach that has been used is aerial seeding of "sterile" cereal crop seeds, which pose no threat of escape and invasion. However, this practice still has negative ecosystem impacts in that aggressive grass growth will diminish native recovery and diversity (Keeley 2004). Since this cereal is a sterile annual, there is no second-year establishment and where there is abundant first-year growth there is a vacuum created in the second year that could attract alien invasions. Regardless of the attributes of seeded species, there is always the potential for contaminants in the seed lot. For example, following the 2000 Cerro Grande Fire in New Mexico, USA, it is estimated that aerial seeding inadvertently sowed over a billion seeds of the highly aggressive alien *Bromus tectorum* (Keeley *et al.* 2006a). Physical barriers created by mulch and hay bales also have the potential for introducing exotic species and as a consequence more and more such projects are requiring weed-free hay.

It is sometimes proposed that seeding of native species is more in line with ecological management objectives and in the Mediterranean Basin natives are routinely used for postfire seeding. However, even this has the potential for impacting communities by upsetting aspect-specific community assemblages. In addition, seeding natives, as with non-natives, tends to result in a dominance of one or a few species and reduce native plant diversity (Dodson *et al.* 2010).

Postfire responses to ecosystem losses such as tree mortality involve replanting seedlings. In crown fire ecosystems this may not be warranted, but in forests with surface fire regimes, where forest regeneration requires survival of parent seed trees, crown fires can greatly set back recovery without active replanting. Many fires in the Mediterranean Basin occur in old-fields, which are dominated by early succession flammable species (see Chapter 4). In such situations, one of the postfire management actions includes planting seedlings of less flammable resprouting species. The objectives of such an option are to increase the resilience to new fires, as well as to increase soil stabilization in the midterm (Pausas *et al.* 2004a; Vallejo *et al.* 2006).

Future Fire Management

Future global changes in MTC regions and associated landscapes with MTV will undoubtedly stress our ability to maintain community safety and sustainable ecosystem structure and processes. Increasing population growth coupled with continued urban sprawl and potential changes in fire regime due to climate change will be some of the most critical factors.

Climate change impacts are intimately tied to fuel structure, both because fuel structure affects how climate changes will alter fire regimes and because climate change impacts will potentially alter fuel structure. Forests where fuels largely comprise understory litter respond differently to climate than forests with herbaceous surface fuels (e.g. see Fig. 2.8) and both respond differently than crown fire shrubland fuels.

There are numerous predictions that in most fire-prone landscapes in western North America, Australia and the Mediterranean Basin fire activity will increase with global change due to expected climatic changes (e.g. Pitman *et al.* 2007;

Lenihan *et al.* 2008; Giannakopoulos *et al.* 2009). Indeed, some contend that climate-driven changes have already occurred in fire regimes of western North American forests (e.g. Westerling *et al.* 2006), although such conclusions may be premature as they are based on short-term datasets and the data are consistent with other explanations. In montane forested environments it has been postulated that increasing temperatures over the twentieth century have caused earlier snow-melt and that this change has increased fire activity. However, as of yet this pattern is not consistent for the western USA (Medler *et al.* 2002; Michaels 2006). Likewise, in the Mediterranean Basin temperatures over the past several decades have risen, but this change has not been correlated with increased area burned (Pausas 2004).

Global climate-forcing mechanisms such as El Niño-Southern Oscillation (ENSO) are known to affect fire activity primarily through effects on precipitation. Predicting future impacts is complicated by the observation that this ENSO effect on fires has changed over time and thus is not always a reliable indicator of fire occurrence (Yocum *et al.* 2010).

Predicting global change impacts on fire is complicated by the complexity of potential interactions. Higher temperatures will create greater atmospheric water deficits and decreased fuel moisture during the fire season. However, these same water deficits may reduce fine fuel production, thus reducing fire frequency in some vegetation types (Pausas & Bradstock 2007). Over time this could result in fewer but more intense fires. In other cases fire frequency may increase and contribute to type conversion of woody vegetation to herbaceous systems, ultimately decreasing fire intensity. In addition, direct effects of elevated CO_2 potentially may confound or reinforce these effects (Bradstock 2010). As CO_2 rises one might expect increased primary productivity and thus increased fuel loads, but as CO_2 increases, direct effects on stomatal behavior may act to decrease drought stress and thus increase fuel moisture.

Predicting future climate and fire interactions is extremely difficult because current models lack the necessary complexity to capture the multitude of ecological interactions that will occur as environments change. This is illustrated by the climate envelope models that attempt to predict future demographic patterns of species based on climate alone (e.g. Loarie *et al.* 2008) and are likely to come to very different conclusions once we have the ability to incorporate other, and often more important, ecosystem factors (e.g. Loehle & LeBlanc 1996; A.J. Davis *et al.* 1998; Pimm 2007).

Climate change is only one of a multitude of global changes for which we need to plan. For example, populations are increasing faster in MTC regions than are temperatures, and since ignitions are largely anthropogenic, and fires are often ignition limited, we should expect population growth to have a huge impact on future fire regimes. Modeling studies show that climate scenarios for southern California are predicted to have minimal impact on persistence of a non-resprouting chaparral shrub relative to the substantial impact of increases in fire frequency (Lawson *et al.* 2010). Also, increasing nitrogen deposition in most MTC regions may increase understory fuel continuity and thus alter fire regimes (Hurteau *et al.* 2009). Fuel continuity may also change with socio-economic changes (Pausas 2004) and with changes in the agricultural policy (e.g. subsidies). In some fire-prone ecosystems the future response to altering fire regimes is not likely to be a monotonic function of increasing temperature, rather *tipping point* thresholds are to be expected (Lenton *et al.* 2008). For example, the potential exists for temperature increases to produce fire regime changes that lead to vegetation shifts, which in turn lead to further fire regime shifts that are not directly predicted by temperature. This line of reasoning leads to the conclusion that better analytical models of future temperature change will not necessarily improve predictive power for fire regime changes and that we should proceed cautiously with the expectation of major and inevitable surprises (Doak *et al.* 2008).

Fire management focus in the future is likely to make the most progress by thinking strategically about locating fuel treatments in ways that can be used to provide direct protection of urban and suburban environments. Much of the future focus must be on preventing ignitions during extreme fire weather and managing home construction and urban fuels to make homes less vulnerable. Perhaps most importantly we need to focus on land planning decisions that put fewer people at risk by following smarter patterns of growth. Increasing our predictive ability through fire hazard indices that capture parameters such as drought-induced dieback may provide some added benefit.

Conclusions

Early European influence in all MTC regions outside the Mediterranean Basin has led to marked similarities in approaches to dealing with wildfires. The combination of annual fire risk and high-intensity crown fires in shrublands and woodlands binds all of the MTC regions. Although all five regions have had very different human histories, there is a convergence in contemporary fire problems. Fire management agencies have benefited from the experiences in other regions and this has often resulted in exchange of ideas and personnel. This exchange of information goes far beyond the MTC regions since many fire-prone landscapes in other climatic regions such as southeastern Australia and western USA have similar fire problems. Two important differences between regions have contributed to different levels of fire danger for human populations. One is the inherent difference in fuel loads between MTC regions that generate very different fire hazards, for example between California chaparral and Western Australian heathland. Another is tied to differences in population density that may vary by more than an order of magnitude between MTC regions. These differences have profound implications for patterns of community vulnerability and fire management options.

Although preventing fires from reaching urban and peri-urban environments is an important goal for fire management agencies, it needs to be recognized that not all fires can be stopped. Indeed, moving away from this model to one that treats wildfires like some other disasters such as earthquakes has value. No one considers stopping earthquakes; rather they are recognized as inevitable and communities develop infrastructure to make living in earthquake country safer. Like flood control in many landscapes, the greatest reduction in community vulnerability is achieved with better planning and the same applies to fire management. These MTC plant communities have adapted over tens of millions of years to frequent fires and now human societies are working toward their own adaptive solutions to living in these fire-prone landscapes.

14 Climate, Fire and Geology in the Convergence of Mediterranean-type Climate Ecosystems

Integrating Climate, Fire and Geology in a Fire-prone World

Fire challenges the long-standing hegemony of ecology, biogeography and paleoecology that climate and soils are sufficient to explain the origin and distribution of plant species. In a world where half of the land surface is fire-prone (Krawchuk *et al.* 2009), understanding the past and predicting the future requires a close integration of climate, fire and geology. The dogma that fire is an anthropogenic phenomenon of little use in understanding paleoecology (Axelrod 1980, 1989), or merely incidental to vegetation development (Hopper 2009), is rapidly being replaced with a better understanding of paleofire's impact on land plant evolution (Scott 2000; Pausas & Keeley 2009). Attempts to model future global vegetation patterns have been demonstrated to be inadequate without including both natural and anthropogenic fire regimes (Bond *et al.* 2005).

Bond and Keeley (2005) outlined the conundrum posed by alternative explanations for the present distribution of vegetation and assembly of communities. Classical explanations have invoked resource-based mechanisms that are driven by climate and soils. There are ecosystems where resource-based mechanisms may be sufficient, but on many seasonally dry landscapes ecosystem processes such as fire play a major role in the organization and evolution of vegetation.

The evidence in favor of resource-based explanations is formidable, but the nexus between theory and evidence is largely correlative and not generally informative about the importance of unrecognized sources of influence. For example, many vegetation boundaries commonly correspond to measured or perceived edaphic and climatic disjunctions such as the close relationship between sclerophyllous vegetation and mediterranean-type climates (MTCs). Although anomalies occur, these are often disregarded as unimportant exceptions, whereas here we maintain that in MTC ecosystems in particular, and many other ecosystems as well, fire is a critical factor that interacts with climate and geology to affect plant traits, community assembly and ecosystem functioning.

There is widespread empirical and theoretical evidence that fire is an essential factor determining species and community dynamics through time. For example, fire regimes at continental and global scales covary strongly in concert with climatic and vegetation patterns (Archibald *et al.* 2009; Bradstock 2010), suggesting that syndromes of productivity and fire weather determined, in part, by

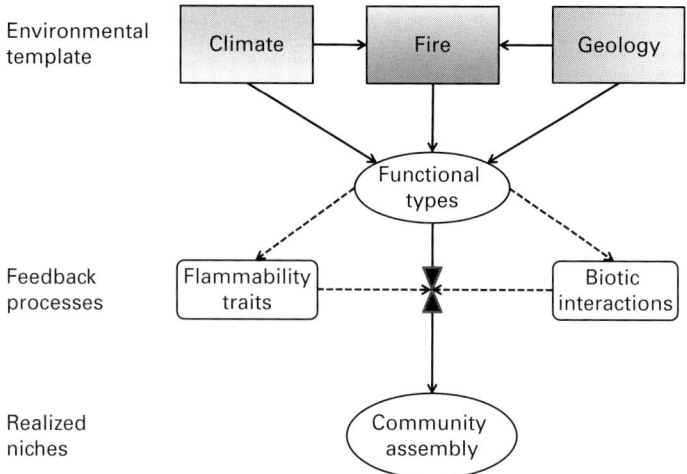

Fig. 14.1 *Schematic model of the environmental template and functional type evolution in fire-prone landscapes.*

climate are major influences on fire regimes (Pausas & Bradstock 2007). However, this is not just a story about how climate shapes fire, but how fire shapes vegetation, which in turn shapes fire regimes. Plant traits that affect flammability are one of the clearest examples of feedback processes that can result in niche construction (Bond & Midgely 1995; Schwilk 2003) and illustrate the inadequacy of a resource-based theory to account for fire regimes and vegetation patterns. While inclusive fitness theory can accommodate flammability, fires can only occur when sufficient numbers of neighbors contribute to fire spread and here is where fire, climate and geology meet in dictating fire regimes. Indeed, the size of fire-prone patches will affect the probability of ignitions, as Wardle *et al.* (1997) has shown on islands, and in this regard, soils and climate are important determinants of vegetation patterns conducive to fire spread.

Thus, a comprehensive approach to understanding fire lies in an integration of the interplay between climate, geological resources, fire-adaptive traits and phylogenetic legacies that place limits on potential plant functional types (Fig. 14.1). A brief examination of the strengths and weaknesses of "resource" and "fire" theories pinpoints the elements needed to develop a more integrated approach.

A strength of resource theory is the demonstrable correlation between patterns of climate and soil variation and vegetation distributions and boundaries. Statistical modeling based on these influences can be used to successfully predict species and community distributions (Franklin 2010b). Although such correlative power is reassuring, mechanistic insights are limited. Models of this kind essentially define realized niches of species and species combinations but give little insight into potential niches. Factors such as fire may play a role in confining species to particular realized niches in concert with other factors such as competition. A long-running debate about the determinants of rainforest distribution in

Australia illustrates the problem. Aspects of the realized niche of rainforest are confounded with attributes of fire refugia. Thus, apparent edaphic/climatic/topographic correlates of rainforest distributions may also reflect boundaries of reduced probability of fire imposed by the same suite of environmental attributes. The potential edaphic/climatic niche of rainforest may be far broader and overlap substantially with sclerophyll species.

Abandoning resource theory entirely for a fire-centric theory of vegetation likewise misses the appropriate balance. Factors responsible for the displacement of rainforest by sclerophyllous eucalypts in Australia spawned the "ecological drift" theory of Jackson (1968). Novel elements of this theory included the influence of vegetation on probability of burning, multiple pathways of vegetation change and the notion that particular fire regimes could eliminate species and vegetation types. This theory postulated a central role for fire in the organization of differing vegetation types. A major drawback of the theory in its original form, however, is omission of varying soil types that control site productivity as a factor that may influence growth rates of species and hence pathways of vegetation dynamics (Bowman 2000). Thus, a merging of elements of resource-based theory with the principles of ecological drift offers the prospect of a more powerful predictive basis for understanding vegetation distributions. Ultimately any future theory of this kind has to account for observable correlations between edaphic and vegetation patterns as well as the successional dynamics that emerge from the interplay between functional types and fire regimes.

Equally problematical are attempts to explain the general lack of forest growth in the Cape region of South Africa on nutrient availability, when almost surely fire has had a heavy hand in determining fynbos/forest boundaries (Bond 2010). The South African MTC region differs from some other MTC regions in lacking as intense a summer drought, and in the widespread extension of shrublands into high-rainfall areas that theoretically should support forests. Extensive conifer and eucalypt plantations in the Cape are testimony to the mismatch between the climate potential for forests and the actual vegetation of extensive shrublands. Mesic Cape fynbos is not at equilibrium with climate, as is the case with much of the global vegetation (Bond *et al.* 2005). Thus, the presence of shrublands over much of the region is not simply a product of a winter-rainfall climate and extended summer drought. The scarcity of forests in the Cape region MTC can be attributed to the interactive effects of frequent fires and slow tree growth rates on nutrient-poor soils, and opportunities for fire afforded by regular annual drought. Unlike Australia, which also has nutrient-poor soils, the Cape region does not have eucalypts with their remarkable fire tolerance and flammable properties, suggesting a potential phylogenetic limitation to forests in the Cape region.

In northern hemisphere shrublands there has likewise been a tradition of interpreting patterns solely in terms of climate, soils and mountain building. Indeed, Axelrod (1989) believed that fire had played no significant role in the evolution of chaparral traits. A huge body of paleoecology work in California was based on the premise that evolution could be understood in terms of climate and geology and thus there was little attempt to consider fire as a factor. However, demographic patterns in the most widespread chaparral species illustrate not only fire resilience but fire dependence. Delaying reproduction to a single risky postfire pulse of recruitment cannot be accounted for by any other reasonable hypothesis. Likewise, serotinous pines found throughout California and the Mediterranean Basin represent as clear an example of fire adaptation (Schwilk & Ackerly 2001) as to be found anywhere. Yet this perspective evaded Axelrod (1980) who apparently could not escape his early training that plant distributions are dictated by climate and geology and these attitudes have been widespread in many disciplines attempting to explain origins and distributions. Some biogeographical problems are highly dependent on understanding fire regimes and their interactions with climate and geology. For example the California endemic tree *Pinus sabiniana* has long perplexed botanists due to its marked absence from certain watersheds. Life history analysis and response to fire points to an interesting interaction between geological uplift, drainage topology and fire intensity as factors that in combination have acted to eliminate this species from watersheds in the middle of its range (Schwilk & Keeley 2006).

One means of affecting community assembly on fire-prone landscapes is by the acquisition of traits that increase resilience to fire but alter fire regimes in ways that promote burning. Bond and Midgely (1995) explored the concept of flammability as a competitive force. They linked flammability to variations in mortality during fires and postfire recovery and demonstrated that selection for more flammable plant types could occur, provided that less flammable competitors suffered high levels of fire mortality and relatively low levels of recovery. A drawback of their approach is that processes of fire spread used in the model are simplistic and variations in flammability among functional types due to exogenous influences of weather are not considered. The model also assumes that space is equally available to all functional types: that is, the possibility of variations in habitat availability among species is not considered. Fire is a particularly interesting phenomenon from an evolutionary perspective because the probability of an individual burning is a function not just of its own flammability but the flammability of the community and the extent of fire-prone landscape to collect lightning ignitions and spread them. Where natural sources of ignition are limiting, the size of fire-prone patches will determine fire probability for individuals.

There have been recent attempts to link observable patterns of variation in diversity and resilience (i.e. resprouting capability) to resource and disturbance syndromes (Bellingham & Sparrow 2000; Clarke & Knox 2002). Attribution of causality is difficult because observed variations in diversity and resprouting often correspond to confounded gradients of resources (i.e. water and soils, which relate to productivity) and fire regimes (Clarke *et al.* 2005; Pausas & Bradstock 2007). Huston (2003) posed a model (Dynamic Equilibrium Model) that attempted to explain how patterns of diversity may emerge from covarying syndromes of productivity and disturbance frequency. A strength of Huston's model is that it links productivity, fire frequency and intensity; a weakness is the lack of feedback of fire on flammability and thus on fire regimes themselves. Also, effects of productivity on resilience are a key element. The model therefore provides some resolution of resource- and disturbance-based perspectives.

A severe limitation of all of these models is that they treat responses as a monotonic function of fire regime parameters such as fire frequency and intensity. They fail to acknowledge that gradients in fire frequency or intensity have very different selective impact in different fuel types such as shrubland vs. forest. As one transitions from a crown fire regime to a surface fire regime, the transition is like a tipping point where trait selection may change radically along gradients of fire frequency or intensity (see Chapter 9). This is an important limitation that affects model building designed to provide globally applicable explanations. Fire is an ecosystem process that, when coupled with climate and geology, forms the environmental template upon which evolution operates to select for functional types (Fig. 14.1). The factors that determine the realized niche and ultimately the community assembly process are competitive sorting through competition for resources on the one hand and, competition via differences in flammability and resilience competition on the other, operating as feedback processes that fine-tune functional type arrangements in communities.

Recognizing that ecosystem processes on fire-prone landscapes involve the interaction of climate, fire and geology is ever more important as we learn how to predict ecological outcomes due to anthropogenically induced climate warming. It is a matter of concern when observed patterns consistent with climate change predictions carry sufficient weight to rule out alternative explanations. The recent demonstration of an upward shift in chaparral species in a southern California mountain range (Kelly & Gouldin 2008), interpreted as a global warming response (Breshears *et al.* 2008), could also be accounted for by different fire histories between the upper and lower portions of the elevational gradient they studied as observed by D.W. Schwilk & J.E. Keeley (unpublished data). Likewise, an impressive study of increased tree mortality over the last several decades in Sierra Nevada forests (van Mantgem *et al.* 2009) has been done in a vacuum devoid of fire. Due to effective fire suppression these highly fire-prone forests are grossly out of balance with the important ecosystem process of fire, not having experienced burning for more than 125 yrs (Schwilk *et al.* 2006), a fire return interval roughly ten times longer than that experienced over at least the last 2000 yrs (Swetnam 1993) and undoubtedly over a much longer time span. In developing a climate change argument, fire exclusion effects were ruled out as a factor because similar mortality patterns were observed in forests with different fire regimes, some of which had not missed fire cycles. In essence their argument is that if fire can't account for mortality patterns everywhere it can't account for them anywhere, but as illustrated throughout this book, the interaction between climate and fire generates qualitatively different responses in different fire regimes. Regardless of the validity of assumptions made in this study of tree mortality, it serves to illustrate the difficulty of trying to parse out fire impacts across different fire

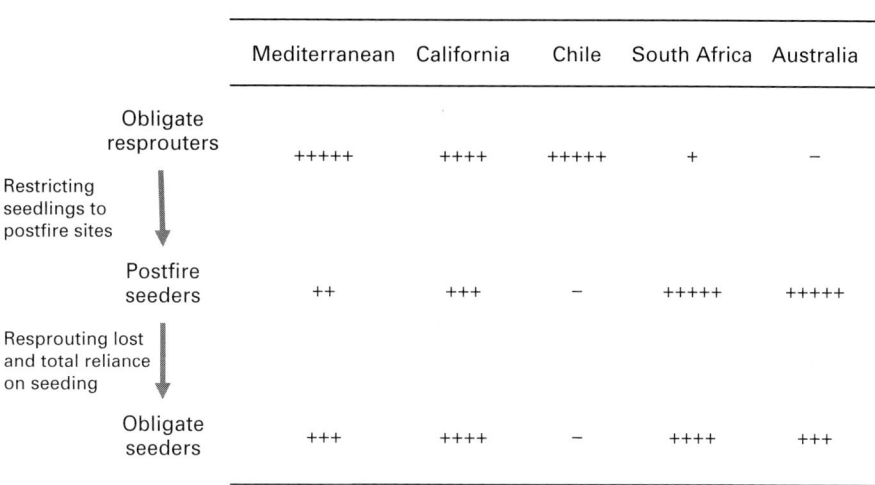

Fig. 14.2 *In the evolution of fire-dependent ecosystems two important transitions are evident (see Fig. 9.5). Mediterranean-type climate ecosystems differ in the importance of these modes.*

regimes. Future global changes will be best anticipated by confronting the issue of fire directly and best management practices require that we think about such systems in terms of climate, fire and geology interactions if we are to anticipate future global changes and adapt.

Convergence in Fire-prone Mediterranean-type Climate Ecosystems

Convergence theory states that similar environments will select for similar biotas through a combination of evolutionary adaptation and ecological sorting. The five MTC regions have long been noted as examples of convergence in both the dominance of evergreen sclerophyllous-leaved woody plant communities of mediterranean-type vegetation (MTV) and their highly fire-prone character. These regions are tied together by their similar seasonal distribution of cool weather growing season sufficiently mesic to produce sufficient biomass capable of spreading fires during the hot, dry fire season. However, no two environments are identical and subtle differences in climate, not so subtle differences in geology, and substantial differences in phylogeographical histories, have all contributed to differences in plant traits (Fig. 14.2), community composition and fire regimes. It should be taken as axiomatic that convergence is to be expected, but since no two environments are identical the question becomes, "which environmental factors have played the greatest role in driving non-convergences?"

In all five regions many MTC plant lineages can be traced to sclerophyllous taxa with origins in the early Tertiary under climates bearing little resemblance to the contemporary winter rains and summer drought. Sclerophylly was of selective value on sites where coarse-grained lower-fertility substrates produced soil-drought stress under some regimen of seasonality, on an annual or longer frequency basis. The MTC in most regions developed gradually through the Miocene with increasing intensity of summer drought, and this coupling of drought with high summer temperatures greatly expanded the drought-prone landscape. Concomitant with this was an increase in fuel continuity and thus fire spread and predictability of fire.

Two features of the MTC have been of immense selective importance: the cool growing season and seasonal soil drought. Both of these predate the MTC and thus much of the present flora persists under conditions with some significant similarities to environments present during their pre-MTC origins. The primary impact of the MTC was a major expansion of fire-prone sclerophyll vegetation. Over the last 10 million years, as the current climate unfolded, and a highly predictable crown fire regime developed, species reassembled, and many fine-tuned their adaptations to this regimen. Some lineages exploited newly generated resource combinations and novel fire regimes and have speciated rapidly since the end of the Tertiary. Other lineages remain ensconced in ecological niches little changed from pre-MTC environments.

An emergent property of this model is that it provides a clearer picture of the relationship of MTV that today persists outside the MTC regions. In many respects the proximal selective factors of soil drought and fire are the same. This is most evident in highly fire-prone evergreen sclerophyllous shrublands and woodlands that persist in winter drought, summer rain climates as well as aseasonal climates with decadal-scale droughts. Noteworthy examples include outcroppings of fynbos in eastern South Africa, the evergreen scrub in the southeastern USA, and sclerophyllous shrublands and woodlands in southeastern Australia. Although these landscapes have similar life forms and functional types adapted to periodic high-intensity crown fires, these are not MTC ecosystems as they typically differ markedly in composition, growth forms and some ecosystem processes.

Likewise, each of the five MTC regions has ecosystems with marked floristic differences and somewhat subtle structural and functional differences. What ties these together is that the landscapes are dominated by sclerophyllous-leaved MTV, and this continuity of fire-prone vegetation increases predictability of fire and regimes dominated by crown fires. Not only does the combination of soil drought and high temperatures greatly enhance flammability, but the expanse of contiguous fuelbeds generated by the mild winters increases the probability of fire spread through communities, and thus predictability of fire for individuals.

Despite the evident structural convergence of all five MTC regions, there are really three different stories with respect to fire. These are the two northern hemisphere systems, the two low soil fertility southern hemisphere systems and the Chilean system.

In the northern hemisphere most of the dominant lineages originated in the early Tertiary in small drought-prone pockets amidst a landscape of mesic forests and woodlands. Seasonal drought necessary for fires was probably not annual and fires could have been as infrequent as roughly once in the life span of these shrubs and small trees and still be of selective importance. These conditions selected for the obligate resprouting mode, suitable for persisting through periodic fires, but a fire regime that did not favor postfire seedling recruitment. Factors responsible for limited selection of postfire seeding would include: fire intensity insufficient to produce widespread resprout mortality (i.e. high adult survivorship) and constraints imposed by drought avoidance strategies of deep root systems. The island-like distribution of these fire-prone pockets selected for plant traits contributing to wider spatial dispersal, including feathery wind-dispersed achenes and fleshy-covered vertebrate-dispersed fruits.

Through a combination of increasing length of drought, and coupling drought with high temperatures, as well as the eroding effect of periodic fires on mesic vegetation, the extent of fire-prone landscape expanded. As islands of fire-prone vegetation expanded within a sea of less flammable woodlands, patch size alone provided feedback by increasing the probability of being struck by lightning and thus increased fire ignitions. In these closed-canopy shrublands and woodlands, high-intensity fire created larger resprout-free gaps. This coupled with the marked shift from closed-canopy resource-limited conditions to more open resource-abundant burned sites increased opportunities for seedling recruitment to the extent that species were selected to delay reproduction to a single postfire pulse. Fire-dependent reproduction selected for niche construction traits such as enhanced flammability created by retention of dead branches in the canopy and other traits. As the predictability of gap formation increased some lineages took the extra hazardous step of eliminating resprouting and becoming obligate seeders. Postfire seeding, including obligate seeding, originated at different times in different lineages and includes apparent origins in the Oligocene, Miocene and Pliocene. Since the late Miocene through a combination of increased drought-prone landscape and fire-favoring adaptations, many lineages have accelerated speciation and contributed to making these MTC regions hotspots of plant diversity.

Southern hemisphere MTC regions share some similarity in the early origins of many contemporary taxa. However, in Western Australia and the Cape region of South Africa, highly leached and infertile soils played a larger role than seasonal drought in the early origins of many sclerophyll taxa. In southwest Australia the early Tertiary sclerophyllous vegetation produced low-growing contiguous expanses of heathlands and this potentially contributed to a much earlier importance for fire. The greater expanse of flammable vegetation would have increased the likelihood of capturing natural ignitions as well as enhancing fire spread and this may have selected for postfire seeders earlier than in northern hemisphere systems. Low-fertility soils in both Australia and South Africa changed the picture in several ways. Fires could more easily drive the replacement of mesic woodlands with fire-prone shrublands due to the longer recovery time of trees on low-fertility sites. In a like manner, low soil fertility would slow resprout recovery and open up sites for seedling recruitment, favoring seeders over obligate resprouters. The absence of the vertebrate-dispersed obligate resprouter mode in southwestern Australian heathlands may be tied to this early development of expansive heathland soils and thus lack of selection for widespread dispersal to isolated habitat islands, as is hypothesized to have been the case in the northern hemisphere. Low-fertility soils, however, produced constraints on reproduction that likely contributed to retention of seeds in serotinous fruits in the canopy and myrmechorous seed dispersal.

Life history characteristics suggest these southern hemisphere communities evolved under much higher fire frequency than their northern hemisphere counterparts. Tolerance to fire return intervals of less than a decade characterize the former but the latter communities require longer intervals. Likewise long fire-free intervals of 50 yrs or more are a threat to the sustainability of serotinous-dominated southern hemisphere heathlands but intervals twice that long are tolerated in northern hemisphere chaparral and matorral.

Higher-fertility soils on Chilean landscapes represent a very different southern hemisphere MTC story. Although lacking much of a fossil record we surmise that early Tertiary evolution of sclerophylls was similar to the history described for the northern hemisphere. As in these northern systems, fire-adaptive traits such as resprouting, lignotubers and postfire seeding would have evolved on marginal sites under various climatic scenarios of seasonality. However, by late Miocene the Andean uplift had blocked summer storms from bringing lightning ignitions into central and northern Chile. This barrier has persisted to this day and explains the presence of certain fire-type traits on a landscape with few natural sources of ignition. Except in the far southeastern corner of the MTC region that still receives regular lightning-ignited fires, Chilean matorral represents an ecosystem in transition from fire-prone to a non-fire-type system. It is conceivable that throughout the geological record other communities have waxed and waned in their fire-prone character due to the cyclical changes in climate and geology.

Another extraordinary convergence feature of MTC regions is the attraction they hold for humans, as all are major contemporary population centers. Today, the combination of moderate winter temperatures and dry summer heat coupled with their coastal setting make these widely attractive environments. Massive high-intensity crown fires are a natural feature of some ecosystems in these regions and when coupled with human populations are a significant threat. However, many aspects of the climate, geology and sociology have combined to produce unique problems in each region. The timing and orientation of high-velocity Santa Ana winds in southern California limit the ability of managers to control fire spread into urban environments. However, the general lack of such conditions in southwest Australia provides greater opportunities for manipulating fire outcomes with fuel treatments. The rural exodus to suburban settings occurring in Spain and other southern European countries in recent decades has removed agricultural controls on wildland fuels and been a significant factor in increased fire size. Anomalously long and severe droughts in southern California and southern Australia are more important factors driving increased fire size in those regions. In the Cape region of South Africa woody plant invasions have contributed to fire problems in recent decades. In southern Chile the widespread replacement of native vegetation with highly flammable pines likewise has contributed to significant fire problems. Thus, despite the remarkable convergence of fire-prone landscapes in all fire regions, human societies have had to adapt to unique features of fires in each region.

References

Aagesen, D. (2004) Burning monkey-puzzle: native fire ecology and forest management in northern Patagonia. *Agriculture and Human Values*, **21**, 233–242.

Abarzúa, A.M. & Moreno, P.I. (2008) Changing fire regimes in the temperate rainforest region of southern Chile over the last 16,000 years. *Quaternary Research*, **69**, 62–71.

Abbott, I. (2003) *Aboriginal fire regimes in south-west Western Australia: evidence from historical documents*. Backhuys Publishers, Leiden, The Netherlands.

Abbott, I. & Burrows, N.D. (2003) *Fire in ecosystems of south-west Western Australia: impacts and management*. Backhuys Publishers, Leiden, The Netherlands.

Abbott, P.L. (1981) Cenozoic paleosols San Diego area, California. *Catena*, **8**, 223–237.

Abdel Malak, D. & Pausas, J.G. (2006) Fire regime and post-fire Normalized Difference Vegetation Index changes in the eastern Iberian peninsula (Mediterranean basin). *International Journal of Wildland Fire*, **15**, 407–413.

Abella, S.R. (2009) Smoke-cued emergence in plant species of ponderosa pine forests: contrasting greenhouse and field results. *Fire Ecology*, **4**, 22–37.

Abella, S.R. & Covington, W.W. (2006) Vegetation–environment relationships and ecological species groups of an Arizona *Pinus ponderosa* landscape, USA. *Plant Ecology*, **185**, 255–268.

Abensperg-Traun, M., Atkins, L., Hobbs, R., & Steven, D. (1998) Exotic plant invasion and understorey species richness: a comparison of two types of eucalypt woodland in agricultural Western Australia. *Pacific Conservation Biology*, **4**, 21–32.

Ackerly, D.D. (2003) Community assembly, niche conservatism, and adaptive evolution in changing environments. *International Journal of Plant Sciences*, **164**, S165–S184.

Ackerly, D.D. (2004a) Adaptation, niche conservatism, and convergence: comparative studies of leaf evolution in the California chaparral. *American Naturalist*, **163**, 654–671.

Ackerly, D.D. (2004b) Functional strategies of chaparral shrubs in relation to seasonal water deficit and disturbance. *Ecological Monographs*, **74**, 25–44.

Ackerly, D.D. (2009) Evolution, origin and age of lineages in the Californian and Mediterranean floras. *Journal of Biogeography*, **36**, 1221–1233.

Adam, P., Stricker, P., & Anderson, D.J. (1989) Species-richness and soil phosphorus in plant communities in coastal New South Wales. *Austral Ecology*, **14**, 189–198.

Agee, J.K. (1991) Fire history along an elevational gradient in the Siskiyou Mountains, Oregon. *Northwest Science*, **65**, 188–199.

Agee, J.K. (1993) *Fire ecology of Pacific Northwest forests*. Island Press, Covelo, California.

Agee, J.K., Bahro, B., Finney, M.A., et al. (2000) The use of shaded fuel breaks in landscape fire management. *Forest Ecology and Management*, **127**, 55–66.

Alados, C.L., ElAich, A., Papanastasis, V.P., *et al.* (2004) Change in plant spatial patterns and diversity along the successional gradient of Mediterranean grazing ecosystems. *Ecological Modelling*, **180**, 523–535.

Albini, F.A. (1979) *Spot fire distance from burning trees: a predictive model*. USDA Forest Service, Intermountain Forest and Range Experiment Station, Ogden, Utah.

Alexander, J.M. & D'Antonio, C.M. (2003) Seed bank dynamics of French broom in coastal California grasslands: effects of stand age and prescribed burning on control and restoration. *Conservation Biology*, **11**, 185–197.

Alexander, M.E. (1982) Calculating and interpreting forest fire intensities. *Canadian Journal of Botany*, **60**, 349–357.

Allan, G.E. & Southgate, R.I. (2002) Fire regimes in the spinifex landscapes of Australia. In *Flammable Australia: the fire regimes and biodiversity of a continent* (eds R.A. Bradstock, J.E. Williams & A.M. Gill). Cambridge University Press, Cambridge.

Allen, C.D., Savage, M., Falk, D.A., *et al.* (2002) Ecological restoration of southwestern ponderosa pine ecosystems: a broad perspective. *Ecological Applications*, **12**, 1418–1433.

Allen, E.B., Eliason, S.A., Marquez, V.J., *et al.* (2000) What are the limits to restoration of coastal sage scrub in southern California? In *2nd Interface between Ecology and Development in California* (eds J.E. Keeley, M.B. Keeley & C.J. Fotheringham), pp. 253–262. U.S. Geological Survey, Sacramento, California.

Allen, P. (1998) Purbeck-Wealden (early Cretaceous) climates. *Proceedings of the Geologists' Association*, **109**, 197–236.

Allen-Diaz, B., Standiford, R., & Jackson, R.D. (2007) Oak woodlands and forests. In *Terrestrial vegetation of California*, third edition (eds M.G. Barbour, T. Keeler-Wolf & A.A. Schoenherr), pp. 313–338. University of California Press, Los Angeles, California.

Álvarez, M. & Maisterrena, J. (1977) Climatological and meteorological characteristics of the Observatorio Astronomico Nacional at San Pedro Mártir, B. C. *Revista Mexicana de Astronomía y Astrofísica*, **2**, 43–52.

Amigo, J. & Ramirez, C. (1998) A bioclimatic classification of Chile: woodland communities in the temperate zone. *Plant Ecology*, **136**, 9–26.

Andersen, A. (1982) Seed removal by ants in the mallee of northwestern Victoria. In *Ant-plant interactions in Australia* (ed. R.C. Buckley), pp. 31–44. Dr. W. Junk, The Hague, The Netherlands.

Andersen, A.N. (1989a) Pre-dispersal seed losses to insects in species of *Leptospermum* (Myrtaceae). *Australian Journal of Ecology*, **14**, 13–18.

Andersen, A.N. (1989b) Impact of insect predation on ovule survivorship in *Eucalyptus baxteri* (Myrtaceae). *Journal of Ecology*, **72**, 62–69.

Andersen, A.N. & Yen, A.L. (1985) Immediate effects of fire on ants in the semi-arid mallee region of north-western Victoria. *Australian Journal of Ecology*, **10**, 25–30.

Anderson, M.K. (2006) The use of fire by Native Americans in California. In *Fire in California's Ecosystems* (eds N.G. Sugihara, J.W. van Wagtendonk, K.E. Shaffer, J. Fites-Kaufman & A.E. Thoede), pp. 417–430. University of California Press, Los Angeles.

Anderson, M.K. & Moratto, M.J. (1996) Native American land-use practices and ecological impacts. In *Sierra Nevada Ecosystem Project. Final Report to Congress. Status of the Sierra Nevada*, Vol. II. *Assessments and Scientific Basis for Management Options*, pp. 187–206. Centers for Water and Wildland Resources, University of California, Davis.

Anderson, R.S. (1989) Development of the southwestern ponderosa pine forests: what do we really know? In *Multiresource management of ponderosa pine forests* (eds A. Tecle,

W.W. Covington & R.H. Hamre). General Technical Report RM-GTR-185, USDS Forest Service, Missoula, Montana.

Anderson, S.A.J. & Anderson, W.R. (2010) Ignition and fire spread thresholds in gorse (*Ulex europaeus*). *International Journal of Wildland Fire*, **19**, 589–598.

Andrews, P.L. (1986) *BEHAVE: fire behavior prediction and fuel modeling system: burn subsystem, Part 1*. USDA Forest Service, Intermountain Forest and Range Experiment Station, Ogden, Utah.

Andrews, P.L., Lofsgaarden, D.O., & Bradshaw, L.S. (2003) Evaluation of fire danger rating indexes using logistic regression and percentile analysis. *International Journal of Wildland Fire*, **12**, 213–226.

Aravena, J.C., Lequesne, C., Jimenez, H., *et al.* (2003) Fire history in central Chile: tree-ring evidence and modern records. In *Fire and climatic change in temperate ecosystems of the western Americas* (ed. W.L. Baker, G. Montenegro, & T.W. Swetnam), pp. 343–356. Springer Verlag, New York.

Araya, S. & Ávila, G. (1981) Rebrote de arbustos afectados por fuego en el matorral chileno. *Anales del Museo de Historia Natural de Valparaiso*, **14**, 107–113.

Archibald, S., Roy, D.P., van Wilgen, B.W., & Scholes, R.J. (2009) What limits fire? An examination of drivers of burnt area in Southern Africa. *Global Change Biology*, **15**, 613–630.

Arianoutsou, M., Delipetrou, P., Celesti-Grapow, L., *et al.* (2010) Comparing naturalized alien plants and recipient habitats across an east-west gradient in the Mediterranean Basin. *Journal of Biogeography*, **37**, 1811–1823.

Arianoutsou, M., Kazanis, D., Kokkoris, Y., & Shourou, P. (2002) Land-use interactions with fire in Mediterranean *Pinus halepensis* landscapes of Greece: patterns of biodiversity. In *Forest Fire Research & Wildland Fire Safety* (ed. D.X. Viegas). Millpress, Rotterdam, The Netherlands.

Arianoutsou, M. & Ne'eman, G. (2000) Post-fire regeneration of natural *Pinus halepensis* forest in the east Mediterranean basin. In *Ecology, biogeography and management of* Pinus halepensis *and P. brutia forest ecosystems in the Mediterranean basin* (eds G. Ne'eman & L. Trabaud), pp. 269–289. Backhuys Publishers, Leiden, The Netherlands.

Arianoutsou-Faraggitaki, M. & Margaris, N.S. (1982) Phryganic (east Mediterranean) ecosystems and fire. *Ecologia Mediterranea*, **VIII**, 473–480.

Arienti, M.C., Cumming, S.G., Krawchuk, M.A., & Boutin, S. (2009) Road network density correlated with increased lightning fire incidence in the Canadian western boreal forest. *International Journal of Wildland Fire*, **18**, 970–982.

Armesto, J.J., Arroyo, M.K., & Villagrán, C. (1980) Altitudinal distribution, cover and size structure of umbelliferous cushion plants in the high Andes of Central Chile. *Acta Oecologica*, **1**, 327–332.

Armesto, J.J., Arroyo, M.T.K., & Hinojosa, L.F. (2007) The mediterranean environment of central Chile. In *The physical geography of South America* (eds T.T. Veblen, K.R. Young & A.R. Orme), pp. 184–199. Oxford University Press, Oxford, UK.

Armesto, J.J. & Gutiérrez, J.R. (1978) El efecto del fuego en la estructura de la vegetación de Chile central. *Anales del Museo de Historia Natural de Valparaiso*, **11**, 43–48.

Armesto, J.J., Manuschevich, D., Mora, A., *et al.* (2010) From the Holocene to the Anthropocene: a historical framework for land cover change in southwestern South America in the past 15,000 years. *Land Use Policy*, **27**, 148–160.

Armesto, J.J. & Pickett, S.T.A. (1985) A mechanistic approach to the study of succession in the Chilean matorral. *Revista Chilena de Historia Natural*, **58**, 9–17.

Arnold, K., Burcham, L.T., Fenner, R.L., & Grah, R.F. (1951) Use of fire in land clearing. *California Agriculture*, **5**(3), 9–11; **5**(4), 4–5,13,15; **5**(5), 11–12; **5**(6), 13–15; **5**(7), 6,15.

Aronne, G. & De Micco, V. (2001) Seasonal dimorphism in the Mediterranean *Cistus incanus* L. subsp. *incanus*. *Annals of Botany*, **87**, 789–794.

Aronson, J., Pereira, J.S., & Pausas, J.G., eds. (2009) *Cork oak woodlands on the edge: ecology, adaptive management, and restoration*. Island Press, Washington, DC.

Aronson, J. & Shmida, A. (1992) Plant species diversity along a Mediterranean-desert gradient and its correlation with interannual rainfall fluctuations. *Journal of Arid Environments*, **23**, 235–247.

Arroyo, M.T.K., Cavieres, L., Marticorena, C., & Munoz-Schick, M. (1995) Convergence in the mediterranean floras in central Chile and California: insights from comparative biogeography. In *Ecology and biogeography of mediterranean ecosystems in Chile, California, and Australia* (eds M.T.K. Arroyo, P.H. Zedler & M.D. Fox), pp. 43–88. Springer, New York.

Aschmann, H. & Bahre, C. (1977) Man's impact on the wild landscape. In *Convergent evolution of Chile and California mediterranean climate ecosystems* (ed. H.A. Mooney), pp. 73–84. Dowden, Hutchinson and Ross, Stroudsburg, Pennsylvania.

Auld, T.D. (1986) Population dynamics of the shrub *Acacia suaveolens* (Sm.) Willd.: fire and the transition to seedlings. *Austral Ecology*, **11**, 373–385.

Auld, T.D. (1987) Population dynamics of the shrub *Acacia suaveolens* (Sm.) Willd.: survivorship throughout the life cycle, a synthesis. *Austral Ecology*, **12**, 139–151.

Auld, T.D. (1994) The role of soil seedbanks in maintaining plant communities in fire-prone habitats. *Proceedings of the Second International Conference on Forest Fire Research*, 1069–1078.

Auld, T.D. & Denham, A.J. (2006) How much seed remains in the soil after a fire? *Plant Ecology*, **187**, 15–24.

Auld, T.D., Keith, D.A., & Bradstock, R.A. (2000) Patterns in longevity of soil seedbanks in fire-prone communities of south-eastern Australia. *Australian Journal of Botany*, **48**, 539–548.

Auld, T.D. & O'Connell, M.A. (1991) Predicting patterns of post-fire germination in 25 eastern Australian Fabaceae. *Australian Journal of Ecology*, **16**, 53–70.

Auld, T.D. & Ooi, M.K.J. (2008) Applying seed germination studies in fire management for biodiversity conservation in south eastern Australia. *Web Ecology*, **8**, 47–54.

Ávila, G., Aljaro, M.E., & Silva, B. (1981) Observaciones en el estrato herbáceo después del fuego. *Anales del Museo de Historia Natural, Valparaíso*, **14**, 99–105.

Ávila, G., Montenegro, G., & Aljaro, M.E. (1988) Incendios en la vegetación mediterranea. In *Ecologia del pasaje en Chile central: estudios sobre sus espacios montanosos* (eds E.R. Fuentes & S. Prenafeta). Ediciones Universidad Católica de Chile, Santiago.

Axelrod, D.I. (1939) *A Miocene flora from the western border of the Mohave Desert,*. Publication 516. Carnegie Institution of Washington, Washington, DC.

Axelrod, D.I. (1950) *The Piru Gorge flora of southern California*. Contributions to Paleontology Publication 590. Carnegie Institution of Washington, Washington, DC.

Axelrod, D.I. (1973) History of the Mediterranean ecosystem in California. In *Mediterranean ecosystems: origin and structure* (eds F. di Castri & H.A. Mooney), pp. 225–277. Springer, New York.

Axelrod, D.I. (1975) Evolution and biogeography of Madrean-Tethyan sclerophyll vegetation. *Annals of the Missouri Botanical Garden*, **62**, 280–334.

Axelrod, D.I. (1980) History of the maritime closed-cone pines, Alta and Baja California. *University of California Publications in Geological Sciences*, **120**, 1–143.

Axelrod, D.I. (1982) Vegetation and climate during middle Ridge Basin deposition. In *Geologic history of Ridge Basin, southern California* (eds J.C. Crowell & M.H. Link), pp. 247–251. Society of Economic Paleontologists and Mineralogists, Pacific Section, Los Angeles, California.

Axelrod, D.I. (1985) Miocene floras from the Middle Basin, west-central Nevada. *University of California Publications in Geological Sciences*, **129**, 1–279.

Axelrod, D.I. (1987) The Late Oligocene Creede flora, Colorado. *University of California Publications in Geological Sciences*, **130**, 1–235.

Axelrod, D.I. (1988) Paleoecology of late Pleistocene Monterey pine at Laguna Niguel, southern California. *Botanical Gazette*, **149**, 458–464.

Axelrod, D.I. (1989) Age and origin of chaparral. In *The California chaparral: paradigms reexamined* (ed. S.C. Keeley), Science Series 34, pp. 7–19. Natural History Museum of Los Angeles County, Los Angeles, California.

Axelrod, D.I., Arroyo, M.T.K., & Raven, P.H. (1991) Historical development of temperate vegetation in the Americas. *Revista Chilena de Historia Natural*, **64**, 413–446.

Axelrod, D.I. & Deméré, T.A. (1984) A Pliocene flora from Chula Vista, San Diego County, California. *Transactions of the San Diego Society of Natural History*, **20**, 277–300.

Axelrod, D.I. & Raven, P.H. (1978) Late Cretaceous and Tertiary vegetation history of Africa. In *The biogeography and ecology of southern Africa* (ed. M.J.A. Werger), pp. 77–130. Dr. W. Junk, The Hague, The Netherlands.

Axelrod, D.I. & Schorn, H.E. (1994) The 15 Ma floristic crisis at Gillam Spring, Washoe County, Northwestern Nevada. *PaleoBios*, **16**, 1–10.

Baeza, M.J., de Luis, M., Raventós, J., & Escarré, A. (2002) Factors influencing fire behaviour in shrublands of different stand ages and the implications for using prescribed burning to reduce wildfire risk. *Journal of Environmental Management*, **65**, 199–208.

Baeza, M.J., Raventós, J., Escarré, A., & Vallejo, V.R. (2006) Fire risk and vegetation structural dynamics in Mediterranean shrubland. *Plant Ecology*, **187**, 189–201.

Baeza, M.J., Santana, V.M., Pausas, J.G., et al. (2011) Successional trends in standing dead biomass in Mediterranean basin species. *Journal of Vegetation Science*, **22**, 467–474.

Bahre, C.J. (1979) Destruction of the natural vegetation of north-central Chile. *University of California Publications in Geography*, **23**, 1–117.

Baird, A.M. (1977) Regeneration after fire in King's Park, Perth, Western Australia. *Journal of the Royal Society of Western Australia*, **60**, 1–22.

Bajocco, S., Pezzatti, G.B., Mazzoleni, S., & Ricotta, C. (2010) Wildfire seasonality and land use: when do wildfires prefer to burn? *Environmental Monitoring and Assessment*, **164**, 445–452.

Baker, W.L. & Ehle, D.S. (2001) Uncertainty in surface-fire history: the case of ponderosa pine forests in the western United States. *Canadian Journal of Forest Research*, **13**, 1205–1226.

Baldwin, I.T., Staszakkozinski, L., & Davidson, R. (1994) Up in smoke. 1. Smoke-derived germination cues for postfire annual, *Nicotiana attenuata* Torr ex Watson. *Journal of Chemical Ecology*, **20**, 2345–2372.

Balfour, D.A. & Howison, O.E. (2002) Spatial and temporal variation in a mesic savanna fire regime: responses to variation in annual rainfall. *African Journal of Range & Forage Science*, **19**, 45–53.

Bands, D.P. (1977) Prescribed burning in Cape fynbos catchments. In *Proceedings of the symposium on environmental consequences of fire and fuel management in Mediterranean ecosystems* (eds H.A. Mooney & C.E. Conrad), General Technical Report WO-3, pp. 245–256. USDA Forest Service, Washington, DC.

Bangert, R. & Huntly, N. (2010) The distribution of native and exotic plants in a naturally fragmented sagebrush-steppe landscape. *Biological Invasions*, **12**, 1627–1640.

Banks, J.C.G. (1989) A history of forest fire in the Australian Alps. In *The scientific significance of the Australian Alps* (ed. R. Good). Australian Academy of Science & AALC, Canberra.

Barbour, M.G. (1987) Community ecology and distribution of California hardwood forests and woodlands. In *Proceedings of the symposium on multiple-use management of California's hardwood resources* (eds T.R. Plumb & N.H. Pillsbury), General Technical Report PSW-100, pp. 18–25. USDA Forest Service, Pacific Southwest Forest and Range Experiment Station, Albany, California.

Barbour, M.G. (2007) Closed-cone pine and cypress forests. In *Terrestrial vegetation of California*, third edition (eds M.G. Barbour, T. Keeler-Wolf & A.A. Schoenherr), pp. 296–312. University of California Press, Los Angeles, California.

Barbour, M.G. & Minnich, R.A. (1990) The myth of chaparral convergence. *Israel Journal of Botany*, **39**, 453–463.

Barker, N.P., Weston, P.H., Rutschmann, F., & Sauquet, H. (2007) Molecular dating of the Gondwanan plant family Proteaceae is only partially congruent with the timing of the break-up of Gondwana. *Journal of Biogeography*, **34**, 2012–2027.

Barlow, B.A. (1994) Phytogeography of the Australian region. In *Australian vegetation* (ed. R.H. Groves), pp. 2–36. Cambridge University Press, Cambridge, UK.

Barrett, J.W., McDonald, P.M., F. Ronco, J., & Russell, A. (1980) Interior ponderosa pine. In *Forest cover types of the United States and Canada* (ed. F.H. Eyer), pp. 114–115. USDA, U.S. Forest Service, Washington, DC.

Barrón, E., Rivas-Carballo, R., Postigo-Miljarra, J.M., *et al.* (2010) The Cenozoic vegetation of the Iberian Peninsula: a synthesis. *Review of Palaeobotany and Palynology*, doi:10.1016/j.revpalbo.2009.11.007.

Barrow, S. (1998) A revision of Phoenix. *Kew Bulletin*, **53**, 513–575.

Barry, J.P. (1989) Reproductive response of a marine annelid to winter storms: an analog to fire adaptation in plants? *Marine Ecology Progress Series*, **54**, 99–107.

Bartolome, J.W., Barry, W.J., Griggs, T., & Hopkinson, P. (2007) Valley grassland. In *Terrestrial vegetation of California*, third edition (eds M.G. Barbour, T. Keeler-Wolf & A.A. Schoenherr), pp. 367–393. University of California Press, Los Angeles, California.

Basanta, A., Vizcaino, E.D., Casal, M., & Morey, M. (1989) Diversity measurements in shrubland communities of Galicia (NW Spain). *Vegetatio*, **82**, 105–112.

Bassi, S., Kettunen, M., & Cavalieri, S. (2008) *Forest fires: causes and contributing factors in Europe.* IP/A/ENVI/ST/2007-15 PE 401.003. Policy Department, Economic and Scientific Policy, European Parliament, Brussels, Belgium.

Batten, D.J. (2002) Palaeoenvironmental setting of the Purbeck limestone group of Dorset, Southern England. In *Life and environments in Purbeck times* (eds A.R. Milner & D.J. Batten), pp. 13–20. The Palaeontological Association, Blackwell Publishing, London.

Battisti, A., Cantini, R., Rouault, G., & Roques, A. (2003) Serotinous cones of *Cupressus sempervirens* provide viable seeds in spite of high seed predation. *Annals of Forest Science*, **60**, 781–787.

Bauer, H.L. (1930) On the flora of the Tehachapi Mountains, California. *Bulletin of the Southern California Academy of Sciences*, **29**, 96–99.

Beadle, N.C.W. (1940) Soil temperatures during forest fires and their effect on the survival of vegetation. *Journal of Ecology*, **28**, 180–192.

Beadle, N.C.W. (1962) Soil phosphate and the delimitation on plant communities in Eastern Australia, II. *Ecology*, **43**, 281–288.

Beadle, N.C.W. (1981) *The vegetation of Australia*. Cambridge University Press, Cambridge, UK.

Beard, J.S. (1977) Tertiary evolution of the Australian flora in the light of latitudinal movements of the continent. *Journal of Biogeography*, **4**, 111–118.

Beard, J.S. (1984) Biogeography of the Kwongan. In *Kwongan plant life of the sandplain: biology of a south-west Australian shrubland ecosystem* (eds J.S. Pate & J.S. Beard), pp. 1–26. University of Western Australia Press, Nedlands, Western Australia.

Beerling, D.J., Woodward, F.I., Lomas, M.R., Wills, M.A., Quick, W.P., & Valdes, P.J. (1998) The influence of Carboniferous palaeo-atmospheres on plant function: an experimetnal and modelling assessment. *Philosophical Transactions of the Royal Society of London, Series B*, **353**, 131–140.

Belcher, C.M., Mander, L., Rein, G., et al. (2010) Increased fire activity at the Triassic/Jurassic boundary in Greenland due to climate-driven floral change. *Nature Geoscience*, **3**, 426–429.

Bell, D.T. (1999) The process of germination in Australian species. *Australian Journal of Botany*, **47**, 475–517.

Bell, D.T. (2001) Ecological response syndromes in the flora of southwestern Western Australia: fire resprouters versus reseeders. *Botanical Review*, **67**, 417–433.

Bell, D.T., Hopkins, A.J.M., & Pate, J.S. (1984) Fire in the kwongan. In *Kwongan plant life of the sandplain: biology of a south-west Australian shrubland ecosystem* (eds J.S. Pate & J.S. Beard), pp. 178–204. University of Western Australia Press, Nedlands, Western Australia.

Bell, D.T. & Koch, J.M. (1980) Post-fire succession in the northern jarrah forest of Western Australia. *Australian Journal of Ecology*, **5**, 9–14.

Bell, D.T. & Loneragan, W.A. (1985) The relationship of fire and soil type to floristic patterns within heathland vegetation near Badgingarra, Western Australia. *Journal of the Royal Society of Western Australia*, **67**, 98–109.

Bell, D.T., McCaw, W.L., & Burrows, N.D. (1989) Influence of fire on jarrah forest vegetation. In *The jarrah forest:– a complex Mediterranean ecosystem* (eds B. Dell, J.J. Havel & N. Malajczuk), pp. 203–215. Kluwer Academic Press, Dordrecht, The Netherlands.

Bell, D.T., Moezel, P.G., van der Delfs, J.C., & Loneragan, W.A. (1987) Northern sandplain kwongan: effect of fire on *Hakea obliqua* and *Beaufortia elegans* population structure. *Journal of the Royal Society of Western Australia*, **69**, 139–143.

Bell, D.T., Plummer, J.A., & Taylor, S.K. (1993) Seed germination ecology in southwestern Western Australia. *Botanical Review*, **59**, 24–73.

Bell, D.T. & Williams, J.E. (1997) Eucalypt ecophysiology. In *Eucalypt ecology: individuals to ecosystems* (eds J.E. Williams & J.C.Z. Woinarski), pp. 168–196. Cambridge University Press, Cambridge, UK.

Bell, T.L. & Ojeda, F. (1999) Underground starch storage in *Erica* species of the Cape Floristic Region – differences between seeders and resprouters. *New Phytologist*, **144**, 143–152.

Bell, T.L. & Pate, J.S. (1996) Growth and fire response of selected Epacridaceae of southwestern Australia. *Australian Journal of Botany*, **44**, 509–526.

Bellingham, P.J. & Sparrow, A.D. (2000) Resprouting as a life history strategy in woody plant communities. *Oikos*, **89**, 409–416.

Bellingham, P.J., Tanner, E.V.J., & Healey, J.R. (1994) Sprouting of trees in Jamaican montane forests, after a hurricane. *Journal of Ecology*, **82**, 747–758.

Bendix, J. (1997) Flood disturbance and the distribution of riparian species diversity. *Geographical Review*, **87**, 468–483.

Bennett, K.D. (1990) Milankovitch cycles and their effects on species in ecological and evolutionary time. *Paleobiology*, **16**, 11–21.

Bennett, K.D. (2004) Continuing the debate on the role of Quaternary environmental change for macroevolution. *Philosophical Transactions of the Royal Society of London, Series B*, **359**, 295–303.

Bennett, L.T. (1994) The expansion of *Leptospermum laevigatum* on the Yanakie Isthmus, Wilson's Promontory, under changes in the burning and grazing regimes. *Australian Journal of Botany*, **42**, 555–564.

Benwell, A.S. (1998) Post-fire seedling recruitment in coastal heathland in relation to regeneration strategy and habitat. *Australian Journal of Botany*, **46**, 75–101.

Beyers, J.L. (2004) Postfire seeding for erosion control: effectiveness and impacts. *Conservation Biology*, **18**, 947–956.

Bhaskar, R., Valiente-Benuet, A., & Ackerly, D.D. (2007) Evolution of hydraulic traits in closely related species pairs from mediterranean and nonmediterranean environments of North America. *New Phytologist*, **176**, 718–726.

Bilgili, E. & Saglam, B. (2003) Fire behavior in maquis fuels in Turkey. *Forest Ecology and Management*, **184**, 201–207.

Black, M.P. & Mooney, S.D. (2006) Holocene fire history from the Greater Blue Mountains World Heritage Area, New South Wales, Australia: the climate, humans and fire nexus. *Regional Environmental Change*, **6**, 41–51.

Black, M.P., Mooney, S.D., & Attenbrow, V. (2008) Implications of a 14 200 year contiguous fire record for understanding human-climate relationships at Goochs Swamp, New South Wales, Australia. *Holocene*, **18**, 437–447.

Blackburn, D.T. & Sluiter, I.R.K. (1994) The Oligo-Miocene coal floras of southeastern Australia. In *Australian vegetation: Cretaceous to recent* (ed. R.S. Hill), pp. 328–367. Cambridge University Press, Cambridge, UK.

Blackburn, G. & Wright, M.J. (1989) Soils. In *Mediterranean landscapes in Australia: mallee ecosystems and their management* (eds J.C. Noble & R.A. Bradstock), pp. 35–54. CSIRO Publications, Victoria, Australia.

Blier, W. (1998) The Sundowner winds of Santa Barbara, California. *Weather and Forecasting*, **13**, 702–716.

Blondel, J. (1991) Assessing convergence at the community-wide level. *Trends in Ecology and Evolution*, **6**, 271–272.

Blondel, J. (2006) The "design" of Mediterranean landscapes: a millennial story of humans and ecological systems during the historic period. *Human Ecology*, **34**, 713–729.

Blondel, J. & Aronson, J. (1999) *Biology and wildlife of the Mediterranean region*. Oxford University Press, Oxford, UK.

Bloom, K.J. & Watson, E.B. (2006) The Arbuckle-Hershey chamisal (*Adenostoma fasciculatum*): a significant anomaly in California plant geography. *Madroño*, **53**, 275–277.

Boatwright, J.S., Savolainen, V., van Wyk, E.B., *et al.* (2008) Systematic position of the anomalous genus *Cadia* and the phylogeny of the tribe Podalyrieae (Fabaceae). *Systematic Botany*, **33**, 133–147.

Boer, M.M., Sadler, R.J., Wittkuhn, R.S., McCaw, L., & Grierson, P.F. (2009) Long-term impacts of prescribed burning on regional extent and incidence of wildfires – evidence from 50 years of active fire management in SW Australian forests. *Forest Ecology and Management*, **259**, 132–142.

Bond, W. & Breytenbach, G.J. (1985) Ants, rodents, and seed predation in Proteaceae. *South African Journal of Zoology*, **20**, 150–154.

Bond, W.J. (1980) Fire and senescent fynbos in the Swartberg, southern Cape. *South African Forestry Journal*, **114**, 68–71.

Bond, W.J. (1983) On alpha diversity and the richness of the Cape flora: a study in southern Cape fynbos. In *Mediterranean-type ecosystems: the role of nutrients* (eds F.J. Kruger, D.T. Mitchell & J.U.M. Jarvis), pp. 337–356. Springer, New York.

Bond, W.J. (1984) Fire survival of Cape Proteaeceae – influence of fire season and seed predators. *Vegetatio*, **56**, 65–74.

Bond, W.J. (1985) Canopy stored seed reserves (serotiny) in Cape Proteaceae. *South African Journal of Botany*, **51**, 181–186.

Bond, W.J. (1988) Proteas as "tumbleseeds": wind dispersal through air and over soil. *South African Journal of Botany*, **54**, 455–460.

Bond, W.J. (1989) The tortoise and the hare: ecology of angiosperm dominance and gymnosperm persistence. *Biological Journal of the Linnean Society*, **36**, 227–249.

Bond, W.J. (1997) Fire. In *Vegetation of Southern Africa* (eds R.M. Cowling, D.M. Richardson & S.M. Pierce), pp. 421–446. Cambridge University Press, Cambridge, UK.

Bond, W.J. (2010) Do nutrient-poor soils inhibit development of forests? A nutrient stock analysis. *Plant and Soil*, **334**, 47–60.

Bond, W.J. & Archibald, S. (2003) Confronting complexity: fire policy choices in South African savanna parks. *International Journal of Wildland Fire*, **12**, 381–389.

Bond, W.J., Honig, M., & Maze, K.E. (1999) Seed size and seedling emergence: an allometric relationship and some ecological implications. *Oecologia*, **120**, 132–136.

Bond, W.J. & Keeley, J.E. (2005) Fire as a global "herbivore": the ecology and evolution of flammable ecosystems. *Trends in Ecology and Evolution*, **20**, 387–394.

Bond, W.J. & Ladd, P.G. (2001) Dynamics of the overstorey and species richness in Australian heathlands. *Journal of Mediterranean Ecology*, **2**, 247–257.

Bond, W.J., Le Roux, D., & Erntzen, R. (1990) Fire intensity and regeneration of myrmecochorous Proteaceae. *South African Journal of Botany*, **56**, 326–330.

Bond, W.J., Maze, K.E., & Desmet, P.G. (1995) Fire life histories and the seeds of chaos. *Ecoscience*, **2**, 252–260.

Bond, W.J., Midgley, G.F., & Woodward, F.I. (2003) What controls South African vegetation, climate or fire? *South African Journal of Botany*, **69**, 79–91.

Bond, W.J. & Midgley, J.J. (1995) Kill thy neighbour: an individualistic argument for the evolution of flammability. *Oikos*, **73**, 79–85.

Bond, W.J. & Midgley, J.J. (2001) Ecology of sprouting in woody plants: the persistence niche. *Trends in Ecology and Evolution*, **16**, 45–51.

Bond, W.J. & Midgley, J.J. (2003) The evolutionary ecology of sprouting in woody plants. *International Journal of Plant Science*, **164**, 103–114.

Bond, W.J. & Scott, A.C. (2010) Fire and the spread of flowering plants in the Cretaceous. *New Phytologist*, **188**, 1137–1150.

Bond, W.J. & Slingsby, P. (1983) Seed dispersal by ants in shrublands of the Cape Province and its evolutionary implications. *South African Journal of Science*, **79**, 231–233.

Bond, W.J. & Stock, W.D. (1989) The costs of leaving home: ants disperse myrmecochorous seeds to low nutrient sites. *Oecologia*, **81**, 412–417.

Bond, W.J. & van Wilgen, B.W. (1996) *Fire and plants*. Chapman & Hall, New York.

Bond, W.J., Vlok, J., & Viviers, M. (1984) Variation in seedling recruitment of Cape Proteaceae after fire. *Journal of Ecology*, **72**, 209–221.

Bond, W.J., Woodward, F.I., & Midgley, G.F. (2005) The global distribution of ecosystems in a world without fire. *New Phytologist*, **165**, 525–538.

Bonet, A. & Pausas, J.G. (2004) Species richness and cover along a 60-year chronosequence in old-fields of southeastern Spain. *Plant Ecology*, **174**, 257–270.

Bonet, A. & Pausas, J.G. (2007) Old field dynamics on the dry side of the Mediterranean Basin: patterns and processes in semiarid SE Spain. In *Old fields: dynamics and restoration of abandoned farmland* (eds V.A. Cramer & R.J. Hobbs), pp. 247–264. Island Press, Washington, DC.

Bonnet, V.H. & Tatoni, T. (2004) Ephemeral establishment of therophytic, ruderal and wind-dispersed species after fire (Marseille, France). In *Ecology, conservation and management of mediterranean climate ecosystems* (eds M. Arianoutsou & V.P. Papanastasis), pp. 1–12. Millpress, Rotterdam, The Netherlands.

Borchert, M. & Tyler, C.M. (2009) Patterns of post-fire flowering and fruiting in *Chlorogalum pomeridianum* var. *pomeridianum* (DC.) Kunth in southern California chaparral. *International Journal of Wildland Fire*, **18**, 623–630.

Botelho, H., Rego, F.C., & Ryan, K.C. (1998) Tree mortality models for *Pinus pinaster* of northern Portugal. In *Proceedings of the 13th Conference on Fire and Forest Meteorology*, pp. 235–240. International Association of Wildland Fire, Fairfield, Washington.

Boucher, C. & Moll, E.J. (1981) South African Mediterranean shrublands. In *Ecosystems of the world*, Vol. 11. *Mediterranean-type shrublands* (eds F. di Castri, D.W. Goodall & R.L. Specht), pp. 233–267. Elsevier Scientific, New York.

Bovio, G. & Camia, A. (1997) Land zoning based on fire history. *International Journal of Wildland Fire*, **7**, 249–258.

Bowman, D. (2003) Wild, tame and feral fire: the fundamental linkage between indigenous fire usage and the conservation of Australian biodiversity. In *3rd international wildland fire conference and exhibition, urban and rural communities living in fire prone environments: managing the future of global problems*, October 3–6, 2003, Sydney Convention and Exhibition Center, Sydney, Australia (CD ROM Proceedings). USDI Bureau of Land Management, Washington, DC.

Bowman, D.M.J.S. (2000) *Australian rainforests: islands of green in a land of fire*. Cambridge University Press, Cambridge, UK.

Bowman, D.M.J.S., Balch, J.K., Artaxo, P., *et al.* (2009) Fire in the earth system. *Science*, **324**, 481–484.

Bowman, D.M.J.S. & Harris, S. (1994) Conifers of Australia's dry forests and open woodlands. In *Ecology of the southern conifers* (eds N.J. Enright & R.S. Hill), pp. 252–270. Melbourne University Press, Carlton, Victoria, Australia.

Boydak, M. (1985) The distribution of *Phoenix theophrasti* in the Datça Peninsula, Turkey. *Biological Conservation*, **32**, 129–135.

Boydak, M., Isik, F., & Dogan, B. (1998) The effect of prescribed fire on the natural regeneration success of Lebanon ceder (*Cedrus libani* A. Rich.) at Antalya-Kas locality. *Turkish Journal of Agriculture and Forestry*, **22**, 399–404.

Brachert, T.C., Reuter, M., Kroeger, K.F. & Lough, J.M. (2006) Coral growth bands: a new and easy to use paleothermometer in paleoenvironment analysis and paeloceanography (late Miocene, Greece). *Paleoceanography*, **21**, doi:10.1029/2006PA001288.

Bradstock, R.A. (1989) Dynamics of a perennial understorey. In *Mediterranean landscapes in Australia: mallee ecosystems and their management* (eds J.C. Noble & R.A. Bradstock), pp. 141–155. CSIRO Publications, Victoria, Australia.

Bradstock, R.A. (1990) Demography of woody plants in relation to fire: *Banksia serrata* Lf. and *Isopogon anemonifolius* (Salisb.) Knight. *Australian Journal of Ecology*, **15**, 117–132.

Bradstock, R.A. (1991) The role of fire in establishment of seedlings of serotinous species from the Sydney region. *Australian Journal of Botany*, **39**, 347–356.

Bradstock, R.A. (1995) Demography of woody plants in relation to fire: *Telopea speciosissima*. *Proceedings of the Linnean Society of New South Wales*, **115**, 25–33.

Bradstock, R.A. (2008) Effects of large fires on biodiversity in south-eastern Australia: disaster or template for diversity? *International Journal of Wildland Fire*, **17**, 809–822.

Bradstock, R.A. (2010) A biogeographic model of fire regimes in Australia: current and future implications. *Global Ecology and Biogeography*, **19**, 145–158.

Bradstock, R.A. & Auld, T.D. (1995) Soil temperatures during experimental bushfires in relation to fire intensity: consequences for legume germination and fire management in south-eastern Australia. *Journal of Applied Ecology*, **32**, 76–84.

Bradstock, R.A., Bedward, M., & Cohn, J.S. (1998c) Weather, ignition and fuel as determinants of fire regimes: investigation of limits to management using a simple, spatial modelling approach. In *III International Conference on Forest Fire Research: 14th Conference on Fire and Forest Meteorology* (ed. D.X. Viegas), Vol. 2, pp. 2365–2378. ADAI, Luso and Coimbra, Portugal.

Bradstock, R.A., Bedward, M., & Cohn, J.S. (2006) The modelled effects of differing fire management strategies on the conifer *Callitris verrucosa* within semi-arid mallee vegetation in Australia. *Journal of Applied Ecology*, **43**, 281–292.

Bradstock, R.A., Bedward, M., Gill, A.M., & Cohn, J.S. (2005) Which mosaic? A landscape ecological approach for evaluating interactions between fire regimes, habitat and animals. *Wildlife Research*, **32**, 409–423.

Bradstock, R.A., Bedward, M., Kenny, B.J., & Scott, J. (1998a) Spatially-explicit simulation of the effect of prescribed burning on fire regimes and plant extinctions in shrublands typical of south-eastern Australia. *Biological Conservation*, **86**, 83–95.

Bradstock, R.A., Bedward, M., Scott, J., & Keith, D.A. (1996) Simulation of the effect of spatial and temporal variation in fire regimes on the population viability of a *Banksia* species. *Conservation Biology*, **10**, 776–784.

Bradstock, R.A. & Cohn, J.S. (2002a) Fire regimes and biodiversity in semi-arid mallee ecosystems. In *Flammable Australia: the fire regimes and biodiversity of a continent* (eds R.A. Bradstock, J.E. Williams & A.M. Gill), pp. 238–258. Cambridge University Press, Cambridge, UK.

Bradstock, R.A. & Cohn, J.S. (2002b) Demographic characteristics of mallee pine (*Callitris verrucosa*) in fire-prone mallee communities of central New South Wales. *Australian Journal of Botany*, **50**, 653–665.

Bradstock, R.A. & Gill, A.M. (1993) Fire in semi-arid, mallee shrublands: size of flames from discrete fuel arrays and their role in the spread of fire. *International Journal of Wildland Fire*, **3**, 3–12.

Bradstock, R.A. & Gill, A.M. (2001) Living with fire and biodiversity at the urban edge: in search of a sustainable solution to the human protection problem in southern Australia. *Journal of Mediterranean Ecology*, **2**, 179–195.

Bradstock, R.A., Gill, A.M., Hastings, S.M., & Moore, P.H.R. (1994) Survival of serotinous seedbanks during bushfires: comparative studies of *Hakea* species from southeastern Australia. *Austral Ecology*, **19**, 276–282.

Bradstock, R.A., Gill, A.M., Kenny, B.J., & Scott, J. (1998b) Bushfire risk at the urban interface estimated from historical weather records: consequences for the use of prescribed fire in the Sydney region of south-eastern Australia. *Journal of Environmental Management*, **52**, 259–271.

Bradstock, R.A., Hammill, K.A., Collins, L., & Price, O. (2010) Effects of weather, fuel and terrain on fire severity in topographically diverse landscapes of south-eastern Australia. *Landscape Ecology*, **25**, 607–619.

Bradstock, R.A., Keith, D.A., & Auld, T.D. (1995) Fire and conservation: imperatives and constraints on managing for diversity. In *Conserving biodiversity: threats and solutions* (eds R.A. Bradstock, T.D. Auld, D.A. Keith, *et al.*), pp. 323–333. Surrey Beatty & Sons, Chipping Norton, New South Wales, Australia.

Bradstock, R.A. & Kenny, B.J. (2003) An application of plant functional types to fire management in a conservation reserve in southeastern Australia. *Journal of Vegetative Science*, **14**, 345–354.

Bradstock, R.A. & Myerscough, P.J. (1988) The survival and population response to frequent fires of two woody resprouters *Banksia serrata* and *Isopogon anemonifolius*. *Australian Journal of Botany*, **36**, 415–431.

Bradstock, R.A. & O'Connell, M.A. (1988) Demography of woody plants in relation to fire: *Banksia ericifolia* L.f. and *Petrophile pulchella* (Schrad) R. Br. *Austral Ecology*, **13**, 505–518.

Bradstock, R.A., Tozer, M.G., & Keith, D.A. (1997) Effects of high frequency fire on floristic composition and abundance in a fire-prone heathland near Sydney. *Australian Journal of Botany*, **45**, 641–655.

Bradstock, R.A., Williams, J.E., & Gill, A.M. (2002) *Flammable Australia: the fire regimes and biodiversity of a continent*, Cambridge University Press, Sydney, Australia.

Breshears, D.D., Huxman, T.E., Adams, H.D., Zou, C.B., & Davison, J.E. (2008) Vegetation synchronously leans upslope as climate warms. *Proceedings of the National Academy of Sciences*, **105**, 11 591–11 592.

Briassoulis, H. (1992) The planning uses of fire: reflections on the Greek experience. *Journal of Environmental Planning and Management*, **35**, 161–173.

Bridgewater, P.B. & Backshall, D.J. (1981) Dynamics of some western Australian ligneous formations with special reference to the invasion of exotic species. *Vegetatio*, **46**, 141–148.

Brits, G.J., Calitz, F.J., Brown, N.A.C., & Manning, J.C. (1993) Desiccation as the active principle in heat-stimulated seed germination of *Leucospermum* R. Br. (Proteaceae) in fynbos. *New Phytologist*, **125**, 397–403.

Brooks, M.L. (1999) Alien annual grasses and fire in the Mojave Desert. *Madroño*, **46**, 13–19.

Brooks, M.L., D'Antonio, C.M., Richardson, D.M., *et al.* (2004) Effects of invasive alien plants on fire regimes. *BioScience*, **54**, 677–688.

Brooks, M.L. & Minnich, R.A. (2006) Southeastern deserts bioregion. In *Fire in California's Ecosystems* (eds N.G. Sugihara, J.W. van Wagtendonk, K.E. Shaffer, J. Fites-Kaufman & A.E. Thoede), pp. 391–414. University of California Press, Los Angeles, California.

Brown, D.E. & Minnich, R.A. (1986) Fire and changes in creosote bush scrub of the western Sonoran Desert, California. *American Midland Naturalist*, **116**, 411–422.

Brown, J.H. & Davidson, D.W. (1977) Competition between seed-eating rodents and ants in desert ecosystems. *Science*, **196**, 880–882.

Brown, N.A.C. (1993) Promotion of germination of fynbos seeds by plant-derived smoke. *New Phytologist*, **123**, 575–584.

Brown, N.A.C., Botha, P.A., & Prosch, D. (1995) Where's the smoke... *The Garden, Journal of the Royal Horticultural Society*, **120**, 402–405.

Brown, N.A.C., van Staden, J., Daws, M.I., & Johnson, T. (2003) Patterns in the seed germination response to smoke in plants from the Cape Floristic Region, South Africa. *South African Journal of Botany*, **69**, 514–525.

Brown, P.J., Manders, P.T., Bands, D.P., Kruger, F.J., & Andrag, R.H. (1991) Prescribed burning as a conservation management practice: a case history from the Cederberg Mountains, Cape Province, South Africa. *Biological Conservation*, **56**, 133–150.

Brown, P.M. & Swetnam, T.W. (1994) A cross-dated fire history from coast redwood near Redwood National Park, California. *Canadian Journal of Forest Research*, **24**, 21–31.

Brown, R.W. (1935) Miocene leaves, fruits, and seeds from Idaho, Oregon, and Washington. *Journal of Paleontology*, **9**, 577–587.

Burcham, L.T. (1957) *California range land: an historico-ecological study of the range resources of California*. State of California Department of Natural Resources, Division of Forestry, Sacramento, California.

Burns, B.R. (1993) Fire-induced dynamics of *Araucaria araucana–Nothofagus antarctica* forest in the southern Andes. *Journal of Biogeography*, **20**, 669–685.

Burrough, P.A., Brown, L., & Morris, E.C. (1977) Variations in vegetation and soil pattern across the Hawkesbury Sandstone plateau from Barren Grounds to Fitzroy Falls, New South Wales. *Austral Ecology*, **2**, 137–159.

Burrows, G.E. (2002) Epicormic strand structure in *Angophora*, *Eucalyptus* and *Lophostemon* (Myrtaceae): implications for fire resistance and recovery. *New Phytologist*, **153**, 111–131.

Burrows, G.E. (2010) *Syncarpia* and *Tristaniopsis* (Myrtaceae) possess specialised fire-resistant epicormic structures. *Australian Journal of Botany*, **56**, 254–264.

Burrows, N.D. (1995) A framework for assessing acute impacts of fire in jarrah forests for ecological studies. *CALM Science Supplement*, **4**, 59–66.

Burrows, N.D. (2001) Flame residence times and rates of weight loss of eucalypt forest fuel particles. *International Journal of Wildland Fire*, **10**, 137–143.

Burrows, N.D. (2008) Linking fire ecology and fire management in south-west Australian forest landscapes. *Forest Ecology and Management*, **255**, 2394–2406.

Burrows, N.D. & Friend, G. (1998) Biological indicators of appropriate fire regimes in southwest Australian ecosystems. In *Fire in ecosystem management: shifting the paradigm from suppression to prescription: Tall Timbers Fire Ecology Conference Proceedings* (eds T.L. Pruden & L.A. Brennan), pp. 413–421. Tall Timbers Research Station, Tallahassee, Florida.

Burrows, N.D. & McCaw, W.L. (1990) Fuel characteristics and bushfire control in *Banksia* low woodlands in Western Australia. *Journal of Environmental Management*, **31**, 229–236.

Burrows, N.D., Ward, B., & Robinson, A. (2010) Fire regimes and tree growth in low rainfall jarrah forest of south-west Australia. *Environmental Management*, **45**, 1332–1343.

Burrows, N.D., Ward, B., & Robinson, A.D. (1995) Jarrah forest fire history from stem analysis and anthropological evidence. *Australian Forestry*, **58**, 7–16.

Burrows, N.D. & Wardell-Johnson, G. (2003) Fire and plant interactions in forested ecosystems of south-west Western Australia. In *Fire in ecosystems of south-west Western Australia: impacts and management* (eds I. Abbott & N.D. Burrows), pp. 225–268. Backhuys Publishers, Leiden, The Netherlands.

Burrows, N.D., Wardell-Johnson, G., & Ward, B. (2008) Post-fire juvenile period of plants in south-west Australia forests and implications for fire management. *Journal of the Royal Society of Western Australia*, **91**, 163–174.

Busby, C.J. & Putirka, K. (2009) Miocene evolution of the western edge of the Nevadaplano in the central and northern Sierra Nevada: paleocanyons, magmatism, and structure. *International Geology Review*, **51**, 670–701.

Busby, P.E., Vitousek, P., & Dirzo, R. (2010) Prevalence of tree regeneration by sprouting and seeding along a rainfall gradient in Hawai'i. *Biotropica*, **42**, 80–86.

Busch, D.E. (1995) Effects of fire on southwestern riparian plant community structure. *Southwestern Naturalist*, **40**, 259–267.

Bustamante, R., Pauchard, A., Marticorena, A.J.A., & Cavieres, L. (2005) Alien plants in Mediterranean-type ecosystems of the Americas: comparing floras at a regional and local scale. In *Invasive plants in mediterranean-type regions of the world* (ed. S. Brunel), pp. 89–97. Council of Europe Publishing, Strasbourg, France.

Bustos-Schindler, C., Quesne, C.L., Gonzalez, M.E., & Solari, M.E. (2010) Historia preliminar de incendios y prácticas (multi) culturales en la cuenca media del río Cachapoal (34° S), Chile central. *Bosque*, **31**, 17–27.

Butler, B. (2009) Efforts to update firefighter safety zone guidelines. *Fire Management Today*, **69**, 15–17.

Byram, G.M. (1959) Combustion of forest fuels. In *Forest fire: control and use* (ed. K.P. Davis), pp. 61–89. McGraw-Hill, New York.

Calder, J.H., Gibling, M.R., Scott, A.C., Davies, S.J., & Hebert, B.L. (2006) A fossil lycopsid forest succession in the classic Joggins section of Nova Scotia: paleoecology of a disturbance-prone Pennsylvanian wetland. In *Wetlands through time* (eds S.F. Greb & W.A. DiMichele), pp. 169–195. Geological Society of America, Boulder, Colorado.

Callaway, R.M., Nadkarni, N.M., & Mahall, B.E. (1991) Facilitation and interference of *Quercus douglasii* on understory productivity in central California. *Ecology*, **72**, 1484–1499.

Calvin, M. & Wettlaufer, D. (2000) Fires in the Southern Cape Peninsula, Western Cape Province, South Africa, January 2000. *International Forest Fire News*, **22**, 69–75.

Camci Çetin, S., Karaca, A., Haktanır, K., & Yildiz, H. (2007) Global attention to Turkey due to desertification. *Environmental Monitoring and Assessment*, **128**, 489–493.

Campbell, M.L. & Clarke, P.J. (2006) Response of montane wet sclerophyll forest understorey species to fire: evidence from high and low intensity fires. *Proceedings of the Linnean Society of New South Wales*, **127**, 63–73.

Canadell, J., Lloret, F., & López-Soria, L. (1991) Resprouting vigour of two Mediterranean shrub species after experimental fire treatments. *Vegetatio*, **95**, 119–126.

Canadell, J. & López-Soria, L. (1998) Lignotuber reserves support regrowth following clipping of two mediterranean shrubs. *Functional Ecology*, **12**, 31–38.

Canadell, J. & Zedler, P.H. (1995) Underground structures of woody plants in Mediterranean ecosystems of Australia, California, and Chile. In *Ecology and biogeography of Mediterranean ecosystems in Chile, California and Australia* (eds M.T.K. Arroyo, P.H. Zedler & M.D. Fox), pp. 177–210. Springer, New York.

Cant, C.M. (1937) Stem structure in the Maddenii series of rhododendrons. *Proceedings of the Botanical Society of Edinburgh*, **22**, 287–291.

Caprio, A.C. (2004) Temporal and spatial dynamics of pre-EuroAmerican fire at a watershed scale, Sequoia and Kings Canyon national parks. In *Proceedings of the symposium: fire management: emerging policies and new paradigms* (eds N.G. Sugihara & M.E. Morales). Miscellaneous Publication 2, pp. 107–125. Association for Fire Ecology, Berkeley, California.

Carbonell, E., Bermúdez de Castro, J.M., Parés, J.M., et al. (2008) The first hominin of Europe. *Nature*, **452**, 465–469.

Carbonell, E., Decastro, J.M.B., Arsuaga, J.L., et al. (1995) Lower pleistocene hominids and artifacts from Atapuerca-TD6 (Spain). *Science*, **269**, 826–830.

Carlson, J.M. & Doyle, J. (2002) Complexity and robustness. *Proceedings of the National Academy of Sciences of the USA*, **99**, 2538–2545.

Carr, D.J., Carr, S.G.M., & Jahnke, R. (1982) The eucalypt lignotuber: a position-dependent organ. *Annals of Botany*, **50**, 481–489.

Carr, D.J., Jahnke, R., & Carr, S.G.M. (1984) Initiation, development and anatomy of lignotubers in some species of *Eucalyptus*. *Australian Journal of Botany*, **32**, 415–437.

Carreira, J.A., Sanchezvazquez, F., & Niell, F.X. (1992) Short-term and small-scale patterns of postfire regeneration in a semiarid dolomite basin of southern Spain. *Acta Oecologica*, **13**, 241–253.

Carrington, M.E. & Keeley, J.E. (1999) Comparison of post-fire seedling establishment between scrub communities in mediterranean and non-mediterranean climate ecosystems. *Journal of Ecology*, **87**, 1025–1036.

Carrión, J.S. (2002) Patterns and processes of Late Quaternary environmental change in a montane region of southwestern Europe. *Quaternary Science Reviews*, **21**, 2047–2066.

Cart, J. (2009) Australia will take a hard look at "Leave Early or Stay and Defend" fire policy. *Los Angeles Times*, February 9, 2009.

Cary, G.J. & Banks, J.C.G. (1999) Fire regime sensitivity to global climate change: an Australian perspective. In *Advances in global change research* (ed. J.L. Innes), pp. 233–246. Kluwer Academic Publishers, Dordrecht, The Netherlands.

Cary, G.J., Keane, R.E., Gardner, R.H., et al. (2006) Comparison of the sensitivity of landscape-fire-succession models to variation in terrain, fuel pattern, climate and weather. *Landscape Ecology*, **21**, 121–137.

Cary, G.J. & Morrison, D.A. (1995) Effects of fire frequency on plant species composition of sandstone communities in the Sydney region: combinations of inter-fire intervals. *Australian Journal of Ecology*, **20**, 418–426.

Cassel, E.J., Graham, S.A., & Chamberlain, C.P. (2009) Cenozoic tectonic and topograpic evolution of the northern Sierra Nevada, California, through stable isotope paleoaltimetry in volcanic glass. *Geology*, **37**, 547–550.

Catchpole, W. (2000) The international scene and its impact on Australia. In *Fire! The Australian experience*, pp. 137–148. Australian Academy of Technological Sciences and Engineering Limited, Canberra.

Catchpole, W. (2002) Fire properties and burn patterns in heterogeneous landscapes. In *Flammable Australia: the fire regimes and biodiversity of a continent* (eds R.A. Bradstock, J.E. Williams & A.M. Gill), pp. 49–75. Cambridge University Press, Cambridge, UK.

Catry, F., Rego, F., Moreira, F., Fernandes, P.M., & Pausas, J.G. (2010) Post-fire tree mortality in mixed forests of central Portugal. *Forest Ecology and Management*, **206**, 1184–1192.

Catry, F.X., Rego, F.C., Bação, F.L., & Moreira, F. (2009) Modeling and mapping wildfire ignition risk in Portugal. *International Journal of Wildland Fire*, **18**, 921–931.

Caturla, R.N., Raventós, J., Guardia, R., & Vallejo, V.R. (2000) Early post-fire regeneration dynamics of *Brachypodium retusum* Pers. (Beauv.) in old fields of the Valencia region (eastern Spain). *Acta Oecologica*, **21**, 1–12.

Cavelier, J. & Tecklin, D. (2005) Conservación de la Cordillera de la Costa: un desafio urgente en la ecoregión valdiviana. In *Historia, biodiversidad y ecología de los bosques costeros de Chile* (eds C. Smith-Ramirez, J.J. Armesto & C. Valdovinos), pp. 632–641. Editorial Universitaria, Santiago, Chile.

Cermak, R.W. (2005) *Fire in the forest: a history of forest fire control on the national forests in California, 1898–1956*, R5-FR-003, USDA Forest Service, Pacific Southwest Region, Albany, California.

Cerrillo, R.M.N., Hayas, A., García-Ferrer, A., et al. (2008) Characteristics of areas affected by fire in 2005 at Parque Nacional de Torres del Paine (Chile) as assessed from multispectral images. *Revista Chilena de Historia Natural*, **81**, 95–110.

Chafer, C.J., Noonan, M., & McNaught, E. (2004) The post-fire measurement of fire severity and intensity in the Christmas 2001 Sydney wildfires. *International Journal of Wildland Fire*, **13**, 227–240.

Chalwell, S.T.S. & Ladd, P.G. (2005) Stem demography and post fire recruitment of *Podocarpus drouynianus*, a resprouting non-serotinous conifer. *Botanical Journal of the Linnean Society*, **149**, 433–449.

Charco, J. (1999) *El bosque mediterráneo en el norte de África*. Ediciones Mundo Árabe e Islam. Agencia Española de Cooperación Internacional, Madrid.

Charnov, E.L. & Schaffer, W.M. (1973) Life-history consequences of natural selection: Cole's result revisited. *American Naturalist*, **107**, 791–793.

Chase, B.M. & Meadows, M.E. (2007) Late Quaternary dynamics of southern Africa's winter rainfall zone. *Earth-Science Reviews*, **84**, 103–138.

Chase, J.M. (2003) Community assembly: when should history matter? *Oecologia*, **136**, 489–498.

Chastain, J. (2007) How environmentalists fanned California fires. *World Net Daily*, Nov 2007.

Cheal, D.C. (1996) Fire succession in heathlands and implications for vegetation management. In *Fire and Biodiversity: the effects and effectiveness of fire management*, pp. 67–80. Department of the Environment, Sport and Territories, Canberra, Australia.

Chen, K. & McAneney, J. (2004) Quantifying bushfire penetration into urban areas in Australia. *Geophysical Research Letters*, **31**, 1–4.

Cheney, N.P. (1979) Bushfire disasters in Australia 1945–1975. In *Natural hazards in Australia* (eds R.L. Heathcote & B.G. Thom), pp. 72–93. Australian Academy of Science, Canberra.

Cheney, N.P. (1981) Fire behavior. In *Fire and the Australian biota* (eds A.M. Gill, R.H. Groves & I.R. Noble), pp. 151–175. Australian Academy of Science, Canberra.

Cheney, P. (1996) The effectiveness of fuel reduction burning for fire management. In *Fire and biodiversity. the effects and effectiveness of fire management*, Biodiversity Paper 8, pp. 9–16. Department of the Environment, Sport and Territories, Canberra, Australia.

Chew, S.J. & Bonser, S.P. (2009) The evolution of growth rate, resource allocation and competitive ability in seeder and resprouter tree seedlings. *Evolutionary Ecology*, **23**, 723–735.

Chiariello, N.R. (1989) Phenology of California grasslands. In *Grassland structure and function: California annual grasslands* (eds L.F. Huenneke & H.A. Mooney), pp. 47–58. Kluwer Academic Publishers, Dordrecht, The Netherlands.

Chiarucci, A., Dominicis, V.D., & Wilson, J.B. (2001) Structure and floristic diversity in permanent monitoring plots in forest ecosystems of Tuscany. *Forest Ecology and Management*, **141**, 201–210.

Childers, A.C. & Piirto, D.D. (1989) Cost-effective fire management for southern California's chaparral wilderness: an analytical procedure. In *Proceedings of the symposium on fire and watershed management* (ed. N.H. Berg), General Technical Report PSW-109, pp. 30–37. USDA Forest Service, Pacific Southwest Forest and Range Experiment Station, Berkeley, California.

Chiwocha, S.D.S., Dixon, K.W., Flematti, G.R., et al. (2009) Karrikins: a new family of plant growth regulators in smoke. *Plant Science*, **177**, 252–256.

Christian, C.E. & Stanton, M.L. (2004) Cryptic consequences of a dispersal mutualism: seed burial, elaiosome removal, and seed-bank dynamics. *Ecology*, **85**, 1101–1110.

Christodoulakis, N.S. (1989) An anatomical study of seasonal dimorphism in the leaves of *Phlomis fruticosa*. *Annals of Botany*, **63**, 389–394.

Christophel, D.C. & Greenwood, D.R. (1989) Changes in climate and vegetation in Australia during the Tertiary. *Review of Palaeobotany and Palynology*, **58**, 95–109.

Chu, P.-S., Yan, W., & Fujioka, F. (2002) Fire-climate relationships and long-leaf seasonal wildfire prediction for Hawaii. *International Journal of Wildland Fire*, **11**, 25–31.

Chuvieco, E., ed. (1999) *Remote sensing of large wildfires in the European Mediterranean Basin*. Springer, Berlin.

Chuvieco, E., ed. (2009) *Earth observation of wildland fires in mediterranean ecosystems*. Springer, Dordrecht, The Netherlands.

Chuvieco, E., Aguado, I., & Dimitrakopoulos, A.P. (2004) Conversion of fuel moisture content values to ignition potential for integrated fire danger assessment. *Canadian Journal of Forest Research*, **34**, 2284–2293.

Chuvieco, E., de Santis, A., Riaño, D., & Halligan, K. (2007) Simulation approaches for burn severity estimation using remotely sensed images. *Fire Ecology*, **3**, 129–150.

Chuvieco, E., Giglio, L., & Justice, C. (2008) Global characterization of fire activity: toward defining fire regimes from Earth observation data. *Global Change Biology*, **14**, 1488–1502.

Chytrý, M., Maskell, L.C., Pino, J., et al. (2008) Habitat invasions by alien plants: a quantitative comparison among Mediterranean, subcontinental and oceanic regions of Europe. *Journal of Applied Ecology*, **45**, 448–458.

Cichocki, O. (1998) Petrified, lignified and carbonized wood remains from the Early Miocene lignite opencast mine Oberdorf (N. Voitsberg, Styria, Austria). *Jahrbuch der Geologischen Bundesanstalt*, **140**, 469–473.

Clark, J.S., Gill, A.M., & Kershaw, A.P. (2002) Spatial variability in fire regimes: its effects on recent and past vegetation. In *Flammable Australia: the fire regimes and biodiversity of a continent* (eds R.A. Bradstock, J.E. Williams & M.A. Gill), pp. 125–141. Cambridge University Press, Cambridge, UK.

Clark, J.S. & Robinson, J. (1993) Paleoecology of fire. In *Fire in the environment* (ed. P.J. Crutzen & J.G. Goldammer), pp. 193–214. Wiley, New York.

Clarke, A.R. & Knox, K.J.E. (2009) Trade-offs in resource allocation that favour resprouting affect the competitive ability of woody seedlings in grassy communities. *Journal of Ecology*, **97**, 1374–1382.

Clarke, P.J. (2002a) Habitat islands in fire-prone vegetation: do landscape features influence community composition? *Journal of Biogeography*, **29**, 1–8.

Clarke, P.J. (2002b) Habitat insularity and fire response traits: evidence from a sclerophyll archipelago. *Oecologia*, **132**, 582–591.

Clarke, P.J. & Dorji, K. (2008) Are trade-offs in plant resprouting manifested in community seed banks? *Ecology*, **89**, 1850–1858.

Clarke, P.J. & Knox, K.J.E. (2002) Post-fire response of shrubs in the tablelands of eastern Australia: do existing models explain habitat differences? *Australian Journal of Botany*, **50**, 53–62.

Clarke, P.J., Knox, K.J.E., Wills, K.E., & Campbell, M. (2005) Landscape patterns of woody plant response to crown fire: disturbance and productivity influence sprouting ability. *Journal of Ecology*, **93**, 544–555.

Clarke, P.J., Kumar, L., Munzo, C., & Knox, K.J.E. (2007) *Burn severity and fire regimes thresholds for plant conservation in New England Tableland National Parks*. University of New England, Armidale.

Clarke, P.J., Myerscough, P.J., & Skelton, N.J. (1996) Plant coexistence in coastal heaths: between- and within-habitat effects of competition, disturbance and predation in the post-fire environment. *Austral Ecology*, **21**, 55–63.

Climent, J., Tapias, R., Pardos, J.A., & Gil, L. (2004) Fire adaptations in the Canary Islands pine (*Pinus canariensis*). *Plant Ecology*, **171**, 185–196.

Close, D.C., Davidson, N.J., Johnson, D.W., *et al.* (2009) Premature decline of *Eucalyptus* and altered ecosystems processes in the absence of fire in some Australian forests. *Botanical Review*, **75**, 191–202.

Cochrane, M.A. (2003) Fire science for rainforests. *Nature*, **421**, 913–919.

Cody, M.L. (1986) Diversity, rarity and conservation in Mediterranean-climate regions. In *Conservation biology: the science of scarcity and diversity*. (ed. M.E. Soulé), pp. 122–152. Sinauer Associates, Sunderland, Massachusetts.

Cody, M.L. (1999) Assembly rules at different scales in plant and bird communities. In *Ecological assembly rules: perspectives, advances, retreats* (eds E. Weiher & P. Keddy), pp. 165–205. Cambridge University Press, Cambridge, UK.

Cody, M.L. & Mooney, H.A. (1978) Convergence versus nonconvergence in Mediterranean-climate ecosystems. *Annual Review of Ecology and Systematics*, **9**, 265–321.

Coetzee, J.A. & Muller, J. (1984) The phytogeographic significance of some extinct Gondwana pollen types from the Tertiary of the southwestern Cape (South Africa). *Annals of the Missouri Botanical Garden*, **71**, 1088–1099.

Coetzee, J.A. & Rogers, J. (1982) Palynological and lithological evidence for the Miocene palaeoenvironment in the Saldanha region (South Africa). *Palaeogeography, Palaeoclimatology, Palaeoecology*, **39**, 71–85.

Coetzee, J.A., Scholtz, A., & Deacon, H.J. (1983) Palynological studies and the vegetation history of the fynbos. In *Fynbos palaeoecology: a preliminary synthesis* (eds H.J. Deacon,

Q.B. Hendey & J.J.N. Lambrechts), Report 75, pp. 156–173. CSIR, South African National Scientific Programmes, Pretoria.

Coffman, G.C., Ambrose, R.F., & Rundel, P.W. (2010) Wildfire promotes dominance of invasive giant reed (*Aundo donax*) in riparian ecosystems. *Biological Invasions*, **12**, 2723–2734.

Cohen, J. (1999) Reducing the wildland fire threat to homes: where and how much? In *Proceedings of the symposium on fire economics, planning, and policy: bottom lines* (ed. A. González-Caban), pp. 189–196. Pacific Southwest Research Station, Albany, California.

Cohen, J. (2008) The wildland-urban interface fire problem: a consequence of the fire exclusion paradigm. *Forest History Today*, Fall, 20–26.

Cohen, J.D. & Stratton, R.D. (2008) *Home destruction examination: Grass Valley Fire, Lake Arrowhead, CA*, Technical Report R5-TP-026b. USDA Forest Service, Region 5, Albany, California.

Cohn, J.S. & Bradstock, R.A. (2000) Factors affecting post-fire seedling establishment of selected mallee understorey species. *Australian Journal of Botany*, **48**, 59–70.

Collier, A.B., Hughes, A.R.W., Lichtenberger, J., & Steinbach, P. (2006) Seasonal and diurnal variation of lightning activity over southern Africa and correlation with European whistler observations. *Annales Geophysicae*, **24**, 529–542.

Collins, B.M. & Stephens, S.L. (2007) Managing natural wildfires in Sierra Nevada wilderness areas. *Frontiers in Ecology and the Environment*, **5**, 523–527.

Collins, B.M. & Stephens, S.L. (2010) Stand-replacing patches within a "mixed severity" fire regime: quantitative characterization using recent fires in a long-established natural fire area. *Landscape Ecology*, **25**, 927–939.

Collins, P. (2006) *Burn: the epic story of bushfire in Australia*. Allen and Unwin, Sydney, Australia.

Collinson, M.E. (2002) The ecology of Cainozoic ferns. *Review of Palaeobotany and Palynology*, **119**, 51–68.

Collinson, M.E., Featherstone, C., Cripps, J.A., Nichols, G.J., & Scott, A.C. (1999) Charcoal-rich plant debris accumulations in the Lower Cretaceous of the Isle of Wight, England. *Acta Palaeobotanica*, Suppl. **2**, 93–105.

Collinson, M.E., Steart, D.C., Scott, A.C., Glasspool, I.J., & Hooker, J.J. (2007) Episodic fire, runoff and deposition at the Palaeocene-Eocene boundary. *Journal of the Geological Society*, **164**, 87.

Colombaroli, D., Vanniѐe, B., Emmanuel, C., Magny, M., & Tinner, W. (2008) Fire-vegetation interactions during the Mesolithic-Neolithic transition at Lago dell'Accesa, Tuscany, Italy. *Holocene*, **18**, 679–692.

Combrink, P., Dwarika, Y., Fowkes, S., Prins, P., & Smith, P. (2003) Challenges of managing fires along an urban-wildland interface: lessons from the Cape Peninsula, South Africa. In *3rd International Wildland Fire Conference and Exhibition. Managing the Future of Global Problems* [no pagination]. Australasian Fire Authorities Council, Sydney, Australia.

CONAF (2003) *Informes finales estatisticos de temporadas*. Programa de Manejo de Fuego, Santiago, Chile.

Conard, S.G. & Radosevich, S.R. (1982) Post-fire succession in white fir (*Abies concolor*) vegetation of the northern Sierra Nevada. *Madroño*, **29**, 42–56.

Conard, S.G., Sukhinin, A.I., Stocks, B.J., *et al.* (2002) Determining effects of area burned and fire severity on carbon cycling and emissions in Siberia. *Climatic Change*, **55**, 197–211.

Conard, S.G. & Weise, D.R. (1998) Management of fire regime, fuels, and fire effects in southern California chaparral: lessons from the past and thoughts for the future. *Tall Timbers Ecology Conference Proceedings*, **20**, 342–350.

Condit, C. (1938) The San Pablo flora of west central California. In *Miocene and Pliocene floras of western North America*, Contributions to Palaeontology 476, pp. 217–268. Carnegie Institution of Washington, Washington, DC.

Condit, R., Hubbell, S.P., Lafranki, J.V., *et al.* (1996) Species-area and species-individual relationships for tropical trees: a comparison of three 50-ha plots. *Journal of Ecology*, **84**, 549–562.

Conroy, R.J. (1996) To burn or not to burn? A description of the history, nature and management of bushfires within Ku-Ring-Gai Chase National Park. *Proceedings of the Linnean Society of New South Wales*, **116**, 79–95.

Cook, S.F., Jr. (1959) The effects of fire on a population of small rodents. *Ecology*, **40**, 102–108.

Cooper, W.S. (1922) *The broad-sclerophyll vegetation of California: an ecological study of the chaparral and its related communities,* Publication 319. Carnegie Institution of Washington, Washington DC.

Corbett, E.S. & Green, L.R. (1965) *Emergency revegetation to rehabilitate burned watersheds in southern California,* Research Paper PSW-22. USDA Forest Service, Pacific Southwest Forest and Range Experiment Station, Berkeley, California.

Corbin, J.D., Dyer, A.R., & Seabloom, E.W. (2007) Competitive interactions. In *California grasslands: ecology and management* (eds M.R. Stromberg, J.D. Corbin & C.M. D'Antonio), pp. 156–168. University of California Press, Los Angeles, California.

Court, A. (1960) *Lightning fire incidence in northeastern California, 1945–1956*, Technical Paper 47. USDA Forest Service, Pacific Southwest Forest and Range Experiment Station, Berkeley, California.

Covington, W.W. & Moore, M.M. (1992) Postsettlement changes in natural fire regimes: implications for restoration of old-growth ponderosa pine forests. In *Old-growth forests in the southwest and Rocky Mountain regions: proceedings of a workshop* (eds M.R. Kaufmann, W.H. Moir & R.L. Bassett), General Technical Report RM-213, pp. 81–99. USDA Forest Service, Rocky Mountain Forest and Range Experiment Station, Fort Collins, Colorado.

Cowling, R.M. (1983a) The occurrence of C_3 and C_4 grasses in fynbos and allied shrublands in the south eastern Cape, South Africa. *Oecologia*, **58**, 121–127.

Cowling, R.M. (1983b) Diversity relations in Cape shrublands and other vegetation in the southeastern Cape, South Africa. *Vegetatio*, **45**, 103–127.

Cowling, R.M. (1987) Fire and its role in coexistence and speciation in Gondwanan shrublands. *South African Journal of Science*, **83**, 106–112.

Cowling, R.M. (1990) Diversity components in a species-rich area of the Cape Floristic Region. *Journal of Vegetation Science*, **1**, 699–710.

Cowling, R.M., Gibbs-Russell, G.E., Hoffmann, M.T., & Hilton-Taylor, C. (1989) Patterns of plant species diversity in southern Africa. In *Biotic diversity in southern Africa: concepts and conservation* (ed. B.J. Huntley), pp. 19–50. Oxford University Press, Cape Town, RSA.

Cowling, R.M. & Gxaba, T. (1990) Effects of a fynbos overstorey shrub on understorey community structure: implications for the maintenance of community-wide species richness. *South African Journal of Ecology*, **1**, 1–7.

Cowling, R.M., Kirkwood, K., Midgley, J.J., & Pierce, S.M. (1997b) Invasion and persistence of bird-dispersed, subtropical thicket and forest species in fire-prone coastal fynbos. *Journal of Vegetation Science*, **8**, 475–488.

Cowling, R.M. & Lamont, B.B. (1987) Post-fire recruitment of four co-occurring *Banksia* species. *Journal of Applied Ecology*, **24**, 645–658.

Cowling, R.M. & Lamont, B.B. (1998) On the nature of Gondwanan species flocks: diversity of Proteaceae in Mediterranean south-western Australia and South Africa. *Australian Journal of Botany*, **46**, 335–355.

Cowling, R.M., Lamont, B.B., & Pierce, S.M. (1987) Seed bank dynamics of four co-occurring *Banksia* species (Proteaceae). *Journal of Ecology*, **75**, 289–302.

Cowling, R.M. & Lombard, A.T. (2002) Heterogeneity, speciation/extinction history and climate: explaining regional plant diversity patterns in the Cape Floristic Region. *Diversity and Distributions*, **8**, 163–179.

Cowling, R.M., Pierce, S.M., & Moll, E.J. (1986) Conservation and utilization of south coast renosterveld, an endangered South African vegetation type. *Biological Conservation*, **37**, 363–377.

Cowling, R.M. & Pressey, R.M. (2001) Rapid plant diversification: planning for an evolutionary future. *Proceedings of the National Academy of Sciences of the USA*, **98**, 5452–5457.

Cowling, R.M., Proches, S., & Partridge, T.C. (2009) Explaining the uniqueness of the Cape flora: incorporating geomorphic evolution as a factor for explaining its diversification. *Molecular Phylogenetics and Evolution*, **51**, 64–74.

Cowling, R.M., Proches, S., & Vlok, J.H.J. (2005) On the origin of southern African subtropical thicket vegetation. *South African Journal of Botany*, **71**, 1.

Cowling, R.M., Richardson, D.M., & Mustart, P.J. (1997a) Fynbos. In *Vegetation of Southern Africa* (eds R.M. Cowling, D.M. Richardson & S.M. Pierce), pp. 99–130. Cambridge University Press, Cambridge, UK.

Cowling, R.M., Rundel, P.W., Lamont, B.B., Arroyo, M.K., & Arianoutsou, M. (1996) Plant diversity in mediterranean-climate regions. *Trends in Ecology and Evolution*, **11**, 362–366.

Cowling, R.M. & Witkowski, E.T.F. (1994) Convergence and non-convergence of plant traits in climatically and edaphically matched sites in Mediterranean Australia and South Africa. *Australian Journal of Ecology*, **19**, 220–232.

Cox, R.D. & Allen, E.B. (2008) Composition of soil seed banks in southern California coastal sage scrub and adjacent exotic grassland. *Plant Ecology*, **198**, 37–46.

Craig, M.D., Hobbs, R.J., Grigg, A.H., et al. (2010) Do thinning and burning sites revegetated after bauxite mining improve habitat for terrestrial vertebrates? *Restoration Ecology*, **18**, 300–310.

Crawley, M.J. & Harral, J.E. (2001) Scale dependence in plant biodiversity. *Science*, **291**, 864–868.

Crayn, D.M., Rosetto, M., & Maynard, D.J. (2006) Molecular phylogeny and dating reveals an Oligo-Miocene radiation of dry-adapted shrubs (former Tremandraceae) from rainforest tree progenitors (Elaeocarpaceae) in Australia. *American Journal of Botany*, **93**, 1328–1342.

Cressler, W.L., III (2001) Evidence of earliest known wildfires. *Palaios*, **16**, 171–174.

Crisp, M.D., Arroyo, M.T.K., Cook, L.G., et al. (2009) Phylogenetic biome conservatism on a global scale. *Nature*, **458**, 754–756.

Crisp, M.D., Burrows, G.E., Cook, L.G., et al. (2011) Flammable biomes dominated by eucalypts originated at the Cretaceous–Palaeogene boundary. *Nature Communications*, **2**, article 193, doi:10.1038/ncomms1191.

Crowder, L.B. (1980) Ecological convergence of community structure: a neutral model analysis. *Ecology*, **61**, 194–204.

Crowley, B.E., Koch, P.L., & Davis, E.B. (2008) Stable isotope constraints on the elevation history of the Sierra Nevada Mountains, California. *Geological Society of America Bulletin*, **120**, 588–598.

Cruz, A. & Moreno, J.M. (2001a) Lignotuber size of *Erica australis* and its relationship with soil resources. *Journal of Vegetation Science*, **12**, 373–384.

Cruz, A. & Moreno, J.M. (2001b) No allocation trade-offs between flowering and sprouting in the lignotuberous, Mediterranean shrub *Erica australis*. *Acta Oecologica*, **22**, 121–127.

Cruz, A., Pérez, B., & Moreno, J.M. (2003a) Resprouting of the Mediterranean-type shrub *Erica australis* with modified lignotuber carbohydrate content. *Journal of Ecology*, **91**, 348–356.

Cruz, A., Pérez, B., & Moreno, J.M. (2003b) Plant stored reserves do not drive resprouting of the lignotuberous shrub *Erica australis*. *New Phytologist*, **157**, 251–261.

CSIRO and Bureau of Meteorology (2007) *Climate Change in Australia*. Online: www.climatechangeinaustralia.gov.au

Curt, T. & Delcros, P. (2010) Managing road corridors to limit fire hazard. A simulation approach in southern France. *Ecological Engineering*, **36**, 457–465.

Cushman, J.H., Tierney, T.A., & Hinds, M.M. (2004) Variable effects of feral pig disturbances on native and exotic plants in a California grassland. *Ecological Applications*, **14**, 1746–1756.

Dallman, P.R. (1998) *Plant life in the world's mediterranean climates: California, Chile, South Africa, Australia, and the Mediterranean Basin*. California Native Plant Society and University of California Press, Los Angeles, California.

Daniau, A.-L., d'Errico, F., Fernanda, M., & Goñi, M.F.S. (2010) Testing the hypothesis of fire use for ecosystem management by Neanderthal and Upper Palaeolithic modern human populations. *PLoS ONE*, **5**, 1–10.

Danin, A. (2001) Near East ecosystems, plant diversity. In *Encyclopedia of Biodiversity* (ed. S.A. Levin), Vol. 4, pp. 353–364. Academic Press, San Diego, California.

D'Antonio, C.M. (2000) Fire, plant invasions, and global changes. In *Invasive species in a changing world* (eds H.A. Mooney & R.J. Hobbs), pp. 65–93. Island Press, Covelo, California.

D'Antonio, C.M. & Vitousek, P.M. (1992) Biological invasions by exotic grasses, the grass/fire cycle, and global change. *Annual Review of Ecology and Systematics*, **23**, 63–87.

Daskalakou, E.N. & Thanos, C.A. (1996) Aleppo pine (*Pinus halepensis*) postfire regeneration: the role of canopy and soil seed banks. *International Journal of Wildland Fire*, **6**, 59–66.

Davies-Vollum, K.S. (1997) Early Palaeocene palaeoclimatic inferences from fossil floras of the western interior, USA. *Palaeogeography, Palaeoclimatology, Palynology*, **136**, 145–164.

Davis, A.J., Jenkinson, L.S., Lawton, J.H., Shorrocks, B., & Wood, S. (1998) Making mistakes when predicting shifts in species range in response to global warming. *Nature*, **391**, 783–786.

Davis, F.W., Borchert, M.I., & Odion, D.C. (1989) Establishment of microscale vegetation pattern in maritime chaparral after fire. *Vegetatio*, **84**, 53–67.

Davis, F.W., Keller, E.A., Parikh, A., & Florsheim, J. (1988) Recovery of the chaparral riparian zone following wildfire. In *Proceedings of California riparian systems conference*, General Technical Report PSW-110, pp. 1–16. USDA Forest Service, Berkeley, California.

Davis, O.K. & Ellis, B. (2010) Early occurrence of sagebrush steppe, Miocene (12 Ma) on the Snake River Plain. *Review of Palaeobotany and Palynology*, **160**, 172–180.

Davis, S.D., Kolb, K.J., & Barton, K.P. (1998) Ecophysiological processes and demographic patterns in the structuring of California chaparral. In *Landscape disturbance and biodiversity in Mediterranean-type ecosystems* (eds P.W. Rundel, G. Montenegro & F.M. Jaksic), pp. 297–310. Springer, New York.

Davy, J.B. (1922) The suffrutescent habit as an adaptation to environment. *Journal of Ecology*, **10**, 211–219.

de Bello, F., Leps, J., & Sebastià, M.T. (2007) Grazing effects on the species-area relationship: variation along a climatic gradient in NE Spain. *Journal of Vegetation Science*, **18**, 25–34.

de Lange, J.H. & Boucher, C. (1990) Autecological studies on *Audouinia capitata* (Bruniaceae), I: Plant-derived smoke as a seed germination cue. *South African Journal of Botany*, **56**, 700–703.

de León, P.V. & Cevallos-Ferriz, S.R.S. (2000) Leaves of *Cercocarpus mixteca* n. sp. (Rosaceae) from Oligocene sediments, near Tepexi de Rodríguez, Puebla. *Review of Palaeobotany and Palynology*, **111**, 285–294.

de Luís, M., Baeza, M.J., Raventós, J., & González-Hidalgo, J.C. (2004) Fuel characteristics and fire behaviour in mature Mediterranean gorse shrublands. *International Journal of Wildland Fire*, **13**, 79–87.

De Santis, A., Chuvieco, E., & Vaughan, P. (2009) Short-term assessment of burn severity using the inversion of PROSPECT and GEOSAIL models. *Remote Sensing of Environment*, **113**, 126–136.

Deacon, H.J. (1983) The peopling of the fynbos region. In *Fynbos palaeoecology: a preliminary synthesis* (eds H.J. Deacon, Q.B. Hendey & J.J.N. Lambrechts), Report 75, pp. 183–204. CSIR, South African National Scientific Programmes, Pretoria.

Deacon, H.J. (1992) Human settlement. In *The ecology of fynbos: nutrients, fire, and diversity* (ed. R.M. Cowling), pp. 260–270. Oxford University Press, Cape Town, RSA.

Deacon, H.J., Hendey, Q.B., & Lambrechts, J.J.N., eds. (1983) *Fynbos palaeoecology: a preliminary synthesis*. Report 75, 216 pp. CSIR, South African National Scientific Programmes, Pretoria.

DeBano, L.F. & Rice, R.M. (1971) Fire in vegetation management: its effect on soil. In *Proceedings of the symposium on interdisciplinary aspects of watershed management*, pp. 327–345. American Society of Civil Engineers, Reston, Virginia.

Debussche, M., Escarré, J., & Lepart, J. (1982) Ornithochory and plant succession in Mediterranean abandoned orchards. *Vegetatio*, **48**, 255–266.

Debussche, M., Escarré, J., Lepart, J., Houssard, C., & Lavorel, S. (1996) Changes in Mediterranean plant succession: old-fields revised. *Journal of Vegetation Science*, **7**, 519–526.

Del Tredici, P. (2001) Sprouting in temperate trees: a morphological and ecological review. *Botanical Review*, **67**, 121–140.

Delitti, W.B.C., Ferran, A., Trabaud, L., & Vallejo, V.R. (2005) Effects of fire recurrence in *Quercus coccifera* L. shrublands of the Valencia Region (Spain), I: plant composition and productivity. *Plant Ecology*, **177**, 57–70.

Dell, B., Havel, J.J., & Malajczuk, N. (1989) *The jarrah forest: a complex mediteranean ecosystem*. Kluwer Academic Publishers, Dordrecht, The Netherlands.

Denham, A.J. & Auld, T.D. (2002) Flowering, seed dispersal, seed predation and seedling recruitment in two pyrogenic flowering resprouters. *Australian Journal of Botany*, **50**, 545–557.

Denham, A.J. & Whelan, R.J. (2000) Reproductive ecology and breeding system of *Lomatia silaifolia* (Proteaceae) following a fire. *Australian Journal of Botany*, **48**, 261–269.

Denk, T., Grímsson, F., & Zetter, R. (2010) Episodic migration of oaks to Iceland: evidence for a North Atlantic "land bridge" in the latest Miocene. *American Journal of Botany*, **97**, 276–287.

Dennison, P.E., Moritz, M.A., & Taylor, R.S. (2008) Evaluating predictive models of critical live fuel moisture in the Santa Monica Mountains, California. *International Journal of Wildland Fire*, **17**, 18–27.

Department of the Environment and Water Resources (2007) *Australia's native vegetation: a summary of Australia's major vegetation groups, 2007*. Australian Government, Canberra.

DeSimone, S.A. & Zedler, P.H. (1999) Shrub seedling recruitment in unburned Californian coastal sage scrub and adjacent grassland. *Ecology*, **80**, 2018–2032.

di Castri, F. (1973) Soil animals in latitudinal and topographical gradients of mediterranean ecosystems. In *Mediterranean type ecosystems: origin and structure* (eds F. di Castri & H.A. Mooney), pp. 171–190. Springer, New York.

di Castri, F. (1981) Mediterranean-type shrublands of the world. In *Ecosystems of the world, Vol. 11: Mediterranean-type shrublands*. (eds F. di Castri, D.W. Goodall & R.L. Specht), pp. 1–52. Elsevier Scientific, New York.

di Castri, F. (1989) History of biological invasions with special emphasis on the Old World. In *Biological invasions: a global perspective* (eds J.A. Drake, H.A. Mooney, F. di Castri, et al.), pp. 1–30. John Wiley & Sons, New York.

di Castri, F. & Mooney, H.A., eds. (1973) *Mediterranean ecosystems: origin and structure*, 405 pp. Springer, New York.

Diamantopoulos, J., Pirintsos, S.A., Margaris, N.S., & Stamou, G.P. (1994) Variation in Greek phrygana vegetation in relation to soil and climate. *Journal of Vegetation Science*, **5**, 355–360.

Diamond, J. (1997) *Guns, germs and steel: the fates of human societies*. Vintage Publications, New York.

Diamond, J.M. (1975) Assembly of species communities. In *Ecology and evolution of communities* (eds M.L. Cody & J.M. Diamond), pp. 342–444. Harvard University Press, Cambridge, Massachusetts.

Díaz-Delgado, R., Lloret, F., Pons, X., & Terradas, J. (2002) Satellite evidence of decreasing resilience in mediterranean plant communities after recurrent wildfires. *Ecology*, **83**, 2293–2303.

Dickinson, K.J.M. & Kirkpatrick, J.B. (1985) The flammability and energy content of some important plant species and fuel components in the forests of southeastern Tasmania. *Journal of Biogeography*, **12**, 121–134.

Diessel, C.F.K. (2010) The stratigraphic distribution of inertinite. *International Journal of Coal Geology*, **81**, 251–268.

Dietenberger, M.A. (2010) Ignition and flame-growth modelling on realistic building and landscape objects in changing environments. *International Journal of Wildland Fire*, **19**, 228–237.

Dieterich, J.H. & Swetnam, T.W. (1984) Dendrochronology of a fire-scarred ponderosa pine. *Forest Science*, **30**, 238–247.

DiMichele, W.A., Behrensmeyer, A.K., Olszewski, T.D., et al. (2004) Long-term stasis in ecological assemblages: evidence from the fossil record. *Annual Review of Ecology and Systematics*, **35**, 285–322.

DiMichele, W.A. & Phillips, T.L. (2002) The ecology of paleozoic ferns. *Review of Palaeobotany and Palynology*, **119**, 143–159.

Dimitrakopoulos, A.P. (2001) A statistical classification of Mediterranean species based on their flammability components. *International Journal of Wildland Fire*, **10**, 113–118.

Dimitrakopoulos, A.P. (2002) Mediterranean fuel models and potential fire behaviour in Greece. *International Journal of Wildland Fire*, **11**, 127–130.

Dimitrakopoulos, A.P. & Panov, P.I. (2001) Pyric properties of some dominant Mediterranean vegetation species. *International Journal of Wildland Fire*, **10**, 23–27.

DiTomaso, J.M., Brooks, M.L., Allen, E.B., et al. (2006) Control of invasive weeds with prescribed burning. *Weed Technology*, **20**, 535–548.

DiTomaso, J.M., Kyser, G.B., & Hastings, M.S. (1999) Prescribed burning for control of yellow starthistle (*Centaurea solstitialis*) and enhanced native plant diversity. *Weed Science*, **47**, 233–242.

Dixon, K. & Barrett, R. (2003) Defining the role of fire in south-west Western Australian plants. In *Fire in ecosystems of south-west Western Australia: impacts and management* (eds I. Abbott & N.D. Burrows), pp. 205–224. Backhuys Publishers, Leiden, The Netherlands.

Dixon, K.W., Roche, S., & Pate, J.S. (1995) The promotive effect of smoke derived from burnt native vegetation on seed germination of Western Australian plants. *Oecologia*, **101**, 185–192.

Doak, D.F., Estes, J.A., Halpern, B.S., et al. (2008) Understanding and predicting ecological dynamics: are major surprises inevitable? *Ecology*, **89**, 952–961.

Dodd, J., Heddle, E.M., Pate, J.S., & Dixon, K.W. (1984) Rooting patterns of sandplain plants and their functional significance. In *Kwongan: plant life of the sandplain* (eds J.S. Pate & J.S. Beard), pp. 146–177. University of Western Australia Press, Nedlands, Australia.

Dodge, J.M. (1975) Vegetational changes associated with land use and fire history in San Diego County. Ph.D. dissertation, University of California, Riverside.

Dodson, E.K., Peterson, D.W., & Harrod, R.J. (2010) Impacts of erosion control treatments on native vegetation recovery after severe wildfire in the Eastern Cascades, USA. *International Journal of Wildland Fire*, **19**, 490–499.

Doerr, S.H., Shakesby, R.A., Blake, W.H., et al. (2006) Effects of differing wildfire severities on soil wettability and implications for hydrological response. *Journal of Hydrology*, **319**, 295–311.

Donoso, C. (1993) *Bosques templados de Chile y Argentina: variación, estructura y dinámica*. Editorial Universitaria, Santiago, Chile.

Downes, K.S., Lamont, B.B., Light, M.E., & van Staden, J. (2010) The fire ephemeral *Tersonia cyathiflora* (Gyrostemonaceae) germinates in response to smoke but not the butenolide 3-methyl-2H-furol[2,3-c]pyran-2-one. *Annals of Botany*, **106**, 381–384.

Drewes, F.E., Smith, M.T., & van Staden, J. (1995) The effect of plant-derived smoke extract on the germination of light-sensitive lettuce seed. *Plant Growth Regulation*, **16**, 205–209.

Driscoll, D.A., Lindenmayer, D.B., Bennett, A.F., et al. (2010) Resolving conflicts in fire management using decision theory: asset-protection versus biodiversity conservation. *Conservation Letters*, **3**, 215–223.

Dufour-Dror, J.-M. (2002) A quantitative classification of Mediterranean mosaic-like landscapes. *Journal of Mediterranean Ecology*, **3**, 3–12.

Dunn, C.P. & Loehl, C. (1988) Species-area parameter estimation: testing the null model of lack of relationship. *Global Ecology and Biogeography*, **9**, 59–74.

Dwire, K.A. & Kauffman, J.B. (2003) Fire and riparian ecosystems in landscapes of the western USA. *Forest Ecology and Management*, **178**, 61–74.

Edgell, M.C.R. (1973) *Nature and perception of the bushfire hazard in southeastern Australia*. Department of Geography, Monash University Melbourne, Australia.

Edinger, J.G., Helvey, R.A., & Baumhefner, D. (1964) *Surface wind patterns in the Los Angeles Basin during "Santa Ana" conditions*. Part I of Final Report on Research Project No. 2606. Supplement No. 49, USFS-UC Contract No. A5fs-16563. Department of Meteorology, University of California, Los Angeles.

Edwards, D. & Axe, L. (2004) Anatomical evidence in the detection of the earliest wildfires. *Palaios*, **19**, 113–128.

Edwards, D. & Hawkins, J.A. (2007) Are Cape floral clades the same age? Contemporaneous origins of two lineages in the genistoids s.l. (Fabaceae). *Molecular Phylogenetics and Evolution*, **45**, 952–970.

Edwards, S.W. (2004) Paleobotany of California. *Four Seasons*, **12**, 3–75.

Ekanayake, D.T. (1962) The anatomy of *Rhododendron zeylanicum* Booth (*R. arboreum* sensu Trim.). *Ceylon Journal of Science (Biological Sciences)*, **4**, 96–111.

Eldredge, N., Thompson, J.N., Brakefield, P.M., et al. (2005) The dynamics of evolutionary stasis. *Paleobiology*, **31**, 133–145.

Eldrett, J.S., Greenwood, D.R., Harding, I.C., & Huber, M. (2009) Increased seasonality through the Eocene to Oligocene transition in northern high latitudes. *Nature*, **459**, 969–974.

Ellis, P. (2003) Spotting and firebrand behaviour in dry eucalypt forest and the implications for fuel management in relation to fire suppression and to "ember" (firebrand) attack on houses. In *3rd international wildland fire conference and exhibition: urban and rural communities living in fire prone environments: managing the future of global problems* (CD ROM) (no pagination). USDI Bureau of Land Management, Washington, DC.

Ellis, S., Kanowski, P., & Whelan, R. (2004) *National inquiry on bushfire mitigation and management*. Commonwealth of Australia, Canberra.

Ellner, S. & Shmida, A. (1981) Why are adaptations for long-range seed dispersal rare in desert plants? *Oecologia*, **51**, 133–144.

Enright, N., Miller, B.P., Johnson, N., Lamont, B.B., & Perry, G.L.W. (2004) Soil seed banks in three contrasting high diversity Mediterranean-type shrublands from SW Australia. In *Ecology, conservation and management of mediterranean climate ecosystems* (eds M. Arianoutsou & V.P. Papanastasis), 33 pp. Millpress, Rotterdam, The Netherlands.

Enright, N.J. & Lamont, B.B. (1989a) Seed banks, fire season, safe sites and seedling recruitment in five co-occurring *Banksia* species. *Journal of Ecology*, **77**, 1111–1122.

Enright, N.J. & Lamont, B.B. (1989b) Fire temperatures and follicle-opening requirements in 10 *Banksia* species. *Australian Journal of Ecology*, **14**, 107–113.

Enright, N.J., Lamont, B.B., & Miller, B.P. (2005) Anomalies in grasstree fire history reconstructions for southwestern Australian vegetation. *Austral Ecology*, **30**, 668–673.

Enright, N.J., Marsula, R., Lamont, B.B., & Wissel, C. (1998) The ecological significance of canopy seed storage in fire-prone environments: a model for non-sprouting shrubs. *Journal of Ecology*, **86**, 946–959.

Enright, N.J., Miller, B.P., & Crawford, A. (1994) Environmental correlates of vegetation patterns and species richness in the northern Grampians, Victoria. *Austral Ecology*, **19**, 159–168.

Enright, N.J., Mosner, E., Miller, B.P., Johnson, N., & Lamont, B.B. (2007) Soil vs. canopy seed storage and plant species coexistence in species-rich Australian shrublands. *Ecology*, **88**, 2292–2304.

Enright, N.J. & Thomas, I. (2008) Pre-European fire regimes in Australian ecosystems. *Geography Compass*, **2**, 979–1011.

Erlandson, J.M. & M.A. Glassow, M.A., eds. (1997) *Archaeology of the California coast during the middle Holocene*, Vol. 4, 187 pp. Institute of Archaeology, Los Angeles, California.

Escudero, A., Sanz, M.V., Pita, J.M., & Pérez-García, F. (1999) Probability of germination after heat treatment of native Spanish pines. *Annals of Forest Science*, **56**, 511–520.

Esplin, B., Gill, A.M., & Enright, N.J. (2003) *Report of the inquiry into the 2002–2003 Victorian bushfires*. State Government of Victoria, Melbourne, Australia.

Estades, C.F. & Escobar, M.A. (2005) Los ecosistemas de las plantaciones de pino de la Cordillera de la Costa. In *Historia, biodiversidad y ecología de los bosques costeros de Chile* (eds C. Smith-Ramirez, J.J. Armesto & C. Valdovinos), pp. 600–616. Editorial Universitaria, Santiago, Chile.

Esther, A., Groeneveld, J., Enright, N.J., *et al.* (2008) Assessing the importance of seed immigration on coexistence of plant functional types in a species-rich ecosystem. *Ecological Modelling*, **213**, 412–416.

Eugenio, M. & Lloret, F. (2006) Effects of repeated burning on Mediterranean communities of the northeastern Iberian Peninsula. *Journal of Vegetation Science*, **17**, 755–764.

Everett, R.G. (2003) Grid-based fire-scar dendrochonology and vegetation sampling in the mixed-conifer forests of the San Bernardino and San Jacinto Mountains of southern California. Ph.D. dissertation. University of California, Riverside.

Evett, R.R., Franco-Vizcaino, E., & Stephens, S.L. (2007) Comparing modern and past fire regimes to assess changes in prehistoric lightning and anthropogenic ignitions in a Jeffrey pine – mixed conifer forest in the Sierra San Pedro Mártir, Mexico. *Canadian Journal of Forest Research*, **37**, 318–330.

Evett, R.R., Woodward, R.A., Harrison, W., *et al.* (2006) Phytolith evidence for the lack of a grass understory in a *Sequoiadendron giganteum* (Taxodiaceae) stand in the central Sierra Nevada, California. *Madroño*, **53**, 351–363.

Falcon-Lang, H.J. (2000) Fire ecology in the Carboniferous tropical zone. *Palaeogeography, Palaeoclimatology, Palaeoecology*, **164**, 355–371.

Falcon-Lang, H.J., Nelson, W.J., Elrick, S., *et al.* (2009) Incised channel fills containing conifers indicate that seasonally dry vegetation dominated Pennsylvanian tropical lowlands. *Geology*, **37**, 923–926.

Falk, D.A. & Swetnam, T.W. (2003) Scaling rules and probability models for surface fire regimes in ponderosa pine forests. In *Fire, fuel treatments, and ecological restoration: conference proceedings* (eds P.N. Omi & L.A. Joyce), Proceedings RMRS-P-29, pp. 301–318. USDA Forest Service, Rocky Mountain Research Station, Fort Collins, Colorado.

Falster, D.S. & Westoby, M. (2005) Tradeoffs between height growth rate, stem persistence and maximum height among plant species in post-fire succession. *Oikos*, **111**, 57–66.

Fernandes, P., Botelho, H., & Bento, J. (1999) Prescribed fire to reduce wildfire hazard: an analysis of management burns in Portuguese pine stands. In *Proceedings of the international symposium on forest fires: needs and innovations, a DELFI Action*, 18–19 November 1999, Athens, Greece, pp. 360–364. CINAR SA, Athens.

Fernandes, P.M. & Botelho, H.S. (2003) A review of prescribed burning effectiveness in fire hazard reduction. *International Journal of Wildland Fire*, **12**, 117–128.

Fernandes, P.M., Catchpole, W.R., & Rego, F.C. (2000) Shrubland fire behaviour modelling with microplot data. *Canadian Journal of Forest Research*, **30**, 889–899.

Fernandes, P.M., Vega, J.A., Jimenez, E., & Rigolot, E. (2008) Fire resistance of European pines. *Forest Ecology and Management*, **256**, 246–255.

Ferrenberg, S.M., Schwilk, D.W., Knapp, E.E., Groth, E., & Keeley, J.E. (2006) Fire decreases arthropod abundance but increases diversity: early and late season prescribed fire effects in a Sierra Nevada mixed-conifer forest. *Fire Ecology*, **2**, 79–102.

Fierro, A., Rutigliano, F.A., De Marco, A., Castaldi, S., & De Santo, A.V. (2007) Post-fire stimulation of soil biogenic emission of CO_2 in a sandy soil of a Mediterranean shrubland. *International Journal of Wildland Fire*, **16**, 573–583.

Figueiral, I., Mosbrugger, V., Rowe, N.P., et al. (1999) The Miocene peat-forming vegetation of northwestern Germany: an analysis of wood remains and comparison with previous palynological interpretations. *Review of Palaeobotany and Palynology*, **104**, 239–266.

Figueiral, I., Mosbrugger, V., Rowe, N.P., et al. (2002) Role of charcoal analysis for interpreting vegetation change and paleoclimate in the Miocene Rhine Embayment (Germany). *Palaios*, **17**, 347.

Figueroa, J.A., Castro, S.A., Marquet, P.A., & Jaksic, F.M. (2004) Exotic plant invasions to the mediterranean region of Chile: causes, history and impacts. *Revista Chilena de Historia Natural* **77**, 465–483.

Figueroa, J.A., Cavieres, L.A., Gómez-González, S., Molina-Montenegro, M.A., & Jaksic, F.M. (2009) Do heat and smoke increase emergence of exotic and native plants in the matorral of central Chile? *Acta Oecologica*, **35**, 335–340.

Figueroa, J.A. & Jaksic, F.M. (2004) Seed bank and dormancy in plants of the mediterranean region of central Chile. *Revista Chilena de Historia Natural*, **77**, 201–215.

Finkelstein, D.B., Pratt, L.M., Curtin, T.M., & Brassell, S.C. (2005) Wildfires and seasonal aridity recorded in Late Cretaceous strata from south-eastern Arizona, USA. *Sedimentology*, **52**, 587–599.

Finney, M.A. (1998) *FARSITE: fire area simulator – model development and valuation*, Research Paper, RMRS-RP-4. USDA Forest Service, Rocky Mountain Research Station, Ogden, Utah.

Finney, M.A. & Martin, R.E. (1992) Short fire intervals recorded by redwoods at Annandel State Park. *Madroño*, **39**, 251–262.

Fisher, J.L., Loneragan, W.A., Dixon, K., Delaney, J., & Veneklaas, E.J. (2009b) Altered vegetation structure and composition linked to fire frequency and plant invasion in a biodiverse woodland. *Biological Conservation*, **142**, 2270–2281.

Fisher, J.L., Loneragan, W.A., Dixon, K., & Veneklaas, E.J. (2009a) Soil seed bank compositional change constrains biodiversity in an invaded species-rich woodland. *Biological Conservation*, **142**, 256–269.

Fites-Kaufman, J.A., Rundel, P., Stephenson, N., & Weixelman, D.A. (2007) Montane and subalpine vegetation of the Sierra Nevada and Cascade ranges. In *Terrestrial vegetation of California*, third edition (eds M.G. Barbour, T. Keeler-Wolf & A.A. Schoenherr), pp. 456–501. University of California Press, Los Angeles.

Fjellstrom, R.G. & Parfitt, D.E. (1995) Phylogenetic analysis and evolution of the genus *Juglans* (Juglandaceae) as determined from nuclear genome RFLPs. *Plant Systematics and Evolution*, **197**, 19–32.

Flematti, G.R., Ghisalberti, E.L., Dixon, K.W., & Trengove, R.D. (2004) A compound from smoke that promotes seed germination. *Science*, **305**, 977.

Fleming, G.M., Diffendorfer, J.E., & Zedler, P.H. (2009) The relative importance of disturbance and exotic-plant abundance in California coastal sage scrub. *Ecological Applications*, **19**, 2210–2227.

Flinn, M.A. & Wein, R.W. (1977) Depth of underground plant organs and theoretical survival during fire. *Canadian Journal of Botany*, **55**, 2550–2554.

Florence, R.G. (1996) *Ecology and silviculture of eucalypt forests*. CSIRO, Collongwood, Australia.

Forest, F., Nänni, I., Chase, M.W., Crane, P.R., & Haskins, J.A. (2007) Diversification of a large genus in a continental biodiversity hotspot: temporal and spatial origin of *Muraltia* (Polygalaceae) in the Cape of South Africa. *Molecular Phylogenetics and Evolution*, **43**, 60–74.

Forsyth, G.G. & van Wilgen, B.W. (2007) *Analysis of the fire history records from protected areas in the Western Cape*. CSIR, Natural Resources and the Environment, Stellenbosch, RSA.

Forsyth, G.G. & van Wilgen, B.W. (2008) The recent fire history of the Table Mountain National Park and implications for fire management. *Koedoe*, **50**, 3–9.

Fosberg, M.A., O'Dell, C.A., & Schroeder, M.J. (1966) *Some characteristics of the three-dimensional structure of Santa Ana winds*, Research Paper PSW-30. USDA Forest Service, Pacific Southwest Forest and Range Experiment Station, Los Angeles, California.

Fotheringham, C.J. (2009) *Postfire recovery in the bimodal rainfall region of Arizona*. University of California, Los Angeles.

Founda, D. & Giannakopoulos, C. (2009) The exceptionally hot summer of 2007 in Athens, Greece: a typical summer in the future climate? *Global and Planetary Change*, **67**, 227–236.

Fowells, H.A., ed. (1965) *Silvics of forest trees of the United States*, Agriculture Handbook 271. USDA, Forest Service, Washington, DC.

Fox, B.J. (1983) Mammal species diversity in Australian heathlands: the importance of pyric succession and habitat diversity. In *Mediterranean-type ecosystems: the role of nutrients* (eds F.J. Kruger, D.T. Mitchell & J.U.M. Jarvis), pp. 473–489. Springer, New York.

Fox, B.J., Fox, M.D., & McKay, G.M. (1979) Litter accumulation after fire in a eucalypt forest. *Australian Journal of Botany*, **27**, 157–165.

Fox, B.J., Quinn, R.D., & Breytenbach, G.J. (1985) A comparison of small mammal succession following fire in shrublands of Australia, California and South Africa. *Proceedings of the Ecological Society of Australia*, **14**, 179–197.

Fox, M.D. (1990) Mediterranean weeds: exchanges of invasive plants between the five mediterranean regions of the world. In *Biological invasions in Europe and the Mediterranean Basin* (eds F. di Castri, A.J. Hansen & M. DeBussche). Kluwer Academic Publishers, Dordrecht, The Netherlands.

Fox, M.D. (1995) Australian Mediterranean vegetation: intra- and intercontinental comparisons. In *Ecology and biogeography of Mediterranean ecosystems in Chile,*

California and Australia (eds M.T.K. Arroyo, P.H. Zedler & M.D. Fox), pp. 137–159. Springer, New York.

Francis, J.E. (1984) The seasonal environment of the Purbeck (Upper Jurassic) fossil forests. *Palaeogeography, Palaeoclimatology, Palynology*, **48**, 285–307.

Franklin, J. (2010a) Vegetation dynamics and exotic plant invasion following high severity crown fire in a southern California conifer forest. *Plant Ecology*, **207**, 281–295.

Franklin, J. (2010b) *Mapping species distributions: spatial inference and prediction (ecology, biodiversity and conservation)*. Cambridge University Press, Cambridge, UK.

Franklin, J., Spears-Lebrun, L.A., Deutschman, D.H., & Marsden, K. (2006) Impact of a high-intensity fire on mixed evergreen and mixed conifer forests in the Peninsular Ranges of southern California, USA. *Forest Ecology and Management*, **235**, 18–29.

Franklin, S.E. (1987) Urban-wildland fire defense strategy, precision prescribed fire: the Los Angeles County approach. In *Proceedings of the symposium on wildland fire 2000*, General Technical Report PSW-101, pp. 22–25. USDA Forest Service, Pacific Southwest Forest and Range Experiment Station, Los Angeles, California.

Frederiksen, N.O. (1991) Pulses of middle Eocene to earliest Oligocene climatic deterioration in southern California and the Gulf Coast. *Palaios*, **6**, 564–571.

French, K. & Westoby, M. (1992) Removal of vertebrate-dispersed fruits in vegetation on fertile and infertile soils. *Oecologia*, **91**, 447–454.

French, K. & Westoby, M. (1996) Vertebrate-dispersed species in a fire-prone environment. *Australian Journal of Ecology*, **21**, 379–385.

Freudenberger, D.O., Fish, B.E., & Keeley, J.E. (1987) Distribution and stability of grasslands in the Los Angeles Basin. *Bulletin of the California Academy of Sciences*, **86**, 13–26.

Fried, J.S., Gilless, J.K., Riley, W.J., et al. (2008) *Predicting the effect of climate change on wildfire severity and outcomes in California: preliminary analysis*, White Paper CEC-500-2005-196-SF. California Climate Change Center, University of California, Berkeley.

Fuentes, E.R., Aviles, R., & Segura, A. (1989) Landscape change under indirect effects of human use: the savanna of central Chile. *Landscape Ecology*, **2**, 73–80.

Fuentes, E.R., Aviles, R., & Segura, A. (1990) The natural vegetation of a heavily man-transformed landscape: the savanna of central Chile. *Interciencia*, **15**, 293–295.

Fuentes, E.R. & Espinosa, G. (1986) Resilience of shrublands in central Chile: a volcanism-related hypothesis. *Interciencia*, **11**, 164–165.

Fuentes, E.R., Otaiza, R.D., Alliende, M.C., Hoffmann, A.J., & Poiani, A. (1984) Shrub clumps of the Chilean matorral vegetation: structure and possible maintenance mechanisms. *Oecologia*, **42**, 405–411.

Fuentes, E.R., Segura, A.M., & Holmgren, M. (1994) Are the responses of matorral shrubs different from those in an ecosystem with a reputed fire history? In *The role of fire in mediterranean-type ecosystems* (eds J.M. Moreno and W.C. Oechel), Ecological Studies 107, pp. 16–25. Springer, New York.

Fulé, P.Z., Ribas, M., Gutiérrez, E., Vallejo, R., & Kaye, M.W. (2008) Forest structure and fire history in an old *Pinus nigra* forest, eastern Spain. *Forest Ecology and Management*, **255**, 1234–1242.

Gadgil, M. & Bossert, W.H. (1970) Life historical consequences of natural selection. *American Naturalist*, **104**, 1–24.

Gagnon, P.R., Passmore, H.A., Platt, W.J., et al. (2010) Does pyrogenicity protect burning plants? *Ecology*, **91**, 3481–3486.

Gajardo, R. (1994) *La vegetación natural de Chile: clasificación y distribución geográfica*. Editorial Universitaria, Santiago, Chile.

Gankin, R. & Major, J. (1964) *Arctostaphylos myrtifolia*, its biology and relationship to the problem of endemism. *Ecology*, **45**, 792–808.

GAO (2009) *Wildland fire management: Federal agencies have taken important steps forward, but additional, strategic action is needed to capitalize on those steps*, GAO-09-877. United States Government Accountability Office, Washington, DC.

García, M., Chuvieco, E., Nieto, H., & Aguado, I. (2008) Combining AVHRR and meteorological data for estimating live fuel moisture content. *Remote Sensing of Environment*, **112**, 3618–3627.

García, M., Litago, J., Palacios-Orueta, A., Pinzón, J.E., & Ustin, S.L. (2010) Short-term propagation of rainfall perturbations on terrestrial ecosystems in central California. *Applied Vegetation Science*, **13**, 146–162.

García-Fayos, P. & Verdú, M. (1998) Soil seed bank, factors controlling germination, and establishment of a Mediterranean shrub: *Pistacia lentiscus* L. *Acta Oecologica*, **19**, 357–366.

Gavin, D.G., Hallett, D.J., Hu, F.S., *et al.* (2007) Forest fire and climate change in western North America: insights from sediment charcoal record. *Frontiers in Ecology and Environment*, **5**, 499–506.

Geldenhuys, C.J. (1994) Bergwind fires and the location pattern of forest patches in the southern Cape landscape, South Africa. *Journal of Biogeography*, **21**, 49–62.

Gent, M.L. & Morgan, J.W. (2007) Changes in the stand structure (1975–2000) of coastal *Banksia* forest in the long absence of fire. *Austral Ecology*, **32**, 239–244.

George, A.S., Hopkins, A.J.M., & Marchant, N.G. (1979) The heathlands of Western Australia. In *Ecosystems of the world*, Vol. 9A: *Heathlands and related shrublands: descriptive studies* (ed. R.L. Specht), pp. 211–230. Elsevier Scientific, New York.

Giannakopoulos, C., Le Sager, P., Bindi, M., *et al.* (2009) Climatic changes and associated impacts in the Mediterranean resulting from a 2°C global warming. *Global and Planetary Change*, **68**, 209–224.

Gibson, A.C. (1983) Anatomy of photosynthetic old stems of nonsucculent dicotyledons from North American deserts. *Botanical Gazette*, **144**, 347–362.

Giglio, L., Csiszar, I., & Justice, C.O. (2006) Global distribution and seasonality of active fires as observed with the Terra and Aqua Moderate Resolution Imaging spectroradiometer (MODIS) sensors. *Journal of Geophysical Research – Biogeosciences*, **111**, G02016, doi:10.1029/2005JG000142.

Gil-Romera, G., Carrion, J.S., McClure, S.B., Schmich, S., & Finlayson, C. (2009) Holocene vegetation dynamics in Mediterranean Iberia: historical contingency and climate-human interactions. *Journal of Anthropological Research*, **65**, 271–285.

Gil-Romera, G., Carrión, J.S., Pausas, J.G., Fernández, S., & Burjachs, F. (2010) Fire regime in southern Iberia: the long-term role of fire as landscape modeller in a western Mediterranean region. *Quaternary Science Reviews*, **29**, 1082–1092.

Gill, A. (1994) *Patterns and processes in open forests of Eucalyptus in southern Australia*. Cambridge University Press, Cambridge, UK.

Gill, A.M. (1973) *Effects of fire on Australia's natural vegetation*, Annual Report, pp. 41–46. CSIRO, Division of Plant Industry, Canberra.

Gill, A.M. (1986) *Research for the fire management of Western Australian state forests and conservation reserves*, Technical Report 12. Department of Conservation and Land Management, Perth, Australia.

Gill, A.M. (1997) Eucalypts and fires: interdependent or independent? In *Eucalypt ecology: individuals to ecosystems* (eds J.E. Williams & J.C.Z. Woinarski), pp. 151–167. Cambridge University Press, Cambridge, UK.

Gill, A.M. (1999) Biodiversity and bushfires: an Australia-wide perspective on plant-species changes after a fire event. In *Australia's biodiversity: responses to fire of plants, birds and invertebrates* (eds A.M. Gill, J.C.Z. Woinarski & A. York), pp. 9–54. Department of the Environment and Heritage, Canberra, Australia.

Gill, A.M. & Allen, G. (2008) Large fires, fire effects and the fire-regime concept. *International Journal of Wildland Fire*, **17**, 688–695.

Gill, A.M. & Bradstock, R.A. (1995) Extinctions of biota by fires. In *Conserving biodiversity: threats and solutions* (eds R.A. Bradstock, T.D. Auld, D.A. Keith, *et al.*), pp. 309–322. Surrey Beatty & Sons, Chipping North, Australia.

Gill, A.M. & Bradstock, R.A. (2003) Fire regimes and biodiversity: a set of postulates. In *Australia burning: fire ecology, policy and management issues* (eds G. Cary, D. Lindenmayer & S. Dovers), pp. 15–25. CSIRO Publishing, Collingwood, Australia.

Gill, A.M., Bradstock, R.A., & Williams, J.E. (2002) Fire regimes and biodiversity: legacy and vision. In *Flammable Australia: the fire regimes and biodiversity of a continent* (eds R.A. Bradstock, J.E. Williams & A.M. Gill), pp. 429–446. Cambridge University Press, Cambridge, UK.

Gill, A.M. & Catling, P.C. (2002) Fire regimes and biodiversity of forested landscapes of southern Australia. In *Flammable Australia: the fire regimes and biodiversity of a continent* (eds R.A. Bradstock, J.E. Williams & M.A. Gill), pp. 351–369. Cambridge University Press, Cambridge, UK.

Gill, A.M. & Ingwersen, F. (1976) Growth of *Xanthorrhoea australis* R. Br. in relation to its fire tolerance. *Journal of Applied Ecology*, **13**, 195–203.

Gill, A.M. & McCarthy, M.A. (1998) Intervals between prescribed fires in Australia: what intrinsic variation should apply? *Biological Conservation*, **85**, 161–169.

Gill, A.M. & McMahon, A. (1986) A postfire chronosequence of cone, collicle and seed production in *Banksia ornata*. *Australian Journal of Botany*, **34**, 425–433.

Gill, A.M. & Moore, P.H.M. (1997) *Contemporary fire regimes in the forests of Southwestern Australia*, Report for Environment Australia. CSIRO Plant Industry, Canberra.

Gill, A.M. & Nicholls, A.O. (1989) Monitoring fire prone flora in reserves for nature conservation. In *Fire management on nature conservation lands* (eds N.D. Burrows, L. McCaw & G. Friend), Occasional Paper 1/89, pp. 137–151. Western Australia Department of Conservation and Land Management, Perth.

Gill, A.M. & Stephens, S.L. (2009) Scientific and social challenges for the management of fire-prone wildland-urban interfaces. *Environmental Research Letters*, **4**, 1–10 (doi:10.1088/1748–9326/4/3/034014).

Gillison, A.N. (1969) Plant succession in an irregularly fired grassland area: Doma Peaks Region, Papua. *Journal of Ecology*, **57**, 415–428.

Ginocchio, R., Holmgren, M., & Montenegro, G. (1994) Effect of fire on plant architecture in Chilean shrubs. *Revista Chilean de Historia Natural*, **67**, 177–182.

Gitas, I.Z., de Santis, A., & Mitri, G.H. (2009) Remote sensing of burn severity. In *Earth observation of wildland fires in Mediterranean ecosystems* (ed. E. Chuvieco), pp. 129–148. Springer, New York.

Givnish, T.J. (1988) Adaptation to sun and shade: a whole plant perspective. *Australian Journal of Plant Physiology*, **15**, 63–92.

Glasspool, I. (2000) A major fire event recorded in the mesofossils and petrology of the Late Permian, Lower Whybrow coal seam, Sydney Basin, Australia. *Palaeogeography, Palaeoclimatology, Palaeoecology*, **164**, 357–380.

Glasspool, I.J., Edwards, D., & Axe, L. (2004) Charcoal in the Silurian as evidence for the earliest wildfire. *Geology*, **32**, 381–383.

Gliessman, S.R. (1978) The establishment of bracken following fire in tropical habitats. *American Fern Journal*, **68**, 41–44.

Glitzenstein, J.S. & Harcombe, P.A. (1988) Effects of the December 1983 tornado on forest vegetation of the big thicket, southeast Texas, U.S.A. *Forest Ecology and Management*, **25**, 269–290.

Goldblatt, P. (1978) Quaternary vegetation changes in southern Africa. In *The biogeography and ecology of Southern Africa* (ed. M.J.A. Werger), pp. 131–143. Dr. W. Junk, The Hague, The Netherlands.

Goldstein, B.E. (2007) The futility of reason: incommensurable differences between sustainability narratives in the aftermath of the 2003 San Diego Cedar Fire. *Journal of Environmental Policy & Planning*, **9**, 227–244.

Gómez-González, S. & Cavieres, L.A. (2009) Litter burning does not equally affect seedling emergence of native and alien species of the Mediterranean-type Chilean matorral. *International Journal of Wildland Fire*, **18**, 213–221.

Gómez-González, S., Sierra-Almeida, A., & Cavieres, L.A. (2008) Does plant-derived smoke affect seed germination in dominant woody species of the mediterranean matorral of central Chile? *Forest Ecology and Management*, **255**, 1510–1515.

González, M.E. & Veblen, T.T. (2006) Climatic influences on fire in *Araucaria araucana–Nothofagus* forests in the Andean cordillera of south-central Chile. *Ecoscience*, **13**, 342–350.

González, M.E. & Veblen, T.T. (2007) Wildfire in *Araucaria araucana* forests and ecological considerations about salvage logging in areas recently burned. *Revista Chilena de Historia Natural*, **80**, 243–253.

González, M.E., Veblen, T.T., & Sibold, J.S. (2005) Fire history of *Araucaria–Nothofagus* forests in Villarrica National Park, Chile. *Journal of Biogeography*, **32**, 1187–1202.

González, M.E., Veblen, T.T., & Sibold, J.S. (2010) Influence of fire severity on stand development of *Araucaria araucana–Nothofagus pumilio* stands in the Andean cordillera of south-central Chile. *Austral Ecology*, **35**, 597–615.

González-Pelayo, O., Andreu, V., Campo, J., Gimeno-García, E., & Rubio, J.L. (2006) Hydrological properties of a Mediterranean soil burned with different fire intensities. *Catena*, **68**, 186–193.

González-Rabanal, F. & Casal, M. (1995) Effect of high temperatures and ash on germination of ten species from gorse shrubland. *Vegetatio*, **116**, 123–131.

González-Tagle, M.A., Schwendenmann, L., Pérez, J.J., & Schulz, R. (2008) Forest structure and woody plant species composition along a fire chronosequence in mixed pine–oak forest in the Sierra Madre Oriental, northeast Mexico. *Forest Ecology and Management*, **256**, 161–167.

Good, R.B. (1996) Fuel dynamics, preplan and future research needs. In *Fire and biodiversity: the effects and effectiveness of fire management: proceedings of the conference held October 1994 in Footscray, Australia*, pp. 253–266. Department of the Environment, Sport and Territories, Canberra, Australia.

Goodman, S.J. & Christian, H.J. (1993) Global observations of lightning. In *Atlas of satellite observations related to global change* (eds R.J. Gurney, J.L. Foster, and C.L. Parkinson), pp. 191–217. Cambridge University Press, Cambridge, UK.

Gordon, D.T. (1970) *Natural regeneration of white and red fir – influence of several factors*, Research Paper PSW-58. USDA Forest Service, Pacific Southwest Forest and Range Experiment Station, Berkeley, California.

Goren-Inbar, N., Alperson, N., Kislev, M.E., et al. (2004) Evidence of hominin control of fire at Gesher Benot Yaaqov, Israel. *Science*, **304**, 725–727.

Gorgens, A.H.M. & van Wilgen, B.W. (2004) Invasive alien plants and water resources in South Africa: current understanding, predictive ability and research challenges. *South African Journal of Science*, **100**, 27–33.

Goubitz, S., Nathan, R., Roitenberg, R., Ne'eman, G., & Shmida, A. (2004) Canopy seed bank structure in relation to: fire, tree size and density. *Plant Ecology*, **173**, 191–201.

Gould, S.J. & Vrba, E.S. (1982) Exaptation: a missing term in the science of form. *Paleobiology* **8**, 4–15.

Grace, J.B. & Keeley, J.E. (2006) A structural equation model analysis of postfire plant diversity in California shrublands. *Ecological Applications*, **16**, 503–514.

Graham, A. (1999) *Late Cretaceous and Cenozoic history of North American vegetation.* Oxford University Press, Oxford, UK.

Graham, R.T., Harvey, A.E., Jain, T.B., & Tonn, J.R. (1999) *The effects of thinning and similar stand treatments on fire behavior in western forests*, General Technical Report, PNW-GTR-463. USDA Forest Service, Pacific Northwest Research Station, Ogden, Utah.

Gratkowski, H.J. (1962) Heat as a factor in germination of seeds of *Ceanothus velutinus* var. *laevigatus* T. & G. Ph.D. dissertation, Oregon State University, Corvallis.

Gray, J.T. (1982) Community structure and productivity in *Ceanothus* chaparral and coastal sage scrub of southern California. *Ecological Monographs*, **52**, 415–435.

Green, L.R. (1981) *Burning by prescription in chaparral*, General Technical Report PSW-51. USDA Forest Service, Pacific Southwest Forest and Range Experiment Station, Berkeley, California.

Greene, D.F. & Johnson, E.A. (2000) Tree recruitment from burn edges. *Canadian Journal of Forest Research*, **30**, 1264–1274.

Greenlee, J.M. & Langenheim, J.H. (1990) Historic fire regimes and their relation to vegetation patterns in the Monterey Bay area of California. *American Midland Naturalist*, **124**, 239–253.

Greenwood, D.R. (1994) Palaeobotanical evidence for Tertiary climates. In *History of the Australian vegetation: Cretaceous to recent* (ed. R.S. Hill), pp. 44–59. Cambridge University Press, Cambridge, UK.

Gregory-Wodzicki, K.M. (2000) Uplift history of the central and northern Andes: a review. *Bulletin of the Geological Society of America*, **112**, 1091–1105.

Griffiths, T. (2002) Environmental history and historical ecology: reviews and meta-analysis. *Australian Journal of Botany*, **50**, 375–389.

Grigulis, K., Lavorel, S., Davies, I.D., et al. (2005) Landscape-scale positive feedbacks between fire and expansion of the large tussock grass, *Ampelodesmos mauritanica* in Catalan shrublands. *Global Change Biology*, **11**, 1042–1053.

Grisebach, A.H.R. (1872) *Die vegetation der Erde nach ihrer klimatishcen Anordnung.* W. Engelmann, Leipzig, Germany.

Gritti, E.S., Smith, B., & Sykes, M.T. (2006) Vulnerability of Mediterranean Basin ecosystems to climate change and invasion by exotic plant species. *Journal of Biogeography*, **33**, 145–157.

Groeneveld, J., Enright, N.J., & Lamont, B.B. (2008) Simulating the effects of different spatio-temporal fire regimes on plant metapopulation persistence in a Mediterranean-type region. *Journal of Applied Ecology*, **45**, 1477–1485.

Groeneveld, J., Enright, N.J., Lamont, B.B., & Wissel, C. (2002) A spatial model of coexistence among three *Banksia* species along a topographic gradient in fire-prone shrublands. *Journal of Ecology*, **90**, 762–774.

Groom, P.K. & Lamont, B.B. (1997) Fruit-seed relations in *Hakea*: serotinous species invest more dry matter in predispersal seed protection. *Australian Journal of Ecology*, **22**, 352–355.

Groom, P.K. & Lamont, B.B. (2010) Phosphorus accumulation in Proteaceae seeds: a synthesis. *Plant and Soil*, **334**, 61–72.

Grove, A.T. & Rackham, O. (2001) *The nature of Mediterranean Europe: an ecological history*. Yale University Press, New Haven, Connecticut.

Groves, R.H. (1991) The biogeography of mediterranean plant invasions. In *Biogeography of mediterranean invasions* (eds R.H. Groves & F. di Castri), pp. 427–438. Cambridge University Press, Cambridge, UK.

Groves, R.H., ed. (1994) *Australian Vegetation*. Cambridge University Press, Cambridge, UK.

Gulmon, S.L. (1977) A comparative study of the grasslands of California and Chile. *Flora*, **166**, 261–278.

Guzmán, B., Lledó, M.D., & Vargas, P. (2009) Adaptive radiation in Mediterranean *Cistus* (Cistaceae). *PLoS ONE*, **4**, e6362.

Guzmán, B. & Vargas, P. (2005) Systematics, character evolution, and biogeography of *Cistus* L. (Cistaceae) based on ITS, trnL-trnF, and matK sequences. *Molecular Phylogenetics and Evolution*, **37**, 644–660.

Guzmán, B. & Vargas, P. (2009a) Historical biogeography and character evolution of Cistaceae (*Malvales*) based on analysis of plastid rbcL and trnL-trnF sequences. *Organisms, Diversity & Evolution*, **9**, 83–99.

Guzmán, B. & Vargas, P. (2009b) Long-distance colonization of the western Mediterranean by *Cistus ladanifer* (Cistaceae) despite the absence of special dispersal mechanisms. *Journal of Biogeography*, **36**, 954–968.

Gworek, J.R., Wall, S.B.V., & Brussard, P.F. (2007) Changes in biotic interactions and climate determine recruitment of Jeffrey pine along an elevation gradient. *Forest Ecology and Management*, **239**, 57–68.

Habrouk, A., Retana, J., & Espelta, J.M. (1999) Role of heat tolerance and cone protection of seeds in the response of three pine species to wildfires. *Plant Ecology*, **145**, 91–99.

Hadar, L., Noy-Meir, I., & Perevolotsky, A. (1999) The effect of shrub clearing and grazing on the composition of a Mediterranean plant community: functional groups versus species. *Journal of Vegetation Science*, **10**, 673–682.

Hahs, A., Enright, N.J., & Thomas, I. (1999) Plant communities, species richness and their environmental correlates in the sandy heaths of Little Desert National Park, Victoria. *Austral Ecology*, **24**, 249–257.

Haidinger, T.L. & Keeley, J.E. (1993) Role of high fire frequency in destruction of mixed chaparral. *Madroño*, **40**, 141–147.

Halpern, C.B. (1989) Early successional patterns of forest species: interactions of life history traits and disturbance. *Ecology*, **70**, 704–720.

Haltenhoff, H. (1991) *Estadísticas de causas de incendios forestales*. Corporación Nacional Forestal, Santiago, Chile.

Hamilton, J.G. (1997) Changing perceptions of pre-European grasslands in California. *Madroño*, **44**, 311–333.

Hammer, R.B., Radeloff, V.C., Fried, J.S., & Stewart, S.I. (2007) Wildland-urban interface housing growth during the 1990s in California, Oregon, and Washington. *International Journal of Wildland Fire*, **16**, 255–265.

Hammill, K.A. & Bradstock, R.A. (2006) Remote sensing of fire severity in the Blue Mountains: influence of vegetation type and inferring fire intensity. *International Journal of Wildland Fire*, **15**, 213–226.

Hampe, A. & Petit, R. (2005) Conserving biodiversity under climate change: the rear edge matters. *Ecology Letters*, **8**, 461–467.

Hanes, T.L. (1971) Succession after fire in the chaparral of southern California. *Ecological Monographs*, **41**, 27–52.

Hanley, M.E. & Lamont, B.B. (2001) Herbivory, serotiny and seedling defense in Western Australian Proteaceae. *Oecologia*, **126**, 409–417.

Hansen, A., Pate, J.S., & Hansen, A.P. (1991) Growth and reproductive performance of a seeder and a resprouter species of *Bossiaea* as a function of plant age after fire. *Annals of Botany*, **67**, 497–510.

Hardig, T.M., Soltis, P.S., & Soltis, D.E. (2000) Diversification of the North American shrub genus *Ceanothus* (Rhamnaceae): conflicting phylogenies from nuclear ribosomal DNA and chloroplast DNA. *American Journal of Botany*, **87**, 108–123.

Hardy, C.R. & Linder, H.P. (2005) Intraspecific variability and timing in ancestral ecology reconstruction: a test case from the Cape flora. *Systematic Biology*, **54**, 299–316.

Harmon, L.J., Kolbe, J.J., Cheverud, J.M., & Losos, J.B. (2005) Convergence and the multidimensional niche. *Evolution*, **59**, 409–421.

Harnis, R.O. & Murray, R.B. (1973) 30 years of vegetal change following burning of sagebrush-grass range. *Journal of Range Management*, **26**, 322–325.

Harper, J.L. (1977) *Population biology of plants*. Academic Press, London.

Harrington, J.P. (1932) *Tobacco among the Karok Indians of California*, Bureau of American Ethnology Bulletin 94. Smithsonian Institution, Washington, DC.

Harris, R.R. (1987) Occurrence of vegetation on geomorphic surfaces in the active floodplain of a California alluvial stream. *American Midland Naturalist*, **118**, 393–405.

Harris, T.M. (1981) Burnt ferns from the English Wealden. *Proceedings of the Geological Association*, **92**, 47–58.

Harrison, S. (1998) Do taxa persist as metapopulations in evolutionary time? In *Biodiversity dynamics: turnover of populations, taxa, and communities* (eds M.L. McKinney & J.A. Drake), pp. 19–30. Columbia University Press, New York.

Harrison, S. & Inouye, B.D. (2002) High B diversity in the flora of Californian serpentine "islands". *Biodiversity and Conservation*, **11**, 1869–1876.

Harrison, S., Inouye, B.D., & Safford, H.D. (2003) Ecological heterogeneity in the effects of grazing and fire grassland diversity. *Conservation Biology*, **17**, 837–845.

Harte, J., McCarthy, S., Taylor, K., Kinzig, A., & Fischer, M.L. (1999) Estimating species-area relationships from plot to landscape scale using species spatial-turnover data. *Oikos*, **86**, 45–54.

Hassell, C.W. & Dodson, J.R. (2003) The fire history of south-west Western Australia prior to European settlement in 1826–1829. In *Fire in ecosystems of south-west Western Australia: impacts and management* (eds I. Abbott & N.D. Burrows), pp. 71–86. Backhuys Publishers, Leiden, The Netherlands.

Hasson, A.E.A., Mills, G.A., Timbal, B., & Walsh, K. (2009) Assessing the impact of climate change on extreme fire weather events over southeastern Australia. *Climate Research*, **39**, 159–172.

Haynes, K., Handmer, J., McAneney, J., Tibbits, A., & Coates, L. (2010) Australian bushfire fatalities 1900–2008: exploring trends in relation to the "Prepare, stay and defend or leave early" policy. *Environmental Science & Policy*, **13**, 185–194.

Heady, H.F. (1977) Valley grasslands. In *Terrestrial vegetation of California* (eds M.G. Barbour & J. Major), pp. 491–514. John Wiley & Sons, New York.

Hédl, R., Kopecký, M., & Komárek, J. (2010) Half a century of succession in a temperate oakwood: from species-rich community to mesic forest. *Diversity and Distributions*, **16**, 267–276.

Heelemann, S., Proches, S., Rebelo, A.G., et al. (2008) Fire season effects on the recruitment of non-sprouting serotinous Proteaceae in the eastern (bimodal rainfall) fynbos biome, South Africa. *Austral Ecology*, **33**, 119–127.

Heinselman, M.L. (1981) Fire intensity and frequency as factors in the distribution and structure of northern ecosystems. In *Proceedings of the conference "Fire regimes and ecosystem properties"* (eds H.A. Mooney, T.M. Bonnicksen, N.L. Christensen, J.E. Lotan & W.A. Reiners), General Technical Report WO-26, pp. 7–57. USDA Forest Service, Washington, DC.

Heizer, R.F. (1978) *Handbook of North American Indians, California*, Vol. 8. Smithsonian Institution, Washington DC.

Henderson, M., Kalabokidis, K., Marmaras, E., Konstantinidis, P., & Marangudakis, M. (2005) Fire and society: a comparative analysis of wildfire in Greece and the United States. *Human Ecology Review*, **12**, 169–182.

Hendey, Q.B. (1983) Palaeontology and palaeoecology of the fynbos region: an introduction. In *Fynbos palaeoecology: a preliminary synthesis* (eds H.J. Deacon, Q.B. Hendey & J.J.N. Lambrechts), Report No. 75, pp. 87–99. CSIR, South African National Scientific Programmes, Pretoria.

Henkin, Z., Seligman, N.G., Noy-Meir, I., & Kafkafi, U. (1999) Secondary succession after fire in a Mediterranean dwarf-shrub community. *Journal of Vegetation Science*, **10**, 503–514.

Hennessy, K., Lucas, C., Nicholls, N., et al. (2006) *Climate change impacts on fire-weather in south-east Australia*. CSIRO Atmospheric Research and Australian Government Bureau of Meteorology, Canberra.

Henry, C.D. (2009) Uplift of the Sierra Nevada, California. *Geology*, **37**, 575–576.

Herendeen, P.S. & Skog, J.E. (1998) *Gleichenia chaloneri*: a new fossil fern from the lower Cretaceous (Albian) of England. *International Journal of Plant Sciences*, **159**, 870–879.

Herranz, J.M., Ferrandis, P., & Martínez-Sánchez, J.J. (1998) Influence of heat on seed germination of seven Mediterranean Leguminosae species. *Plant Ecology*, **136**, 95–103.

Herranz, J.M., Ferrandis, P., & Martínez-Sánchez, J.J. (1999) Influence of heat on seed germination of nine woody Cistaceae species. *International Journal of Wildland Fire*, **9**, 173–182.

Herranz, J.M., Martínez-Sánchez, J.J., Marín, A., & Ferrandis, P. (1997) Postfire regeneration of *Pinus halepensis* Miller in a semi-arid area in Albacete province (southeastern Spain). *Ecoscience*, **4**, 86–90.

Herrera, C.M. (1992) Historical effects and sorting processes as explanations for contemporary ecological patterns: character syndromes in Mediterranean woody plants. *American Naturalist*, **140**, 421–446.

Herring, J.R. (1985) Charcoal fluxes into sediments of the North Pacific Ocean: the Cenozoic record of burning. In *The carbon cycle and atmospheric CO_2: natural variations Archean to present* (eds E.T. Sundquist & W.S. Broecker), pp. 419–442. American Geophysical Union, Washington, DC.

Hessl, A.E., McKenzie, D., & Schellhaas, R. (2004) Drought and Pacific Decadal Oscillation linked to fire occurrence in the inland Pacific Northwest. *Ecological Applications*, **14**, 425–442.

Heusser, L. (1978) Pollen in the Santa Barbara Basin, California: a 12,000-yr record. *Geological Society of America Bulletin*, **89**, 673–678.

Higuera, P.E., Brubaker, L.B., Anderson, P.M., *et al.* (2008) Frequent fires in ancient shrub tundra: implications of paleorecords for Arctic environmental change. *PLoS ONE*, **3**, doi:10.1371/journal.pone.0001744.

Hileman, L.C., Vasey, M.C., & Parker, V.T. (2001) Phylogeny and biogeography of the Arbutoideae (Ericaceae): implications for the Madrean-Tethyan hypothesis. *Systematic Botany*, **26**, 131–143.

Hill, K.D. (1989) Mallee eucalypt communities: their classification and biogeography. In *Mediterranean landscapes in Australia: mallee ecosystems and their management* (eds J.C. Noble & R.A. Bradstock), pp. 93–109. CSIRO Publications, Melbourne, Australia.

Hill, K.D. (1990) Biogeography of the mallee eucalypts. In *The mallee lands: a conservation perspective*, Proceedings of the National Mallee Conference, Adelaide, April 1989 (eds J.C. Noble, P.J. Joss & G.K. Jones), pp. 16–20. CSIRO, Melbourne, Australia.

Hill, K.D. & Johnson, L.A.S. (1995) Systematic studies in the eucalypts, 7: A revision of the bloodwoods genus *Corymbia* (Myrtaceae). *Telopea*, **6**, 185–504.

Hill, R.S. (1994) The history of selected Australian taxa. In *History of the Australian vegetation: Cretaceous to recent* (ed. R.S. Hill), pp. 390–420. Cambridge University Press, Cambridge, UK.

Hill, R.S. (2004) Origins of the southeastern Australian vegetation. *Philosophical Transactions of the Royal Society of London, Series B*, **359**, 1537–1549.

Hill, R.S. & Brodribb, T.J. (2001) Macrofossil evidence for the onset of xeromorphy in Australian Casuarinaceae and tribe Banksieae (Proteaceae). *Journal of Mediterranean Ecology*, **2**, 127–136.

Hill, R.S. & Christophel, D.C. (1988) Tertiary leaves of the tribe Banksieae (Proteaceae) from south-eastern Australia. *Botanical Journal of the Linnean Society*, **97**, 205–227.

Hill, R.S. & Read, J. (1984) Post-fire regeneration of rainforest and mixed forest in Western Tasmania. *Australian Journal of Botany*, **32**, 481–493.

Hill, R.S., Truswell, E.M., McLoughlin, S., & Dettmann, M. (1999) Evolution of the Australian flora: fossil evidence. In *Flora of Australia*, Vol. 1, second edition (ed. A.E. Orchard), pp. 251–320. CSIRO, Melbourne, Australia.

Hinojosa, L.F. (2005) Cambios climáticos y vegetacionales inferidos a partir de paleofloras cenozoicas del sur de Sudamérica. *Revista Geológica de Chile*, **32**, 95–115.

Hinojosa, L.F., Armesto, J.J., & Villagran, C. (2006) Are Chilean coastal forests pre-pleistocene relicts? Evidence from foliar physiognomy, palaeoclimate, and phytogeography. *Journal of Biogeography*, **33**, 331–341.

Hirsch, K.G. & Martell, D.L. (1996) A review of initial attack fire crew productivity and effectiveness. *International Journal of Wildland Fire*, **6**, 199–215.

Hobbs, R. (2005) Landscapes, ecology and wildlife management in highly modified environments: an Australian perspective. *Wildlife Research*, **32**, 389–398.

Hobbs, R.J. (2002) Fire regimes and their effects in Australian temperate woodlands. In *Flammable Australia: the fire regimes and biodiversity of a continent* (eds R.A. Bradstock, J.E. Williams & A.M. Gill), pp. 305–326. Cambridge University Press, Cambridge, UK.

Hobbs, R.J. & Atkins, L. (1990) Fire-related dynamics of a *Banksia* woodland in south-western Western Australia. *Australian Journal of Botany*, **38**, 97–110.

Hobbs, R.J. & Atkins, L. (1991) Interactions between annuals and woody perennials in a Western Australian nature reserve. *Journal of Vegetation Science*, **2**, 643–654.

Hobbs, R.J. & Mooney, H.A. (1985) Vegetative regrowth following cutting in the shrub *Baccharis pilularis* ssp. *consanguinea* (DC) C.B. Wolf. *American Journal of Botany*, **72**, 514–519.

Hoffman, M.T. (1997) Human impacts on vegetation. In *Vegetation of southern Africa* (eds R.M. Cowling, D.M. Richardson & S.M. Pierce), pp. 507–534. Cambridge University Press, Cambridge, UK.

Hoffmann, A. et al. (2001) *Enciclopedia de los bosques chilenos*. Coleccion Voces del Bosque, Santiago, Chile.

Hoffmann, A. & Kummerow, J. (1978) Root studies in the Chilean matorral. *Oecologia*, **32**, 57–69.

Hoffmann, A.J., Teillier, S., & Fuentes, E.R. (1989) Fruit and seed characteristics of woody species in Mediterranean-type regions of Chile and California. *Revista Chilena de Historia Natural*, **62**, 43–60.

Holden, Z.A., Morgan, P., Smith, A.M.S., & Vierling, L. (2010) Beyond Landsat: a comparison of four satellite sensors for detecting burn severity in ponderosa pine forests of the Gila Wilderness, NM, USA. *International Journal of Wildland Fire*, **19**, 449–458.

Holdgate, G.R., Cartwright, I., Blackburn, D.T., et al. (2007) The middle Miocene Yallourn coal seam: the last coal in Australia. *International Journal of Coal Geology*, **70**, 95–115.

Holland, P.G. (1986) Mallee vegetation: steady state or successional? *Australian Geography*, **17**, 113–20.

Holland, V.L. & Keil, D.J. (1995) *California vegetation*. Kendall/Hunt Publishing Company, Dubuque, Iowa.

Holmes, P.M. (2002) Depth distribution and composition of seed-banks in alien-invaded and uninvaded fynbos vegetation. *Austral Ecology*, **27**, 110–120.

Holmes, P.M. & Cowling, R.M. (1997a) The effects of invasion by *Acacia saligna* on the guild structure and regeneration capabilities of South African fynbos shrublands. *Journal of Applied Ecology*, **34**, 317–332.

Holmes, P.M. & Cowling, R.M. (1997b) Diversity, composition and guild structure relationships between soil-stored seed banks and mature vegetation in alien plant-invaded South African fynbos shrublands. *Plant Ecology*, **133**, 107–122.

Holmes, P.M. & Foden, W. (2001) The effectiveness of post-fire soil disturbance in restoring fynbos after alien clearance. *South African Journal of Botany*, **67**, 533–539.

Holmes, P.M. & Newton, R.J. (2004) Patterns of seed persistence in South African fynbos. *Plant Ecology*, **172**, 143–158.

Holmes, P.M., Richardson, D.M., vanWilgen, B.W., & Gelderblom, C. (2000) Recovery of South African fynbos vegetation following alien woody plant clearing and fire: implications for restoration. *Austral Ecology*, **25**, 631–639.

Holmes, W.B.K., Holmes, F.M., & Martin, H.A. (1982) Fossil *Eucalyptus* remains from the Middle Miocene Chalk Mountain Formation, Warrumbungle Mountains, New South Wales. *Proceedings of the Linnean Society N.S.W.*, **106**, 299–310.

Holmgren, M., Aviles, R., Sierralta, L., Segura, A.M., & Fuentes, E.R. (2000a) Why have European herbs so successfully invaded the Chilean matorral? Effects of herbivory, soil nutrients, and fire. *Journal of Arid Environments*, **44**, 197–211.

Holmgren, M., Segura, A.M., & Fuentes, E.R. (2000b) Limiting mechanisms in the regeneration of the Chilean matorral: experiments on seedling establishment in burned and cleared mesic sites. *Plant Ecology*, **147**, 49–57.

Hoover, R.F. (1936) Character and distribution of the primitive vegetation of the San Joaquin Valley, M.A. thesis. University of California, Berkeley.

Hopkins, A. & Griffin, E. (1984) Floristic patterns. In *Kwongan: Plant life of the sandplain: biology of a south-west Australian shrubland ecosystem* (eds J. Pate & J. Beard), pp. 69–83. University of Western Australia Press, Nedlands, Australia.

Hopkins, A.J.M. & Griffin, E.A. (1989) Fire in *Banksia* woodlands of the Swan Coastal Plain. *Journal of the Royal Society of Western Australia*, **71**, 93–94.

Hopkins, A.J.M., Keighery, G.J., & Marchant, N.G. (1983) Species-rich uplands of south-western Australia. *Proceedings of the Ecological Society of Australia*, **12**, 15–26.

Hopkins, A.J.M. & Robinson, C.J. (1981) Fire induced structural change in a Western Australian woodland. *Australian Journal of Ecology*, **6**, 177–188.

Hopper, S.D. (1979) Biogeographic aspects of speciation in the southwest Australian flora. *Annual Review of Ecology and Systematics*, **10**, 399–422.

Hopper, S.D. (2003) An evolutionary perspective on south-west Western Australia landscapes, biodiversity and fire: a review and management implications. In *Fire in ecosystems of south-west Western Australia: impacts and management* (eds I. Abbott & N. Burrows), pp. 9–35. Backhuys Publishers, Leiden, The Netherlands.

Hopper, S.D. (2009) OCBIL theory: towards an integrated understanding of the evolution, ecology and conservation of biodiversity on old, climatically buffered, infertile landscapes. *Plant and Soil*, **322**, 49–86.

Hopper, S.D. & Gioia, P. (2004) The southwest Australian floristic region: evolution and conservation of a global hot spot of biodiversity. *Annual Review of Ecology and Systematics*, **35**, 623–650.

Hopper, S.D., Smith, R.J., Fay, M.F., Manning, J.C., & Chase, M.W. (2009) Molecular phylogenetics of Haemodoraceae in the greater Cape and Southwest Australian floristic regions. *Molecular Phylogenetics and Evolution*, **51**, 19–30.

Horne, I.P. (1981) The frequency of veld fires in the Groot Swartberg Mountain Catchment Area, Cape Province. *South African Forestry Journal*, **118**, 56–60.

Howell, J., Humphreys, G.S., & Mitchell, P.B. (2006) Changes in soil water repellence and its distribution in relation to surface microtopographic units after a low severity fire in eucalypt woodland, Sydney, Australia. *Australian Journal of Soil Research*, **44**, 205–217.

Hren, M.T., Pagani, M., Erwin, D.M., & Brandon, M. (2010) Biomarker reconstruction of the early Eocene paleotopography and paleoclimate of the northern Sierra Nevada. *Geology*, **38**, 7–10.

Huenneke, L.F. (1989) Distribution and regional patterns of Californian grasslands. In *Grassland structure and function: California annual grasslands* (eds L.F. Huenneke & H.A. Mooney), pp. 1–12. Kluwer Academic Publishers, Dordrecht, The Netherlands.

Huenneke, L.F. & Mooney, H.A., eds. (1989) *Grassland structure and function: California annual grasslands*, 218 pp. Kluwer Academic Publishers, Dordrecht, The Netherlands.

Hunter, J.C. & Parker, V.T. (1993) The disturbance regime of an old-growth forest in coastal California. *Journal of Vegetation Science*, **4**, 19–24.

Hurteau, M.D., North, M., & Foin, T. (2009) Modeling the influence of precipitation and nitrogen deposition on forest understory fuel connectivity in Sierra Nevada mixed-conifer forest. *Forest Ecology and Management*, **220**, 2460–2468.

Huston, J.M. (1964) *The Western Mediterranean World: an introduction to its landscapes*. Longman, London.

Huston, M. (2003) Understanding the effects of fire and other mortality-causing disturbances on species diversity. In *Fire in ecosystems of south west Western Australia: impacts and management* (eds I. Abbott & N.D. Burrows). Backhuys Publishers, Leiden, The Netherlands.

Hutchinson, M.F., McIntyre, S., Hobbs, R.J., Stein, J.L., Garnett, S., & Kinlock, J. (2005) Integrating a global agro-climatic classification with bioregional boundaries in Australia. *Global Ecology and Biogeography*, **14**, 197–212.

Ibáñez, J.J., Lledó, M.J., Sánchez, J.R., & Rodà, F. (1999) Stand structure, aboveground biomass and production. In *Ecology of Mediterranean evergreen oak forests* (eds F. Rodà, J. Retana, C.A. Gracia & J. Bellot), pp. 31–46. Springer, Berlin.

IBHS (2009) *Mega fires: the case for mitigation: the Witch Creek Wildfire, October 21–31, 2007*. Institute for Business & Home Safety, Tampa, Florida.

Insole, A.N. & Hutt, S. (1994) The palaeoecology of the dinosaurs of the Wessex Formation (Wealden Group, Early Cretaceous), Isle of Wight, Southern England. *Zoological Journal of the Linnean Society*, **112**, 197–215.

Jackson, W.D. (1968) Fire, air, water and earth: an elemental ecology of Tasmania. *Proceedings of the Ecological Society of Australia*, **3**, 9–16.

Jacobsen, A.L., Davis, S.D., & Babritus, S.L. (2004) Fire frequency impacts non-sprouting chaparral shubs in the Santa Monica Mountains of southern California. In *Ecology and conservation of mediterranean climate ecosystems* (eds M. Arianoutsou & V.P. Panastasis). Millpress, Rotterdam, The Netherlands.

Jacobsen, A.L., Pratt, R.B., Davis, S.D., & Ewers, F.W. (2008) Comparative community physiology: nonconvergence in water relations among three semi-arid shrub communities. *New Phytologist*, **180**, 100–113.

Jacobsen, A.L., Pratt, R.B., Moe, L.M., & Ewers, F.W. (2009) Plant community water use and invasibility of semi-arid shrublands by woody species in southern California. *Madroño*, **56**, 213–220.

Jacobson, A.L., Davis, S.D., & Babritius, S.L. (2004) Fire frequency impacts non-sprouting chaparral shrubs in the Santa Monica Mountains of southern California. In *Ecology, conservation and management of mediterranean climate ecosystems* (eds M. Arianoutsou & V.P. Panastasis), CD ROM. Millpress, Rotterdam, The Netherlands.

Jain, T., Pilliod, D., & Graham, R. (2004) *Tongue-tied. Wildfire*, **4**, 22–36.

James, S. (1984) Lignotubers and burls: their structure, function and ecological significance in Mediterranean ecosystems. *Botanical Review*, **50**, 225–266.

Jefferson, L.V., Pennacchio, M., Havens, K., *et al.* (2008) Ex situ germination responses of midwestern USA prairie species to plant-derived smoke. *American Midland Naturalist*, **159**, 251–256.

Jeong, S.C., Liston, A., & Myrold, D.D. (1997) Molecular phylogeny of the genus *Ceanothus* (Rhamnaceae) using rbcL and ndhF sequences. *Theoretical and Applied Genetics*, **94**, 852–857.

Jia, G., Peng, P., Zhao, Q., & Jian, Z. (2003) Changes in terrestrial ecosystem since 30 Ma in East Asia: stable isotope evidence from black carbon in the South China Sea. *Geology*, **31**, 1093–1096.

Jimenez, A., Pauchard, A., Cavieres, L.A., Marticorena, A., & Bustamante, R.O. (2008) Do climatically similar regions contain similar alien floras? A comparison between the mediterranean areas of central Chile and California. *Journal of Biogeography*, **35**, 614–624.

Jiménez, H.E. & Armesto, J.J. (1992) Importance of the soil seed bank of disturbed sites in Chilean matorral in early secondary succession. *Journal of Vegetation Science*, **3**, 579–586.

Johansson, M., Rooke, T., Fetene, M., & Granström, A. (2010) Browser selectivity alters post-fire competition between *Erica arborea* and *E. trimera* in the sub-alpine heathlands of Ethiopia. *Plant Ecology*, **207**, 149–160.

Johnston, R.D. & Lacey, C.J. (1983) Multi-stemmed trees in rainforest. *Australian Journal of Botany*, **31**, 189–195.

Johnstone, J.F. & F.S. Chapin, I. (2006) Effects of soil burn severity on post-fire tree recruitment in boreal forest. *Ecosystems*, **9**, 14–31.

Johnstone, M.H., Lowry, D.C., & Quilty, P.G. (1973) The geology of southwestern Australia: a review. *Journal of the Royal Society of Western Australia*, **56**, 5–15.

Jones, J.H. (2008) SDG&E's fire plan would cut service. *San Diego Union–Tribune*, October 3, 2008.

Jones, R. (1969) Fire-stick farming. *Australian Natural History*, 224–228.

Jones, T.P., Ash, S., & Figueiral, I. (2002) Late Triassic charcoal from Petrified Forest National Park, Arizona, USA. *Palaeogeography, Palaeoclimatology, Palynology*, **188**, 127–139.

Jordan, G.J., Weston, P.H., Carpenter, R.J., Dillon, R.A., & Brodribb, T.J. (2008) The evolutionary relations of sunken covered, and encrypted stomata to dry habitats in Proteaceae. *American Journal of Botany*, **95**, 521–530.

Jordano, P., García, C., Godoy, J.A., & García-Castaño, J.L. (2007) Differential contribution of frugivores to complex seed dispersal patterns. *Proceedings of the National Academy of Sciences*, **104**, 3278–3282.

Jurado, E. & Westoby, M. (1992) Seedling growth in relation to seed size among species of arid Australia. *Journal of Ecology*, **80**, 407–416.

Kaniewski, D., Paulissen, E., De Laet, V., & Waelkens, M. (2008) Late Holocene fire impact and post-fire regeneration from the Bereket basin, Taurus Mountains, southwest Turkey. *Quaternary Research*, **70**, 228–239.

Kauffman, J.B. (1991) Survival by sprouting following fire in tropical forests of the eastern Amazon. *Biotropica* **23**, 219–224.

Kauffman, J.B. & Martin, R.E. (1990) Sprouting shrub response to different seasons and fuel consumption levels of prescribed fire in Sierra Nevada mixed conifer ecosystems. *Forest Science*, **36**, 748–764.

Kauffman, J.B. & Martin, R.E. (1991) Factors affecting the scarification and germination of three montane Sierra Nevada shrubs. *Northwest Science*, **65**, 180–187.

Kavgaci, A., Carni, A., Basaran, S., *et al.* (2010) Long-term post-fire succession of *Pinus brutia* forest in the east Mediterranean. *International Journal of Wildland Fire*, **19**, 599–605.

Kazanis, D. & Arianoutsou, M. (2004) Long-term post-fire vegetation dynamics in *Pinus halepensis* forests of central Greece: a functional group approach. *Plant Ecology*, **171**, 101–121.

Keane, R.E., Cary, G.J., & Parsons, R. (2003) Using simulation to map fire regimes: an evaluation of approaches, strategies, and limitations. *International Journal of Wildland Fire*, **12**, 309–322.

Keeley, J.E. (1977) Seed production, seed populations in the soil and seedling production after fire for two congeneric pairs of sprouting and nonsprouting chaparral shrubs. *Ecology*, **58**, 820–829.

Keeley, J.E. (1981) Reproductive cycles and fire regimes. In *Proceedings of the conference "Fire regime and ecosystems properties", December 11–15, 1978, Honolulu, Hawaii* (eds H.A. Mooney, T.M. Bonnicksen, N.L. Christensen, J.E. Lotan & W.A. Rainers), General Technical Report WO-26, pp. 231–277. USDA Forest Service, Washington, DC.

Keeley, J.E. (1982) Distribution of lightning and man-caused wildfires in California. In *Proceedings of the symposium on dynamics and management of mediterranean-type ecosystems* (eds C.E. Conrad & W.C. Oechel), General Technical Report PSW-58, pp. 431–437. USDA Forest Service, Pacific Southwest Forest and Range Experiment Station, Berkeley, California.

Keeley, J.E. (1986) Resilience of Mediterranean shrub communities to fire. In *Resilience in mediterranean-type ecosystems* (eds B. Dell, A.J.M. Hopkins & B.B. Lamont), pp. 95–112. Dr. W. Junk, Dordrecht, The Netherlands.

Keeley, J.E. (1987) Role of fire in seed germination of woody taxa in California chaparral. *Ecology*, **68**, 434–443.

Keeley, J.E. (1990a) The California valley grassland. In *Endangered plant communities of southern California* (ed. A.A. Schoenherr), Special Publication 3, pp. 2–23. Southern California Botanists, Fullerton, California.

Keeley, J.E. (1990b) Demographic structure of California black walnut (*Juglans californica*; Juglandaceae) woodlands in southern California. *Madroño*, **37**, 237–248.

Keeley, J.E. (1991) Seed germination and life history syndromes in the California chaparral. *Botanical Review*, **57**, 81–116.

Keeley, J.E. (1992a) Demographic structure of California chaparral in the long-term absence of fire. *Journal of Vegetation Science*, **3**, 79–90.

Keeley, J.E. (1992b) A Californian's view of fynbos. In *The ecology of fynbos* (ed. R.M. Cowling), pp. 372–388. Oxford University Press, Cape Town, RSA.

Keeley, J.E. (1992c) Recruitment of seedlings and vegetative sprouts in unburned chaparral. *Ecology*, **73**, 1194–1208.

Keeley, J.E. (1992d) Temporal and spatial dispersal syndromes. In *MEDECOS VI: Proceedings of the 6th international conference on Mediterranean climate ecosystems, "Plant-animal interactions in mediterranean-type ecosystems"* (ed. C.A. Thanos), pp. 251–256. University of Athens, Greece.

Keeley, J.E. (1993a) Assessing suitable sites for grassland restoration. In *Interface between ecology and land development in California* (ed. J.E. Keeley), pp. 277–281. Southern California Academy of Sciences, Los Angeles.

Keeley, J.E. (1993b) Smoke-induced flowering in the fire-lily *Cyrtanthus ventricosus*. *South African Journal of Botany*, **59**, 638.

Keeley, J.E. (1995a) Seed germination patterns in fire-prone Mediterranean-climate regions. In *Ecology and biogeography of Mediterranean ecosystems in Chile, California and Australia* (eds M.T.K. Arroyo, P.H. Zedler & M.D. Fox), pp. 239–273. Springer, New York.

Keeley, J.E. (1995b) Future of California floristics and systematics: wildfire threats to the California flora. *Madroño*, **42**, 175–179.

Keeley, J.E. (1996) Postfire vegetation recovery in the Santa Monica mountains under two alternative management programs. *Bulletin of the Southern California Academy of Sciences*, **95**, 103–119.

Keeley, J.E. (1998) Coupling demography, physiology and evolution in chaparral shrubs. In *Landscape disturbance and biodiversity in Mediterranean-type ecosystems* (eds P.W. Rundel, G. Montenegro & F.M. Jaksic), pp. 257–264. Springer, New York.

Keeley, J.E. (2000) Chaparral. In *North American terrestrial vegetation* (eds M.G. Barbour & W.D. Billings), pp. 203–253. Cambridge University Press, Cambridge, UK.

Keeley, J.E. (2001) Fire and invasive species in Mediterranean-climate ecosystems of California. In *Proceedings of the invasive species workshop: the role of fire in the control and spread of invasive species* (eds K.E.M. Galley & T.P. Wilson), Miscellaneous Publication 11, pp. 81–94. Tall Timbers Research Station, Tallahassee, Florida.

Keeley, J.E. (2002a) Fire management of California shrubland landscapes. *Environmental Management*, **29**, 395–408.

Keeley, J.E. (2002b) Native American impacts on fire regimes of the California coastal ranges. *Journal of Biogeography*, **29**, 303–320.

Keeley, J.E. (2003) Relating species abundance distributions to species-area curves in two Mediterranean-type shrublands. *Diversity and Distributions*, **9**, 253–259.

Keeley, J.E. (2004) Ecological impacts of wheat seeding after a Sierra Nevada wildfire. *International Journal of Wildland Fire*, **13**, 73–78.

Keeley, J.E. (2005) Fire history of the San Francisco East Bay region and implications for landscape patterns. *International Journal of Wildland Fire*, **14**, 285–296.

Keeley, J.E. (2006a) South coast bioregion. In *Fire in California's ecosystems* (eds N.G. Sugihara, J.W. van Wagtendonk, K.E. Shaffer, J. Fites-Kaufman & A.E. Thoede), pp. 350–390. University of California Press, Los Angeles.

Keeley, J.E. (2006b) Fire management impacts on invasive plant species in the western United States. *Conservation Biology*, **20**, 375–384.

Keeley, J.E. (2006c) Fire severity and plant age in postfire resprouting of woody plants in sage scrub and chaparral. *Madroño*, **53**, 373–379.

Keeley, J.E. (2009) Fire intensity, fire severity and burn severity: a brief review and suggested usage. *International Journal of Wildland Fire*, **18**, 116–126.

Keeley, J.E., Allen, C.D., Betancourt, J., et al. (2006a) A 21st century perspective on postfire seeding. *International Journal of Wildland Fire*, **104**, 103–104.

Keeley, J.E., Aplet, G.H., Christensen, N.L., et al. (2009a) *Ecological foundations for fire management in North American forest and shrubland ecosystems*, General Technical Report PNW-GTR-779, pp. 99. USDA Forest Service, Pacific Northwest Research Station, Portland, Oregon.

Keeley, J.E., Baer-Keeley, M., & Fotheringham, C.J. (2005d) Alien plant dynamics following fire in Mediterranean-climate California shrublands. *Ecological Applications*, **15**, 2109–2125.

Keeley, J.E. & Bond, W.J. (1997) Convergent seed germination in South African fynbos and Californian chaparral. *Plant Ecology*, **133**, 153–167.

Keeley, J.E. & Bond, W.J. (1999) Mast flowering and semelparity in bamboos: the bamboo fire cycle hypothesis. *American Naturalist*, **154**, 383–391.

Keeley, J.E., Brennan, T., & Pfaff, A.H. (2008) Fire severity and ecosystem responses following crown fires in California shrublands. *Ecological Applications* **18**, 1530–1546.

Keeley, J.E. & Davis, F.W. (2007) Chaparral. In *Terrestrial vegetation of California*, third edition (eds M.G. Barbour, T. Keeler-Wolf & A.A. Schoenherr), pp. 339–366. University of California Press, Los Angeles, California.

Keeley, J.E. & Fotheringham, C.J. (1997) Trace gas emissions in smoke-induced seed germination. *Science*, **276**, 1248–1250.

Keeley, J.E. & Fotheringham, C.J. (1998) Smoke-induced seed germination in California chaparral. *Ecology*, **79**, 2320–2336.

Keeley, J.E. & Fotheringham, C.J. (2000) Role of fire in regeneration from seed. In *Seeds: the ecology of regeneration in plant communities*, second edition (ed. M. Fenner), pp. 311–330. CAB International, Wallingford, UK.

Keeley, J.E. & Fotheringham, C.J. (2001a) Historic fire regime in Southern California shrublands. *Conservation Biology*, **15**, 1536–1548.

Keeley, J.E. & Fotheringham, C.J. (2001b) History and management of crown-fire ecosystems: a summary and response. *Conservation Biology*, **15**, 1561–1567.

Keeley, J.E. & Fotheringham, C.J. (2003a) Impact of past, present, and future fire regimes on North American Mediterranean shrublands. In *Fire and climatic change in temperate ecosystems of the western Americas* (eds T.T. Veblen, W.L. Baker, G. Montenegro & T.W. Swetnam), pp. 218–262. Springer, New York.

Keeley, J.E. & Fotheringham, C.J. (2003b) Species-area relationships in Mediterranean-climate plant communities. *Journal of Biogeography*, **30**, 1629–1657.

Keeley, J.E. & Fotheringham, C.J. (2005) Plot shape effects on plant species diversity measurements. *Journal of Vegetation Science*, **16**, 249–256.

Keeley, J.E., Fotheringham, C.J., & Baer-Keeley, M. (2005a) Determinants of postfire recovery and succession in mediterranean-climate shrublands of California. *Ecological Applications*, **15**, 1515–1534.

Keeley, J.E., Fotheringham, C.J., & Baer-Keeley, M. (2005c) Factors affecting plant diversity during post-fire recovery and succession of mediterranean-climate shrublands in California, USA. *Diversity and Distributions*, **11**, 525–537.

Keeley, J.E., Fotheringham, C.J., & Baer-Keeley, M. (2006b) Demographic patterns of postfire regeneration in mediterranean-climate shrublands of California. *Ecological Monographs*, **76**, 235–255.

Keeley, J.E., Fotheringham, C.J., & Morais, M. (1999a) Reexamining fire suppression impacts on brushland fire regimes. *Science*, **284**, 1829–1832.

Keeley, J.E., Fotheringham, C.J., & Moritz, M.A. (2004) Lessons from the 2003 wildfires in southern California. *Journal of Forestry*. **102**, 26–31.

Keeley, J.E., Keeley, M.B., & Bond, W.J. (1999b) Stem demography and post-fire recruitment of a resprouting serotinous conifer. *Journal of Vegetation Science*, **10**, 69–76.

Keeley, J.E. & Keeley, S.C. (1984) Postfire recovery of California coastal sage scrub. *American Midland Naturalist*, **111**, 105–117.

Keeley, J.E. & Keeley, S.C. (1987) The role of fire in the germination of chaparral herbs and suffrutescents. *Madroño*, **34**, 240–249.

Keeley, J.E. & Keeley, S.C. (1988) Chaparral. In *North American terrestrial vegetation* (eds M.G. Barbour & W.D. Billings). Cambridge University Press, Cambridge, UK.

Keeley, J.E., Lubin, D., & Fotheringham, C.J. (2003) Fire and grazing impacts on plant diversity and alien plant invasions in the southern Sierra Nevada. *Ecological Applications*, 13, 1355–1374.

Keeley, J.E. & McGinnis, T.W. (2007) Impact of prescribed fire and other factors on cheatgrass persistence in a Sierra Nevada ponderosa pine forest. *International Journal of Wildland Fire*, 16, 96–106.

Keeley, J.E. & Nitzberg, M.E. (1984) The role of charred wood in the germination of the chaparral herbs *Emmenanthe penduliflora* (Hydrophyllaceae) and *Eriophyllum confertiflorum* (Asteraceae). *Madroño*, 31, 208–218.

Keeley, J.E., Pausas, J.G., Rundel, P.W., et al. (2011) Fire as an evolutionary pressure shaping plant traits. *Trends in Plant Science*, 16, 406–411.

Keeley, J.E., Pfaff, A.H., & Safford, H.D. (2005b) Fire suppression impacts on postfire recovery of Sierra Nevada chaparral shrublands. *International Journal of Wildland Fire*, 14, 255–265.

Keeley, J.E. & Rundel, P.W. (2005) Fire and the Miocene expansion of C_4 grasslands. *Ecology Letters*, 8, 683–690.

Keeley, J.E., Safford, H., Fotheringham, C.J., Franklin, J., & Moritz, M. (2009b) The 2007 southern California wildfires: lessons in complexity. *Journal of Forestry*, 107, 287–296.

Keeley, J.E. & Stephenson, N.L. (2000) Restoring natural fire regimes in the Sierra Nevada in an era of global change. In *Wilderness science in a time of change conference* (eds D.N. Cole, S.F. McCool & J. O'Loughlin), RMRS-P-15, Vol. 5, pp. 255–265. USDA Forest Service, Rocky Mountain Research Station, Missoula, Montana.

Keeley, J.E. & Swift, C.C. (1995) Biodiversity and ecosystem functioning in Mediterranean-climate California. In *Biodiversity and function in mediterranean-type ecosystems* (eds G.W. Davis & D.M. Richardson), pp. 121–183. Springer, New York.

Keeley, J.E. & van Mantgem, P.J. (2008) Community ecology of seedlings. In *Seedling ecology and evolution* (eds M.A. Leak, V.T. Parker & R.L. Simpson), pp. 253–270. Cambridge University Press, Cambridge, UK.

Keeley, J.E., Vasey, M.C., & Parker, V.T. (2007) Subspecific variation in the widespread burl-forming *Arctostaphylos glandulosa*. *Madroño* 54, 42–62.

Keeley, J.E. & Zedler, P.H. (1978) Reproduction of chaparral shrubs after fire: a comparison of sprouting and seeding strategies. *American Midland Naturalist*, 99, 142–161.

Keeley, J.E. & Zedler, P.H. (1998) Evolution of life histories in *Pinus*. In Ecology and biogeography of *Pinus* (ed. D.M. Richardson), pp. 219–250. Cambridge University Press, Cambridge, UK.

Keeley, J.E. & Zedler, P.H. (2009) Large, high intensity fire events in southern California shrublands: debunking the fine-grained age-patch model. *Ecological Applications*, 19, 69–94.

Keeley, J.E., Zedler, P.H., Zammit, C.A., & Stohlgren, T.J. (1989) Fire and demography. In *The California chaparral: paradigms reexamined* (ed. S.C. Keeley), Science Series 34, pp. 151–153. Natural History Museum of Los Angeles County, Los Angeles, California.

Keeley, S.C. & Johnson, A.W. (1977) A comparison of the pattern of herb and shrub growth in comparable sites in Chile and California. *American Midland Naturalist*, 97, 120–132.

Keeley, S.C. & Pizzorno, M. (1986) Charred wood stimulated germination of two fire-following herbs of the California chaparral and the role of hemicellulose. *American Journal of Botany*, 73, 1289–1297.

Keifer, M.B. (1998) Fuel load and tree density changes following prescribed fire in the giant sequoia-mixed conifer forest: the first 14 years of fire effects monitoring. In *Fire in ecosystem management: shifting the paradigm from suppression to prescription* (eds T.L. Pruden & L.A. Brennan), Tall Timbers Fire Ecology Conference Proceedings 20, pp. 306–309. Tall Timbers Research Station, Tallahassee, Florida.

Keifer, M.B., van Wagtendonk, J.W., & Buhler, M. (2006) Long-term surface fuel accumulation in burned and unburned mixed-conifer forests of the central and southern Sierra Nevada, CA (USA). *Fire Ecology*, **2**, 53–72.

Keith, D. (1996) Fire-driven extinction of plant populations: a synthesis of theory and review of evidence from Australian vegetation. *Proceedings of the Linnean Society of New South Wales*, **116**, 37–78.

Keith, D.A. & Bradstock, R.A. (1994) Fire and competition in Australian heath: a conceptual model and field investigations. *Journal of Vegetation Science*, **5**, 347–354.

Keith, D.A., Holman, L., Rodoreda, S., Lemmon, J., & Bedward, M. (2007) Plant functional types can predict decade-scale changes in fire-prone vegetation. *Journal of Ecology*, **95**, 1324–1337.

Keith, D.A. & Myerscough, P.J. (1993) Floristics and soil relations of upland swamp vegetation near Sydney. *Austral Ecology*, **18**, 325–344.

Keith, D.A., Williams, J.E., & Woinarski, J.C.Z. (2002) Fire management and biodiversity conservation: key approaches and principles. In *Flammable Australia: the fire regimes and biodiversity of a continent* (eds R.A. Bradstock, J.E. Williams & A.M. Gill), pp. 401–428. Cambridge University Press, Cambridge, UK.

Kelly, A.E. & Goulden, M.L. (2008) Rapid shifts in plant distribution with recent climate change. *Proceedings of the National Academy of Sciences*, **105**, 11 823–11 826.

Kemp, E.M. (1981) Pre-quaternary fire in Australia. In *Fire and the Australian biota* (eds A.M. Gill, R.H. Groves & I.R. Noble), pp. 3–21. Australian Academy of Science, Canberra.

Kemper, J., Cowling, R.M., Richardson, D.M., Forsyth, G.G., & McKelly, D.H. (2000) Landscape fragmentation in South Coast Renosterveld, South Africa, in relation to rainfall and topography. *Austral Ecology*, **25**, 179–186.

Kerr, B., Schwilk, D.W., Bergman, A., & Feldman, M.W. (1999) Rekindling an old flame: a haploid model for the evolution and impact of flammability in resprouting plants. *Evolutionary Ecology Research*, **1**, 807–833.

Kerr, L.R. (1925) The lignotubers of eucalypt seedlings. *Proceedings of the Royal Society of Victoria*, **37**, 79–97.

Kershaw, A.P., Clark, J.S., Gill, A.M., & D'Costa, D.M. (2002) A history of fire in Australia. In *Flammable Australia: the fire regimes and biodiversity of a continent* (eds R.A. Bradstock, J.E. Williams & A.M. Gill), pp. 3–25. Cambridge University Press, Cambridge, UK.

Key, C.H. & Benson, N.C. (2006) *Landscape assessment: sampling and anlaysis methods*. USDA Forest Service, Rocky Mountain Research Station, Ogden, Utah.

Keyes, C.R., Acker, S.A., & Greene, S.E. (2001) Overstory and shrub influences on seedling recruitment patterns in an old-growth ponderosa pine stand. *Northwest Science*, **75**, 204–210.

Keyes, C.R., Maguire, D.A., & Tappeiner, J.C. (2007) Observed dynamics of ponderosa pine (*Pinus ponderosa* var. *ponderosa* Dougl. ex Laws.) seedling recruitment in the Cascade Range, USA. *New Forests*, **34**, 95–105.

Kilgore, B.M. & Briggs, G.S. (1972) Restoring fire to high elevation forests in California. *Journal of Forestry*, **70**, 267–271.

Kilgore, B.M. & Taylor, D. (1979) Fire history of a sequoia-mixed conifer forest. *Ecology*, **60**, 129–142.

Kimber, R. (1983) Black lightning: Aborigines and fire in Central Australia and the Western Desert. *Archaeology in Oceania*, **18**, 38–44.

King, K.J. (2004) *Simulating the effects of anthropogenic burning on patterns of biodiversity*. Australian National University, Canberra.

King, K.J., Bradstock, R.A., Cary, G., Chapman, C., & Marsden-Smedley, J.B. (2008) The relative importance of fine-scale fuel mosaics on reducing fire risk in south-west Tasmania, Australia. *International Journal of Wildland Fire*, **17**, 421–430.

King, K.J., Cary, G.J., Bradstock, R.A., Chapman, J., Pyrke, A., & Marsden-Smedley, J.B. (2006) Simulation of prescribed burning strategies in south-west Tasmania, Australia: effects on unplanned fires, fire regimes, and ecological management values. *International Journal of Wildland Fire*, **15**, 527–540.

Kirkpatrick, J.B. (1997) Vascular plant-eucalypt interactions. In *Eucalypt ecology: individuals to ecosystems* (eds J.E. Williams & J.C.Z. Woinarski), pp. 227–245. Cambridge University Press, Cambridge, UK.

Kirkpatrick, J.B. & Dickinson, K.J.M. (1984) The impact of fire on Tasmanian alpine vegetation and soils. *Australian Journal of Botany*, **32**, 613–629.

Kitzberger, T. & Veblen, T.T. (1997) Influences of humans and ENSO on fire history of *Austrocedrus chilensis* woodlands in northern Patagonia, Argentina. *Ecoscience*, **4**, 508–520.

Kitzberger, T. & Veblen, T.T. (1999) Fire-induced changes in northern Patagonian landscapes. *Landscape Ecology*, **14**, 1–15.

Kitzberger, T. & Veblen, T.T. (2003) Influences of climate on fire in Northern Patagonia, Argentina. In *Fire and climatic change in temperate ecosystems of the western Americas* (ed. T.T. Veblen, W.L. Baker, G. Montenegro & T.W. Swetnam), pp. 296–321. Springer, New York.

Kitzberger, T., Veblen, T.T., & Villalba, R. (1997) Climatic influences on fire regimes along a rain forest to xeric woodland gradient in northern Patagonia, Argentina. *Journal of Biogeography*, **24**, 35–47.

Klaus, W. (1989) Mediterranean pines and their history. *Plant Systematics and Evolution*, **162**, 133–163.

Klein, J. (1920) *The Mesta: a study in Spanish economic history 1273–1836*. Harvard University Press, Cambridge, Massachusetts.

Knapp, E.E., Brennan, T.J., Ballenger, E.A., & Keeley, J.E. (2005) Fuel reduction and coarse woody debris dynamics with early season and late season prescribed fires in a Sierra Nevada mixed conifer forest. *Forest Ecology and Management*, **208**, 383–397.

Knapp, E.E. & Keeley, J.E. (2006) Heterogeneity in fire severity within early season and late season prescribed burns in a mixed conifer forest. *International Journal of Wildland Fire.*, **15**, 1–9.

Knapp, E.E., Schwilk, D.W., Kane, J.M., & Keeley, J.E. (2007) Role of burning season on initial understory vegetation response to prescribed fire in a mixed conifer forest. *Canadian Journal of Forest Research*, **37**, 11–22.

Knox, K.J.E. & Clarke, P.J. (2005) Nutrient availability induces contrasting allocation and starch formation in resprouting and obligate seeding shrubs. *Functional Ecology*, **19**, 690–698.

Kolcon, I. & Sachsenhofer, R.F. (1999) Petrography, palynology and depositional environments of the early Miocene Oberdorf lignite seam (Syrian Basin, Austria). *International Journal of Coal Geology*, **41**, 275–308.

Koperberg, P. (2003) The politics of fire management. In *Proceedings 3rd international wildland fire conference and exhibition: urban and rural communities living in fire prone environments: managing the future of global problems, 3–6 October 2003, Sydney, Australia* (CD ROM) (no pagination). USDI Bureau of Land Management, Washington, DC.

Kraaij, T. (2010) Changing the fire management regime in the renosterveld and lowland fynbos of the Bontebok National Park. *South African Journal of Botany*, **76**, 550–557.

Krannitz, P.G. & Duralia, T.E. (2004) Cone and seed production in *Pinus ponderosa*: a review. *Western North American Naturalist*, **64**, 208–218.

Krawchuk, M.A., Moritz, M.A., Parisien, M.-A., Van Dorn, J., & Hayhoe, K. (2009) Global pyrogeography: the current and future distribution of wildfire. *PLoS ONE*, **4**, e5102.

Krech, S.I. (1999) *The ecological Indian: myth and history*. W.W. Norton, New York.

Krick, I.P. (1933) Foehn winds of southern California. *Beitrage zur Geophysik*, **39**, 399–407.

Krock, B., Schmidt, S., Hertweck, C., & Baldwin, I.T. (2002) Vegetation-derived abscisic acid and four terpenes enforce dormancy in seeds of the post-fire annual, *Nicotiana attenuata*. *Seed Science Research*, **12**, 239–252.

Krug, R.M., Krug, C.B., Iponga, D.M., et al. (2004) Reconstructing west coast renosterveld: past and present ecological processes in a Mediterranean shrubland of South Africa. In *Ecology, conservation and management of mediterranean climate ecosystems* (eds M. Arianoutsou & V.P. Papanastasis) [no pagination]. Millpress, Rotterdam, The Netherlands.

Kruger, F.J. (1977) A preliminary account of aerial plant biomass in fynbos communities of the Mediterranean-type climate zone of the Cape Province. *Bothalia*, **12**, 301–307.

Kruger, F.J. (1979) South African heathlands. In *Ecosystems of the world. Vol. 9A: Heathlands and related shrublands: descriptive studies* (ed. R.L. Specht), pp. 19–80. Elsevier Scientific, New York.

Kruger, F.J., Reid, P., Mayet, M., et al. (2000) *A review of the veld fires in the Western Cape during 15 to 25 January 2000*. Department of Water Affairs and Forestry, Pretoria, RSA.

Kruger, L.M., Midgley, J.J., & Cowling, R.M. (1997) Resprouters vs. reseeders in South African forest trees – a model based on forest canopy height. *Functional Ecology*, **11**, 101–105.

Kunst, C., Bravo, K., Moscovich, F., Herrera, J., Godoy, J., & Velez, S. (2003) Date of prescribed fire and herbaceous diversity of an *Elionorus muticus* (Spreng.) O. Kuntze savanna. *Revista Chilena de Historia Natural*, **76**, 105–115.

Kutiel, H. & Kutiel, P. (1991) The distribution of autumnal easterly wind spells favoring rapid spread of forest wildfires on Mount Carmel, Israel. *GeoJournal*, **23**, 147–152.

Kutiel, P. (1997) Spatial and temporal heterogeneity of species diversity in a Mediterranean ecosystem following fire. *International Journal of Wildland Fire*, **7**, 307–315.

Kvaček, Z., Manchester, S.R., & Schorn, H.E. (2000) Cones, seeds, and foliage of *Tetraclinis salicornioides* (Cupressaceae) from the Oligocene and Miocene of western North America: a geographic extension of the European Tertiary species. *International Journal of Plant Sciences*, **161**, 331–344.

Kyser, G.B. & DiTomaso, J.M. (2002) Instability in a grassland community after the control of yellow starthistle (*Centaurea solstitialis*) with prescribed burning. *Weed Science*, **50**, 648–657.

Lacey, C.J. (1974) Rhizomes in tropical eucalypts and their role in recovery from fire damage. *Australian Journal of Botany*, **22**, 29–38.

Lacey, C.J. (1983) The development of large plate-like lignotubers in *Eucalyptus botryoides* in relation to environmental factors. *Australian Journal of Botany*, **31**, 105–118.

Lacey, C.J. & Jahnke, R. (1984) The occurrence and nature of lignotubers in *Notelaea longifolia* and *Elaeocarpus reticulatus*. *Australian Journal of Botany*, **32**, 311–321.

Ladd, P.G., Crosti, R., & Pignatti, S. (2005) Vegetative and seedling regeneration after fire planted Sardinian pinewood compared with that in other areas of Mediterranean-type climate. *Journal of Biogeography*, **32**, 85–98.

Lambdon, P.W., Lloret, F., & Hulme, P.E. (2008) Do non-native species invasions lead to biotic homogenization at small scales? The similarity and functional diversity of habitats compared for alien and native components of Mediterranean floras. *Diversity and Distributions*, **14**, 774–785.

Lambers, H., Raven, J.A., Shaver, G.R., & Smith, S.E. (2008) Plant nutrient-acquisition strategies change with soil age. *Trends in Ecology and Evolution*, **23**, 95–103.

Lamont, B.B. (1982) Mechanisms for enhancing nutrient uptake in plants, with special reference to Mediterranean South Africa and Western Australia. *Botanical Review*, **48**, 597–689.

Lamont, B.B. & Enright, N.J. (2000) Adaptive advantages of aerial seed banks. *Plant Species Biology*, **15**, 157–166.

Lamont, B.B., Enright, N.J., Groeneveld, J., & He, T. (2004a) Coping with fire in species-rich heathlands: is there an optimum fire interval for management? In *Ecology, conservation and management of mediterranean climate ecosystems* (eds M. Arianoutsou & V.P. Papanastasis) [no pagination]. Millpress, Rotterdam, The Netherlands.

Lamont, B.B., Enright, N.J., Witkowski, E.T.F., & Groeneveld, J. (2007) Conservation biology of banksias: insights from natural history to simulation modelling. *Australian Journal of Botany*, **55**, 280–292.

Lamont, B.B. & Groom, P.K. (1998) Seed and seedling biology of the woody-fruited Proteaceae. *Australian Journal of Botany*, **46**, 387–406.

Lamont, B.B., Groom, P.K., Richards, M.B., & Witkowski, E.T.F. (1999) Recovery of *Banksia* and *Hakea* communities after fire in mediterranean Australia: the role of species identity and functional attributes. *Diversity and Distributions*, **5**, 15–26.

Lamont, B.B., Le Maitre, D.C., Cowling, R.M., & Enright, N.J. (1991) Canopy seed storage in woody plants. *Botanical Review*, **57**, 277–317.

Lamont, B.B. & Markey, A. (1995) Biogeography of fire-killed and resprouting *Banksia* species in south-western Australia. *Australian Journal of Botany*, **43**, 283–303.

Lamont, B.B., Olesen, J.M., & Briffa, P.J. (1998) Seed production, pollinator attractants and breeding system in relation to fire response: are there reproductive syndromes among co-occurring Proteaceous shrubs? *Australian Journal of Botany*, **46**, 377–385.

Lamont, B.B., Ward, D.J., Eldridge, J., *et al.* (2003) Believing the Balga: a new method for gauging the fire history of vegetation using grasstrees. In *Fire in ecosystems of south-west Western Australia: impacts and management* (eds I. Abbott & N.D. Burrows), pp. 147–170. Backhuys Publishers, Leiden, The Netherlands.

Lamont, B.B. & Wiens, D. (2003) Are seed set and speciation rates always low among species that resprout after fire, and why? *Evolutionary Ecology*, **17**, 277–292.

Lamont, B.B. & Witkowski, E.T.F. (1995) A test for lottery recruitment among four *Banksia* species based on their demography and biological attributes. *Oecologia*, **101**, 299–308.

Lamont, B.B., Wittkuhn, R., & Korczynskyj, D. (2004b) Ecology and ecophysiology of grasstrees. *Australian Journal of Botany*, **52**, 561–582.

LaMotte, R.S. (1952) *Catalogue of the Cenozoic plants of North America through 1950*. Geological Society of America, Washington, DC.

Landrum, L.R. (1981) The phylogeny and geography of *Myrceugenia* (Myrtaceae). *Brittonia*, **33**, 105–129.

Lara, A., Fraver, S., Aravena, J.C., & Wolodarsky-Franke, A. (1999) Fire and the dynamics of *Fitzroya cupressoides* (*alerce*) forests of Chile's Cordillera Pelada. *Ecoscience*, **6**, 100–109.

Lara, A. & Veblen, T.T. (1993) Forest plantations in Chile: a successful model? *Afforestation: Policies, Planning and Progress*, 118–139.

Lara, A., Villalba, R., Wolodarsky-Franke, A., et al. (2005) Spatial and temporal variation in *Nothofagus pumillio* growth at tree line along its latitudinal range (35°40′ – 55°S) in the Chilean Andes. *Journal of Biogeography*, **32**, 879–893.

Lara, A., Wolodarsky-Franke, A., Aravena, J.C., et al. (2003) Fire regimes and forest dynamics in the Lake Region of south-central Chile. In *Fire and climatic change in temperate ecosystems of the western Americas* (ed. T.T. Veblen, W.L. Baker, G. Montenegro & T.W. Swetnam), pp. 322–342. Springer, New York.

Lauder, G.V., Leroi, A.M., & Rose, M.R. (1993) Adaptations and history. *Trends in Ecology and Evolution*, **8**, 294–297.

Laurie, H. & Cowling, R.M. (1994) Lottery coexistence models extended to plants with disjoint generations. *Journal of Vegetation Science*, **5**, 161–168.

Lawrence, G.E. (1966) Ecology of vertebrate animals in relation to chaparral fire in the Sierra Nevada foothills. *Ecology*, **47**, 278–291.

Lawson, D.M., Regan, H.M., Zedler, P.H., & Franklin, J. (2010) Cumulative effects of land use, altered fire regime and climate change on persistence of *Ceanothus verrucosus*, a rare, fire-dependent plant species. *Global Change Biology*, **16**, 2518–2529.

Lawson, G.W., Jenik, J., & Armstrong-Mensah, K.O. (1968) A study of a vegetation catena in Guinea savanna at Mole Game Reserve (Guinea). *Journal of Ecology*, **56**, 505–522.

Le Brocque, A.F. & Buckney, R.T. (2003) Species richness–environment relationships within coastal sclerophyll and mesophyll vegetation in Ku-ring-gai Chase National Park, New South Wales, Australia. *Austral Ecology*, **28**, 404–412.

Le Floc'h, E., Houérou, H.N.L., & Mathez, J. (1990) History and patterns of plant invasion in northern Africa. In *Biological invasions in Europe and the Mediterranean Basin* (eds F. di Castri, A.J. Hansen & M. Debussche), pp. 105–133. Kluwer Academic Publishers, Dordrecht, The Netherlands.

Le Houerou, H.N. (2004) An agro-bioclimatic classification of arid and semiarid lands in the isoclimatic mediterranean zones. *Arid Land Research and Management*, **18**, 301–346.

Le Maitre, D.C. (1984) A short note on seed predation in *Watsonia pyramidata* (Andr.) Stapf in relation to season of burn. *Journal of South African Botany*, **50**, 407–415.

Le Maitre, D.C. (1988) Effects of season of burn on the regeneration of two Proteaceae with soil-stored seed. *South African Journal of Botany*, **54**, 581–584.

Le Maitre, D.C. & Brown, P.J. (1992) Life cycles and fire-stimulated flowering in geophytes. In *Fire in South African mountain fynbos* (eds B.W. van Wilgen, D.M. Richardson, F.J. Kruger & H.J. van Hensbergen), pp. 145–160. Springer, Berlin.

Le Maitre, D.C. & Midgley, J.J. (1992) Plant reproductive ecology. In *The ecology of fynbos: nutrients, fire and diversity* (ed. R.M. Cowling), pp. 135–174. Oxford University Press, Oxford, UK.

Le Maitre, D.C., van Wilgen, B.W., Chapman, R.A., & McKelly, D.H. (1996) Invasive plants and water resources in the Western Cape Province, South Africa: modelling the consequences of a lack of management. *Journal of Applied Ecology*, **33**, 161–172.

Le Maitre, D.C., van Wilgen, B.W., Gelderblom, C.M., *et al.* (2002) Invasive alien trees and water resources in South Africa: case studies of the costs and benefits of management. *Forest Ecology and Management*, **160**, 143–159.

League, K. & Veblen, T. (2006) Climatic variability and episodic *Pinus ponderosa* establishment along the forest-grassland ecotones of Colorado. *Forest Ecology and Management*, **228**, 98–107.

Lecomte, N., Simard, M., Fenton, N., & Bergeron, Y. (2006) Fire severity and long-term ecosystem biomass dynamics in coniferous boreal forests of eastern Canada. *Ecosystems*, **9**, 1215–1230.

Leisz, D.R. & Wilson, C.R. (1980) To burn or not to burn: fire and chaparral management in southern California. *Journal of Forestry*, **78**, 107–108.

Lengyel, S., Gove, A.D., Latimer, A.M., Majer, J.D., & Dunn, R.R. (2010) Convergent evolution of seed dispersal by ants, and phylogeny and bigeography in flowering plants: a global survey. *Perspectives in Plant Ecology, Evolution and Systematics*, **12**, 43–55.

Lenihan, J.M., Bachelet, D., Neilson, R.P., & Drapek, R. (2008) Response of vegetation distribution, ecosystem productivity, and fire to climatic change scenarios for California. *Climatic Change*, **87**, S215–S230.

Lentile, L.B., Holden, Z.A., Smith, A.M.S., *et al.* (2006) Remote sensing techniques to assess active fire characteristics and post-fire effects. *International Journal of Wildland Fire*, **18**, 319–345.

Lenton, T.M., Held, H., Kriegler, E., *et al.* (2008) Tipping elements in the Earth's climate system. *Proceedings of the National Academy of Sciences of the USA*, **105**, 1786–1793.

Leone, V., Borghetti, M., & Saracino, A. (2000) Ecology of post-fire recovery in *Pinus halepensis* in southern Italy. In *Life and environment in the Mediterranean* (ed. L. Trabaud), pp. 129–154. Wit Press, Southampton, UK.

Leroi, A.M., Rose, M.R., & Lauder, G.V. (1994) What does the comparative method reveal about adaptation? *American Naturalist*, **143**, 381–402.

Lescure, R. (2009) San Diego City Council passes SDG&E power shut-off plan. *East County Magazine*, July 27, 2009.

Lessard, A.G. (1988) The Santa Ana wind of southern California. *Weatherwise*, **41**, 100–104.

Levin, N. & Saaroni, H. (1999) Fire weather in Israel: synoptic climatological analysis. *GeoJournal*, **47**, 523–538.

Lev-Yadun, S. (1995) Living serotinous cones in *Cupressus sempervirens*. *International Journal of Plant Sciences*, **156**, 50–54.

Levyns, M.R. (1929) Veld burning experiments at Ida's Valley, Stellenbosch. *Transactions of the Royal Society of South Africa*, **17**, 61–92.

Lewis, S.A., Wu, J.Q., & Robichaud, P.R. (2006) Assessing burn severity and comparing soil water repellency, Hayman Fire, Colorado. *Hydrological Processes*, **20**, 1–16.

Light, M.E., Burger, B.V., Staerk, D., Kohout, L., & van Staden, J. (2010) Butenolides from plant-derived smoke: natural plant-growth regulators with antagonistic actions on seed germination. *Journal of Natural Products*, **73**, 267–269.

Linder, H.P. (2003) The radiation of the Cape flora, southern Africa. *Biological Reviews*, **78**, 597–638.

Linder, H.P. (2005) Evolution of diversity: the Cape flora. *Trends in Plant Science*, **10**, 536–541.

Linder, H.P., Eldenás, P., & Briggs, B.G. (2003) Contrasting patterns of radiation in African and Australian Restionaceae. *Evolution*, **57**, 2688–2702.

Linder, H.P. & Ellis, R.P. (1990) Vegetative morphology and interfire survival strategies in the Cape fynbos grasses. *Bothalia*, **20**, 91–103.

Linder, H.P. & Hardy, C.R. (2004) Evolution of the species-rich Cape flora. *Philosophical Transactions of the Royal Society of London, Series B*, **359**, 1623–1632.

Lindon, H.L. & Menges, E. (2008) Scientific note: effects of smoke on seed germination of twenty species of fire-prone habitats in Florida. *Castanea*, **73**, 106–110.

Linhart, Y.B. (1988) Ecological and evolutionary studies of ponderosa pine in the Rocky Mountains. In *Ponderosa pine: the species and its management: symposium proceedings* (eds D.M. Baumgarner & J.E. Lotan), pp. 77–89. USDA Forest Service, Ogden, Utah.

Linstädter, A. & Zielhofer, C. (2010) Regional fire history shows abrupt responses of Mediterranean ecosystems to centennial-scale climate change (*Olea–Pistacia* woodlands, NE Morocco). *Journal of Arid Environments*, **74**, 101–110.

Litton, C.M. & Santelices, R. (2002) Early post-fire succession in a *Nothofagus glauca* forest in the Coastal Cordillera of south-central Chile. *International Journal of Wildland Fire*, **11**, 115–125.

Litton, C.M. & Santelices, R. (2003) Effect of wildfire on soil physical and chemical properties in a *Nothofagus glauca* forest, Chile. *Revista Chilena de Historia Natural*, **76**, 529–542.

Liu, K.B., Lu, H.Y., & Shen, C.M. (2008) A 1200-year proxy record of hurricanes and fires from the Gulf of Mexico coast: testing the hypothesis of hurricane-fire interactions. *Quaternary Research*, **69**, 29–41.

Lledó, M.J., Sánchez, J.R., Bellot, J., et al. (1992) Structure, biomass and production of a resprouted holm-oak (*Quercus ilex* L.) forest in NE Spain. *Vegetatio*, **99–100**, 51–59.

Lloret, F. (1998) Fire, canopy cover and seedling dynamics in Mediterranean shrubland of northeastern Spain. *Journal of Vegetation Science*, **9**, 417–430.

Lloret, F., Calvo, E., Pons, X., & Díaz-Delgado, R. (2002) Wildfires and landscape patterns in the Eastern Iberian Peninsula. *Landscape Ecology*, **17**, 745–759.

Lloret, F., Casanovas, C., & Peñuelas, J. (1999b) Seedling survival of Mediterranean shrubland species in relation to root: shoot ratio, seed size and water and nitrogen use. *Functional Ecology*, **13**, 210–216.

Lloret, F. & López-Soria, L. (1993) Resprouting of *Erica multiflora* after experimental fire treatments. *Journal of Vegetation Science*, **4**, 367–374.

Lloret, F., Medail, F., Brundu, G., & Hulme, P.E. (2004) Local and regional abundance of exotic plant species on Mediterranean islands: are species traits important? *Global Ecology and Biogeography*, **13**, 37–45.

Lloret, F., Pausas, J.G., & Vilà, M. (2003) Response of Mediterranean plant species to different fire regimes in Garraf Natural Park (Catalonia, Spain): field observations and modelling predictions. *Plant Ecology*, **167**, 223–235.

Lloret, F., Verdú, M., Flores-Hernández, N., & Valiente-Banuet, A. (1999a) Fire and resprouting in Mediterranean ecosystems; insights from an external biogeographical region, the mexical shrubland. *American Journal of Botany*, **86**, 1655–1661.

Loarie, S.R., Carter, B.E., Hayhoe, K., *et al.* (2008) Climate change and the future of California's endemic flora. *Public Library of Science (PLoS)*, **3**, 1–10.

Loehle, C. & LeBlanc, D. (1996) Model-based assessments of climate change effects on forests: a critical review. *Ecological Modelling*, **90**, 1–31.

Loepfe, L., Martinez-Vilata, J., Oliveres, J., Piñol, J., & Lloret, F. (2010) Feedbacks between fuel reduction and landscape homogenisation determine fire regimes in three Mediterranean areas. *Forest Ecology and Management*, **259**, 2366–2374.

Lombardo, K.J., Swetnam, T.W., Baisan, C.H., & Borchert, M.I. (2009) Using bigcone Douglas-fir fire scars and tree rings to reconstruct interior chaparral fire history. *Fire Ecology*, **5**, 35–56.

Lomolino, M.V. (2001) The species-area relationship: new challenges for an old pattern. *Progress in Physical Geography*, **25**, 1–21.

Long, S.P., Moya, E.G., Imbamba, S.K., *et al.* (1989) Primary productivity of natural grass ecosystems of the tropics: a reappraisal. *Plant and Soil*, **115**, 155–166.

Longhurst, W.M. (1956) Stump sprouting of oaks in response to seasonal cutting. *Journal of Range Management*, **9**, 194–196.

Loveless, A.R. (1961) A nutritional interpretation of sclerophylly based on differences in the chemical composition of sclerophyllous and mesophytic leaves. *Annals of Botany*, **25**, 168–184.

Low, A.B. & Lamont, B.B. (1990) Aerial and below-ground phytomass of *Banksia* scrub-heath at Eneabba, south-western Australia. *Australian Journal of Botany*, **38**, 351–359.

Luck, G.W., Smallbone, L., & O'Brien, R. (2009) Socio-economics and vegetation change in urban ecosystems: patterns in space and time. *Ecosystems*, **12**, 604–620.

Luger, A.D. & Moll, E.J. (1993) Fire protection and afromontane forest expansion in Cape fynbos. *Biological Conservation*, **64**, 51–56.

Luis-Calabuig, E., Tarrega, R., & Valbuena, L. (2000) Ten years of recovery of *Cistus ladanifer* after experimental disturbances. *Israel Journal of Plant Sciences*, **48**, 271–276.

Luke, R.H. & McArthur, A.G. (1978) *Bushfires in Australia*. Australian Government Publishing Service, Canberra.

Luly, J.G. (2001) On the equivocal fate of late pleistocene *Callitris* Vent. (Cupressaceae) woodlands in arid south Australia. *Quaternary International*, **83–85**, 155–168.

Lunt, I.D. (1998) Two hundred years of land use and vegetation change in a remnant coastal woodland in southern Australia. *Australian Journal of Botany*, **46**, 629–647.

Lunt, I.D. & Morgan, J.W. (2002) The role of fire regimes in temperate lowland grasslands of south-eastern Australia. In *Flammable Australia: the fire regimes and biodiversity of a continent* (eds R.A. Bradstock, J.E. Williams & A.M. Gill), pp. 177–198. Cambridge University Press, Cambridge, UK.

Lynch, A.H., Beringer, J., Kershaw, P., *et al.* (2007) Using the paleorecord to evaluate climate and fire interactions in Australia. *Annual Review of Earth and Planetary Science*, **35**, 215–239.

MacGinitie, H.D. (1969) The Eocene Green River flora of northwestern Colorado and northeastern Utah. *University of California Publications in Geological Sciences*, **83**, 1–202.

Mack, M.C. & D'Antonio, C.M. (1998) Impacts of biological invasions on disturbance regimes. *Trends in Ecology and Evolution*, **13**, 195–198.

Mack, R.N. (1981) Invasion of *Bromus tectorum* L. into western North America: an ecological chronicle. *Agro-Ecosystems*, **7**, 145–165.

Magri, D., Fineschi, S., Bellarosa, R., *et al.* (2007) The distribution of *Quercus suber* chloroplast haplotypes matches the palaeogeographical history of the western Mediterranean. *Molecular Ecology*, **16**, 5259–5266.

Mahall, B.E., Thwing, L.K., & Tyler, C.M. (2010) A quantitative comparison of two extremes in chaparral shrub phenology. *Flora*, **205**, 513–526.

Main, A.R. (1981) Fire tolerance of heathland animals. In *Ecosystems of the world, Vol. 9B: Heathlands and related shrublands: analytical studies* (ed. R.L. Specht), pp. 85–90. Elsevier Scientific, New York.

Mairota, P., Thornes, J.B., & Geeson, N. (1998) *Atlas of Mediterranean Environments in Europe: the desertification context*. John Wiley & Sons, Chichester, UK.

Malamud, B.D., Morein, G., & Turcotte, D.L. (1998) Forest fires: an example of self-organized critical behavior. *Science*, **281**, 1840–1842.

Malamud, B.D. & Turcotte, D.L. (1999) Self-organized criticality applied to natural hazards. *Natural Hazards*, **20**, 93–116.

Malanson, G.P. & Trabaud, L. (1988) Vigour of post-fire resprouting by *Quercus coccifera* L. *Journal of Ecology*, **76**, 351–365.

Mallik, A.U. (1993) Ecology of a forest weed of Newfoundland: vegetative regeneration strategy of *Kalmia angustifolia*. *Canadian Journal of Botany*, **71**, 161–166.

Malo, J.E. & Suarez, F. (1997) Dispersal mechanism and transcontinental naturalization proneness among Mediterranean herbaceous species. *Journal of Biogeography*, **24**, 391–394.

Manders, P.T. (1986) An assessment of the current status of the Clanwilliam cedar (*Widdringtonia cedarbergensis*) and the reasons for its decline. *South African Forest Journal*, **139**, 48–53.

Manders, P.T. (1990) Fire and other variables as determinants of forest/fynbos boundaries in the Cape Province. *Journal of Vegetation Science*, **1**, 483–490.

Manders, P.T. (1991) The relationships between forest and mountain fynbos communities in the southwestern Cape Province of South Africa, Ph.D. dissertation, Unversity of Cape Town, Rondebosch, RSA.

Manders, P.T., Richardson, D.M., & Masson, P.H. (1992) Is fynbos a stage in succession to forest? Analysis of the perceived ecological distinction between two communities. In *Fire in South African mountain fynbos* (eds B.W. van Wilgen, D.M. Richardson, F.J. Kruger & H.J. van Hensbergen), pp. 81–107. Springer, Berlin.

Manry, D.E. & Knight, R.S. (1986) Lightning density and burning frequency in South African vegetation. *Vegetatio*, **66**, 67–76.

Marais, C. & Wannenburgh, A.M. (2008) Restoration of water resources (natural capital) through the clearing of invasive alien plants from riparian areas in South Africa: costs and water benefits. *South African Journal of Botany*, **74**, 526–537.

Margaris, N.S. (1976) Structure and dynamics in a phryganic (east Mediterranean) ecosystem. *Journal of Biogeography*, **3**, 249–259.

Margaris, N.S. (1977) Physiological and biochemical observations in seasonal dimorphic leaves of *Sarcopoterium spinosum* and *Phlomis fruticosa*. *Oecologia Plantarum*, **12**, 343–350.

Markgraf, V., Whitlock, C., Anderson, R.S., & Garcia, A. (2009) Late Quaternary vegetation and fire history in the northernmost *Nothofagus* forest region: Mallin Vaca Lauquen, Neuquen Province, Argentina. *Journal of Quaternary Science*, **24**, 248–258.

Markgraf, V., Whitlock, C., & Haberle, S. (2007) Vegetation and fire history during the last 18,000 cal yr B.P. in southern Patagonia: Mallín Pollux, Coyhaique, Province Aisén. *Palaeogeography, Palaeoclimatology, Palaeoecology*, **254**, 492–507.

Marlon, J.R., Bartlein, P.J., Walsh, M.K., *et al.* (2009) Wildfire responses to abrupt climate change in North America. *Proceedings of the National Academy of Sciences of the USA*, **106**, 2519–2524.

Marrinan, M.J., Edwards, W., & Landsberg, J. (2005) Resprouting of saplings following a tropical rainforest fire in north-east Queensland, Australia. *Austral Ecology*, **30**, 817–826.

Marsh, G. (1864) *Man and Nature; or, Physical Geography as Modified by Human Action*. Charles Scribner, New York.

Marshall, S. (2007) Lawsuit alleges fire insurance wrongdoing. *North County Times*, April 17, 2007.

Martin, H.A. (1990) Tertiary climate and phytogeography in southeastern Australia. *Palaeobotany and Palynology*, **65**, 47–55.

Martin, H.A. (1994) Australian tertiary phytogeography: evidence from palynology. In *History of the Australian vegetation: Cretaceous to recent* (ed. R.S. Hill), pp. 104–142. Cambridge University Press, Cambridge, UK.

Martin, H.A. (1996) Wildfires in past ages. *Proceedings of the Linnean Society of New South Wales*, **116**, 3–18.

Martin, H.A. (2006) Cenozoic climatic change and the development of the arid vegetation in Australia. *Journal of Arid Environments*, **66**, 533–563.

Martin, M.P., Flasse, S., Downey, I., & Ceccato, P. (1999) Fire detection and fire growth monitoring using satellite data. In *Remote Sensing of Large Wildfires in the European Mediterranean Basin* (ed. E. Chuvieco). Springer, Berlin.

Martin, P.S. (1984) Prehistoric overkill: the global model. In *Quaternary extinctions: a prehistoric revolution* (eds P.S. Martin & R.G. Klein), pp. 354–403. University of Arizona Press, Tucson.

Martin, P.S. & Klein, R.G., eds. (1984) *Quaternary extinctions: a prehistoric revolution*. University of Arizona Press, Tucson.

Martin, R.E. & Sapsis, D.B. (1995) A synopsis of large or disastrous wildland fires. In *The Biswell symposium: fire issues and solutions in urban interface and wildland ecosystems* (eds D.R. Weise & R.E. Martin), General Technical Report PSW-GTR-158, pp. 35–38. USDA Forest Service, Pacific Southwest Research Station, Albany, California.

Marynowski, L. & Simoneit, B.R.T. (2009) Widespread Upper Triassic to Lower Jurassic wildfire records from Poland: evidence from charcoal and pyrolytic polycyclic aromatic hydrocarbons. *Palaios*, **24**, 785–798.

Mazarakis, N., Kotroni, V., Lagouvardos, K., & Argiriou, A.A. (2008) Storms and lighning activity in Greece during the warm periods of 2003–06. *Journal of Applied Meteorology and Climatology*, **47**, 3089–3098.

Masson, P.H. & Moll, E.J. (1987) The factors affecting forest colonisation of fynbos in the absence of recurrent fire at Orange Kloof, Cape Province, South Africa. *South African Forestry Journal*, **143**, 5–10.

Mast, A.R. & Givnish, T.J. (2002) Historical biogeography and the origin of stomatal distributions in *Banksia* and *Dryandra* (Proteaceae) based on their cpDNA phylogeny. *American Journal of Botany*, **89**, 1311–1323.

Mast, A.R., Jones, E.H., & Havery, S.P. (2005) An assessment of old and new DNA sequence evidence for the paraphyly of *Banksia* with respect to *Dryandra* (Proteaceae). *Australian Systematic Botany*, **18**, 75–88.

Mast, A.R. & Thiele, K. (2007) The transfer of *Dryandra* R.Br. to *Banksia* L.f. (Proteaceae). *Australian Systematic Botany*, **20**, 63–71.

McAneney, J., Chen, K., & Pitman, A. (2009) 100-years of Australian bushfire property losses: is the risk significant and is it increasing? *Journal of Environmental Management*, **90**, 2819–2822.

McArthur, A.G. (1972) Fire control in the arid and semi-arid lands of Australia. In *The use of trees and shrubs in the dry country of Australia* (eds N. Hall et al.), pp. 488–515. Government Publishing Service, Canberra, Australia.

McCaw, L. (1997) *Predicting fire spread in Western Australian mallee-heath shrubland.* University of New South Wales, Australia.

McCaw, L., Cheney, P., & Sneeuwjagt, R. (2003) *Development of a scientific understanding of fire behaviour and use in south-west Western Australia.* Backhuys Publishers, Leiden, The Netherlands.

McCaw, L. & Hanstrum, B. (2003) Fire environment of Mediterranean south-west Western Australia. In *Fire in ecosystems of south-west Western Australia: impacts and management* (eds I. Abbott & N.D. Burrows), pp. 87–106. Backhuys Publishers, Leiden, The Netherlands.

McCaw, L., Maher, T., & Gillen, K. (1992) *Wildfires in the Fitzgerald River National Park, Western Australia, December 1989.* Department of Conservation and Land Management, Como, Australia.

McCaw, W.L., Smith, R.H., & Neal, J.E. (1997) Prescribed burning of thinning slash in regrowth stands of karri (*Eucalyptus diversicolor*), 1: Fire characteristics, fuel consumption and tree damage. *International Journal of Wildland Fire*, **7**, 29–40.

McDonald, P.M. (1980) *Seed dissemination in small clearcuttings in north-central California*, Research Paper PSW-150. USDA Forest Service, Pacific Southwest Research Station, Albany, California.

McDonald, P.M. & Abbott, C.S. (1994) *Seedfall, regeneration, and seedling development in group-selection openings,* Report PSW-RP-220. USDA Forest Service, Pacific Southwest Research Station, Albany, California.

McDonald, P.M. & Fiddler, G.O. (1990) *Ponderosa pine seedlings and competing vegetation: ecology, growth, and cost,* Research Paper PSW-199. USDA Forest Service, Pacific Southwest Research Station, Albany, California.

McIntosh, P.D., Laffan, M.D., & Hewitt, A.E. (2005) The role of fire and nutrient loss in the genesis of the forest soils of Tasmania and southern New Zealand. *Forest Ecology and Management*, **220**, 185–215.

McIvor, J.G. & McIntyre, S. (2005) Understanding grassy woodland ecosystems. In *Managing and conserving grassy woodlands* (eds S. McIntyre, J.G. McIvor & K.M. Heard), pp. 1–24. CSIRO Publishing, Collingwood, Australia.

McMahon, A.R.G., Carr, G., Bedggood, S.E., Hill, R.J., & Pritchard, A.M. (1996) Prescribed fire and control of coast wattle (*Acacia sophorae* (Labill) R. Br.). In *Fire and biodiversity: the effects and effectiveness of fire management*, pp. 87–96. Department of Environment, Sport and Territories, Canberra, Australia.

McNaughton, S.J. (1968) Structure and function in California grasslands. *Ecology*, **49**, 962–972.

McParland, L.C., Collinson, M.E., Scott, A.C., et al. (2007) Ferns and fires: experimental charring of ferns compared to wood and implications for paleobiology, paleoecology, coal petrology, and isotope geochemistry. *Palaios*, **22**, 528–538.

Medler, M.J., Montesano, P., & Robinson, D. (2002) Examining the relationship between snowfall and wildfire patterns in the western United States. *Physical Geography*, **23**, 335–342.

Meentemeyer, R.K. & Moody, A. (2002) Distribution of plant history types in California chaparral: the role of topographically-determined drought severity. *Journal of Vegetation Science*, **13**, 67–78.

Mellmann-Brown, S. & Barbour, M.G. (1995) Understory/overstory species patterns through a Sierra Nevada ecotone. *Phytocoenologia*, **25**, 89–106.

Menges, E.S. (2007) Integrating demography and fire management: an example from Florida scrub. *Australian Journal of Botany*, **55**, 261–272.

Menges, E.S. & Kohfeld, N. (1995) Life history strategies of Florida scrub plants in relation to fire. *Bulletin of the Torrey Botanical Club*, **122**, 282–297.

Mensing, S.A., Michaelsen, J., & Byrne, R. (1999) A 560-year record of Santa Ana fires reconstructed from charcoal deposited in the Santa Barbara Basin, California. *Quaternary Research*, **51**, 295–305.

Meredith, C.W. (1996) Is fire management effective? In *Fire and biodiversity: the effects and effectiveness of fire management*, Biodiversity Paper 8, pp. 227–232. Department of the Environment, Sport and Territories, Canberra, Australia.

Mermoz, M., Kitzberger, T., & Veblen, T.T. (2005) Landscape influences on occurrence and spread of wildfires in Patagonian forests and shrublands. *Ecology*, **86**, 2705–2715.

Merriam, K.E., Keeley, J.E., & Beyers, J.L. (2006) Fuel breaks affect nonnative species abundance in Californian plant communities. *Ecological Applications*, **16**, 515–527.

Merritt, D.J., Turner, S.R., Clarke, S., & Dixon, K.W. (2007) Seed dormancy and germination stimulation syndromes for Australian temperate species. *Australian Journal of Botany*, **55**, 335–344.

Mesléard, F. & Lepart, J. (1989) Continuous basal sprouting from lignotuber. *Arbutus unedo* L. and *Erica arborea* L. as woody Mediterranean examples. *Oecologia*, **80**, 127–131.

Mezquida, E.T. & Benkman, C.W. (2004) The geographic selection mosaic for squirrels, crossbills and Aleppo pine. *Journal of Evolutionary Biology*, **18**, 348–357.

Mibus, R. & Sedgley, M. (2000) Early lignotuber formation in *Banksia*: investigations into the anatomy of the cotyledonary node of two *Banksia* (Proteaceae) species. *Annals of Botany*, **86**, 575–587.

Michaels, P.J. (2006) Drought scare goes up in smoke. *Environment & Climate News*, October 2006, p. 11.

Midgley, J.J. (1996) Why the world's vegetation is not totally dominated by resprouting plants; because resprouters are shorter than reseeders. *Ecography*, **19**, 92–95.

Midgley, J.J. & Enright, N.J. (2000) Serotinous species show correlation between retention time for leaves and cones. *Journal of Ecology*, **88**, 348–351.

Midgley, J.J. & Seydack, A. (2006) No adverse signs of the effect of environmental change on tree biomass in the Knysna forest during the 1990s. *South African Journal of Science*, **102**, 96–97.

Milberg, P. & Lamont, B.B. (1995) Fire enhances weed invasion of roadside vegetation in southwestern Australia. *Biological Conservation*, **73**, 45–49.

Milewski, A.V. & Bond, W.J. (1982) Convergence of myrmecochory in mediterranean Australia and South Africa. In *Ant-plant interactions in Australia* (ed. R.C. Buckley), pp. 89–98. Dr. W. Junk Publishers, The Hague.

Millán, M.M. (2002) El cíclo hídrico en el Mediterráneo: un estudio del efecto de las masas forestales. In *La regeneración natural del bosque mediterráneo en la Peninsula Ibérica: evaluación de problemas y propuesta de soluciones* (ed. J. Charco). Ministerio de Medio Ambiente, Madrid.

Millán, M.M., Estrela, M.J., & Badenas, C. (1998) Meteorological processes relevant to forest fire dynamics on the Spanish Mediterranean coast. *Journal of Applied Meteorology*, **37**, 83–100.

Millar, C.I. (1996) Historical variation and desired condition: review and critique. In *What is watershed stability? Sixth Biennial Watershed Management Conference* (ed. S. Sommarstrom), Water Resources Center Report 92, pp. 105–132. University of California, Lake Tahoe, California/Nevada.

Millar, C.I. (1998) Early evolution of pines. In *Ecology and biogeography of* Pinus (ed. D.M. Richardson), pp. 69–91. Cambridge University Press, Cambridge, UK.

Millar, C.I. (1999) Evolution and biogeography of *Pinus radiata*, with a proposed revision of its Quaternary history. *New Zealand Journal of Forestry Science*, **29**, 335–365.

Miller, A.J., Young, D.A., & Wen, J. (2001) Phylogeny and biogeography of *Rhus* (Anacardiaceae) based on its sequence data. *International Journal of Plant Science*, **162**, 1401–1407.

Miller, B.P., Walshe, T., Enright, N.J., & Lamont, B.B. (2007) Error in the inference of fire history from grasstrees. *Austral Ecology*, **32**, 908–916.

Miller, J.D. & Thode, A.E. (2007) Quantifying burn severity in a heterogeneous landscape with a relative version of the delta Normalized Burn Ratio (dNBR). *Remote Sensing of Environment*, **109**, 66–80.

Miller, J.D. & Yool, S.R. (2002) Mapping forest post-fire canopy consumption in several overstory types using multi-temporal Landsat TM and ETM data. *Remote Sensing of Environment*, **82**, 481–496.

Mills, A.J. & Cowling, R.M. (2006) Rate of carbon sequestration at two thicket restoration sites in the Eastern Cape, South Africa. *Restoration Ecology*, **14**, 38–49.

Millspaugh, S.H., Whitlock, C., & Bartlein, P.J. (2004) Postglacial fire, vegetation, and climate history of the Yellowstone-Lamar and central plateau provinces, Yellowstone National Park. In *After the fires: the ecology of change in Yellowstone National Park* (ed. L.L. Wallace). Yale University Press, New Haven, Connecticut.

Milne, R.I. (2006) Northern Hemisphere plant disjunctions: a window on Tertiary land bridges and climate change? *Annals of Botany*, **98**, 465–472.

Milton, S.J. (2004) Grasses as invasive alien plants in South Africa. *South African Journal of Science*, **100**, 69–75.

Minnich, R.A. (1978) The geography of fire and conifer forests in the eastern Transverse Ranges, California. Ph.D. dissertation, University of California, Los Angeles.

Minnich, R.A. (1983) Fire mosaics in southern California and northern Baja California. *Science*, **219**, 1287–1294.

Minnich, R.A. (1987a) Fire behavior in southern California chaparral before fire control: the Mount Wilson burns at the turn of the century. *Annals of the Association of American Geographers*, **77**, 599–618.

Minnich, R.A. (1987b) The distribution of forest trees in northern Baja California, Mexico. *Madroño* **34**, 98–127.

Minnich, R.A. (1995) Fuel-driven fire regimes of the California chaparral. In *Brushfires in California wildlands: ecology and resource management* (eds J.E. Keeley & T. Scott), pp. 21–27. International Association of Wildland Fire, Fairfield, Washington.

Minnich, R.A. (1998) Landscapes, land-use and fire policy: where do large fires come from? In *Large forest fires* (ed. J.M. Moreno), pp. 133–158. Backhuys Publishers, Leiden, The Netherlands.

Minnich, R.A. (2001) An integrated model of two fire regimes. *Conservation Biology*, **15**, 1549–1553.

Minnich, R.A. (2003) Fire is inevitable but we can mitigate the damage. *San Diego Union–Tribune*, Nov 2, 2003, G–1.

Minnich, R.A. (2007) Southern California conifer forests. In *Terrestrial vegetation of California*, third edition (eds M.G. Barbour, T. Keeler-Wolf & A.A. Schoenherr), pp. 501–538. University of California Press, Los Angeles, California.

Minnich, R.A. (2008) *California's fading wildflowers: lost legacy and biological invasions.* University of California Press, Los Angeles, California.

Minnich, R.A. & Chou, Y.H. (1997) Wildland fire patch dynamics in the chaparral of southern California and northern Baja California. *International Journal of Wildland Fire*, **7**, 221–248.

Minnich, R.A. & Dezzani, R.J. (1991) Suppression, fire behavior, and fire magnitudes in Californian chaparral at the urban/wildland interface. In *California watersheds at the urban interface:, proceedings of the third biennial watershed conference* (ed. J.J. DeVries), Water Resources Center Report 75, pp. 67–83. University of California, Davis.

Minnich, R.A. & Dezzani, R.J. (1998) Historical decline of coastal sage scrub in the Riverside-Perris Plain, California. *Western Birds*, **29**, 366–391.

Minnich, R.A. & Everett, R.G. (2002) What unmanaged fire regimes in Baja California tell us about presuppression fire in Californian mediterranean ecosystems. In *Proceedings of the symposium: fire in California ecosytems: integrating ecology, prevention and management*, Miscellaneous Publication 1, pp. 325–338. Association for Fire Ecology, Berkeley, California.

Minnich, R.A. & Franco-Vizcaino, E. (1998) Land of chamise and pines: historical accounts and current statusof northern Baja California's vegetation. *University of California Publications in Botany*, **80**, 1–166.

Minnich, R.A. & Franco-Vizcaino, E. (1999) Prescribed mosaic burning in California chaparral? In *Proceedings of the symposium on fire economics, planning, and policy: bottom lines* (ed. A. González-Caban), PSW-GTR-173, pp. 247–254. USDA Forest Service Pacific Southwest Research Station, Albany, California.

Minore, D. & Weatherly, H. (1996) *Stump sprouting of the Pacific Yew*. USDA Forest Service Pacific Northwest Research Station, Portland, Oregon.

Mitchell, V.L. (1969) The regionalization of climate in montane areas, Ph.D. dissertation, University of Wisconsin, Madison.

Mitri, G.H. & Gitas, I.Z. (2008) Mapping the severity of fire using object-based classification of IKONOS imagery. *International Journal of Wildland Fire*, **17**, 431–442.

Miyanishi, K. (2001) Duff consumption. In *Forest fires: behavior and ecological effects* (eds E.A. Johnson & K. Miyanishi), pp. 437–475. Academic Press, San Diego, California.

Mladenoff, D.J. (2004) LANDIS and forest landscape models. *Ecological Modelling*, **180**, 7–19.

Moles, A.T., Falster, D.S., Leishman, M.R., & Westoby, M. (2004) Small-seeded species produce more seeds per square metre of canopy per year, but not per individual per lifetime. *Journal of Ecology*, **92**, 384–396.

Molinas, M.L. & Verdaguer, D. (1993) Lignotuber ontogeny in the cork-oak (*Quercus suber*; Fagaceae). *American Journal of Botany*, **80**, 172–181, 182–191.

Moll, E.J., McKenzie, B., & McLachlan, D. (1980) A possible explanation for the lack of trees in the fynbos, Cape Province, South Africa. *Biological Conservation*, **17**, 221–228.

Montenegro, G., Ávila, G., & Schatte, P. (1983) Presence and development of lignotubers in shrubs of the Chilean matorral. *Canadian Journal of Botany*, **61**, 1804–1808.

Montenegro, G., Ginocchio, R., Segura, A., Keeley, J.E., & Gómez, M. (2004) Fire regimes and vegetation responses in two Mediterranean-climate regions. *Revista Chilena de Historia Natural*, **77**, 455–464.

Montenegro, G., Gómez, M., Diaz, F., Ginocchio, R., & Veblen, T.T. (2003) Regeneration potential of Chilean matorral after fire: an updated view. In *Fire and climatic change in temperate ecosystems of the western Americas* (ed. T.T. Veblen, W.L. Baker, G. Montenegro, & T.W. Swetnam), pp. 381–409. Springer, New York.

Montenegro, G., Rivera, O., & Bas, F. (1978) Herbaceous vegetation in the Chilean matorral: dynamics of growth and evaluation of allelopathic effects of some dominant shrubs. *Oecologia*, **36**, 237–244.

Moody, J.A. & Martin, P.A. (2001) Initial hydrologic and geomorphic response following a wildfire in the Colorado front range. *Earth Surface Processes and Landforms*, **26**, 1049–1070.

Mooney, H.A. (1972) The carbon balance of plants. *Annual Review of Ecology and Evolution*, **3**, 315–346.

Mooney, H.A., ed. (1977a) *Convergent evolution of Chile and California Mediterranean climate ecosystems*, 224 pp. Dowden, Hutchinson and Ross, Stroudsburg, Pennsylvania.

Mooney, H.A. (1977b) Frost sensitivity and resprouting behavior of analogous shrubs of California and Chile. *Madroño*, **24**, 74–78.

Mooney, H.A. (1989) Chaparral physiological ecology: paradigms revisited. In *The California chaparral: paradigms reexamined* (ed. S.C. Keeley), Science Series 34, pp. 85–90. Natural History Museum of Los Angeles County, Los Angeles.

Mooney, H.A. & Bartholomew, B. (1974) Comparative carbon balance and reproductive modes of two Californian *Aesculus* species. *Botanical Gazette*, **135**, 306–313.

Mooney, H.A. & Dunn, E.L. (1970) Convergent evolution of mediterranean-climate evergreen sclerophyll shrubs. *Evolution*, **24**, 292–303.

Mooney, H.A., Hamburg, S.P., & Drake, J.A. (1986) The invasions of plants and animals into California. In *Ecology of biological invasions of North America and Hawaii* (eds H.A. Mooney & J.A. Drake), pp. 250–272. Springer, New York.

Mooney, H.A. & Hays, R.I. (1973) Carbohydrate storage cycles in two Californian Mediterranean-climate trees. *Flora*, **162**, 295–304.

Mooney, H.A., Kummerow, J., Johnson, A.W., *et al.* (1977) The producers: their resources and adaptive responses. In *Convergent evolution of Chile and California Mediterranean climate ecosystems* (ed. H.A. Mooney), pp. 85–143. Dowden, Hutchinson and Ross, Stroudsburg, Pennsylvania.

Mooney, S.D. & Maltby, E.L. (2006) Two proxy records revealing the late Holocene fire history at a site on the central coast of New South Wales, Australia. *Austral Ecology*, **31**, 682–695.

Mooney, S.D., Radford, K.L., & Hancock, G. (2001) Clues to the "burning question": pre-European fire in the Sydney coastal region from sedimentary charcoal and palynology. *Ecological Management and Restoration*, **2**, 203–212.

Moore, A.D. & Noble, I.R. (1990) An individualistic model of vegetation stand dynamics. *Journal of Environmental Management*, **31**, 61–81.

Moravec, J. (1990) Regeneration of NW African *Pinus halepensis* forests following fire. *Vegetatio*, **87**, 29–36.

Moreira, B., Tormo, J., Estrelles, E., & Pausas, J.G. (2010) Disentangling the role of heat and smoke as germination cues in Mediterranean Basin flora. *Annals of Botany*, **105**, 627–635.

Moreira, F., Duarte, I., Catry, F., & Acácio, V. (2007) Cork extraction as a key factor determining post-fire cork oak survival in a mountain region of southern Portugal. *Forest Ecology and Management*, **253**, 30–37.

Moreira, F., Rego, F.C., & Ferreira, P.G. (2001) Temporal (1958–1995) pattern of change in a cultural landscape of northwestern Portugal: implications for fire occurrence. *Landscape Ecology*, **16**, 557–567.

Moreira, F. & Russo, D. (2007) Modelling the impact of agricultural abandonment and wildfires on vertebrate diversity in Mediterranean Europe. *Landscape Ecology*, **22**, 1461–1476.

Moreno, J.M., Cruz, A., Fernández, F., *et al.* (2004) Ecología del monte mediterráneo en relación con el fuego: el jaral-brezal de Quintos de Mora (Toledo). In *Avances en el estudio de la gestión del monte mediterráneo* (eds V.R. Vallejo & J.A. Alloza), pp. 17–45. Fundación Centro de Estudios Ambientales del Mediterráneo (CEAM), Valencia, Spain.

Moreno, J.M. & Oechel, W.C. (1989) A simple method for estimating fire intensity after a burn in California chaparral. *Acta Oecologica*, **10**, 57–68.

Moreno, J.M. & Oechel, W.C. (1994) Fire intensity as a determinant factor of postfire plant recovery in southern California chaparral. In *The role of fire in mediterranean-type ecosystems* (eds J.M. Moreno & W.C. Oechel), pp. 26–45. Springer, New York.

Moreno, J.M., Vázquez, A., & Vélez, R. (1998) Recent history of forest fires in Spain. In *Large fires*, pp. 159–185. Backhuys Publishers, Leiden, The Netherlands.

Moreno, P.I. (2000) Climate, fire, and vegetation between about 13,000 and 9200 C-14 yr BP in the Chilean lake district. *Quaternary Research*, **54**, 81–89.

Morgan, P., Hardy, C.C., Swetnam, T.W., Rollins, M.G., & Long, D.G. (2001) Mapping fire regimes across time and space: understanding coarse and fine-scale fire patterns. *International Journal of Wildland Fire*, **10**, 329–342.

Moritz, M.A. (1997) Analyzing extreme disturbance events: fire in the Los Padres National Forest. *Ecological Applications*, **7**, 1252–1262.

Moritz, M.A., Keeley, J.E., Johnson, E.A., & Schaffner, A.A. (2004) Testing a basic assumption of shrubland fire management: does the hazard of burning increase with the age of fuels? *Frontiers in Ecology and the Environment*, **2**, 67–72.

Moritz, M.A., Krawchuk, M.A., & Parisien, M.-A. (2010a) Pyrogeography: understanding the ecological niche of fire. *PAGES news*, **18**, 83–85.

Moritz, M.A., Moody, T.J., Krawchuk, M.A., Huges, M., & Hall, A. (2010b) Spatial variation in extreme winds predicts large wildfire locations in chaparral ecosystems. *Geophysical Research Letters*, **37**, L04801, doi:10.1029/2009GL041735.

Moritz, M.A., Morais, M.E., Summerell, L.A., Carlson, J.M., & Doyle, J. (2005) Wildfires, complexity, and highly optimized tolerance. *Proceedings of the National Academy of Sciences*, **102**, 17 912–17 917.

Moritz, M.A. & Stephens, S.L. (2008) Fire and sustainability: considerations for California's altered future climate. *Climatic Change*, **87** (Suppl 1), S265–S271.

Morrison, D.A. (1995) Some effects of low-intensity fires on populations of co-occurring small trees in the Sydney region. *Proceedings of the Linnean Society of New South Wales*, **115**, 109–119.

Morrison, D.A., Buckney, R.T., Bewick, B.J., & Cary, G.J. (1996) Conservation conflicts over burning bush in south-eastern Australia. *Biological Conservation*, **76**, 167–175.

Morrison, D.A., Gary, G.J., Pengelly, S.M., *et al.* (1995) Effects of fire frequency on plant species composition of sandstone communities in the Sydney region: inter-fire interval and time-since-fire. *Austral Ecology*, **20**, 239–247.

Morvan, D. & Dupuy, J.L. (2004) Modeling the propagation of a wildfire through a Mediterranean shrub using a multiphase formulation. *Combustion and Flame*, **138**, 199–210.

Mott, J.J. & Groves, R.H. (1994) Natural and derived grasslands. In *Australian vegetation*, second edition (ed. R.H. Groves), pp. 369–392. Cambridge University Press, Melbourne, Australia.

Mucina, L. & Rutherford, M.C. (2006) *The vegetation of South Africa, Lesotho and Swaziland*. South African National Biodiversity Institute, Pretoria.

Müller, M.J. (1982) *Selected climatic data for a global set of standard stations for vegetation science*. Dr. W. Junk Publishers, The Hague, The Netherlands.

Muller, C.H. (1939) Relations of the vegetation and climatic types in Nuevo Leon, Mexico. *American Midland Naturalist*, **21**, 687–729.

Muller, C.H. (1947) Vegetation and climate of Coahuila, Mexico. *Madroño*, **9**, 33–57.

Mummenhoff, K., Al-Shehbaz, I.A., Bakker, F.T., Linder, H.P., & Mühlhousen, A. (2005) Phylogeny, morphological evolution, and speciation of endemic Brassicaceae genera in the Cape flora of southern Africa. *Annals of the Missouri Botanical Garden*, **92**, 400–424.

Muñoz, M.R. & Fuentes, E.R. (1989) Does fire induce shrub germination in the Chilean matorral? *Oikos*, **56**, 177–181.

Murphy, B.P. & Bowman, D.M.J.S. (2007) Seasonal water availability predicts the relative abundance of C_3 and C_4 grasses in Australia. *Global Ecology and Biogeography*, **16**, 160–169.

Murphy, K.A., Reynolds, J.H., & Koltun, J.M. (2008) Evaluating the ability of the differenced Normalized Burn Ratio (dNBR) to predict ecologically significant burn severity in Alaskan boreal forests. *International Journal of Wildland Fire*, **17**, 490–499.

Musil, C.F., Milton, S.J., & Davis, G.W. (2005) The threat of alien invasive grasses to lowland Cape floral diversity: an empirical appraisal of the effectiveness of practical control strategies. *South African Journal of Science*, **101**, 337–344.

Mutch, L.S. & Parsons, D.J. (1998) Mixed conifer forest mortality and establishment before and after prescribed fire in Sequoia National Park, California. *Forest Science*, **44**, 341–355.

Mutch, R.W. (1970) Wildland fires and ecosystems: a hypothesis. *Ecology*, **51**, 1046–1051.
Mutch, R.W. (2003) Safety considerations for plume dominated fires. USDA Forest Service. Online: www.fs.fed.us/nwacfire/quick/docs/plumeage.htm.
Myers, N., Mittermeier, R.A., Mittermeier, C.G., da Fonseca, G.A.B., & Kent, J. (2000) Biodiversity hotspots for conservation priorities. *Nature*, **403**, 853–858.
Myerscough, P.J. & Clarke, J.S. (2007) Burnt to blazes: landscape fires, resilience and habitat interaction in frequently burnt coastal heath. *Australian Journal of Botany*, **55**, 91–102.
Myerscough, P.J., Whelan, R.J., & Bradstock, R.A. (2000) Ecology of Proteaceae with special reference to the Sydney region. *Cunninghamia*, **6**, 951–1015.
Nagel, T.A. & Taylor, A.H. (2005) Fire and persistence of montane chaparral in mixed conifer forest landscapes in the northern Sierra Nevada, Lake Tahoe Basin, California, USA. *Journal of the Torrey Botanical Society*, **132**, 442–457.
Naiman, J. (2009) SDG&E explains countersuit threat to establish rights. *Fallbrook Village News,* March **5**, 2009.
Nano, C.E.M. & Clarke, P.J. (2010) Woody-grass ratios in a grassy arid system are limited by multi-causal interactions of abiotic constraint, competition and fire. *Oecologia*, **162**, 719–732.
Naumburg, E. & DeWald, L.E. (1999) Relationships between *Pinus ponderosa* forest structure, light characteristics, and understory graminoid species presence and abundance. *Forest Ecology and Management*, **124**, 205–215.
Naveh, Z. (1974) Effect of fire in the Mediterranean region. In *Fire and ecosystems* (eds T.T. Kozlowski & C.E. Ahlgren), pp. 401–434. Academic Press, New York.
Naveh, Z. (1975) The evolutionary significance of fire in the Mediterranean region. *Vegetatio*, **29**, 199–208.
Naveh, Z. & Whittaker, R.H. (1979) Measurements and relationships of plant species diversity in Mediterranean shrublands and woodlands. In *Ecological diversity in theory and practice* (eds J.F. Grassie, G.P. Patil, W. Smith & C. Taillie), pp. 219–239. International Co-operative Publishing House, Fairfield, Maryland.
Naveh, Z. & Whittaker, R.H. (1980) Structural and floristic diversity of shrublands and woodlands in northern Israel and other Mediterranean areas. *Vegetatio*, **41**, 171–190.
Neary, D.G., Klopatek, C.C., DeBano, L.F., & Ffolliott, P.F. (1999) Fire effects on belowground sustainability: a review and synthesis. *Forest Ecology and Management*, **122**, 51–71.
Ne'eman, G. (1997) Regeneration of natural pine forest: review of work done after the 1989 fire in Mount Carmel, Israel. *International Journal of Wildland Fire*, **7**, 295–306.
Ne'eman, G. (2000) The effect of burned pine trees on post-fire regeneration. In *Ecology, biogeography and management of* Pinus halepensis *and* P. brutia *forest ecosystems in the Mediterranean Basin* (eds G. Ne'eman & L. Trabaud), pp. 303–320. Backhuys Publishers, Leiden, The Netherlands.
Ne'eman, G., Fotheringham, C.J., & Keeley, J.E. (1999) Patch to landscape patterns in post fire recruitment of a serotinous conifer. *Plant Ecology*, **145**, 235–242.
Ne'eman, G., Goubitz, S., & Nathan, R. (2004) Reproductive traits of *Pinus halepensis* in the light of fire: a critical review. *Plant Ecology*, **171**, 69–79.
Ne'eman, G., Lahav, H., & Izhaki, I. (1992) Spatial pattern of seedlings one year after fire in a Mediterranean pine forest. *Oecologia*, **91**, 365–370.
Nichols, G. & Jones, T. (1992) Fusain in Carboniferous shallow marine sediments, Donegal, Ireland: the sedimentological effects of fire. *Sedimentology*, **39**, 487–502.

Nicolle, D. (2006) A classification and census of regenerative strategies in the eucalypts (*Angophora*, *Corymbia* and *Eucalyptus* – Myrtaceae), with special reference to the obligate seeders. *Australian Journal of Botany*, **54**, 391–407.

Nieuwenhuis, A. (1987) The effect of fire frequency on the sclerophyll vegetation of the West Head, New South Wales. *Australian Journal of Ecology*, **12**, 373–385.

Niklasson, M., Zin, E., Zielonka, T., *et al.* (2010) A 350-year tree-ring fire record from Biaøowieza Primeval Forest, Poland: implications for Central European lowland fire history. *Journal of Ecology*, **98**, 1319–1329.

Noble, I.R. (1984) Mortality of lignotuberous seedlings of *Eucalyptus* species after an intense fire in montane forest. *Australia Journal of Ecology*, **9**, 47–50.

Noble, I.R. & Slatyer, R.O. (1980) The use of vital attributes to predict successional changes in plant communities subject to recurrent disturbances. *Vegetatio*, **43**, 5–21.

Norris, J.R., Jackson, S.T., & Betancourt, J.L. (2006) Classification tree and minimum-volume ellipsoid analyses of the distribution of ponderosa pine in the western USA. *Journal of Biogeography*, **33**, 342–360.

North, M., Hurteau, M., Fiegener, R., & Barbour, M. (2005a) Influence of fire and El Niño on tree recruitment varies by species in Sierran mixed conifer. *Forest Science*, **51**, 187–197.

North, M., Oakley, B., Chen, J., *et al.* (2002) *Vegetation and eoclogical characteristics of mixed-conifer and red fir forests at the Teakettle Experimental Forest*, General Technical Report PSW-GTR-186. USDA Forest Service, Albany, California.

North, M., Oakley, B., Fiegener, R., Gray, A., & Barbour, M. (2005b) Influence of light and soil moisture on sierran mixed-conifer understory communities. *Plant Ecology*, **177**, 13–24.

NSWRFS (2006) *Planning for bush fire protection*. New South Wales Rural Fire Service, Lidcombe, Australia.

Núñez, M.R. & Calvo, L. (2000) Effect of high temperatures on seed germination of *Pinus sylvestris* and *Pinus halepensis*. *Forest Ecology and Management*, **131**, 183–190.

O'Connor, T.G. & Bredenkamp, G.J. (1997) Grassland. In *Vegetation of Southern Africa* (eds R.M. Cowling, D.M. Richardson & S.M. Pierce), pp. 215–257. Cambridge University Press, Cambridge, UK.

Odion, D.C., Frost, E.J., Strittholt, J.R., *et al.* (2004) Patterns of fire severity and forest conditions in the western Klamath mountains, California. *Conservation Biology*, **18**, 927–936.

Odion, D.C. & Haubensak, K.A. (2002) Response of French broom to fire. In *Proceedings of the symposium: fire in California ecosystems: integrating ecology, prevention and management* (eds N.G. Sugihara, M.E. Morales & T.J. Morales), Miscellaneous Publication 1, pp. 296–307. Association for Fire Ecology, Davis, California.

O'Donnell, A.J., Cullen, L.E., McCaw, W.L., Boer, M.M., & Grierson, P.F. (2010) Dendroecological potential of *Callistris preissii* for dating historical fires in semi-arid shrublands of southern Western Australia. *Dendrochronologia*, **28**, 37–48.

O'Dowd, D.J. & Gill, A.M. (1984) Predator satiation and site alteration following fire: mass reproduction of alpine ash (*Eucalyptus delegatensis*) in southeastern Australia. *Ecology*, **65**, 1052–1066.

OIG (2006) *Audit Report: Forest Service large fire suppression costs*, Report No. 08601–44-SF, United States Department of Agriculture, Office of Inspector General, Western Region, Washington, DC.

Ojeda, F. (1998) Biogeography of seeder and resprouter *Erica* species in the Cape Floristic Region: where are the resprouters? *Biological Journal of the Linnean Society*, **63**, 331–347.

Ojeda, F., Brun, F.G., & Vergara, J.J. (2005) Fire, rain and the selection of seeder and resprouter life-histories in fire-recruiting, woody plants. *New Phytologist*, **168**, 155–165.

Ojeda, F., Marañón, T., & Arroyo, J. (1996) Postfire regeneration of a mediterranean heathland in southern Spain. *International Journal of Wildland Fire*, **6**, 191–198.

Ojeda, F., Pausas, J.G., & Verdú, M. (2010) Soil shapes community structure through fire. *Oecologia*, **163**, 729–735.

O'Leary, J.F. & Westman, W.E. (1988) Regional disturbance effects on herb succession patterns in coastal sage scrub. *Journal of Biogeography*, **15**, 775–786.

Oliver, E.G.H. & Fellingham, A.C. (1994) A new serotinous species of *Cliffortia* (Rosaceae) from the southwestern Cape with notes on *Cliffortia arborea*. *Bothalia*, **24**, 153–162.

Omi, P.N., Rideout, D.B., & Botti, S.J. (1999) An analytical approach for assessing cost-effectiveness of landscape prescribed fires. In *Proceedings of the symposium on fire economics, planning, and policy: bottom lines* (ed. A. González-Caban), pp. 237–242. Pacific Southwest Research Station, Albany, California.

Ooi, M.K., Whelan, R.J., & Auld, T.D. (2006) Persistence of obligate-seeding species at the population scale: effects of fire intensity, fire patchiness and long fire-free intervals. *International Journal of Wildland Fire*, **15**, 261–269.

Orians, G.H. & Milewski, A.V. (2007) Ecology of Australia: the effects of nutrient-poor soils and intense fires. *Biological Reviews*, **82**, 393–423.

Orians, G.H. & Solbrig, O.T., eds. (1977) *Convergent evolution in warm deserts*, 352 pp. Dowden, Hutchinson & Ross, Stroudsburg, Pennsylvania.

Orshan, G. (1989) *Plant phenomorphological studies in Mediterranean type ecosystems*. Kluwer Academic Publishers, Dordrecht, The Netherlands.

Ovalle, C., Aronson, J., & Del Pozo, A. (1990) The espinal: systems of the mediterranean-type climate region of Chile. *Agroforestry Systems*, **10**, 213–249.

Owen-Smith, N. (1987) Pleistocene extinctions: the pivotal role of megaherbivores. *Paleobiology*, **13**, 351–362.

Paciorek, C.J., Condit, R., Hubbell, S.P., & Foster, R.B. (2000) The demographics of resprouting in tree and shrub species of a moist tropical forest. *Journal of Ecology*, **88**, 765–777.

Padgett, P.E. & Allen, E.B. (1999) Differential responses to nitrogen fertilization in native shrubs and exotic annuals common to Mediterranean coastal sage scrub of California. *Plant Ecology*, **144**, 93–101.

Palamarev, E. (1989) Paleobotanical evidences of the Tertiary history and origin of the Mediterranean sclerophyll dendroflora. *Plant Systematics and Evolution*, **162**, 93–107.

Palfner, G., Canseco, M.I., & Casanova-Katny, A. (2008) Post-fire seedlings of *Nothofagus alpina* in southern Chile show strong dominance of a single ectomycorrhizal fungus and a vertical shift in root architecture. *Plant and Soil*, **313**, 237–250.

Pannell, J.R. & Myerscough, P.J. (1993) Canopy-stored seed banks of *Allocasuarina distyla* and *A. nana* in relation to time since fire. *Australian Journal of Botany*, **41**, 1–9.

Papanastasis, V.P. (1980) Effects of season and frequency of burning on a phryganic rangeland in Greece. *Journal of Range Management*, **33**, 251–255.

Parker, V.T. (1984) Correlation of physiological divergence with reproductive mode in chaparral shrubs. *Madroño*, **31**, 231–242.

Parker, V.T. (1987) Effects of wet-season management burns on chaparral vegetation: implications for rare species. In *Conservation and management of rare and endangered plants* (ed. T.S. Elias), pp. 233–237. California Native Plant Society, Sacramento.

Parker, V.T. & Kelly, V.R. (1989) Seed banks in California chaparral and other Mediterranean climate shrublands. In *Ecology of soil seed banks* (eds M.A. Leck, V.T. Parker & R.L. Simpson), pp. 231–255. Academic Press, New York.

Parker, V.T., Vasey, M.C., & Keeley, J.E. (2009) Arctostaphylos. In *Flora of North America north of Mexico*, Vol. 8 (ed. Flora of North America Committee), pp. 406–445. Oxford University Press, New York.

Parker, V.T., Vasey, M.C., & Keeley, J.E. (2010) *Arctostaphylos*. In *The Jepson manual: vascular plants of California* (eds B.G. Baldwin, S. Boyd, D.J. Keil, et al.). University of California Press, Los Angeles.

Parker-Allie, F., Musil, C.F., & Thuiller, W. (2009) Effects of climate warming on the distributions of invasive Eurasian annual grasses: a South African perspective. *Climatic Change*, **94**, 87–103.

Parr, C.L. & Andersen, A.N. (2006) Patch mosaic burning for biodiversity conservation: a critique of the pyrodiversity paradigm. *Conservation Biology*, **20**, 1610–1619.

Parrish, J.T. (1998) *Interpreting pre-Quaternary climate from the geologic record*. Columbia University Press, New York.

Parrish, J.T., Ziegler, A.M., & Scotese, C.H. (1982) Rainfall patterns and the distribution of coals and evaporites in the Mesozoic and Cenozoic. *Palaeogeography, Palaeoclimatology, Palynology*, **40**, 67–101.

Parsons, D.J. (1976) Vegetation structure in the Mediterranean climate scrub communities of California and Chile. *Journal of Ecology*, **64**, 435–447.

Parsons, D.J. (1995) Restoring fire to giant sequoia groves: what have we learned in 25 years? In *Proceedings: symposium on fire in wilderness and park management* (eds J.K. Brown, R.W. Mutch, C.W. Spoon & R.H. Wakimotos), General Technical Report INT-GTR-320, pp. 256–258. USDA Forest Service, Missoula, Montana.

Parsons, R.F. (1981) Eucalyptus scrubs and shrublands. In *Australian vegetation* (ed. R.H. Groves), pp. 227–252. Cambridge University Press, Cambridge, UK.

Parsons, R.F. & Hopper, S.D. (2003) Monocotyledonous geophytes: comparison of south-western Australia with other areas of mediterranean climate. *Australian Journal of Botany*, **51**, 129–133.

Pärtel, M., Moora, M., & Zobel, M. (2001) Variation in species richness within and between calcareous (alvar) grassland stands: the role of core and satellite species. *Plant Ecology*, **157**, 203–211.

Pate, J. & Beard, J. (1984) *Kwongan: plant life of the sandplain: biology of a south-west Australian shrubland ecosystem*. University of Western Australia Press, Nedlands, Australia.

Pate, J. & Dell, B. (1984) Economy of mineral nutrients in sandplain species. In *Kwongan: plant life of the sandplain: biology of a south-west Australian shrubland ecosystem* (eds J. Pate & J. Beard), pp. 227–252. University of Western Australia Press, Nedlands, Australia.

Pate, J.S., Casson, N.E., Rullo, J., & Kuo, J. (1985) Biology of fire ephemerals of the sandplains of the kwongan of south-western Australia. *Australian Journal of Plant Physiology*, **12**, 641–655.

Pate, J.S., Froend, R.H., Bowen, B.J., Hansen, A., & Kuo, J. (1990) Seedling growth and storage characteristics of seeder and resprouter species of Mediterranean-type ecosystems of south western Australia. *Annuals of Botany*, **65**, 585–601.

Pate, J.S., Meney, K.A., & Dixon, K. (1991) Contrasting growth and morphological characteristics of fire-sensitive (obligate-seeder) and fire-resistant (resprouter) species of

Restionaceae (S. Hemisphere restiads) from south-western Western Australia. *Australian Journal of Botany*, **39**, 505–525.

Pauchard, A., Cavieres, L.A., & Bustamante, R.O. (2004) Comparing alien plant invasions among regions with similar climates: where to from here? *Diversity and Distributions*, **10**, 371–375.

Pauchard, A., García, R.A., Pena, E., *et al.* (2008) Positive feedbacks between plant invasions and fire regimes: *Teline monspessulana* (L.) K. Koch (Fabaceae) in central Chile. *Biological Invasions*, **10**, 547–553.

Paula, S., Arianoutsou, M., Kazanis, D., *et al.* (2009) Fire-related traits for plant species of the Mediterranean Basin. *Ecology*, **90**, 1420.

Paula, S. & Pausas, J.G. (2006) Leaf traits and resprouting ability in the Mediterranean basin. *Functional Ecology*, **20**, 941–947.

Paula, S. & Pausas, J.G. (2008) Burning seeds: germinative response to heat treatments in relation to resprouting ability. *Journal of Ecology*, **96**, 543–552.

Paula, S. & Pausas, J.G. (2009) *BROT: a plant trait database for Mediterranean Basin species*, Version 2009.01. Online: www.uv.es/jgpausas/brot.htm.

Paula, S. & Pausas, J.G. (2011) Root traits explain different foraging strategies between resprouting life histories. *Oecologia* **165**, 321–331.

Paull, R. & Hill, R.S. (2010) Early Oligocene *Callitris* and *Fitzroya* (Cupressaceae) from Tasmania. *American Journal of Botany*, **97**, 809–820.

Pausas, J.G. (1997) Resprouting of *Quercus suber* in NE Spain after fire. *Journal of Vegetation Science*, **8**, 703–706.

Pausas, J.G. (1999) The response of plant functional types to changes in the fire regime in Mediterranean ecosystems: a simulation approach. *Journal of Vegetation Science*, **10**, 717–722.

Pausas, J.G. (2001) Resprouting vs. seedling: a Mediterranean perspective. *Oikos*, **94**, 193–194.

Pausas, J.G. (2004) Changes in fire and climate in the eastern Iberian Peninsula (Mediterranean basin). *Climatic Change*, **63**, 337–350.

Pausas, J.G. (2006) Simulating mediterranean landscape pattern and vegetation dynamics under different fire regimes. *Plant Ecology*, **187**, 249–259.

Pausas, J.G., Bladé, C., Valdecantos, A., *et al.* (2004a) Pines and oaks in the restoration of Mediterranean landscapes in Spain: new perspectives for an old practice: a review. *Plant Ecology*, **171**, 209–220.

Pausas, J.G., Bonet, A., Maestre, F.T., & Climent, A. (2006a) The role of the perch effect on the nucleation process in Mediterranean semi-arid oldfields. *Acta Oecologica*, **29**, 346–352.

Pausas, J.G. & Bradstock, R.A. (2007) Fire persistence traits of plants along a productivity and disturbance gradient in Mediterranean shrublands of SE Australia. *Global Ecology and Biogeography*, **16**, 330–340.

Pausas, J.G., Bradstock, R.A., Keith, D.A., Keeley, J.E., & the GCTE Fire Network (2004b) Plant functional traits in relation to fire in crown-fire ecosystems. *Ecology*, **85**, 1085–1100.

Pausas, J.G., Carbó, E., Caturla, R.N., Gil, J.M., & Vallejo, R. (1999) Post-fire regeneration patterns in the eastern Iberian Peninsula. *Acta Oecologica*, **20**, 499–508.

Pausas, J.G. & Fernández-Muñoz, S. (2011) Fire regime changes in the Western Mediterranean Basin: from fuel-limited to drought-driven fire regime. *Climatic Change*, doi: 10.1007/s10584-011-0060-6.

Pausas, J.G., Gimeno, T., & Vallejo, R. (2002) Fire severity and pine regeneration in the eastern Iberian Peninsula. *In Forest fire research and wildland fire safety* (ed. D.X. Viegas), 8 pp. Millpress, Rotterdam, The Netherlands.

Pausas, J.G. & Keeley, J.E. (2009) A burning story: the role of fire in the history of life. *BioScience*, **59**, 593–601.

Pausas, J.G., Keeley, J.E., & Verdú, M. (2006b) Inferring differential evolutionary processes of plant persistence traits in Northern Hemisphere Mediterranean fire-prone ecosystems. *Journal of Ecology*, **94**, 31–39.

Pausas, J.G. & Lloret, F. (2007) Spatial and temporal patterns of plant functional types under simulated fire regimes. *International Journal of Wildland Fire*, 484–492.

Pausas, J.G., Lloret, F., & Vilà, M. (2006c) Simulating the effects of different disturbance regimes on *Cortaderia selloana* invasion. *Biological Conservation*, **128**, 128–135.

Pausas, J.G., Llovet, J., Rodrigo, A., & Vallejo, R. (2008) Are wildfires a disaster in the Mediterranean basin? A review. *International Journal of Wildland Fire*, **17**, 713–723.

Pausas, J.G., Ouadah, N., Ferrán, A., Gimeno, T., & Vallejo, R. (2003) Fire severity and seedling establishment in *Pinus halepensis* woodlands, eastern Iberian Peninsula. *Plant Ecology*, **169**, 205–213.

Pausas, J.G., Pereira, J.S., & Aronson, J. (2009) The tree. In *Cork oak woodlands on the edge: ecology, adaptive management, and restoration* (eds J. Aronson, J.S. Pereira & J.G. Pausas), pp. 11–21. Island Press, Washington, DC.

Pausas, J.G., Ribeiro, E., Dias, S.G., Pons, J., & Beseler, C. (2006d) Regeneration of a marginal cork oak (*Quercus suber*) forest in the eastern Iberian Peninsula. *Journal of Vegetation Science*, **17**, 729–738.

Pausas, J.G., Ribeiro, E., & Vallejo, R. (2004c) Post-fire regeneration variability of *Pinus halepensis* in the eastern Iberian peninsula. *Forest Ecology and Management*, **203**, 251–259.

Pausas, J.G. & Vallejo, R. (2008) Bases ecológicas para convivir con los incendios forestales en la Región Mediterránea: decálogo. *Ecosistemas*, **17**, 128–129.

Pausas, J.G. & Verdú, M. (2005) Plant persistence traits in fire-prone ecosystems of the Mediterranean basin: a phylogenetic approach. *Oikos*, **109**, 196–202.

Pausas, J.G. & Verdú, M. (2010) The jungle of methods for evaluating phenotypic and phylogenetic structure of communities. *BioScience*, **60**, 614–625.

Paysen, T.E. & Cohen, J.D. (1990) Chamise chaparral dead fuel fraction is not reliably predicted by age. *Western Journal of Applied Forestry*, **5**, 127–131.

Pearson, G.A. (1942) Herbaceous vegetation a factor in natural regeneration of ponderosa pine in the southwest. *Ecological Monographs*, **12**, 316–338.

Pearson, G.A. (1951) A comparison of the climate in four ponderosa pine regions. *Journal of Forestry*, **49**, 256–258.

Peet, R.K. (1978) Ecosystem convergence. *American Naturalist*, **112**, 441–444.

Pekin, B.K., Boer, M.M., Macfarlane, C., & Grierson, P.F. (2009) Impacts of increased fire frequency and aridity on eucalypt forest structure, biomass and composition in southwest Australia. *Forest Ecology and Management*, **258**, 2136–2142.

Pelton, J. (1984) *An ecological monograph of christmasberry*, Heteromeles arbutifolia *M. Roem, in the chaparral and woodland ecosystems of California* [unknown binding].

Penman, T. & Towerton, A.L. (2008) Soil temperatures during autumn prescribed burning: implications for the germination of fire responsive species? *International Journal of Wildland Fire*, **17**, 572–578.

Penman, T.D., Binns, D.L., & Kavanagh, R.P. (2008b) Quantifying successional changes in response to forest disturbances. *Applied Vegetation Science*, **11**, 261–268.

Penman, T.D., Binns, D.L., Sheils, R.J., Allen, R.M., & Kavanagh, R.P. (2008a) Changes in understorey plant species richness following logging and prescribed burning in shrubby dry sclerophyll forests of south-eastern Australia. *Austral Ecology*, **33**, 197–210.

Peppin, D., Fulé, P.Z., Sieg, C.H., Beyers, J.L., & Hunter, M.E. (2010) Post-wildfire seeding in forest of the western United States: an evidence-based review. *Forest Ecology and Management*, **260**, 573–586.

Pérez, B., Cruz, A., Fernández-González, F., & Moreno, J.M. (2003) Effects of the recent land-use history on the postfire vegetation of uplands in Central Spain. *Forest Ecology and Management*, **182**, 273–283.

Pérez, B. & Moreno, J.M. (1998) Methods for quantifying fire severity in shrubland-fires. *Plant Ecology*, **139**, 91–101.

Pérez, B., Sánchez-Mata, D., & Moreno, J.M. (1997) Effects of past and current land-use on post-fire vegetation in Sierra de Gredos, Spain. In *Forest fire risk and management* (eds P. Balabanis, G. Eftichidis & R. Fantechi), pp. 419–424. European Commission, Brussels.

Perry, T. (2009) SDG&E to pay the state $14.3 million after brush fires in 2007. *Los Angeles Times*, October 31, 2009.

Pessi, A., Businger, S., & Turner, K.L.C.T. (2004) On the relationship between lightning and convective rainfall over the central Pacific Ocean. In *18th International lightning detection conference*, Reference 21, pp. 1–9. Online: www.vaisala.com/ILDC2004.

Peters, D.P., Pielke, R.A., Bestelmeyer, B.T., et al. (2004) Cross-scale interactions, nonlinearities, and forecasting catastrophic events. *Proceedings of the National Academy of Sciences of the USA*, **101**, 15 130–15 135.

Peterson, D.L., Evers, L., Gravenmier, R.A., & Eberhardt, E. (2007) *A consumer guide: tools to manage vegetation and fuels*, General Technical Report PNW-GRR-690. USDA Forest Service, Pacific Northwest Research Station, Portland, Oregon.

Peterson, D.L. & Ryan, K.C. (1986) Modeling postfire conifer mortality for long-range planning. *Environmental Management*, **10**, 797–808.

Peterson, G.L. & Abbott, P.L. (1979) Mid-Eocene climatic change, southwestern California and northwestern Baja California. *Palaeogeography, Palaeoclimatology, Palynology*, **26**, 73–87.

Pfeiffer, M., Huttenlocher, H., & Ayasse, M. (2010) Myrmecochorous plants use chemical mimicry to cheat seed-dispersing ants. *Functional Ecology*, **24**, 545–555.

Phillips, J.F.V. (1931) Forest succession and ecology in the Knysna region. *Memoirs of the Botanical Survey of South Africa*, **14**, 1–327.

Phipps, J.B. (1992) *Heteromeles* and *Photinia* (Rosaceae, subfam. Maloideae) of Mexico and Central America. *Canadian Journal of Botany*, **70**, 2138–2162.

Pierce, S.M., Esler, K., & Cowling, R.M. (1995) Smoke-induced germination of succulents (Mesembryanthemaceae) from fire-prone and fire-free habitats in South Africa. *Oecologia*, **102**, 520–522.

Pillans, N.S. (1924) Destruction of indigenous vegetation by burning on the Cape Peninsula. *South African Journal of Science*, **21**, 348–350.

Pimm, S.L. (2007) Biodiversity: climate change or habitat loss: which will kill more species? *Current Biology*, **18**, R117–R119.

Pinaya, I., Soto, B., Arias, M., & Diaz-Fierros, F. (2000) Revegetation of burnt areas: relative effectiveness of native and commercial seed mixtures. *Land Degradation & Development*, **11**, 93–98.

Pincetl, S.S. (1999) *Transforming California: a political history of land use and development*. Johns Hopkins University Press, Baltimore, Maryland.

Piñol, J., Beven, K., & Viegas, D.X. (2005) Modelling the effect of fire-exclusion and prescribed fire on wildfire size in Mediterranean ecosystems. *Ecological Modelling*, **183**, 397–409.

Piñol, J., Castellnou, M., & Beven, K.J. (2007) Conditioning uncertainty in ecological models: assessing the impact of fire management strategies. *Ecological Modelling*, **207**, 34–44.

Piñol, J., Terradas, J., & Lloret, F. (1998) Climate warming, wildfire hazard, and wildfire occurrence in coastal eastern Spain. *Climatic Change*, **38**, 345–357.

Pinter, N., Fiedel, S., & Keeley, J.E. (2011) Fire and vegetation shifts at the vanguard of Paleoindian migrations. *Quaternary Science Reviews*, **30**, 269–272.

Pitman, A.J., Narisma, G.T., & McAneney, J. (2007) The impact of climate change on the risk of forest and grassland fires in Australia. *Climatic Change*, **84**, 383–401.

Platt, W.J. (1999) Southeastern pine savannas. In *The savanna, barren, and rock outcrop communities of North America* (eds R.C. Anderson, J.S. Fralish & J. Baskin), pp. 23–51. Cambridge University Press, Cambridge, UK.

Platt, W.J., Carr, S.M., Reilly, M., & Fahr, J. (2006) Pine savanna overstorey influences on ground-cover biodiversity. *Applied Vegetation Science*, **9**, 37–50.

Plucinski, M.P. & Anderson, W.R. (2008) Laboratory determination of factors influencing successful point ignition in the litter layer of shrubland vegetation. *International Journal of Wildland Fire*, **17**, 628–637.

Pole, M. (2003) New Zealand climate in the Neogene and implications for global atmospheric circulation. *Palaeogeography, Palaeoclimatology, Palaeoecology*, **193**, 269–284.

Pons, A. & Thinon, M. (1987) The role of fire from paleoecological data. *Ecologia Mediterranea*, **13**, 3–11.

Pons, J. & Pausas, J.G. (2006) Oak regeneration in heterogeneous landscapes: the case of fragmented *Quercus suber* forests in the eastern Iberian peninsula. *Forest Ecology and Management*, **231**, 196–204.

Pons, J. & Pausas, J.G. (2007) Acorn dispersal estimated by radio-tracking. *Oecologia*, **153**, 903–911.

Potter, A. (2002) Wildlife in the big smoke: there's more to the Sydney bushfires than saving property and fiscal assets. *Habitat Australia*, **30**, 11–14.

Potts, J.B. & Stephens, S.L. (2009) Invasive and native plant responses to shrubland fuel reduction: comparing prescribed fire, mastication, and treatment season. *Biological Conservation*, **142**, 1657–1664.

Powell, A., Shennan, S., & Thomas, M.G. (2009) Late Pleistocene demography and the appearance of modern human behavior. *Science*, **324**, 1298–1301.

Power, M.J., Marlon, J., Ortiz, N., *et al.* (2008) Changes in fire regimes since the Last Glacial Maximum: an assessment based on a global synthesis and analysis of charcoal data. *Climate Dynamics*, **30**, 887–907.

Pratt, R.B., Jacobsen, A.L., Ewers, F.W., & Davis, S.D. (2008) Linkage between water stress tolerance and life history type in seedlings of nine chaparral species (Rhamnaceae). *Journal of Ecology*, **96**, 1252–1265.

Pratt, R.B., Jacobsen, A.L., Golgotiu, K.A., *et al.* (2007) Life history type and water stress tolerance in nine California chaparral species (Rhamnaceae). *Ecological Monographs*, **77**, 239–253.

Premoli, A.C. & Steinke, L. (2008) Genetics of sprouting: effects of long-term persistence in fire-prone ecosystems. *Molecular Ecology*, **17**, 3827–3835.

Pretorius, M.R., Esler, K.J., Holmes, P.M., & Prins, N. (2008) The effectiveness of active restoration following alien clearance in fynbos riparian zones and resilience of treatments to fire. *South African Journal of Botany*, **74**, 517–525.

Price, O.F. & Bradstock, R.A. (2010) The effect of fuel age on the spread of fire in sclerophyll forest in the Sydney region of Australia. *International Journal of Wildland Fire*, **19**, 35–45.

Prober, S.M., Lunt, I.D., & Thiele, K.R. (2002) Determining reference conditions for management and restoration of temperate grassy woodlands: relationships among trees, topsoil and understorey flora in little-grazed remnants. *Australian Journal of Botany*, **50**, 687–697.

Prober, S.M., Thiele, K.R., & Lunt, I.D. (2007) Fire frequency regulates tussock grass competition, structure and resilience in endangered temperate woodlands. *Austral Ecology*, **32**, 808–824.

Proches, S. & Cowling, R.M. (2004) Cape geophytes: putting the pieces together. In *Ecology, conservation and management of mediterranean climate ecosystems* (eds M. Arianoutsou & V.P. Panastasis) [no pagination]. Millpress, Rotterdam, The Netherlands.

Proches, S., Cowling, R.M., Goldblatt, P., Manning, J.C., & Snijman, D.A. (2006) An overview of the Cape geophytes. *Biological Journal of the Linnean Society*, **87**, 27–43.

Pulido, F.J., Campos, P., & Montero, G. (2003) *La gestión forestal de las dehesas*. IPROCOR, Mérida, Spain.

Putz, F.E. & Brokaw, N.V.L. (1989) Sprouting of broken trees on Barro Colorado Island, Panama. *Ecology*, **70**, 508–512.

Pyke, G.H. & Paton, D.C. (1983) Why are waratah inflorescences so high and so large? In *Pollination '82* (eds E.G. Williams, R.B. Knox, J.H. Gilbert & P. Bernhardt), pp. 57–68. Botany Department, University of Melbourne, Australia.

Pyne, S.J. (1982) *Fire in America: a cultural history of wildland and rural fire*. Princeton University Press, Princeton, New Jersey.

Pyne, S.J. (1991) *Burning bush: a fire history of Australia*. Henry Holt and Company, New York.

Pyne, S.J. (1995) *World fire: the culture of fire on earth*. Henry Holt and Company, New York.

Pyne, S.J. (2000) Where have all the fires gone? *Fire Management Today*, **60**, 4–6.

Pyne, S.J., Andrews, P.L., & Laven, R.D. (1996) *Introduction to wildland fire*. John Wiley & Sons, New York.

Pyšek, P. (1998) Alien and native species in central European urban floras: a quantitative comparison. *Journal of Biogeography*, **25**, 155–163.

Quézel, P. (1981) Floristic composition and phytosociological structures of sclerophyllous matorrals around the Mediterranean. In *Ecosystems of the world, Vol. 11: Mediterranean-type shrublands* (eds F. di Castri, D.W. Goodall & R.L. Specht), pp. 107–122. Elsevier Scientific, New York.

Quézel, P. & Médail, F. (2003) *Écologie et biogeographie des forêts du bassin méditerranéen*. Elsevier, Paris.

Quick, C.R. (1961) How long can a seed remain alive? In *Seed: the yearbook of agriculture* (ed. A. Stefferud), pp. 93–99. U.S. Government Printing Office, Washington, DC.

Quinn, R.D. (1979) Effects of fire on small mammals in the chaparral. In *Cal-Neva Wildlife Transactions 1979*, pp. 125–133. Western Section of the Wildlife Society and the California-Nevada Chapter of the American Fisheries Society, Smartsville, California.

Quint, M. & Claßen-Bockhoff, R. (2008) Ancient or recent? Insights into the temporal evolution of the Bruniaceae. *Organisms, Diversity & Evolution*, **8**, 293–304.

Quintana, J. (2000) The drought in Chile and La Niña. *Drought Network News*, **12**, 1–6.

Quintana, J.R., Cruz, A., Fernández-González, F., & Moreno, J.M. (2004) Time of germination and establishment success after fire of three obligate seeders in a Mediterranean shrubland of central Spain. *Journal of Biogeography*, **31**, 241–249.

Radeloff, V.C., Hammer, R.B., Stewart, S.I., et al. (2005) The wildland-urban interface in the United States. *Ecological Applications*, **15**, 799–805.

Raison, R.J., Woods, P.V., & Khanna, P.K. (1983) Dynamics of fine fuels in recurrently burnt eucalypt forests. *Australian Forestry*, **46**, 294–302.

Ramírez, C. & San Martin, C. (2005) Asociaciones vegetales de la Cordillera de la Costa de la region de Los Lagos. In *Historia, biodiversidad y ecologia de los bosques costeros de Chile* (eds C. Smith-Ramirez, J.J. Armesto & C. Valdovinos), pp. 206–224. Editorial Universitaria, Santiago, Chile.

Ramsay, G.C., McArthur, N.A., & Rudolph, L. (1995) Towards an integrated model for designing for building survival in bushfires. CALM Science, *Supplement* **4**, 101–108.

Raphael, M.N. (2003) The Santa Ana winds of California. *Earth Interactions*, **7**, 1–13.

Raven, P.H. (1973) The evolution of Mediterranean floras. In *Mediterranean ecosystems: origin and structure* (eds F. di Castri & H.A. Mooney), pp. 213–223. Springer, New York.

Raven, P.H. & Axelrod, D.I. (1972) Plate tectonics and Australasian paleobiogeography. *Science*, **176**, 1379–1386.

Raven, P.H. & Axelrod, D.I. (1974) Angiosperm biogeography and past continental movements. *Annals of the Missouri Botanical Garden*, **61**, 539–673.

Raven, P.H. & Axelrod, D.I. (1978) Origin and relationships of the California flora. *University of California Publications in Botany*, **72**, 1–134.

Rebelo, A.G., Boucher, C., Helme, N., Mucina, L., & Rutherford, M.C. (2006) Fynbos biome. In *The vegetation of South Africa, Lesotho and Swaziland*, Vol. 19: Strelitzia, pp. 53–219. South African National Biodiversity Institute, Pretoria.

Regan, H.M., Auld, T.D., Keith, D.A., & Burgman, M.A. (2003) The effects of fire and predators on the long-term persistence of an endangered shrub, *Grevillea caleyi*. *Biological Conservation*, **109**, 73–83.

Regan, H.M., Crookston, J.B., Swab, R., Franklin, J., & Lawson, D.M. (2010) Habitat fragmentation and altered fire regime create trade-offs for an obligate seeding shrub. *Ecology*, **91**, 1114–1123.

Regelbrugge, J. & Conrad, S.G. (2002) Biomass and fuel characteristics of chaparral in southern California. In *Fire in California ecosystems: integrating ecology, prevention and management* (eds M. Morales & T. Morales), Vol. 1, pp. 308–317. Association for Fire Ecology, San Diego, California.

Rego, F.C., Botelho, H., & Bunting, S.C. (1987) Prescribed fire effects on soils and vegetation in *Pinus pinaster* forests in northern Portugal. *Ecologia Mediterranea*, **13**, 189–195.

Reich, P.B., Walters, M.B., & Ellsworth, D.S. (1997) From tropics to tundra: global convergence in plant functioning. *Proceedings of the National Academy of Sciences of the USA*, **94**, 13 730–13 734.

Reiner, R.J. (2007) Fire in California grasslands. In *California grasslands: ecology and management* (eds M.R. Stromberg, J.D. Corbin & C.M. D'Antonio), pp. 207–217. University of California Press, Los Angeles, California.

Rejmánek, M. (1989) Invasibility of plant communities. In *Biological invasions: a global perspective* (eds J.A. Drake, F. di Castri, R.H. Groves, *et al.*), pp. 369–388. John Wiley & Sons, New York.

Retana, J., Espelta, J.M., Habrouk, A., Ordoñez, J.L., & de Solà-Morales, F. (2002) Regeneration patterns of three Mediterranean pines and forest changes after a large wildfire in northeastern Spain. *Écoscience*, **9**, 89–97.

Reynolds, J.H., Jordan, T.E., Johnson, N.M., Damanti, J.F., & Tabbutt, K.D. (1990) Neogene deformation of the flat-subduction segment of the Argentine-Chilean Andes: magnetostratigraphic constraints from Las Juntas, La Rioja province, Argentina. *Geological Society of America Bulletin*, **102**, 1607–1622.

Riaño, D., Chuvieco, E., Salas, J., Palacios-Orueta, A., & Bastarrika, A. (2002) Generation of fuel type maps from Landsat TM images and ancillary data in Mediterranean ecosystems. *Canadian Journal of Forest Research*, **32**, 1301–1315.

Rice, B. & Westoby, M. (1983) Plant species richness and the 0.1 ha scale in Australian vegetation compared to other continents. *Vegetatio*, **52**, 129–140.

Rice, B. & Westoby, M. (1986) Evidence against the hypothesis that ant-dispersed seeds reach nutrient-enriched microsites. *Ecology*, **67**, 1270–1274.

Rice, B. & Westoby, M. (1999) Regeneration after fire in *Triodia* R. Br. *Australian Journal of Ecology*, **24**, 563–572.

Rice, B., Westoby, M., Griffin, G.F., & Friedel, M.H. (1994) Effects of supplementary soil nutrients on hummock grasses. *Australian Journal of Botany*, **42**, 687–703.

Rich, F.J. (1989) A review of the taphonomy of plant remains in lacustrine sediments. *Review of Palaeobotany and Palynology*, **58**, 33–46.

Richards, R. (2000) The sensitivity of snow gum to fire scarring in relation to Aboriginal landscape burning. BS Honours thesis. Australian National University, Canberra.

Richardson, D.M. (1998) *Ecology and biogeography of* Pinus. Cambridge University Press, Cambridge, UK.

Richardson, D.M., Bond, W.J., Dean, W.R.J., *et al.* (2000a) Invasive alien species and global change: a South African perspective. In *Invasive species in a changing world* (eds H.J. Mooney & R.J. Hobbs), pp. 303–349. Island Press, Washington, DC.

Richardson, D.M., Cowling, R.M., Lamont, B.B., & van Hensbergen, H.J. (1995) Coexistence of *Banksia* species in southwestern Australia: the role of regional and local processes. *Journal of Vegetation Science*, **6**, 329–342.

Richardson, D.M. & Kluge, R.L. (2008) Seed banks of invasive Australian *Acacia* species in South Africa: role in invasiveness and options for management. *Perspectives in Plant Ecology Evolution and Systematics*, **10**, 161–177.

Richardson, D.M. & Kruger, F.J. (1990) Water relations and photosynthetic characteristics of selected trees and shrubs of riparian and hillside habitats in the southwestern Cape Province, South Africa. *South African Journal of Botany*, **56**, 214–225.

Richardson, D.M., Rouget, M., Henderson, L., & Nel, J.L. (2004) Invasive alien plants in South Africa: macroecological patterns, with special emphasis on the Cape Floristic

Region. In *Ecology, conservation and management of mediterranean climate ecosystems* (eds M. Arianoutsou & V.P. Papanastasis), 132 pp. Millpress, Rotterdam, The Netherlands.

Richardson, D.M. & van Wilgen, B.W. (1986) The effects of fire in felled *Hakea sericea* and natural fynbos and implications for weed control in mountain catchments. *South African Forestry Journal*, **139**, 4–14.

Richardson, J.E., Fay, M.F., Cronk, Q.C.B., Bowman, D., & Chase, M.W. (2000b) A phylogenetic analysis of Rhamnaceae using rbcL and trnL-F plastid DNA sequences. *American Journal of Botany*, **87**, 1309–1324.

Ricklefs, R.E. & Latham, R.E. (1992) Intercontinental correlation of geographical ranges suggests stasis in ecological traits of relict genera of temperate perennial herbs. *American Naturalist*, **139**, 1305–1321.

Rideout, D.B., Loomis, J.B., & Omi, P.N. (1999) Incorporating non-market values in fire management planning. In *Proceedings of the symposium on fire economics, planning, and policy: bottom lines* (eds A. González-Cabán & P.N. Omi), General Technical Report PSW-GTR-173, pp. 217–225. USDA Forest Service, Pacific Southwest Research Station, Albany, California.

Riggan, P.J. & Dunn, P.H. (1982) Harvesting chaparral biomass for energy: an environmental assessment. In *Proceedings of the symposium on dynamics and management of Mediterranean-type ecosystems* (eds C.E. Conrad & W.C. Oechel), General Technical Report PSW-58, pp. 149–157. USDA Forest Service, Pacific Southwest Forest and Range Experiment Station, Albany, California.

Riggan, P.J., S. Goode, Jacks, P.M., & Lockwood, R.W. (1988) Interaction of fire and community development in chaparral of southern California. *Ecological Monographs*, **58**, 155–175.

Rigolot, E. (2004) Predicting postfire mortality of *Pinus halepensis* Mill. and *Pinus pinea* L. *Plant Ecology*, **171**, 139–151.

Rigolot, E., Dupuy, J.L., Lambert, B., Coudor, R., Larini, M., & Etienne, M. (1999) Fire prevention management plans in the French mediterranean region: technical and economic assessment of fuel-break networks. In *Proceedings of the international symposium on forest fires: needs and innovations, a DELFI Action, November 18–19, 1999, Athens, Greece*, pp. 120–124. CINAR, SA, Athens.

Rigolot, E., Etienne, M., & Lambert, B. (1998) Different fire regime effects on a *Cytisus purgans* community. In *Fire management and landscape ecology* (ed. L. Trabaud), pp. 137–146. International Association of Wildland Fire, Fairfield, Washington.

Ripple, W.J. & van Valkenburgh, B. (2010) Linking top-down forces to the Pleistocene megafaunal extinctions. *BioScience*, **60**, 516–526.

Roads, J., Tripp, P., Juang, H., *et al.* (2010) NCEP-ECPC monthly to seasonal U.S. fire danger forecasts. *International Journal of Wildland Fire*, **19**, 399–414.

Roalson, E.H. (2007) C_4 photosynthesis: convergence upon convergence upon. *Current Biology*, **17**, R776–R778.

Roberts, R.G., Flannery, T.F., Ayliffe, L.K., *et al.* (2001) New ages for the last Australian megafauna: continent-wide extinction about 46,000 years ago. *Science*, **292**, 1888–1892.

Robichaud, P.R., Beyers, J.L., & Neary, D.G. (2000) *Evaluating the effectiveness of postfire rehabilitation treatments*, General Technical Report RMRS-GTR-63, USDA Forest Service, Rocky Mountain Research Station, Ogden, Utah.

Robinson, J.M. (1989) Phanerozoic O_2 variation, fire, and terrestrial ecology. *Palaeogeography, Palaeoclimatology, Palynology*, **75**, 223–240.

Rocca, M.E. (2009) Fine-scale patchiness in fuel load can influence initial post-fire understory composition in a mixed conifer forest, Sequoia National Park, California. *Natural Areas Journal*, **29**, 126–132.

Rodà, F., Retana, J., Gracia, C.A., & Bellot, J., eds. (1999) *Ecology of Mediterranean evergreen oak forests*, 373 pp. Springer, Berlin.

Rodrigo, A., Quintana, V., & Retana, J. (2007) Fire reduces *Pinus pinea* L. distribution in the north-eastern Iberian Peninsula. *Écoscience*, **14**, 23–30.

Rodrigo, A., Retana, J., & Picó, F.X. (2004) Direct regeneration is not the only response of Mediterranean forests to large fires. *Ecology*, **85**, 716–729.

Rodríguez-Trejo, D.A. (2008) Fire regimes, fire ecology, and fire management in Mexico. *Ambio*, **37**, 548–556.

Rogan, J. & Franklin, J. (2001) Mapping wildfire burn severity in southern California forests and shrublands using enhanced thematic mapper imagery. *Geocarto International*, **16**, 89–99.

Rogers, C., Parker, V.T., Kelly, V., & Wood, M.K. (1989) Maximizing chaparral vegetation response to prescribed burns: experimental considerations. In *Proceedings of the symposium on fire and watershed management* (ed. N.H. Berg), General Technical Report PSW-109, 158 pp. USDA Forest Service, Pacific Southwest Forest and Range Experiment Station, Albany, California.

Rogers, G.F. & Vint, M.K. (1987) Winter precipitation and fire in the Sonoran Desert. *Journal of Arid Environments*, **13**, 47–52.

Rogers, K. & Biggs, H. (1999) Integrating indicators, endpoints and value systems in strategic management of the rivers of the Kruger National Park. *Freshwater Biology*, **41**, 439–451.

Rogers, M.J. (1982) Fire management in southern California. In *Proceedings of the symposium on dynamics and management of Mediterranean-type ecosystems* (eds C.E. Conrad & W.C. Oechel), General Technical Report PSW-58, pp. 496–497. USDA Forest Service, Pacific Southwest Forest and Range Experiment Station, Albany, California.

Roldán-Zamarron, A., Merino-de-Miguel, S., González-Alonso, F., García-Gigorro, S., & Cuevas, J.M. (2006) Minas de Riotinto (south Spain) forest fire: burned area assessment and fire severity mapping using Landsat 5-TM, Envistat-MERIS, and Terra MODIS postfire images. *Journal of Geophysical Research*, **111**, G04S11, 1 of 9.

Román-Cuesta, R.M., Gracia, M., & Retana, J. (2003) Environmental and human factors influencing fire trends in ENSO and non-ENSO years in tropical Mexico. *Ecological Applications*, **13**, 1177–1192.

Romme, W.H. (1982) Fire and landscape diversity in subalpine forests of Yellowstone National Park. *Ecological Monographs*, **52**, 199–221.

Rosenzweig, M.L. (1995) *Species diversity in space and time*. Cambridge University Press, Cambridge, UK.

Ross, K.A., Fox, B.J., & Fox, M.D. (2002) Changes to plant species richness in forest fragments: fragment age, distribution and fire history may be as important as area. *Journal of Biogeography*, **29**, 749–765.

Ross, K.A., Taylor, J.E., Fox, M.D., & Fox, B.J. (2004) Interaction of multiple disturbances: importance of disturbance interval in the effects of fire on rehabilitating mined areas. *Austral Ecology*, **29**, 508–529.

Rothermel, R.C. (1972) *A mathematical model for predicting fire spread in wildland fuels*, Research Paper INT-115. USDA Forest Service, Intermountain Forest and Range Experiment Station, Ogden, Utah.

Roura-Pascual, N., Richardson, D.M., Krug, R.M., *et al.* (2009) Ecology and management of alien plant invasions in South African fynbos: accommodating key complexities in objective decision making. *Biological Conservation*, **142**, 1595–1604.

Rowe, N.P. & Jones, T.P. (2000) Devonian charcoal. *Palaeogeography, Palaeoclimatology, Palaeoecology*, **164**, 347–354.

Roy, D.P., Boschetti, L., & Ju, J. (2008) The collection 5 MODIS burned area product: global evaluation by comparison with the MODIS active fire product. *Remote Sensing of Environment*, **12**, 3690–3707.

Roy, D.P., Boschetti, L., & Trigg, S.N. (2006) Remote sensing of fire severity: assessing the performance of the normalized burn ratio. *IEEE Geoscience and Remote Sensing Letters*, **3**, 112–116.

Roy, J. & Sonié, L. (1992) Germination and population dynamics of *Cistus* species in relation to fire. *Journal of Applied Ecology*, **29**, 647–655.

Royce, E.B. & Barbour, M.G. (2001) Mediterranean climate effects, II: Conifer growth phenology across a Sierra Nevada ecotone. *American Journal of Botany*, **88**, 919–932.

Ruiz-Gallardo, J.R., Castano, S., & Calera, A. (2004) Application of remote sensing and GIS to locate priority intervention areas after wildland fires in Mediterranean systems: a case study from south-eastern Spain. *International Journal of Wildland Fire*, **13**, 241–252.

Rundel, P.W. (1981a) The matorral zone of central Chile. In *Ecosystems of the world, Vol. 11: Mediterranean-type shrublands* (eds F. di Castri, D.W. Goodall & R.L. Specht), pp. 175–201. Elsevier Scientific, New York.

Rundel, P.W. (1981b) Structural and chemical components of flammability. In *Proceedings of the conference "Fire regime and ecosystems properties", December 11–15, 1978, Honolulu, Hawaii* (eds H.A. Mooney, T.M. Bonnicksen, N.L. Christensen, J.E. Lotan & W.A. Rainers), General Technical Report WO-26, pp. 183–207. USDA Forest Service, Washington, DC.

Rundel, P.W. (1996) Monocotyledonous geophytes in the California flora. *Madroño*, **43**, 355–368.

Rundel, P.W. (1998) Landscape disturbance in Mediterranean-type ecosystems: an overview. In *Landscape disturbance and biodiversity in Mediterranean-type ecosystems* (eds P.W. Rundel, G. Montenegro & F.M. Jaksic), pp. 3–22. Springer, New York.

Rundel, P.W. (2004) Mediterranean-climate ecosystems: defining their extent and community dominance. In *Ecology, conservation and management of mediterranean climate ecosystems* (eds M. Arianoutsou & V.P. Panastasis), pp. 1–12. Millpress, Rotterdam, The Netherlands.

Rundel, P.W. (2007) Sage scrub. In *Terrestrial vegetation of California*, third edition (eds M.G. Barbour, T. Keeler-Wolf & A.A. Schoenherr), pp. 208–228. University of California Press, Los Angeles, California.

Rundel, P.W. (2010) Convergence and divergence in mediterranean-climate ecosystems: what we can learn by comparing similar places. In *The ecology of place* (eds M. Price & I. Billick). University of Chicago Press, Chicago, Illinois.

Rundel, P.W., Baker, G.A., & Parsons, D.J. (1981) Productivity and nutritional responses of *Chamaebatia foliolosa* (Rosaceae) to seasonal burning. In *Components of productivity of Mediterranean climate regions: basic and applied aspects* (eds N.S. Margaris & H.A. Mooney), pp. 191–196. Dr. W. Junk, The Hague, The Netherlands.

Rundel, P.W., Baker, G.A., Parsons, D.J., & Stohlgren, T.J. (1987) Postfire demography of resprouting and seedling establishment by *Adenostoma fasciculatum* in the California chaparral. In *Plant response to stress: functional analysis in Mediterranean ecosystems* (eds J.D. Tenhunen, F.M. Catarino, O.L. Lange & W.C. Oechel), pp. 575–596. Springer, Berlin.

Rundel, P.W., Montenegro, G., & Jaksic, F.M., eds. (1998) *Landscape disturbance and biodiversity in mediterranean-type ecosystems*. Springer Verlag, Berlin.

Rundel, P.W. & Parsons, D.J. (1979) Structural changes in chamise (*Adenostoma fasciculatum*) along a fire induced age gradient. *Journal of Range Management*, **32**, 462–466.

Rundel, P.W., Parsons, D.J., & Baker, G.A. (1980) The role of shrub structure and chemistry in the flammability of chaparral shrubs. In *Proceedings of the second conference on scientific research in national parks*, Vol. 10: *Fire ecology*, pp. 248–260. USDI National Park Service, Washington, DC.

Rundel, P.W., Parsons, D.J., & Gordon, D.T. (1977) Montane and subalpine vegetation of the Sierra Nevada and Cascade Ranges. In *Terrestrial vegetation of California* (eds M.G. Barbour & J. Major), pp. 559–599. John Wiley & Sons, New York.

Rundel, P.W. & Sturmer, S.B. (1998) Native plant diversity in riparian communities of the Santa Monica mountains, California. *Madroño*, **45**, 93–100.

Rundel, P.W., Villagra, P.E., Dillon, M.O., Roig-Juñent, S., & Debandi, G. (2007) Arid and semi-arid ecosystems. In *The physical geography of South America* (eds T.T. Veblen, K.R. Young & A.R. Orme), pp. 158–183. Oxford University Press, Oxford, UK.

Russell, R.P. & Parsons, R.F. (1978) Effects of time since fire on heath floristics at Wilson's Promontory, South Australia. *Australian Journal of Botany*, **26**, 53–61.

Russell, W. & Tompkins, R. (2005) Estimating biomass in coastal *Baccharis pilularis* dominated plant communities. *Fire Ecology*, **1**, 20–27.

Russell-Smith, J. & Stanton, P. (2002) Fire regimes and fire management of rainforest communities across northern Australia. In *Flammable Australia: the fire regimes and biodiversity of a continent* (eds R.A. Bradstock, J.E. Williams & A.M. Gill), pp. 329–350. Cambridge University Press, Cambridge, UK.

Ryan, B.C. (1969) *A vertical perspective of Santa Ana winds in a canyon*, Research Paper PSW-52. USDA Forest Service, Pacific Southwest Forest and Range Experiment Station, Berkeley, California.

Ryan, K.C. (2002) Dynamic interactions between forest structure and fire behavior in boreal ecosystems. *Silva Fennica*, **36**, 13–39.

Ryan, K.C. & Frandsen, W.H. (1991) Predicting postfire mortality of seven western conifers. *Canadian Journal of Forest Research*, **18**, 1291–1297.

Sackett, S.S. & Haase, S.M. (1998) Two case histories for using prescribed fire to restore ponderosa pine ecosystems in northern Arizona. In *Fire in ecosystem management: shifting the paradigm from suppression to prescription* (eds T.L. Pruden & L.A. Brennan), Tall Timbers Fire Ecology Conference Proceedings 20, pp. 380–389. Tall Timbers Research Station, Tallahassee, Florida.

Sadori, L. & Giardini, M. (2007) Charcoal analysis, a method to study vegetation and climate of the Holocene: the case of Lago di Pergusa (Sicily, Italy). *Geobios*, **40**, 173–180.

Safford, H.D. & Harrison, S. (2004) Fire effects on plant diversity in serpentine vs. sandstone chaparral. *Ecology*, **85**, 539–548.

Safford, H.D. & Schmidt, D. (2010) *Southern California National Forest fire regime departure maps*. The Nature Conservancy and U.S. Forest Service, Pacific Southwest Region, University of California, Davis.

Safford, H.D., Schmidt, D.A., & Carlson, C.H. (2009) Effects of fuel treatments on fire severity in an area of wildland-urban interface, Angora Fire, Lake Tahoe Basin, California. *Forest Ecology and Management*, **258**, 773–787.

Safford, H.D., Viers, J.H., & Harrison, S.P. (2005) Serpentine endemism in the California flora: a database of serpentine affinity. *Madroño*, **52**, 222–257.

Sağlam, B., Kűcűk, O., Dilgili, E., Durmaz, B.D., & Baysal, I. (2008) Estimating fuel biomass of some shrub species (maquis) in Turkey. *Turkish Journal of Agricultural Forestry*, **32**, 349–356.

Sakamoto, S. & Radford, S.J.E. (2004) *Lightning near Cerro Chascón, ALMA Memo 487*, pp. 1–22. National Radio Astronomy Observatory, Charlottesville, Virginia.

San Martin, J. (2005) Vegetación y biodiversidad florística en la Cordillera de la Costa de Chile central (34°44′–35°50′ S). In *Historia, biodiversidad y ecología de los bosques costeros de Chile* (eds C. Smith-Ramirez, J.J. Armesto & C. Valdovinos), pp. 178–196. Editorial Universitaria, Santiago, Chile.

San Martin, J. & Donoso, C. (1995) Estructura florísitica e impacto antrópico en los bosques nativos de Chile. In *Ecología de los bosques nativos de Chile* (eds J.J. Armesto, C. Villagran & M.K. Arroyo), pp. 153–168. Editorial Universitaria, Santiago, Chile.

Sapsis, D. (1999) *Development patterns and fire suppression*. Department of Forestry and Fire Protection, Fire and Resource Assessment Program, Sacramento, California.

Sardoy, N., Consalvi, J.-L., Porterie, B., & Fernandez-Pello, A.C. (2007) Modeling transport and combustion of firebrands from burning trees. *Combustion and Flame*, **150**, 151–169.

Sauer, J.D. (1979) Living fences in Costa Rican agriculture. *Turrialba*, **29**, 255–261.

Sauquet, H., Weston, P.H., Anderson, C.L., et al. (2009b) Contrasted patterns of hyperdiversification in Mediterranean hotspots. *Proceedings of the National Academy of Sciences of the USA*, **106**, 221–225.

Sauquet, H., Weston, P.H., Barker, N.P., Anderson, C.L., & Cantrill, D.J. (2009a) Using fossils and molecular data to reveal the orgins of the Cape proteas (subfamily Proteoideae). *Molecular Phylogenetics and Evolution*, **51**, 31–43.

Saura-Mas, S. & Lloret, F. (2007) Leaf and shoot water content and leaf dry matter content of Mediterranean woody species with different post-fire regenerative strategies. *Annals of Botany*, **99**, 545–554.

Savage, M. (1997) The role of anthropogenic influences in a mixed-conifer forest mortality episode. *Journal of Vegetation Science*, **8**, 95–104.

Savage, M. & Swetnam, T.W. (1990) Early 19th-century fire decline following sheep pasturing in a Navajo ponderosa pine forest. *Ecology*, **71**, 2374–2378.

Sawyer, J.O., Thornburgh, D.A., & Griffin, J.R. (1977) Mixed evergreen forest. In *Terrestrial vegetation of California* (eds M.G. Barbour & J. Major), pp. 359–381. John Wiley & Sons, New York.

Sax, D.F. (2002) Native and naturalized plant diversity are positively correlated in scrub communities of California and Chile. *Diversity and Distributions*, **8**, 193–210.

Scarcella, M. (2008) Homeowners seek damages against government for California wildfire. In *The Blog of Legal Times*, Oct 6, 2008.

Scarff, F.R. & Westoby, M. (2006) Leaf litter flammability in some semi-arid Australian woodlands. *Functional Ecology*, **20**, 745–752.

Schatral, A., Kailis, S.G., & Fox, J.E.D. (1994) Seed dispersal of *Hibbertia hypericoides* (Dilleniaceae) by ants. *Journal of the Royal Society of Western Australia*, **77**, 81–85.

Scher, H.D. & Martin, E.E. (2006) Timing and climatic consequences of the opening of Drake Passage. *Science*, **312**, 428–430.

Schiffman, P.M. (2007) Pleistocene and pre-European grassland ecosystems: species composition at the time of first European settlement. In *California grasslands: ecology and management* (eds M.R. Stromberg, J.D. Corbin & C.M. D'Antonio), pp. 52–56. University of California Press, Los Angeles, California.

Schimper, A.F.W. (1903) *Plant-geography upon a physiological basis*. Clarendon Press, Oxford, UK.

Schlesinger, W.H., Gray, J.T., Gill, D.S., & Mahall, B.E. (1982) *Ceanothus megacarpus* chaparral: a synthesis of ecosystem properties during development and annual growth. *Botanical Review*, **48**, 71–117.

Schluter, D. (1986) Tests for similarity and convergence of finch communities. *Ecology*, **67**, 1073–1085.

Schluter, D. & Ricklefs, R.E. (1993) Convergence and the regional component of species diversity. In *Species diversity in ecological communities: historical and geographical perspectives* (eds R.E. Ricklefs & D. Schluter), pp. 230–249. University of Chicago Press, Chicago, Illinois.

Schoenberg, F.P., Chang, C.-H., Keeley, J.E., Pompa, J., Woods, J., & Xu, H. (2007) A critical assessment of the burning index in Los Angeles County, California. *International Journal of Wildland Fire*, **16**, 473–483.

Schoenberg, F.P., Peng, R., Huang, Z., & Rundel, P. (2003) Detection of non-linearities in the dependence of burn area on fuel age and climate variables. *International Journal of Wildland Fire*, **12**, 1–6.

Scholze, M., Knorr, W., Arnell, N.W., & Prentice, I.C. (2006) A climate-change risk analysis for world ecosystems. *Proceedings of the National Academy of Sciences of the USA*, **103**, 13 116–13 120.

Schroeder, M.J. & Buck, C.C. (1970) *Fire weather: A guide for application of meteorological information to forest fire control operations*, Agricultural Handbook 360. USDA Forest Service, Washington, DC.

Schurr, F.M., Bond, W.J., Midgley, G.F., & Higgins, S.I. (2005) A mechanistic model for secondary seed dispersal by wind and its experimental validation. *Journal of Ecology*, **93**, 1017–1028.

Schutte, A.L., Vlok, J.H.J., & van Wyk, B.E. (1995) Fire-survival strategy: a character of taxonomic, ecological and evolutionary importance in fynbos legumes. *Plant Systematics and Evolution*, **195**, 243–259.

Schwartz, M.W., Porter, D.J., Randall, J.M., & Lyons, K.E. (1996) Impact of nonindigenous plants. In *Sierra Nevada ecosystem project, final report to Congress: status of the Sierra Nevada*, Vol. II: *Assessments and scientific basis for management options*, pp. 1203–1226. Centers for Water and Wildland Resources, University of California, Davis.

Schwilk, D. & Ackerly, D. (2005) Is there a cost to resprouting? Seedling growth rate and drought tolerance in sprouting and nonsprouting *Ceanothus* (Rhamnaceae). *American Journal of Botany*, **92**, 404–410.

Schwilk, D.W. (2003) Flammability is a niche construction trait: canopy architecture affects fire intensity. *American Naturalist*, **162**, 725–733.

Schwilk, D.W. & Ackerly, D.D. (2001) Flammability and serotiny as strategies: correlated evolution in pines. *Oikos*, **94**, 326–336.

Schwilk, D.W. & Keeley, J.E. (2006) The role of fire refugia in the distribution of *Pinus sabiniana* (Pinaceae) in the southern Sierra Nevada. *Madroño*, **53**, 364–372.

Schwilk, D.W., Keeley, J.E., & Bond, W.J. (1997) The intermediate disturbance hypothesis does not explain fire and diversity pattern in fynbos. *Plant Ecology*, **132**, 77–84.

Schwilk, D.W. & Kerr, B. (2002) Genetic niche-hiking: an alternative explanation for the evolution of flammability. *Oikos*, **99**, 431–442.

Schwilk, D.W., Knapp, E.E., Ferrenberg, S.M., Keeley, J.E., & Caprio, A.C. (2006) Tree mortality from fire and bark beetles following early and late season prescribed fires in a Sierra Nevada mixed-conifer forest. *Forest Ecology and Management*, **232**, 36–75.

Scott, A.C. (2000) The Pre-Quaternary history of fire. *Palaeogeography, Palaeoclimatology, Palaeoecology*, **164**, 297–345.

Scott, A.C. (2010) Charcoal recognition, taphonomy and uses in palaeoenvironmental analysis. *Palaeogeography, Palaeoclimatology, Palaeoecology*, **291**, 11–39.

Scott, A.C. & Glasspool, I.J. (2006) *The diversification of Paleozoic fire systems and fluctuations in atmospheric oxygen concentration. Proceedings of the National Academy of Sciences of the USA, 103, 10* **861–10** 865.

Scott, J.H. & Burgan, R.E. (2005) *Standard fire behavior fuel models: a comprehensive set for use with Rothermel's surface fire spread model*, General Technical Report, RMRS-GTR-153. USDA Forest Service, Rocky Mountain Research Station, Ogden, Utah.

Scott, J.H. & Reinhardt, E.D. (2001) *Assessing crown fire potential by linking models of surface and crown fire behavior.* USDA Forest Service, Rocky Mountain Research Station, Fort Collins, Colorado.

Scott, L. (2002) Microscopic charcoal in sediments: Quaternary fire history of the grassland and savanna regions in South Africa. *Journal of Quaternary Science*, **17**, 77–86.

Segarra-Moragues, J.G. & Ojeda, F. (2010) Post-fire response and genetic diversity in *Erica coccinea*: connecting population dynamics and diversification in a biodiversity hotspot. *Evolution*, doi: 10.1111/j.1558–5646.2010.01064.x.

Segura, A.M., Holmgren, M., Anabalon, J.J., & Fuentes, E.R. (1998) The significance of fire intensity in creating local patchiness in the Chilean matorral. *Plant Ecology*, **139**, 259–264.

Seijo, F. (2009) Who framed the forest fire? State framing and peasant counter-framing of anthropogenic forest fires in Spain since 1940. *Journal of Environmental Policy & Planning*, **11**, 103–128.

Seligman, N.G. & Henkin, Z. (2000) Regeneration of a dominant Mediterranean dwarf-shrub after fire. *Journal of Vegetation Science*, **11**, 893–902.

Seydack, A.H.W., Bekker, S.J., & Marshall, A.H. (2007) Shrubland fire regime scenarios in the Swartsberg Mountain range, South Africa: implications for management. *International Journal of Wildland Fire*, **16**, 81–95.

Shakesby, R.A., Wallbrink, P.J., Doerr, S.H., *et al.* (2007) Distinctiveness of wildfire effects on soil erosion in south-east Australian eucalypt forests assessed in a global context. *Forest Ecology and Management*, **238**, 347–364.

Sharples, J.J., Mills, G.A., McRae, R.H.D., & Weber, R.O. (2010) Foehn-like winds and elevated fire danger conditions in southeastern Australia. *Journal of Applied Meteorology and Climatology*, **49**, 1067–1095.

Shearer, R.C. & Schmidt, W.C. (1970) *Natural regeneration in ponderosa pine forests of western Montana.* USDA Forest Service, Intermountain Forest and Range Experiment Station, Ogden, Utah.

Shmida, A. & Wilson, M.V. (1985) Biological determinants of species diversity. *Journal of Biogeography*, **12**, 1–20.

Show, S.B. & Kotok, E.I. (1923a) The occurrence of lightning storms in relation to forest fires in California. *Monthly Weather Review*, **51**, 175–182.

Show, S.B. & Kotok, E.I. (1923b) *Forest fires in California 1911–1920: an analytical study*. USDA, Washington, DC.

Shreve, F. (1936) The transition from desert to chaparral in Baja California. *Madroño*, **3**, 257–264.

Shreve, F. (1939) Observations on the vegetation of Chihuahua. *Madroño*, **5**, 1–12.

Sikes, K., Witter, M., Christian, J., Taylor, R.S., & Keeley, J.E. (2006) Field validation of NPS-USGS burn severity mapping techniques in chaparral and coastal sage scrub. In *3rd international fire ecology and management congress: changing fire regimes* [poster]. Association for Fire Ecology, San Diego, California.

Silla, F., Fraver, S., Lara, A., Allnutt, T.R., & Newton, A. (2002) Regeneration and stand dynamics of *Fitzroya cupressoides* (Cupressaceae) forests of southern Chile's central depression. *Forest Ecology and Management*, **165**, 213–224.

Simard, A.J. (1991) Fire severity, changing scales, and how things hang together. *International Journal of Wildland Fire*, **1**, 23–34.

Singh, G., Kershaw, A.P., & Clark, R. (1981) Quaternary vegetation and fire history in Australia. In *Fire and the Australian biota* (eds A.M. Gill, R.H. Groves & I.R. Noble), pp. 23–54. Australian Academy of Science, Canberra.

Skinner, C.N. & Chang, C.-R. (1996) *Fire regimes, past and present. In Sierra Nevada ecosystem project: final report to congress*, Vol. II (ed. SNEP). Centers for Water and Wildland Resources, University of California, Davis, California.

Skinner, C.N., Taylor, A.H., & Agee, J.K. (2006) Klamath mountains bioregion. In *Fire in California's ecosystems* (eds N.G. Sugihara, J.W. van Wagtendonk, K.E. Shaffer, J. Fites-Kaufman & A.E. Thoede), pp. 170–194. University of California Press, Los Angeles, California.

Slingsby, P. & Johns, A. (2009) *T.P. Stokoe, the man, the myths, the flowers*. Baardskeerders, Cape Town, RSA.

Sloan, L.C. & Barrón, E.J. (1990) "Equable" climates during Earth history? *Geology* **18**, 489–492.

Slocum, M.G., Beckage, B., Platt, W.J., Orzell, S.L., & Taylor, W. (2010) Effect of climate on wildfire size: a cross-scale analysis. *Ecosystems*, doi: 10.1007/S10021–010–9357–Y.

Smith, A.M.S., Wooster, M.J., Drake, N.A., et al. (2005) Testing the potential of mulit-spectral remote sensing for retrospectively estimating fire severity in African savannahs. *Remote Sensing of Environment*, **97**, 92–115.

Smith, M.A., Grant, C.D., Loneragan, W.A. & Koch, J.M. (2004) Fire management implications of fuel loads and vegetation structure in jarrah forest restoration on bauxite mines in Western Australia. *Forest Ecology and Management*, **187**, 247–266.

Smith, S.A. & Beaulieu, J.M. (2009) Life history influences rates of climatic niche evolution in flowering plants. *Proceedings of the Royal Society of London, Series B*, **276**, 4345–4352.

Smith, S.A. & Donoghue, M.J. (2008) Rates of molecular evolution are linked to life history in flowering plants. *Science*, **322**, 86–89.

SMNRA (2004) *Santa Monica Mountains National Recreation Area fire management plan*. USDI National Park Service, Thousand Oaks, California.

Snyder, J.R. (1984) The role of fire: mutch ado about nothing? *Oikos*, **43**, 404–405.

Sorensen, F.C., Mandel, N.L., & Aagaard, J.E. (2001) Role of selection versus historical isolation in racial differentiation of ponderosa pine in southern Oregon: an investigation of alternative hypotheses. *Canadian Journal of Forest Research*, **31**, 1127–1139.

Soriano, L.R.S., de Pablo, F., & Tomas, C. (2005) Ten-year study of cloud-to-ground lightning activity in the Iberian Peninsula. *Journal of Atmospheric and Solar-Terrestrial Physics*, **67**, 1632–1639.

Soto, L. (1995) Estadísticas de ocurrencia y daño de incendios forestales, temporadas 1964–1995. Departamento de Manejo del Fuego, CONAF, Santiago, Chile.

Soto, O.R. (2010) SDG&E reaches accord in wildfires: utility agrees to pay $14.8 million to state. In *San Diego Union–Tribune*, April 22, 2010.

Southey, D. (2009) *Wildfires in the Cape Floristic Region: exploring vegetation and weather as drivers of fire frequency*, Department of Botany, University of Cape Town, RSA.

Southgate, R. & Carthew, S. (2007) Post-fire ephemerals and spinifex-fuelled fires: a decision model for bilby habitat management in the Tanami Desert, Australia. *International Journal of Wildland Fire*, **16**, 741–754.

Sparrow, A. (1989) Mallee vegetation in South Australia. In *Mediterranean landscapes in Australia: mallee ecosystems and their management* (eds J.C. Noble & R.A. Bradstock), pp. 109–124. CSIRO Publications, Collingwood, Australia.

Sparrow, A.D. & Bellingham, P.J. (2001) More to resprouting than fire. *Oikos*, **94**, 195–197.

Specht, R.L. (1979) Heathlands and related shrublands of the world. In *Ecosystems of the world*, Vol. 9A: *Heathlands and related shrublands: descriptive studies* (ed. R.L. Specht), pp. 1–18. Elsevier Scientific, New York.

Specht, R.L. (1981) Responses to fires in heathlands and related shrublands. In *Fire and the Australian biota* (eds A.M. Gill, R.H. Groves & I.R. Noble), pp. 394–415. Australian Academy of Science, Canberra.

Specht, R.L. & Moll, E.J. (1983) Mediterranean-type heathlands and sclerophyllous shrubland of the world: an overview. In *Mediterranean-type ecosystems: the role of nutrients* (eds F.J. Kruger, D.T. Mitchell & J.U.M. Jarvis), pp. 41–65. Springer, New York.

Specht, R.L. & Morgan, D.G. (1981) The balance between the foliage projective covers of overstorey and understorey strata in Australian vegetation. *Austral Ecology*, **6**, 193–202.

Specht, R.L. & Specht, A.S. (1999) *Australian plant communities: dynamics of structure, growth and biodiversity*. Oxford University Press, Melbourne, Australia.

Sperry, J.S. & Hacke, U.G. (2002) Desert shrub water relations with respect to soil characteristics and plant functional type. *Functional Ecology*, **16**, 367–378.

Spicer, R.A. & Wolfe, J.A. (1987) Plant taphonomy of late Holocene deposits in Trinity (Clair Engle) Lake, northern California. *Paleobiology*, **13**, 227–245.

Spyratos, V., Bourgeron, P.S., & Ghil, M. (2007) Development at the wildland–urban interface and the mitigation of forest-fire risk. *Proceedings of the National Academy of Sciences of the USA*, **104**, 14 272–14 276.

Stadum, C.J. & Weigand, P.W. (1999) Fossil wood from the middle Miocene Conejo volcanics, Santa Monica Mountains, California. *Bulletin of the Southern California Academy of Sciences*, **98**, 15–25.

Stahli, M., Finsinger, W., Tinner, W., & Allgower, B. (2006) Wildfire history and fire ecology of the Swiss National Park (Central Alps): new evidence from charcoal, pollen and plant macrofossils. *The Holocene*, **16**, 805–817.

Steele, J. (2004) CDF, city, county are target of fire claims: insurer seeks reparation; class-action suit pursued. *San Diego Union–Tribune*, July 18, 2004, Local, B1.

Steele, R. (1988) Ecological relationships of ponderosa pine. In *Ponderosa pine. the species and its management symposium proceedings* (eds D.M. Baumgarner & J.E. Lotan), pp. 71–76. USDA Forest Service, Ogden, Utah.

Steinke, L.R., Premoli, A.C., Souto, C.P., & Hedrén, M. (2008) Adaptive and neutral variation of the resprouter *Nothofagus antarctica* growing in distinct habitats in north-western Patagonia. *Silva Fennica*, **42**, 177–188.

Stephens, S.L. (2001) Fire history differences in adjacent Jeffrey pine and upper montane forests in the eastern Sierra Nevada. *International Journal of Wildland Fire*, **10**, 161–167.

Stephens, S.L., Adams, M.A., Handmer, J., et al. (2009) Urban-wildland fires: how California and other regions of the US can learn from Australia. *Environmental Research Letters* **4**, 1–5.

Stephens, S.L. & Fry, D.L. (2005) Fire history in coast redwood stands in the northeastern Santa Cruz Mountains, California. *Fire Ecology*, **1**, 2–19.

Stephens, S.L., Skinner, C.N., & Gill, S.J. (2003) Dendrochonology-based fire history of Jeffrey pine–mixed conifer forests in the Sierra San Pedro Mártir, Mexico. *Canadian Journal of Forest Research*, **33**, 1090–1101.

Stephenson, N., Parsons, D.J., & Swetnam, T.W. (1991) Restoring natural fire to the *Sequoia*-mixed conifer forest: should intense fire play a role? In *Proceedings 17th Tall Timbers Fire Ecology Conference, High Intensity Fire in Wildlands: Management Challenges and Options*, pp. 321–337. Tall Timbers Research Station, Tallahassee, Florida.

Stephenson, N.L. (1990) Climatic control of vegetation distribution: the role of the water balance. *American Naturalist*, **135**, 649–670.

Stephenson, N.L. (1998) Actual evapotranspiration and deficit biologically meaningful correlates of vegetation distribution across spatial scales. *Journal of Biogeography*, **25**, 855–870.

Stevenson, A.C. & Harrison, R.J. (1992) Ancient forests in Spain: a model for land-use and dry forest management in south-west Spain from 4000 BC to 1900 AD. *Proceedings of the Prehistoric Society*, **58**, pp. 227–247.

Stewart, O.C. (1956) Fire as a first great force employed by man. In *Man's role in changing the face of the earth* (ed. W.L. Thomas), pp. 115–133. University of Chicago, Illinois.

Stock, W.D. & Allsopp, N. (1992) Functional perspective of ecosystems. In *The ecology of fynbos* (ed. R.M. Cowling), pp. 241–259. Oxford University Press, Cape Town, RSA.

Stock, W.D., Pate, J.S., & Delfs, J. (1990) Influence of seed size and quality on seedling development under low nutrient conditions in five Australian and South African members of the Proteaceae. *Journal of Ecology*, **78**, 1005–1020.

Stock, W.D., Pate, J.S., & Rasins, E. (1991) Seed developmental patterns in *Banksia attenuata* R. Br. and *B. laricina* C. Gardner in relation to mechanical defense costs. *New Phytologist*, **117**, 109–114.

Stohlgren, T.J., Bull, K.A., & Otsuki, Y. (1998) Comparison of rangeland vegetation sampling techniques in the central grasslands. *Journal of Range Management*, **51**, 164–172.

Stohlgren, T.J., Falkner, M.B., & Schell, L.D. (1995) A modified-Whittaker nested vegetation sampling method. *Vegetatio*, **117**, 113–121.

Stone, E.C. (1951) The stimulative effect of fire on the flowering of the golden brodiaea (*Brodiaea ixiodes* Wats. var. *lugens* Jeps.). *Ecology*, **32**, 534–537.

Stone, E.L. & Cornwall, S.M. (1968) Basal bud burls in *Betula populifolia*. *Forest Science*, **14**, 64–65.

Stone, E.L., Jr. & Stone, M.H. (1954) Root collar sprouts in pine. *Journal of Forestry*, **52**, 487–491.

Strauss, D., Dednar, L., & Mees, R. (1989) Do one percent of forest fires cause ninety-nine percent of the damage? *Forest Science*, **35**, 319–328.

Stringer, C. & McKie, R. (1996) *African exodus: the origin of modern humanity* (eds C. Stringer and R. McKie). Jonathan Cape, London.

Stuart, J.D. (1987) Fire history of an old-growth forest of *Sequoia sempervirens* (Taxodiaceae) forest in Humboldt Redwoods State Park, California. *Madroño*, **34**, 128–141.

Stylinski, C.D. & Allen, E.B. (1999) Lack of native species recovery following severe exotic disturbance in southern Californian shrublands. *Journal of Applied Ecology*, **36**, 544–554.

Suc, J.P. (1984) Origin and evolution of Mediterranean vegetation and climate in Europe. *Nature*, **307**, 429–432.

Sugihara, N.G., van Wagtendonk, J.W., Shaffer, K.E., Fites-Kaufman, J., & Thode, A.E. (2006) *Fire in California's ecosystems*. University of California Press, Los Angeles, California.

Swank, S.E. & Oechel, W.C. (1991) Interactions among the effects of herbivory, competition, and resource limitation on chaparral herbs. *Ecology*, **72**, 104–115.

Swetnam, T.W. (1993) Fire history and climate change in giant sequoia groves. *Science*, **262**, 885–889.

Swetnam, T.W., Allen, C.D., & Betancourt, J.L. (1999) Applied historical ecology: using the past to manage for the future. *Ecological Application*, **9**, 1189–1206.

Swetnam, T.W. & Baisan, C.H. (2003) Tree-ring reconstructions of fire and climate history in the Sierra Nevada and southwestern United States. In *Fire and climatic change in temperate ecosystems of the western Americas* (eds T.T. Veblen, W.L. Baker, G. Montenegro & T.W. Swetnam), pp. 158–195. Springer, New York.

Swetnam, T.W. & Betancourt, J.L. (1998) Mesoscale disturbance and ecological response to decadal climatic variability in the American Southwest. *Journal of Climate*, **11**, 3128–3147.

Swezy, M. & Odion, D.C. (1998) Fire on the mountain: a land manager's manifesto for broom control. In *Proceedings of California Exotic Pest Plant Council symposium, Vol. 3: 1997* (eds M. Kelly, E. Wagner & P. Warner), pp. 76–81. California Exotic Pest Plant Council, Sacramento, California.

Syphard, A.D., Franklin, J., & Keeley, J.E. (2006) Simulating the effects of frequent fires on southern California coastal shrublands. *Ecological Applications*, **16**, 1744–1756.

Syphard, A.D., Radeloff, V.C., Hawbaker, T.J., & Stewart, S.I. (2009) Conservation threats due to human-caused increases in fire frequency in Mediterranean-climate ecosystems. *Conservation Biology*, **23**, 758–769.

Syphard, A.D., Radeloff, V.C., Keeley, J.E., *et al.* (2007) Human influence on California fire regimes. *Ecological Applications*, **17**, 1388–1402.

Talluto, M.V. & Suding, K.N. (2008) Historical change in coastal sage scrub in southern California, USA in relation to fire frequency and air pollution. *Landscape Ecology*, **23**, 803–815.

Tapias, R., Climent, J., Pardos, J.A., & Gil, L. (2004) Life histories of Mediterranean pines. *Plant Ecology*, **171**, 53–68.

Tapias, R., Gil, L., Fuentes-Utrilla, P., & Pardos, J.A. (2001) Canopy seed banks in Mediterranean pines of southeastern Spain: a comparison between *Pinus halepensis* Mill., *P. pinaster* Ait., *P. nigra* Arn. and *P. pinea* L. *Journal of Ecology*, **89**, 629–638.

Tappeiner, J.C., II & Helms, J.A. (1971) Natural regeneration of Douglas fir and white fir on exposed sites in the Sierra Nevada of California. *American Midland Naturalist*, **86**, 358–370.

Taylor, A.H. & Beaty, R.M. (2005) Climatic influences on fire regimes in the northern Sierra Nevada mountains, Lake Tahoe Basin, Nevada, USA. *Journal of Biogeography*, **32**, 425–438.

Taylor, A.H. & Skinner, C.N. (1998) Fire history and landscape dynamics in a late-successional reserve, Klamath Mountains, California, USA. *Forest Ecology and Management*, **111**, 285–301.

Taylor, H.C. (1978) Capensis. In *The biogeography and ecology of Southern Africa* (ed. M.J.A. Werger), pp. 171–229. Dr. W. Junk, The Hague, The Netherlands.

Taylor, J.L.S. & van Staden, J. (1998) Plant-derived smoke solutions stimulate the growth of *Lycopersicon esculentum* roots in vitro. *Plant Growth Regulation*, **26**, 77–84.

Terradas, J., Piñol, J., & Lloret, F. (1998) Risk factors in wildfires along the Mediterranean coast of Iberian Peninsula. In *Fire Management and Landscape Ecology* (ed. L. Trabaud), pp. 297–304. International Association of Wildland Fire, Fairfield, Washington.

Thanos, C.A. & Daskalakou, E.N. (2000) Reproduction in *Pinus halepensis* and *Pinus brutia* In *Ecology, biogeography and management of* Pinus halepensis *and* P. brutia *forest ecosystems in the Mediterranean Basin* (eds G. Ne'eman & L. Trabaud), pp. 79–90. Backhuys Publishers, Leiden, The Netherlands.

Thanos, C.A., Daskalakou, E.N., & Nikolaidou, S. (1996) Early post-fire regeneration of a *Pinus halepensis* forest on Mount Parnis, Greece. *Journal of Vegetation Science*, **7**, 273–280.

Thanos, C.A. & Doussi, M.A. (2000) Post-fire regeneration of *Pinus brutia* forests In *Ecology, biogeography and management of* Pinus halepensis *and* P. brutia *forest ecosystems in the Mediterranean Basin* (eds G. Ne'eman & L. Trabaud), pp. 291–302. Backhuys Publishers Leiden, The Netherlands.

Thirgood, J.V. (1981) *Man and the Mediterranean forest: a history of resource depletion*. Academic Press, London.

Thomas, C.M. & Davis, S.D. (1989) Recovery patterns of three chaparral shrub species after wildfire. *Oecologia*, **80**, 309–320.

Thomas, I., Enright, N.J., & Kenyon, C.E. (2001) The Holocene history of mediterranean-type plant communities, Little Desert National Park, Victoria, Australia. *The Holocene*, **11**, 691–697.

Thomas, P.B., Morris, C.E., & Auld, T.D. (2007) Response surfaces for the combined effects of heat shock and smoke on germination of 16 species forming soil seed banks in south-east Australia. *Austral Ecology*, **32**, 605–616.

Thompson, J.D. (2005) *Plant evolution in the Mediterranean*. Oxford University Press, Oxford, UK.

Thompson, J.R. & Spies, T.A. (2010) Factors associated with crown damage following recurring mixed-severity wildfires and post-fire management in southwestern Oregon. *Landscape Ecology*, **25**, 775–789.

Thomsen, C.D., Williams, W.A., Vayssiéres, M., Bell, F.L., & George, M. (1993) Controlled grazing on annual grassland decreases yellow starthistle. *California Agriculture*, **47**, 36–40.

Thorne, R.F. (1977) Montane and subalpine forests of the Transverse and Peninsular Ranges. In *Terrestrial vegetation of California* (eds M.G. Barbour & J. Major), pp. 537–557. John Wiley & Sons, New York.

Thuiller, W., Slingsby, J.A., Privett, S.D.J., & Cowling, R.M. (2007) Stochastic species turnover and stable coexistence in a species-rich, fire-prone plant community. *PLoS ONE*, **2**, e938.

Timbrook, J., Johnson, J.R., & Earle, D.D. (1982) Vegetation burning by the Chumash. *Journal of California and Great Basin Anthropology*, **4**, 163–186.

Titshall, L.W., O'Connor, T.G., & Morris, C.D. (2000) Effect of long-term exclusion of fire and herbivory on the soils and vegetation of sour grassland. *African Journal of Range & Forage Science*, **17**, 70–80.

Torrejon, F., Cisternas, M., & Araneda, A. (2004) Environmental effects of the Spanish colonization from de Maullin river to the Chiloe archipelo southern Chile. *Revista Chilena de Historia Natural*, **77**, 661–677.

Torres-Rojo, J.M., Magaña-Torres, O.S., & Ramírez-Fuentes, G.A. (2007) Índice de peligro de incendios forestales de largo plazo. *Agrociencia*, **41**, 663.

Tozer, M.G. & Auld, T.D. (2006) Soil heating and germination: investigations using leaf scorch on graminoids and experimental seed burial. *International Journal of Wildland Fire*, **15**, 509–516.

Tozer, M.G. & Bradstock, R.A. (1998) Factors influencing the establishment of seedlings of the mallee, *Eucalyptus leuhmanniana* (Myrtaceae). *Australian Journal of Botany*, **45**, 997–1008.

Tozer, M.G. & Bradstock, R.A. (2003) Fire-mediated effects of overstorey on plant species diversity and abundance in an eastern Australian heath. *Plant Ecology*, **164**, 213–223.

Trabaud, L. (1990) Fire as an agent of plant invasion? A case study in the French Mediterranean vegetation. In *Biological invasions in Europe and the Mediterranean Basin* (eds F. di Castri, A.J. Hansen & M. Debussche), pp. 417–437. Kluwer Academic Publishers, Dordrecht, The Netherlands.

Trabaud, L. (1991) Fire regimes and phytomass growth dynamics in a *Quercus coccifera* garrigue. *Journal of Vegetation Science*, **2**, 307–314.

Trabaud, L. & Campant, C. (1991) Difficulté de recolonisation naturelle du pin de Salzmann *Pinus nigra* Arn. ssp. *salzmanii* (Dunal) Franco après incendie. *Biological Conservation*, **58**, 329–343.

Trabaud, L., Christensen, N.L., & Gill, A.M. (1993) Historical biogeography of fire in temperate and Mediterranean ecosystems. In *Fire in the environment: the ecological, atmospheric, and climatic importance of vegetation fires* (eds P.J. Crutzen & J.G. Goldammer), pp. 277–295. John Wiley & Sons, New York.

Trabaud, L. & Lepart, J. (1980) Diversity and stability in garrigue ecosystems after fire. *Vegetatio*, **43**, 49–57.

Trabaud, L. & Lepart, J. (1981) Changes in the floristic composition of a *Quercus coccifera* L. garrigue in relation to different fire regimes. *Vegetatio*, **46**, 105–116.

Trabaud, L., Michels, C., & Grosman, J. (1985) Recovery of burnt *Pinus halepensis* Mill. forests, 2: Pine reconstitution after wildfire. *Forest Ecology and Management*, **13**, 167–179.

Trigo, R.M., Pereira, J.M.C., Pereira, M.G., et al. (2006) Atmospheric conditions associated with the exceptional fire season of 2003 in Portugal. *International Journal of Climatology*, **26**, 1741–1757.

Trollope, W.S.W. (1973) Fire as a method of controlling macchia (fynbos) vegetation on the Amatole mountains of the eastern Cape. *Proceedings of the Grassland Society of Southern Africa*, **8**, 35–41.

Trouet, V., Taylor, A.H., Carleton, A.M., & Skinner, C.N. (2009) Interannual variations in fire weather, fire extent, and synoptic-scale circulation patterns in northern California and Oregon. *Theoretical and Applied Climatology*, **95**, 349–360.

Troumbis, A.Y. & Trabaud, L. (1989) Some questions about flammability in fire ecology. *Acta Oecologica*, **10**, 167–175.

Tsitsoni, T. (1997) Conditions determining natural regeneration after wildfires in the *Pinus halepensis* (Miller, 1768) forests of Kassandra Peninsula (North Greece). *Forest Ecology and Management*, **92**, 199–208.

Tuomisto, H. (2010) A diversity of beta diversities: straightening up a concept gone awry, 1: Defining beta diversity as a function of alpha and gamma diversity. *Ecography*, **33**, 2–22.

Turner, M.G. & Romme, W.H. (1994) Landscape dynamics in crown fire ecosystems. *Landscape Ecology*, **9**, 59–77.

Turner, M.G., Romme, W.H., & Gardner, R.H. (1999) Prefire heterogeneity, fire severity, and early postfire plant reestablishment in subalpine forests of Yellowstone National Park, Wyoming. *International Journal of Wildland Fire*, **9**, 21–36.

Turner, R., Roberts, N., Eastwood, W.J., Jenkins, E., & Rosen, A. (2010) Fire, climate and the origins of agriculture: micro-charcoal records of biomass burning during the last glacial-interglacial transition in Southwest Asia. *Journal of Quaternary Science*, **25**, 371–386.

Turner, R., Roberts, N., & Jones, M.D. (2008) Climatic pacing of Mediterranean fire histories from lake sedimentary microcharcoal. *Global and Planetary Change*, **63**, 317–324.

Tyler, C. & Borchert, M. (2002) Reproduction and growth of the chaparral geophyte, *Zigadenus fremontii* (Liliaceae), in relation to fire. *Plant Ecology*, **165**, 11–20.

Tzedakis, P.C. (2007) Seven ambiguities in the Mediterranean palaeoenvironmental narrative. *Quaternary Science Reviews*, **26**, 2042–2066.

Udeze, C.U. & Oboh-Ikuenobe, F.E. (2005) Neogene palaeoceanographic and palaeoclimatic events inferred from palynological data: Cape Basin off South Africa, ODP Leg 175. *Palaeogeography, Palaeoclimatology, Palaeoecology*, **219**, 199–223.

Uhl, D., Hamad, A.A., Kerp, H., & Bandel, K. (2007) Evidence for palaeo-wildfire in the Late Permian palaeotropics: charcoalified wood from the Um Irna formation of Jordan. *Review of Palaeobotany and Palynology* **144**, 221–230.

Uhl, D. & Kerp, H. (2003) Wildfires in the Late Palaeozoic of Central Europe: the Zechstein (Upper Permian) of NW-Hesse (Germany). *Palaeogeography, Palaeoclimatology, Palaeoecology*, **199**, 1–15.

Underwood, E.C., Klausmeyer, K.R., Cox, R.L., et al. (2009) Expanding the global network of protected areas to save the imperiled Mediterranean biome. *Conservation Biology*, **23**, 43–52.

Urretavizcaya, M.F. & Defosse, G.E. (2004) Soil seed bank of *Austrocedrus chilensis* (D. Don) Pic. Serm. et Bizarri related to different degrees of fire disturbance in two sites of southern Patagonia, Argentina. *Forest Ecology and Management*, **187**, 361–372.

Urretavizcaya, M.F., Defosse, G.E., & Gonda, H.E. (2006) Short-term effects of fire on plant cover and soil conditions in two *Austrocedrus chilensis* (cypress) forests in Patagonia, Argentina. *Annals of Forest Science*, **63**, 63–71.

USFS (2005) *Land management plan, Part 1: Southern California national forests vision*, R5-MB-075. USDA Forest Service, Pacific Southwest Region, Albany, California.

Ustin, S.L., Riaño, D., Koltunov, A., Roberts, D.A., & Dennison, P.E. (2009) Mapping fire risk in mediterranean ecosystems of California: vegetation type, density, invasive species, and fire frequency. In *Earth observation of wildland fires in mediterranean ecosystems* (ed. E. Chuvieco), pp. 41–53. Springer, Berlin.

Uys, R.G., Bond, W.J., & Everson, T.M. (2004) The effect of different fire regimes on plant diversity in southern African grasslands. *Biological Conservation*, **118**, 489–499.

Vaghti, M.G. & Greco, S.E. (2007) Riparian vegetation of the Great Valley. In *Terrestrial vegetation of California*, third edition (eds M.G. Barbour, T. Keeler-Wolf & A.A. Schoenherr), pp. 425–455. University of California Press, Los Angeles, California.

Vaillant, N.M., Fites-Kaufman, J.A., & Stephens, S.L. (2009) Effectiveness of prescribed fire as a fuel treatment in Californian coniferous forests. *International Journal of Wildland Fire*, **18**, 165–175.

Valente, L.M., Reeves, G., Schnitzler, J., *et al.* (2009) Diversification of the African genus *Protea* (Proteaceae) in the Cape biodiversity hotspot and beyond: equal rates in different biomes. *Evolution*, **64**, 745–760.

Valiente-Banuet, A., Rumebe, A.V., Verdú, M., & Callaway, R.M. (2006) Modern Quaternary plant lineages promote diversity through facilitation of ancient Tertiary lineages. *Proceedings of the National Academy of Sciences of the USA*, **103**, 16 812–16 817.

Vallejo, R., Aronson, J., Pausas, J.G., & Cortina, J. (2006) Restoration of mediterranean woodlands. In *Restoration ecology from a European perspective* (eds J. van Andel & J. Aronson), pp. 193–207. Blackwell Science, Oxford, UK.

van der Moezel, P.G. & Bell, D.T. (1989) Plant species richness in the mallee region of Western Australia. *Australian Journal of Ecology*, **14**, 221–226.

van Horne, M.L. & Fúle, P.Z. (2006) Comparing methods of reconstructing fire history usng fire scars in a southwestern United States ponderosa pine forest. *Canadian Journal of Forest Research*, **36**, 855–867.

Van Konijnenburg-Van Cittert, J.H.A. (2002) Ecology of some Late Triassic to Early Cretaceous ferns in Eurasia. *Review of Palaeobotany and Palynology*, **119**, 113–124.

van Laar, A. (1982) Sampling for above-ground biomass for *Pinus radiata* in the Bosbouklof catchment at Jonkershoek. *South African Forestry Journal*, **123**, 8–13.

van Mantgem, P.J., Stephenson, N.L., Byrne, J.C., *et al.* (2009) Widespread increase of tree mortality rates in the western United States. *Science*, **323**, 521–524.

van Mantgem, P.J., Stephenson, N.L., & Keeley, J.E. (2006) Forest reproduction along a climatic gradient in the Sierra Nevada, California. *Forest Ecology and Management*, **225**, 391–399.

Van Staden, J., Brown, N.A.C., Jager, A.K., & Johnson, T.A. (2000) Smoke as a germination cue. *Plant Species Biology*, **15**, 167–178.

van Wagner, C.E. (1977) Conditions for the start and spread of crown fire. *Canadian Journal of Forest Research*, **7**, 23–34.

van Wagtendonk, J.W. (1993) Spatial patterns of lightning strikes and fires in Yosemite National Park. In *Proceedings of the 12th international conference on fire and forest meteorology*, pp. 223–231. Society of American Foresters, Jekyll Island, Georgia.

van Wagtendonk, J.W. (1996) Use of a deterministic fire growth model to test fuel treatments. In *Sierra Nevada Ecosystem Project final report to Congress: status of the Sierra Nevada, Vol. II: Assessments and scientific basis for management options*, pp. 1155–1166. Centers for Water and Wildland Resources, University of California, Davis.

van Wagtendonk, J.W. & Cayan, D.R. (2008) Temporal and spatial distribution of lightning strikes in California in relation to large-scale weather patterns. *Fire Ecology*, **4**, 34–56.

van Wagtendonk, J.W. & Fites-Kaufman, J.A. (2006) Sierra Nevada bioregion. In *Fire in California's ecosystems* (eds N.G. Sugihara, J.W. van Wagtendonk, K.E. Shaffer, J. Fites-Kaufman & A.E. Thoede), pp. 264–294. University of California Press, Los Angeles, California.

van Wilgen, B.W. (1982) Some effects of post-fire age on the above-ground biomass of fynbos (macchia) vegetation in South Africa. *Journal of Ecology*, **70**, 217–225.

van Wilgen, B.W. (1984) Fire climates in the southern and western Cape Province and their potential use in fire control and management. *South African Journal of Science*, **80**, 358–362.

van Wilgen, B.W. (1986) A simple relationship for estimating the intensity of fires in natural vegetation. *South African Journal of Botany*, **52**, 384–386.

van Wilgen, B.W. (2009) The evolution of fire and invasive alien plant management practices in fynbos. *South African Journal of Science*, **105**, 335–342.

van Wilgen, B.W., Bond, W.J., & Richardson, D.M. (1992a) Ecosystem management. In *The ecology of fynbos* (ed. R.M. Cowling), pp. 345–371. Oxford University Press, Cape Town, RSA.

van Wilgen, B.W. & Burgan, R.E. (1984) Adaptation of the United States fire danger rating system to fynbos conditions, II: Historic fire danger in the fynbos biome. *South African Forestry Journal*, **129**, 66–78.

van Wilgen, B.W., Cowling, R.M., & Burgers, C.J. (1996) Valuation of ecosystem services: a case study from South African fynbos ecosystems. *BioScience*, **46**, 184–190.

van Wilgen, B.W., Everson, C.S., & Trollope, W.S.W. (1990a) Fire management in Southern Africa: some examples of current objectives, practices, and problems. In *Fire in the tropical biota* (ed. J.G. Goldammer), pp. 179–215. Springer, Berlin.

van Wilgen, B.W. & Forsyth, G.G. (1992) Regeneration strategies in fynbos plants and their influence on the stability of community boundaries after fire. In *Fire in South African mountain fynbos* (eds B.W. van Wilgen, D.M. Richardson, F.J. Kruger & H.J. van Hensbergen), pp. 54–80. Springer, Berlin.

van Wilgen, B.W., Forsyth, G.G., de Klerk, H., et al. (2010) Fire management in Mediterranean-climate shrublands: a case study from the Cape fynbos, South Africa. *Journal of Applied Ecology*, **47**, 631–638.

van Wilgen, B.W., Govender, N., Biggs, H.C., Ntsala, D., & Funda, X.N. (2004) Response of savanna fire regimes to changing fire-management policies in a large African national park. *Conservation Biology*, **18**, 1533–1540.

van Wilgen, B.W., Higgins, K.B., & Bellstedt, D.U. (1990b) The role of vegetation structure and fuel chemistry in excluding fire from forest patches in the fire-prone fynbos shrublands of South Africa. *Journal of Ecology*, **78**, 210–222.

van Wilgen, B.W., Maitre, D.C.L., & Kruger, F.J. (1985) Fire behaviour in South African fynbos (macchia) vegetation and predictions from Rothermel's fire model. *Journal of Applied Ecology*, **22**, 207–216.

van Wilgen, B.W. & Richardson, D.M. (1985) The effects of alien shrub invasions on vegetation structure and fire behaviour in South African fynbos shrublands: a simulation study. *Journal of Applied Ecology*, **22**, 955–966.

van Wilgen, B.W., Richardson, D.M., Kruger, F.J., & Hensbergen, H.J.v. (1992b) *Fire in South African mountain fynbos*. Springer, Berlin.

van Wilgen, B.W., Richardson, D.M., & Seydack, A.H.W. (1994) Managing fynbos for biodiversity: constraints and options in a fire-prone environment. *South African Journal of Science*, **90**, 322–329.

van Wilgen, B.W. & Scott, D.F. (2001) Managing fires on the Cape Peninsula, South Africa: dealing with the inevitable. *Journal of Mediterranean Ecology*, **2**, 197–208.

van Wilgen, B.W. & van Hensbergen, H.J. (1992) Fuel properties of vegetation in Swartboskloof. In *Fire in South African mountain fynbos* (eds B.W. van Wilgen, D.M. Richardson, F.J. Kruger & H.J. van Hensbergen), pp. 37–53. Springer, Berlin.

van Wilgen, B.W. & Viviers, M. (1985) The effect of season of fire on serotinous Proteaceae in the Western Cape and the implications for management. *South African Forestry Journal*, **133**, 49–53.

Vankat, J. (1989) Water stress in chaparral shrubs in summer rain versus summer drought climates: whither the Mediterranean type climate paradigm. In *The California chaparral: paradigms reexamined* (ed. S.C. Keeley), Science Series 34, pp. 117–124. Natural History Museum of Los Angeles County, Los Angeles, California.

Vankat, J.L. (1985) General patterns of lightning ignitions in Sequoia National Park, California. In *Proceedings of the symposium and workshop on wilderness fire* (eds J.E. Lotan, B.M. Kilgore, W.C. Fischer & R.W. Mutch), General Technical Report INT-182, pp. 408–411. USDA Forest Service, Intermountain Forest and Range Experiment Station, Ogden, Utah.

Vaughton, G. & Ramsey, M. (2001) Relationships between seed mass, seed nutrients, and seedling growth in *Banksia cunninghamii* (Proteaceae). *International Journal of Plant Sciences*, **162**, 599–606.

Vázquez, A. & Moreno, J.M. (1998) Patterns of lightning- and people-caused fires in peninsular Spain. *International Journal of Wildland Fire*, **8**, 103–115.

Veblen, T.T. (2003) Key issues in fire regime research for fuels management and ecological restoration. In *Fire, fuel treatments, and ecological restoration: conference proceedings* (eds P.N. Omi & L.A. Joyce), Proceedings RMRS-P-29, pp. 319–333. USDA Forest Service, Rocky Mountain Research Station, Fort Collins, Colorado.

Veblen, T.T., Ashton, D.H., & Schlegel, F.M. (1979) Tree regeneration strategies in a lowland *Nothofagus*-dominated forest in south-central Chile. *Journal of Biogeography*, **6**, 329–340.

Veblen, T.T., Kitzberger, T., Raffaele, E., & Lorenz, D.C. (2003) Fire history and vegetation changes in northern Patagonia, Argentina. In *Fire and climatic change in temperate ecosystems of the western Americas* (eds T.T. Veblen, W.L. Baker, G. Montenegro & T.W. Swetnam), pp. 265–295. Springer, New York.

Veblen, T.T., Kitzberger, T., Raffaele, E., *et al.* (2008) The historical range of variability of fires in the Andean-Patagonian *Nothofagus* forest region. *International Journal of Wildland Fire*, **17**, 724–741.

Veblen, T.T., Kitzberger, T., Villalba, R., & Donnegan, J. (1999) Fire history in northern Patagonia: the roles of humans and climatic variation. *Ecological Monographs*, **69**, 47–67.

Veblen, T.T., Schlegel, F.M., & Escobar, B. (1980) Structure and dynamics of old-growth *Nothofagus* forests in the Valdivian Andes, Chile. *Journal of Ecology*, **68**, 1–31.

Vega, J.A. (2000) Resistencia vegetativa ante el fuego a través de la historia de los incendios. In *La defensa contra incendios forestales* (ed. R. Vélez), pp. 4.66–4.85. McGraw Hill, Madrid.

Vega, J.A., Fernández, C., Pérez-Gorostiaga, P., & Fonturbel, T. (2010) Response of maritime pine (*Pinus pinaster* Ait.) recruitment to fire severity and post-fire management in a coastal burned area in Galicia (NW Spain). *Plant Ecology*, **206**, 297–308.

Veirs, S.D., Jr. (1979) The role of fire in northern coast redwood forest dynamics. In *Proceedings of the second conference on scientific research in the national parks*, Vol. 10: *Fire ecology*, pp. 190–209. USDI National Park Service, Washington, DC.

Veirs, S.D., Jr. (1982) Coast redwood forest: stand dynamics, successional status, and the role of fire. In *Proceedings of the symposium on forest succession and stand development research in the Pacific Northwest* (ed. J.E. Means), pp. 119–141. USDA Forest Service, Pacific Northwest Forest and Range Experiment Station, Corvalis, Oregon.

Veraverbeke, S., Verstraeten, W.W., Lhermitte, S., & Goossens, R. (2010) Evaluating Landsat Thematic Mapper spectral indices for estimating burn severity of the 2007 Peloponnese wildfires in Greece. *International Journal of Wildland Fire*, **19**, 558–569.

Verboom, G.A., Archibald, J.K., Bakker, F.T., *et al.* (2009) Origin and diversification of the Greater Cape flora: ancient species repository, hot-bed of recent radiation, or both. *Molecular Phylogenetics and Evolution*, **51**, 44–53.

Verboom, G.A., Linder, H.P., & Stock, W.D. (2003) Phylogenetics of the grass genus *Ehrharta*: evidence for radiation in the summer-arid zone of the South African Cape. *Evolution*, **57**, 1008–1021.

Verboom, G.A., Linder, H.P., & Stock, W.D. (2004) Testing the adaptive nature of radiation: growth form and life history divergence in the African grass genus *Ehrharta* (Poaceae: Ehrhartoideae). *American Journal of Botany*, **91**, 1364–1370.

Verboom, G.A., Stock, W.D., & Linder, H.P. (2002) Determinants of postfire flowering in the geophytic grass *Ehrharta capensis*. *Functional Ecology*, **16**, 705–713.

Verdaguer, D. & Ojeda, F. (2002a) Evolutionary transition from resprouter to seeder life history in two *Erica* (Ericaceae) species: insights from seedling axillary buds. *Annals of Botany*, **95**, 593–599.

Verdaguer, D. & Ojeda, F. (2002b) Root starch storage and allocation patterns in seeder and resprouter seedlings of two Cape *Erica* (Ericaceae) species. *American Journal of Botany*, **89**, 1189–1196.

Verdon, D.C., Kiem, A.S., & Franks, S.W. (2004) Multi-decadal variability of forest fire risk: eastern Australia. *International Journal of Wildland Fire*, **13**, 165–171.

Verdú, J.R., Crespo, M.B., & Galante, E. (2000) Conservation strategy of a nature reserve in Mediterranean ecosystems: the effects of protection from grazing on biodiversity. *Biodiversity and Conservation*, **9**, 1707–1721.

Verdú, M. (2000) Ecological and evolutionary differences between Mediterranean seeders and resprouters. *Journal of Vegetation Science*, **11**, 265–268.

Verdú, M., Barron-Sevilla, J.A., Valiente-Banuet, A., Fores-Hernandez, N., & Garcia-Fayos, P. (2002) Mexical plant phenology: is it similar to Mediterranean communities? *Botanical Journal of the Linnean Society*, **138**, 297–303.

Verdú, M., P. Dávila, P.G.-F., Flores-Hernández, N., & Valiente-Banuet, A. (2003) "Convergent" traits of mediterranean woody plants belong to pre-mediterranean lineages. *Biological Journal of the Linnean Society*, **78**, 415–427.

Verdú, M. & Pausas, J.G. (2007) Fire drives phylogenetic clustering in Mediterranean Basin woody plant communities. *Journal of Ecology*, **95**, 1316–1323.

Verdú, M., Pausas, J.G., Segarra-Moragues, J.G., & Ojeda, F. (2007) Burning phylogenies: fire, molecular evolutionary rates, and diversification. *Evolution*, **61**, 2195–2204.

Versfeld, D.B., Richardson, D.M., van Wilgen, B.W., Chapman, R.A., & Forsyth, G.G. (1992) The climate of Swartboskloof. In *Fire in South African mountain fynbos* (eds B.W. van Wilgen, D.M. Richardson, F.J. Kruger & H.J. van Hensbergen), pp. 21–36. Springer, Berlin.

Vesk, P.A. & Westoby, M. (2004) Sprouting ability across diverse disturbances and vegetation types worldwide. *Journal of Ecology*, **92**, 310–320.

Viegas, D.X. (2004) High mortality. *Wildfire*, **13**, 22–26.

Viegas, D.X., Cruz, M.G., Ribeiro, L.M., *et al.* (2002) Gestosa fire spread experiments. In *Forest fire research and wildland fire safety*. Millpress, Rotterdam, The Netherlands.

Vilà, M., Lloret, F., Ogheri, E., & Terradas, J. (2001) Positive fire-grass feedback in Mediterranean Basin woodlands. *Forest Ecology and Management*, **147**, 3–14.

Vilà, M. & Terradas, J. (1995) Sprout recruitment and self-thinning of *Erica multiflora* after clipping. *Oecologia*, **102**, 64–69.

Villagrán, C. & Armesto, J.J. (2005) Fitogeografía histórica de la Cordillera de la Costa de Chile. In *Biodiversidad y ecología de los bosques de la Cordillera de la costa de Chile* (eds C. Smith, J.J. Armesto & C. Valdovinos), pp. 99–116. Editorial Universitaria, Santiago, Chile.

Vines, R.G. (1974) *Weather patterns and bushfire cycles in southern Australia*, Technical Paper 2. CSIRO, Australian Division of Chemical Technology, Melbourne, Australia.

Vives, R. & Boxall, B. (2009) Setting fires lets crews fight on their own terms. *Los Angeles Times*, September 4, 2009.

Vivian, L.M., Cary, G.J., Bradstock, R.A., & Gill, A.M. (2008) Influence of fire severity on the regeneration, recruitment and distribution of eucalypts in the Cotter River Catchment, Australian Capital Territory. *Austral Ecology*, **33**, 55–67.

Vlok, J.H.J. (1988) Alpha diversity of lowland fynbos herbs at various levels of infestation by alien annuals. *South African Journal of Botany*, **54**, 623–628.

Vlok, J.H.J., Euston-Brown, D.I.W., & Cowling, R.M. (2003) Acocks' valley bushveld 50 years on: new perspective on the delimitation, characterisation, and origin of subtropical thicket vegetation. *South African Journal of Botany*, **69**, 27–51.

Vlok, J.H.J. & Yeaton, R.I. (1999) The effect of overstorey proteas on plant species richness in South African mountain fynbos. *Diversity and Distributions*, **5**, 213–222.

Vlok, J.H.J. & Yeaton, R.I. (2000) The effect of short fire cycles on the cover and density of understorey sprouting species in South African mountain fynbos. *Diversity and Distributions*, **6**, 233–242.

Vogl, R.J., Armstrong, W.P., White, K.L., & Cole, K.L. (1977) The closed-cone pines and cypresses. In *Terrestrial vegetation of California* (eds M.G. Barbour & J. Major), pp. 295–358. John Wiley & Sons, New York.

Wade, D.D. (1993) Thinning young loblolly pine stands with fire. *International Journal of Wildland Fire*, **3**, 169–178.

Wake, D.B. (1991) Homoplasy: the result of natural selection, or evidence of design limitations? *American Naturalist*, **138**, 543–567.

Walker, J. (1981) Fuel dynamics in Australian vegetation. In *Fire and the Australian biota* (eds A.M. Gill, R.H. Groves & I.R. Noble), pp. 101–128. Australian Academy of Science, Canberra.

Walker, L.R. & Boneta, W. (1995) Plant and soil responses to fire on a fern-covered landslide in Puerto Rico. *Journal of Tropical Ecology*, **11**, 473–479.

Walter, H. (1977) Effects of fire on wildlife communities. In *Proceedings of the symposium on environmental consequences of fire and fuel management in Mediterranean ecosystems* (eds H.A. Mooney & C.E. Conrad), General Technical Report WO-3, pp. 183–192. USDA Forest Service, Washington, DC.

Walters, J.R., Bell, T.L., & Read, S. (2005) Intra-specific variation in carbohydrate reserves and sprouting ability in *Eucalyptus obliqua* seedlings. *Australian Journal of Botany*, **53**, 195–203.

Ward, D.J., Lamont, B.B., & Burrows, C.L. (2001) Grasstrees reveal contrasting fire regimes in eucalypt forest before and after European settlement of southwestern Australia. *Forest Ecology and Management*, **150**, 323–329.

Wardell-Johnson, G.W., Williams, M.R., Mellican, A.E., & Annells, A. (2007) Floristic patterns and disturbance history in karri (*Eucalyptus diversicolor*: Myrtaceae) forest, south-western Australia, 2: Origin, growth form and fire response. *Acta Oecologica*, **31**, 137–150.

Wardle, D.A., Zackrisson, O., Hörnberg, G., & Gallet, C. (1997) The influence of island area on ecosystem properties. *Science*, **277**, 1296–1299.

Warman, L. & Moles, A.T. (2009) Alternative stable states in Australia's wet tropics: a theoretical framework for the field data and a field-case for the theory. *Landscape Ecology*, **24**, 1–13.

Warter, J.K. (1976) Late Pleistocene plant communities: evidence from the Rancho La Brea tar pits. In *Symposium proceedings: plant communities of southern California* (ed. J. Latting), Special Publication 2, pp. 32–39. California Native Plant Society, Sacramento, California.

Wasson, R.J. (1982) Landform development in Australia. In *Evolution of the flora and fauna of arid Australia* (eds W.R. Barker & P.J.M. Greenslade), pp. 23–34. Peacock Publications, Adelaide, Australia.

Wasson, R.J. (1989) Landforms. In *Mediterranean landscapes in Australia: mallee ecosystems and their management* (eds J.C. Noble & R.A. Bradstock), pp. 13–34. CSIRO Publications, Collingwood, Australia.

Waters, D.A., Burrows, G.E., & Harper, J.D.I. (2010) *Eucalyptus regnans* (Myrtaceae): a fire-sensitive eucalypt with a resprouter epicormic structure. *American Journal of Botany*, **97**, 545–556.

Watson, J. & Alvin, K.L. (1996) An English Wealden floral list, with comments on possible environmental indicators. *Cretaceous Research*, **17**, 5–26.

Watson, L.H. & Cameron, M.J. (2002) Forest tree and fern species as indicators of an unnatural fire event in a southern Cape mountain forest. *South African Journal of Botany*, **68**, 357–361.

Watson, P. & Wardell-Johnson, G. (2004) Fire frequency and time-since-fire effects on the open-forest and woodland flora of Girraween National Park, south-east Queensland, Australia. *Austral Ecology*, **29**, 225–236.

Weber, R.O. (2001) Wildland fire spread models. In *Forest fires: behavior and ecological effects* (eds E.A. Johnson & K. Miyanishi), pp. 151–169. Academic Press, San Diego, California.

Weber, R.O. & Kaufmann, P. (1998) Relationship of synoptic winds and complex terrain flows during the MISTRAL field experiment. *Journal of Applied Meteorology*, **37**, 1486–1496.

Weiher, E. & Keddy, P., eds. (1995) *Ecological assembly rules: perspectives, advances, retreats*. Cambridge University Press, Cambridge, UK.

Weise, D., White, R.H., & Frommer, S. (2003) Seasonal changes in selected combustion characteristics of ornamental vegetation. In *Second international wildland fire ecology and fire management congress and fifth symposium on fire and forest meteorology*, pp. 1E.4. American Meteorological Society, Orlando, Florida.

Weise, D.R. & Biging, G.S. (1996) Effects of wind velocity and slope on flame properties. *Canadian Journal of Forest Research*, **26**, 1849–1858.

Weise, D.R., Hartford, R.A., & Mahaffey, L. (1998) Assessing live fuel moisture for fire management applications. In *Tall Timbers fire ecology conference* (eds T.L. Pruden & L.A. Brennan), pp. 49–55. Tall Timbers Research Station, Tallahassee, Florida.

Weise, D.R., Stephens, S.L., Fujioka, F.M., Moody, T.J., & Benoit, J. (2010) Estimation of fire danger in Hawai'i using limited weather data and simulation. *Pacific Science*, **64**, 199–220.

Wellington, A.B. (1989) Seedling regeneration and the population dynamics of eucalypts. In *Mediterranean landscapes in Australia: mallee ecosystems and their management* (eds J.C. Noble & R.A. Bradstock), pp. 155–167. CSIRO, Melbourne, Australia.

Wellington, A.B. & Noble, J.C. (1985) Seed dynamics and factors limiting the recruitment of the mallee *Eucalyptus incrassata* in semi-arid, south-eastern Australia. *Journal of Ecology*, **73**, 657–666.

Wells, P.V. (1962) Vegetation in relation to geological substratum and fire in the San Luis Obispo quadrangle, California. *Ecological Monographs*, **32**, 79–103.

Wells, P.V. (1965) Scarp woodlands, transported grassland soils, and concept of grassland climate in the Great Plains region. *Science*, **148**, 246–249.

Wells, P.V. (1969) The relation between mode of reproduction and extent of speciation in woody genera of the California chaparral. *Evolution*, **23**, 264–267.

Wells, P.V. (2000) Pleistocene macrofossil records of four-needled pinyon or juniper encinal in the northern Vizcaino Desert, Baja California del Norte. *Madroño*, **47**, 189–194.

Westerling, A.L., Gershunov, A., Cayan, D.R., & Barnett, T.P. (2002) Long lead statistical forecasts of area burned in western U.S. wildfires by ecosystem province. *International Journal of Wildland Fire*, **11**, 257–266.

Westerling, A.L., Hidalgo, H.G., Cayan, D.R., & Swetnam, T.W. (2006) Warming and earlier spring increase western U.S. forest wildfire activity. *Science*, **313**, 940–943.

Westman, W.E. (1979) Oxidant effects on Californian coastal sage scrub. *Science*, **205**, 1001–1003.

Westman, W.E. (1988) Vegetation, nutrition and climate: data-tables. In *Mediterranean-type ecosystems: data source book* (ed. R.L. Specht), pp. 81–91. Kluwer Academic Publishers, Boston, Maryland.

Westman, W.E., O'Leary, J.F., & Malanson, G.P. (1981) The effects of fire intensity, aspect, and substrate on post-fire growth of California coastal sage scrub. In *Components of productivity of Mediterranean climate regions: basic and applied aspects* (eds N.S. Margaris & H.A. Mooney), pp. 151–179. Dr. W. Junk, The Hague, The Netherlands.

Westoby, M., French, K., Hughes, L., Rice, B., & Rodgerson, L. (1991) Why do more plant species use ants for dispersal on infertile compared with fertile soils? *Australian Journal of Ecology*, **16**, 445–456.

Westoby, M., Leishman, M.R., & Lord, J.M. (1995) Further remarks on phylogenetic correction. *Journal of Ecology*, **83**, 727–730.

Weston, P.H. & Barker, N.P. (2006) A new suprageneric classification of the Proteaceae, with an annotated checklist of genera. *Telopea*, **11**, 314–344.

Whelan, R.J. (1986) Seed dispersal in relation to fire. In *Seed dispersal* (ed. D.R. Murray), pp. 237–271. Academic Press, Sydney, Australia.

Whelan, R.J. (1995) *The ecology of fire*. Cambridge University Press, Cambridge, UK.

Whelan, R.J. (2002) Managing fire regimes for conservation and property protection: an Australian response. *Conservation Biology*, **16**, 1659–1661.

Whelan, R.J., de Jong, N., & von der Burg, S. (1998) Variation in bradyspory and seedling recruitment without fire among populations of *Banksia serrata* (Proteaceae). *Australian Journal of Ecology*, **23**, 121–128.

Whelan, R.J., Rogerson, L., Dickman, C.R., & Sutherland, E.F. (2002) Critical lifecycles of plants and animals: developing a process-based understanding of population changes in fire-prone landscapes. In *Flammable Australia: the fire regimes and biodiversity of a continent* (eds R.A. Bradstock, J.E. Williams & A.M. Gill), pp. 94–124. Cambridge University Press, Cambridge, UK.

Whight, S. & Bradstock, R. (1999) Indices of fire characteristics in sandstone heath near Sydney, Australia. *International Journal of Wildland Fire*, **9**, 145–153.

White, F. (1976) The underground forests of Africa: a preliminary review. *Garden's Bulletin, Singapore*, **29**, 57–71.

White, P.S. & Pickett, S.T.A. (1985) Natural disturbance and patch dynamics: an introduction. In *The ecology of natural disturbance and patch dynamics* (eds S.T.A. Pickett & P.S. White), pp. 1–13. Academic Press, San Diego, California.

White, R.H. & Zipperer, W.C. (2010) Testing and classification of individual plants for fire behaviour: plant selection for the wildland-urban interface. *International Journal of Wildland Fire*, **19**, 213–227.

Whiteman, C.D. (2000) *Mountain meteorology: fundamentals and applications*. Oxford University Press, New York.

Whitlock, C., Skinner, C.N., Bartlein, P.J., Minckley, T., & Mohr., J.A. (2004) Comparison of charcoal and tree-ring records of recent fires in the eastern Klamath Mountains, California, USA. *Canadian Journal of Forest Research*, **34**, 2110–2121.

Whittaker, R.H. (1972) Evolution and measurement of species diversity. *Taxon*, **21**, 213–251.

Wicklow, D.T. (1977) Germination response in *Emmenanthe penduliflora* (Hydrophyllaceae). *Ecology*, **58**, 201–205.

Wiens, D., Calvin, C.L., Wilson, C.A., et al. (1987) Reproductive success, spontaneous embryo abortion, and genetic load in flowring plants. *Oecologia*, **71**, 501–509.

Wieslander, A.E. & Schreiber, B.O. (1939) Notes on the genus *Arctostaphylos*. *Madroño*, **5**, 42–43.

Wilken, D.H. (2010) *Ceanothus*. In *The Jepson manual: vascular plants of California* (eds B.G. Baldwin, S. Boyd, D.J. Keil, et al.). University of California Press, Los Angeles, California.

Williams, J.E., Whelan, R.J., & Gill, A.M. (1994) Fire and environmental heterogeneity in southern temperate forest ecosystems: implications for management. *Australian Journal of Botany*, **42**, 125–137.

Williams, K.S., Davis, S.D., Gartner, B.L., & Karlsson, S. (1991) Factors limiting the establishment of a chaparral oak, *Quercus durata* Jeps., in grassland. In *Proceedings of the symposium on oak woodlands and hardwood rangeland management, October 31–November 2,*

1990, Davis, California (ed. R.B. Standiford), PSW-126, pp. 70–73. USDA Forest Service, Pacific Southwest Research Station, Berkeley, California.

Williams, P.R. (2000) Fire-stimulated rainforest seedling recruitment and vegetative regeneration in a densely grassed wet sclerophyll forest of north-eastern Australia. *Australian Journal of Botany*, **48**, 651–658.

Williams, R.J., Bradstock, R.A. & Cary, G.J. (2009) *Interactions between climate change, fire regimes and biodiversity in Australia: a preliminary assessment*. CSIRO, Canberra.

Williams, R.J., Griffiths, A.D., & Allan, G.E. (2002) Fire regimes and biodiversity in the savannas of northern Australia. In *Flammable Australia: the fire regimes and biodiversity of a continent* (eds R.A. Bradstock, J.E. Williams & A.M. Gill), pp. 281–304. Cambridge University Press, Cambridge, UK.

Williamson, G.B. & Black, E.M. (1981) High temperature of forest fires under pines as a selective advantage over oaks. *Nature*, **293**, 643–644.

Williamson, T.N., Graham, R.C., & Shouse, P.J. (2004) Effects of a chaparral-to-grass conversion on soil physical and hydrologic properties after four decades. *Geoderma*, **123**, 99–114.

Willis, K.J. & Niklas, K.J. (2004) The role of Quaternary environmental change in plant macroevolution: the exception or the rule? *Philosophical Transactions of the Royal Society of London, Series B*, **359**, 159–172.

Wills, K.E. & Clarke, P.J. (2008) Plant trait-environmental linkages among contrasting landscapes and climate regimes in temperate eucalypt woodlands. *Australian Journal of Botany*, **56**, 422–432.

Wills, T.J. & Read, J. (2007) Soil seed bank dynamics in post-fire heathland succession in south-eastern Australia. *Plant Ecology*, **190**, 1–12.

Willyard, A., Syring, J., Gernandt, D.S., Liston, A., & Cronn, R. (2007) Fossil calibration of molecular divergence infers a moderate mutation rate and recent radiations for *Pinus*. *Molecular Biology and Evolution*, **24**, 90–101.

Wilson, A.A.G. & Ferguson, I.S. (1986) Predicting the probability of house survival during bushfires. *Journal of Environmental Management*, **23**, 259–270.

Wilson, A.M., Latimer, A.M., Silander, Jr, J.A., Gelfand, A.E., & de Klerk, H. (2010) A hierarchical Bayesian model of wildfire in a Mediterranean biodiversity hotspot: implications of weather variability and global circulation. *Ecological Modelling*, **221**, 106–112.

Wilson, C.C. (1979a) Management problems and solutions at the interface between man and mediterranean wildlands. In *Fire and fuel management in mediterranean-climate ecosystems: research priorities and programmes* (ed. J.K. Agee), Technical Notes 11, pp. 33–37. UNESCO, Paris.

Wilson, C.C. (1979b) Roadsides: corridors with high fire hazard and risk. *Journal of Forestry*, **77**, 576–578.

Wilson, C.J., Carey, J.W., Beeson, P.C., Gard, M.O., & Lane, L.J. (2001) A GIS-based hillslope erosion and sediment delivery model and its application in the Cerro Grande burn area. *Hydrological Processes*, **15**, 2995–3010.

Wilson, J.B. (1999) Assembly rules in plant communities. In *Ecological assembly rules: perspectives, advances, retreats* (eds E. Weiher & P. Keddy), pp. 130–164. Cambridge University Press, Cambridge, UK.

Wing, S.L. (1987) Eocene and Oligocene floras and vegetation of the Rocky Mountains. *Annals of the Missouri Botanical Garden*, **74**, 748–784.

Wing, S.L. & Greenwood, D.R. (1993) Fossil and fossil climate: the case for equable continental interiors in the Eocene. *Philosophical Transactions of the Royal Society of London, Series B*, **341**, 243–252.

Wisheu, I.C., Rosenzweig, M.L., Olsvig-Whittaker, L., & Shmida, A. (2000) What makes nutrient-poor mediterranean heathlands so rich in plant diversity? *Evolutionary Ecology Research*, **2**, 935–955.

Wittenberg, L. & Malkinson, D. (2009) Spatio-temporal perspectives of forest fires regimes in a maturing Mediterranean mixed pine landscape. *European Journal of Forest Research*, **128**, 297–304.

Witkowski, E.T.F., Lamont, B.B., Connell, S.J. (1991) Seed bank dynamics of three co-occuring banksias in south coastal Western Australia: the role of plant age, cockatoos, senescence and interfire establishment. *Australian Journal of Botany*, **39**, 385–397.

Wolfe, J. (1964) *Miocene floras from Fingerrock Wash southwestern Nevada*. U.S. Government Printing Office, Washington, DC.

Wolfe, J.A. (1975) Some aspects of plant geography of the Northern Hemisphere during the Late Cretaceous and Tertiary. *Annals of the Missouri Botanical Garden*, **62**, 264–279.

Wolfe, J.A. (1994) Tertiary climatic changes at middle latitudes of western North America. *Palaeogeography, Palaeoclimatology, Palaeoecology*, **108**, 195–205.

Wolfe, J.A. (1995) Paleoclimatic estimates from Tertiary leaf assemblages. *Annual Review of Earth and Planetary Sciences*, **23**, 119–142.

Wolfe, J.A., Forest, C.E., & Molnar, P. (1998) Paleobotanical evidence of Eocene and Oligocene paleoaltitudes in midlatitude western North America. *Geological Society of America Bulletin*, **110**, 664–678.

Wolfe, J.A. & Schorn, H.E. (1989) Paleoecologic, paleoclimatic, and evolutionary significance of the Oligocene Creede flora, Colorado. *Paleobiology*, **15**, 180–198.

Woodall, C.W., Charney, J.J., Liknes, G.C., & Potter, B.E. (2005) What is the fire danger now? Linking fuel inventories with atmospheric data. *Journal of Forestry*, **103**, 293–298.

Wooller, S.J., Wooller, R.D., & Brown, K.L. (2002) Regeneration by three species of *Banksia* on the south coast of Western Australia in relation to fire interval. *Australian Journal of Botany*, **50**, 311–317.

Wrathall, J.E. (1985) The hazard of forest fires in southern France. *Disasters*, **9**, 104–114.

Wright, B.R. & Clarke, P.J. (2007) Fire regime (recency, interval and season) changes the composition of spinifex (*Triodia* spp.)-dominated desert dunes. *Australian Journal of Botany*, **55**, 709–724.

Wright, S. (1932) The roles of mutation, inbreeding, crossbreeding and selection in evolution. In *Proceedings of the sixth international congress on genetics*, Vol. 1 (ed. D.F. Jones), pp. 356–366. Brooklyn Botanic Garden, New York.

Xanthopoulos, G. (2007) Forest fire related deaths in Greece: confirming what we already know. In *Proceedings of Wildfire 2007: 4th international wildland fire conference, May 13–17, 2007, Seville, Spain*. Ministerio Environment, Madrid and Junta de Andalucia, Seville, Spain.

Xiang, Q.Y., Crawford, D.J., Wolfe, A.D., Tang, Y.-C., & DePamphilis, C.W. (1998) Origin and biogeography of *Aesculus* L. (Hippocastanaceae): a molecular phylogenetic perspective. *Evolution*, **52**, 988–997.

Yates, C.J., Abbott, I., Hopper, S.D., & Coates, D.J. (2003a) Fire as a determinant of rarity in the south-west Western Australian global biodiversity hotspot. In *Fire in ecosystems of south-west Western Australia: impacts and management* (eds I. Abbott & N.D. Burrows), pp. 395–420. Backhuys Publishers, Leiden, The Netherlands.

Yates, C.J., Hopper, S.D., Brown, S., & Leeuwen, S. (2003b) Impact of two wildfires on endemic granite outcrop vegetation in Western Australia. *Journal of Vegetation Science*, **14**, 185–194.

Yates, C.J., Ladd, P.G., Coates, D.J., & McArthur, S. (2007) Hierarchies of cause: understanding rarity in an endemic shrub *Verticordia staminosa* (Myrtaceae) with a highly restricted distribution. *Australian Journal of Botany*, **55**, 194–205.

Yates, M.J., Verboom, G.A., Rebelo, A.G., & Cramer, M.D. (2010) Ecophysiological significance of leaf size variation in Proteaceae from the Cape Floristic Region. *Functional Ecology*, **24**, 485–492.

Yeaton, R.I. & Bond, W.J. (1991) Competition between two shrub species: dispersal differences and fire promote coexistence. *American Naturalist*, **138**, 328–341.

Yelenik, S.G., Stock, W.D., & Richardson, D.M. (2004) Ecosystem level impacts of invasive *Acacia saligna* in the South African fynbos. *Restoration Ecology*, **12**, 44–51.

Yocum, L.L., Fulé, P.Z., Brown, P.M., *et al.* (2010) El Niño – Southern Oscillation effect on a fire regime in northeastern Mexico has changed over time. *Ecology*, **91**, 1660–1671.

Young, J.A. (2006) *Hakeas of Western Australia: a field and identification guide*. (Self-published), Perth, Australia.

Zammit, C. & Westoby, M. (1988) Pre-dispersal seed losses, and the survival of seeds and seedlings of two serotinous *Banksia* shrubs in burnt and unburnt heath. *Journal of Ecology*, **76**, 200–214.

Zammit, C.A. & Zedler, P.H. (1988) The influence of dominant shrubs, fire, and time since fire on soil seed banks in mixed chaparral. *Vegetatio*, **75**, 175–187.

Zedler, P.H. (1981) Vegetation change in chaparral and desert communities in San Diego County, California. In *Forest succession: concepts and applications* (eds D.C. West, H.H. Shugart & D. Botkin), pp. 406–430. Springer, New York.

Zedler, P.H. (1986) Closed-cone conifers of the chaparral. *Fremontia*, **14**, 14–17.

Zedler, P.H. (1995a) Are some plants born to burn? *Trends in Ecology and Evolution*, **10**, 393–395.

Zedler, P.H. (1995b) Fire frequency in southern California shrublands: biological effects and management options. In *Brushfires in California: ecology and resource management* (eds J.E. Keeley & T. Scott), pp. 101–112. International Association of Wildland Fire, Fairfield, Washington.

Zedler, P.H., Gautier, C.R., & McMaster, G.S. (1983) Vegetation change in response to extreme events: the effect of a short interval between fires in California chaparral and coastal scrub. *Ecology*, **64**, 809–818.

Zedler, P.H. & Seiger, L.A. (2000) Age mosaics and fire size in chaparral: a simulation study. In *2nd Interface between ecology and land development in California*. U.S. Geological Survey open-file report 00–62 (eds J.E. Keeley, M. Baer-Keeley & C.J. Fotheringham), pp. 9–18. U.S. Geological Survey, Sacramento, California.

Zhang, J. & Cregg, B.M. (2005) Growth and physiological responses to varied environments among populations of *Pinus ponderosa*. *Forest Ecology and Management*, **219**, 1–12.

Ziemer, R.R. (1964) Summer evapotranspiration trends as related to time after logging of forests in Sierra Nevada. *Journal of Geophysical Research*, **69**, 615–620.

Zimmerman, G.T. & Omi, P.N. (1998) Fire restoration options in lodgepole pine ecosystems. In *Fire in ecosystem management: shifting the paradigm from suppression to prescription: proceedings of the 20th Tall Timbers fire ecology conference* (eds T.L. Pruden & L.A. Brennan), pp. 285–297. Tall Timbers Research Station, Tallahassee, Florida.

Zohary, M. (1962) *Plant life of Palestine*. Israel and Jordan Ronald Press, New York.

Zunino, S. & Riveros, G. (1990) Cartografia de los incendios forestales en la 5 region. *Anales Museo de Historia Natural de Chile*, **21**, 89–94.

Zylstra, P. (2006) *Fire history of the Australian Alps: prehistory to 2003*. Australian Alps Liaison Committee, Canberra.

Index

Abies concolor, 48, 79, 128, 129, 131, 133
Aborigines, 350, 351, 352
Acacia, 335, 342
Acacia caven, 154, 155
Acacia gummifera, 108
active crown fires, 33
adaptations vs. exaptations, 238, 240, 285, 289
adaptive evolution, 271, 272
adaptive radiation, 177
Adenostoma, 48, 117
Adenostoma fasciculatum, 51, 127, 135, 294
Aesculus californica, 280, 281
Aesculus parryi, 282
age effects, and resprouting, 65
agriculture, 208, *see also* grazing
   renosterveld, 187
   South Africa, 195
agroforestry, 107, 147, 165, 209
air pollution, 338
alien plants, 160, 182, 187, 320, 330–348, 332–339
   Australia, 346–347
   California, 335–340
   Chile, 341–342
   fire management, 196
   Mediterranean Basin, 331–335
   prescription burning, 366
   South Africa, 342–346
Allocasuarina, 217
Ampelodesmos mauritanica, 102, 335
Andean Cordillera, 150, 151, 154, 155, 164
Andean uplift, 282, 396
angiosperms, 253
   serotiny, 74
Angophora, 205
animal communities, chaparral, 128
anisohydric behavior, 246
annual fire season, 157, 353
annual grass invasion, 332, 333
   Australia, 346
   chaparral and sage scrub, 336–339
   Mediterranean Basin, 334
   South Africa, 342

annual plants
   California, 126, 127
   grassland, 135, 136
   matorral, 159, 160
   Mediterranean Basin, 93
   origins, 295
   pyro-endemics, 319
ant dispersers, 185, 256
antecedent climate, 51–54
Anthyllis, 101
apparent colonizers, 68
Aqua satellite, 44
Araucaria, 235
Araucaria araucana, 25, 156, 164
Arbutus, 107, 293
archaeophytes, 334
Arctostaphylos, 48, 117, 127, 134
   community structure, 272
   diversification, 270
   obligate seeders, 259, 260, 261
   obligate vs. facultative seeders, 266
   origins, 293, 302
Arctostaphylos glandulosa, 62
Arctostaphylos pungens, 250
Arctostaphylos rainbowensis, 263
area burned
   Chile, 157, 158, 163, 165
   Mediterranean Basin, 88, 92
   South Africa, 175, 192, 198, 199
area treated, 374
Argania spinosa, 107
arid lands, alien plants in, 340–341
Arizona, 140, 141, 250
   chaparral, 315, 321
aromatic species, 84, 102, 108
arson fires, 159, 367
Artemisia, 293
Artemisia tridentata, 340
Arundo donax, 332
aseasonal climate, 14, 45, 276
Asparagus, 96, 97
asset protection zone, 375
Audoinia, 176
austral forests, 156, 163–165

Australia. *See* Southern Australia, *see* Western Australia
Austrocedrus chilensis, 155, 156, 163
auto-successional, 125

Baccharis, 48
Baja California, 124
    fire regimes, 120–124
bamboos, 258
Banksia, 49, 177, 178, 269, 296, 315, 346–347
bark, thick, 76, 99
bark stripping, 99
basal burls, 59, 60, 62
basal resprouts, 59, 63
batha, 98
BEHAVE model, 55, 116
bergwinds, 19, 26, 174, 188
beta diversity, 326
bimodal rainfall regime, 285
bimodal rainfall region, 287
biodiversity
    and fire regimes, 211–221
    and flammability, 221–225
    and life history, 268–273
    Australia, 209–225
    chaparral, 316, 327
    declines in, 216
    fire and community diversity, 310–327
    fire and regional diversity, 326–328
    fire regime heterogeneity, 223–225
    MTV, 209–225
    postfire, 125, 316
    resource cost, 365
    sampling design measures, 313
biodiversity hotspot, 83, 108, 109, 268, 310
biomass
    and fire, 30, 32
    and fire regimes, 192
    chaparral, 117
    dead, 257, 258
    fynbos, 169
    loss, 381
    South Africa ecosystems, 171
bole scorch, 35
bottom-up effects, 226, 241
BP (years before present), 14, 145
Brachypodium, 89
Brachypodium retusum, 102
bradyspory, 73
Brassica nigra, 343, 383
brickfielder winds, 19
broadleaf evergreen shrublands, 95–99
broadleaf forests, 10, 18
Bromus diandrus, 343
Bromus madritensis, 340
Bromus tectorum, 340, 384

Bronze Age, 85
Bruniaceae, 306
brush removal, 344
buffer zones, 374, 377
bunchgrasses, 135
burn severity. *See* fire severity
burning index, 192, 368
bushfires, 357, 359, 360
butenolide, 252

C_3 grasses, 332, 346
C_4 grasses, 228–229, *see also* grassland
    invasion, 332
    origins, 237, 301
cacti, 152, 161
Calicotome, 101
California, 122, *see also* northern hemisphere, chaparral
    alien plants, 335–341
    biodiversity measures, 313
    biogeography, 311
    climate characteristics, 7, 113, 141
    climate diagrams, 6, 288
    community diversity, 326
    conifer forests, 128–135, 138–139
    drought, 53
    Eocene climate, 278
    fire frequency, 241
    fire frequency records, 41, 42
    fire history, 301
    fire ignitions, 118
    fire impact, 359
    fire management, 144–149
    fire regimes, 120–124, 364–365
    fire seasonality, 44
    fire weather winds, 19, 20
    fire–diversity interactions, 319–321
    floristic comparisons, 24
    fuel characteristics of vegetation, 48, 113
    human impact, 350, 352, 354–356
    landscapes, 10, 11, 123
    lightning strike density, 41
    lignotubers, 63
    major fire events, 4, 355, 359, 360, 361
    Miocene climate, 279, 280, 282
    MTC region, 23–25
    MTV origins, 291–294
    obligate resprouters, 59
    Oligocene climate, 279
    population density, 355
    postfire annuals, 126
    postfire seeding, 69, 382
    socio-political aspects of fire, 379, 380–384
    vegetation, 11, 113, 135–140
    winter growing season, 289

California Fair Access to Insurance
    Plan, 367
California Floristic Province, 23, 24, 113
California Santa Monica Mountains National
    Recreation Area (SMNRA), 374
Callitris, 217, 225
Calocedrus decurrens, 128
Canary Islands, 106
canopy closure, 128
canopy coverage, and biodiversity, 316
canopy porosity, 51
canopy recovery, shrub, 127–128
canopy seedbanks, 73–76, 213–214, 253–256
    seedling/parent ratios, 249
Cape Floristic Region, 25, 168, 181,
    see also South Africa
carbohydrate reserves, 64
carbon dioxide levels, 236, 385
Carduus pycnocephalus, 331
Castanea sativa, 108
Casuarinaceae, 297
catastrophic fires. *See* major fire events
Ceanothus, 117, 127, 134
    community structure, 271, 272
    diversification, 270
    obligate seeders, 259, 266
    origins, 293, 302
    phylogeny, 272
Ceanothus greggii, 250
Ceanothus megacarpus, 48
Ceanothus oliganthus, 48
Cedar Fire, 379
Cenozoic, 236, 237, 275
    climate, 277–278
Centaurea solstitialis, 344, 345
Central Valley, 150
Cerastes, 271, 272, 302
Ceratonia siliqua, 96
Cercocarpus, 292
Cerro Grande Fire, 384
Chamaebatia foliolosa, 134
Chamaebatiaria, 294
chamaephytes, 15
Chamaerops humilis, 97
chaotic dynamics, 185
chaparral, 24, 32, *see also* California
    alien plants and type conversion,
        335–339
    biodiversity, 316, 327
    biomass and fuels in, 117
    crown fire regimes, 114–128
    enhanced flammability, 257
    evolution, 233
    fire history, 301, 302, 303
    fire regimes, 365
    fire return intervals, 360
    fire suppression, 147
    fuel characteristics, 48
    global warming response, 392
    Miocene distribution, 292
    mosaic, 115
    MTV origins, 290, 292–294
    non-resprouters, 259
    obligate vs. facultative seeders, 266
    outside MTC regions, 140–142
    postfire annuals, 126
    postfire diversity, 327
    postfire seeding, 383
    postfire succession, 320, 321
    prescription burning, 370
    seedling recruitment, 67, 254
    seral stage, 116
    soil moisture, 288
    soil stored seed, 70
    species–area relationships, 323
    species richness, 315
    trait evolution, 390
    understory, 79
charcoal, 207, 217, 305
    deposits, 300, 301
    deposits and fire activity, 351
    deposits and fire frequency, 42–43
    flux, 301
cheatgrass, 340
Chile, 151–167, *see also* southern hemisphere,
    matorral
    alien plants, 341–342
    biogeography, 311
    climate characteristics, 7, 150, 151
    community diversity, 314
    fire history, 307, 396
    fire impact, 359
    fire management, 165–166
    fire regimes, 156–165
    fire weather winds, 20
    fire–diversity interactions, 321–322
    floristic comparisons, 24
    fuel characteristics of vegetation, 49
    human impact, 350, 352, 356
    landscape, 10, 150
    lignotubers, 63, 159
    Miocene climate, 282
    MTC region, 25, 150–152
    MTV origins, 299, 309
    obligate resprouters, 59
    population density, 355
    postfire seeding woody genera, 69
    species richness, 315
    summer precipitation, 5
    vegetation, 11, 152–156
Cistaceae, 294
Cistus, 303

diversification, 270
Mediterranean Basin, 101
Cistus ladanifer, 91
clearance zone, 377
Clematis, 96
Cleveland National Forest, 148
Cliffortia ruscifolia, 181
climate. *See also* fire weather, winds, precipitation, drought
  and seed dormancy, 215
  antecedent, 51–53
  aseasonal, 276
  Australia, 7, 204, 210
  California, 7, 113, 141
  Chile, 7, 150, 151
  fire and geology convergence in MTEs, 388–397
  fire weather, 54
  major fire events, 360–361
  Mediterranean Basin, 7, 83
  Mediterranean Basin vegetation patterns, 94–107
climate change, 384–385
climate diagrams
  California, 6, 288
climate envelope models, 385
climate seasonality, 276, 277, 283, 286, 287, 288, 308
climax vegetation, 188
closed-canopy forests, 129, 339
closed-canopy shrublands, 66, 79
coal petrological analyses, 237
Coastal Cordillera, 150
coastal dunes, 109
coexistence, species, 221
colonization. *See also* European colonization
  postfire, 78, 126, 320
community assembly, 9
community assembly rules, 271
community diversity
  and fire, 310–326
community dynamics
  and fire, 182–186, 217–221
community species–area relationships, 322–325
community structure
  and life history, 271–273
community vulnerability, 366–380
community's invasibility, 330
competitive exclusion, 320
competitive height disadvantage, 262, 263
competitive sorting, 392
Composite Burn Index (CBI), 38, 381
cones, 253
  mast, 77
conifer forests, 13, 18
  alien plants, 339–340
  California, 128–134, 138–139

canopy seedbanks, 75
Chile, 164
fire regimes, 109, 131–132, 264, 364
fire suppression, 46
fuel structure, 52
Mediterranean Basin, 107
outside MTC regions, 142–144
precipitation, 53
prescription burning, 356
surface fuels, 130
conservation reserves, 209, 357
convergence theory, 393
convergent evolution, 252
coppicing, 60, 99
core habitat, 226
Coris monspeliensis, 94
Corporación Nacional Forestal (CONAF), 158, 166
Corymbia, 205
Cretaceous landscapes, 236
crown fires, 32–33, 34, 39, 45, 46
  and non-random sorting, 271
  canopy seed beds, 73
  chaparral, 114–128
  fuel characteristics, 48
  fynbos, 174
  Mediterranean Basin, 109
  MTC regions, 394
  Pinus, 103
  postfire seedling recruitment, 69
  regeneration from, 58
  seedling density after, 104
  serotiny, 253
  severity assessment, 380
crown volume scorch, 35
Cryptocarya alba, 25, 154
cryptophytes, 15, 60
Cunonia capensis, 189
Cyrtanthus, 176
Cyrtanthus ventricosus, 240, 322
Cytisus, 101
Cytisus scoparius, 343

D (risk of destruction), 366, 378
Daphne, 96
dead biomass, 257, 258
dead fuels, 50, 51, 52, 53
  and drought, 53
  and major fire events, 361
  live/dead ratio, 115
decadal oscillations, climate, 278
defensible space, 375
deforestation, 156
  Mediterranean Basin, 85, 86
dehesa, 107
delta diversity, 326

density-dependent effects, 185
dependent crown fires, 33
desert winds, 22
Devonian fires, 234
dicotyledons, 59, 93
differenced Normalized Burn Ratio (dNBR), 36, 37, 39, 381
disturbance cycles, 340
disturbance frequency, 45, 46, 211, 264
disturbances
  and grazing, 331, 335
  fire, 31, 163
diversity. See biodiversity
dNBR
  absolute, 381
  relative, 36, 381
dominance diversity curves, 324
dormancy
  and climate, 215
  smoke-stimulated species, 72
Dorycnium, 101
drought. See also summer drought
  and community diversity, 326
  and fire activity, 53
  and major fire events, 361
  avoidance, 246, 250
  selective factor, 289
  tolerance, 246
drought-prone landscape, 275, 276
Dynamic Equilibrium Model, 391

E (probability of fire encroaching into the urban environment), 374, 376, 377
eastern Mediterranean Basin, biodiversity, 319
Echinopsis chilensis, 152, 161
ecological convergence, 17
ecological drift theory, 390
ecological Indian model, 350
ecological sorting, 16, 271, 289
ecosystem convergence, 10–18
  and MEDECOS, 27–29
ecosystem response, fire intensity/severity, 39
ecosystem sustainability, 362–366
Ehrharta, 270
Ehrharta calycina, 176, 347
El Niño, 14, 45, 157, 209
Elytropappus, 187
Elytropappus rhinocerotis, 170, 186, 187, 195
ember-generated fires, 368
emergent property, 210
Emmenanthe penduliflora, 73
endogenous postfire regeneration, 58–66, 125
ENSO events, 14, 45, 158, 164, 385
Eocene, 277–278, 291
Eocene–Oligocene transition, 292
ephemeral flora, 115, 125–127, 319, 322

epicormic resprouts, 59, 61, 63, 64, 181
eucalyptus, 240
Erica, 48
  Mediterranean Basin, 94, 97, 101, 102
  obligate seeders, 260
  South Africa, 181
Erica arborea, 61
erosion. See soil erosion
espinal, 154
Etesian winds, 20
eucalypts, 204, 205, 209
  flammability, 222
  resilience, 225
Eucalyptus, 26, 203, 205
  and alien plants, 346
  Chile, 165
  crown fires, 33
  early evidence, 237
  epicormic resprouting, 240
  epicormic sprouting, 64
  fire history, 305
  fire–diversity interactions, 322
  fuel characteristics, 49
  origins, 297
  prescription burning, 370
  resprouting, 61
Eucalyptus globulus, 335
Eucalyptus marginata, 208
European colonization, 352, 357
evacuation orders, 368
evapotranspiration, 286, 287
event-dependent effects
  postfire recovery, 183–184
evergreen broadleaf shrublands, 95–99
evergreen oak woodlands, 99–100
evergreen species. See also sclerophyllous woody plants, conifer forests
  Mediterranean Basin, 93
evolutionary conservatism, 272
evolutionary convergence, 16, 256
evolutionary history, MTV, 289–290
exaptations, 238, 240, 285, 289
exponential model, 323, 325

facultative resprouters, 213
facultative seeders, 59, 67, 68, 69, 180
  Australia, 212
  iteroparity, 265
  life histories, 266
  resprouting levels, 260
  seedling/parent ratios, 248, 249
  vs. obligate seeders, 261, 266
fading fire regime, 81
FARSITE model, 55
FATELAND model, 55, 56
ferns, resprouting, 239

Index

financial responsibilities, 380
fire. *See also* major fire events, area burned
  accidental fires, 367
  and alien plants, 330–348
  and community diversity, 310–326
  and community dynamics, 182–186, 217–221
  and herbivores, 30
  and maintenance of MTV, 225–230
  and regional diversity, 326–328
  climate and geology convergence in MTEs, 388–397
  five parameters, 32
  impact on human populations, 357–360
fire-adapted species, 362
fire-adaptations, 238–268
  life history and diversification, 268–273
  non-resprouting obligate seeders, 258–267
  obligate resprouters and seeders, 247
  postfire resprouting and seeding, 267–268
  postfire resprouting vs. seeding, 241–258
  resprouting, 238–241
fire behavior, 47
  and antecedent climate, 51–53
  and topography, 55
  landscape models, 55–56
  Mediterranean Basin, 91–92
  Paleozoic, 236
fire cycle, 223
  seedling/parent ratios, 248
fire danger indices, 52, 361, 366
fire departure map, 148
fire-dependent ecosystems
  fynbos, 182
fire-dependent recruitment, 14, 250–256
  seedling/parent ratios, 248
fire-dependent reproducers, 66, 76
fire diamond, 31, 40
fire–diversity interactions, 321
fire ephemerals, 217
fire exclusion, 356
fire exclusion experiments, 190
fire-free landscapes, 8
fire frequency, 38–43, 119
  and biodiversity, 220, 221
  and population density, 357
  and resprouting success, 264
  and seed storage mode, 264
  Australia, 208
  California, 131, 241
  Chile, 156, 158, 164
  fynbos, 174, 194
  gradients, 392
  grasslands, 190
  South Africa, 191
fire hazard, 366–380
fire hazard reduction treatment, 362, 364
fire hazard severity maps, 378
fire history, 207, 208, 220, 308
  Chile, 396
  from the Paleozoic, 233–238
  MTV, 300–308
  northern hemisphere, 395
  southern hemisphere, 395–396
fire-independent recruitment, 14, 78–80, 160, 243–250
  seedling/parent ratios, 248
fire intensity, 14, 34–38
  and community diversity, 326
  fynbos, 174, 342
  gradients, 392
  inhibition of establishment, 215
  vs. fire severity, 38
fire losses
  reducing, 366–367
fire management, 349–387, 358–383, *see also* fire suppression, prescription burning
  and alien plants, 338–339
  Australia, 370
  California, 144–149
  Chile, 165–166
  conifer forests, 340
  fire hazard and community vulnerability, 366–380
  future, 384–386
  goals, 361
  Mediterranean Basin, 109–111
  natural resources and ecosystem sustainability, 362–366
  postfire, 380–384
  South Africa, 196–199
  wildland, 366–374
fire management strategies, 361–380
fire numbers
  Chile, 157
  Mediterranean Basin, 88
  South Africa, 175
fire patch size and distribution, 43–44
fire persisters, 260
fire potential, 366
fire prevention, 120, 121, 367, *see also* fire management, fire regimes
  Australia, 209
fire-promoting capacity
  alien plants, 330
fire-prone communities, 362, 363
fire-prone ecosystems, 168, 172
  convergence in, 388–393
  resprouting and serotiny, 259
fire-prone landscapes, 3, 237, 366
  climate change, 384
  environmental template, 9
  global, 3
  global distribution, 8

Index

fire recruiters, 260
fire refugia, 225
fire regime heterogeneity, 221, 223–225
fire regime model, 219
fire regimes, 31–32, *see also* surface fires, crown fires, human impacts
  and biodiversity, 211–221
  and fire-adapted species, 362
  and fuel consumption, 265
  and vegetation patterns, 388
  Australia, 204, 207–209, 357
  California, 364–365
  chaparral, 119–125
  Chile, 156–165
  classes of, 32
  conifer forests, 109, 131–132, 264, 364
  emergent properties, 45–47
  fynbos, 174–186, 189–194
  human impacts on, 353–357, 363
  Mediterranean Basin, 87–89
  MTC ecosystems, 56–57
  MTC regions, 353–361
  Paleozoic, 234
  plant adaptations to, 58
  productivity and disturbance frequency, 46
  sage scrub, 136
  South Africa, 189–194
  target species, 365
fire-resistant ecosystems, 172
  South Africa, 188–189
fire return interval, 14, 39, 41, 68, 130, 132, 392
  and seed storage mode, 261
  and species richness, 218
  and type conversion, 336
  Australia, 208
  chaparral, 360
  fynbos, 194, 356
  obligate seeders, 217–219, 220
  South Africa, 197
fire risk maps, 111, 174
fire rotation interval, 14, 38, 119
fire-scar dendrochronology, 42, 119, 131, 133, 139
  Australia, 352
fire seasonality, 44–45, 56, 132
fire-stick cultivation, 350
fire-stimulated recruitment, 214, 215
  fynbos, 175, 176, 177, 182
  renosterveld, 187
  annual, 353
  Chile, 156, 158
  fynbos, 183
  South Africa, 174
fire severity, 14, 37–38
  assessment, 39, 380–382
  Pinus, 105
  vs. fire intensity, 38

fire size. *See also* major fire events
  California, 117, 120, 122
  Chile, 158
  fynbos, 174
fire size classes, 92
fire spread, 224
  chaparral, 116
  models, 55
  patterns, 32–34
fire suppression, 46, *see also* fire management
  California, 120, 121, 146, 147, 340, 354
  Mediterranean Basin, 110
  South Africa, 198
  strategies, 366
fire triangle, 30
fire weather, 54, 222–223
Fire Weather Index, 368
fire weather winds, 18–19, 20–21
fire world, 225, 227, 228–230
firebelts, 196, 197
firebrands, 54, 371, 374
  transport, 376
FIRESCAPE model, 55
firestorms, 19
Fitzroya cupressoides, 156, 164
flame length, 34, 35
flammability, 211
  and biodiversity, 221–225
  and postfire recruitment, 257–258
  as competitive force, 391
  chaparral, 116
  within MTV, 225
flooding cycles
  annual, 137
flora
  ephemeral, 115, 125–127, 319, 322
  fossil flora, 277, 280, 305
  Mediterranean Basin, 93–94
foehn winds, 18, 19–23
foothill coniferous trees, 138–139
forest clearance, 158, 163, 165
forests. *See also* rainforests, conifer forests
  agroforestry, 165, 209
  area burned in Chile, 157, 158, 163, 165
  austral, 156, 163–164
  Chile, 153, 155, 163
  crown fires, 33
  fire frequency records, 41–42
  fire intensity, 35, 36
  fire severity, 35
  fire size and distribution, 43
  fire suppression, 146
  fuel loads, 50
  MTC regions, 11, 13, 18

prescription burning, 369–370
resprouting, 238
scarcity in South Africa, 390
South Africa, 171, 188–189
surface fires, 32
tree mortality, 392
fossil floras, 277, 280, 305
fossil record, 93, 283, 290, 291
   bias, 250, 290, 294, 295
   fire, 300
France
   fire management, 110
   fire weather winds, 20
   prescription burning, 354
Freeway Fire, 376
fuel breaks, 110, 338, 339, 372
   benefits of, 372
   costs to, 372
fuel characteristics, 47–51
   chaparral crown fires, 115–117
   conifer forest surface fires, 128–130
   Mediterranean Basin, 48, 89–91
   vegetation types, 48
fuel connectivity, 191
fuel consumption, 32–34
fuel loads, 89, 115, 146, 359
   in MTC regions, 386
fuel mapping, 111
   regional, 368
fuel mass, 189
fuel moisture, 51, 52, 137, 138, 189, 221, 223
fuel structure, 30, 32, 194, 353
   and climate change, 384
   fire weather and flammability, 222–223
fuel treatments, 110, 361
   buffer zones, 375
   cost-benefit analysis, 372
   strategic planning, 372–374
   wildland, 369
Fumana, 101
fynbos, 25, 26, 32, 168, 170, *see also* South Africa
   alien plants, 342–344
   biodiversity, 322
   biomass, 169
   fire history, 306
   fire regimes, 174–186, 189–194
   fire return interval, 356
   fuel characteristics, 49
   graminoids, 181
   human impacts, 194, 195, 196
   origins, 297–298
   prescription burning, 370
   soil, 170
   species richness, 315
   taxa in, 173–200

G (chance of fire propagating), 378
gamma diversity, 326
gap size, 146, 147, 251
   conifer forests, 132, 133
garrigue, 48, 95–97
   definition, 98
   fire–diversity interactions, 319
   species richness, 315
genetic advantages of resprouters, 269
Genista, 101
geochemical mass balance model, 236
geological timescale, 234
geology, climate and fire convergence, 388–397
geophytes, 13, 18, 60, 194
   biodiversity, 322
   Mediterranean Basin, 93
germination. *See also* smoke-stimulated seed germination, heat-stimulated germination
   soil seedbanks, 69, 70, 71–73, 251–253
GIS tools, 111
Gleichenia, 239
global warming, 124, 346, 392
Gondwana, 177, 209, 296–300
graminoids, 181, 182, 192
granivory, 253, 256
grass fire cycles, 330, 332, 333, 334, 335, 336
grass fire feedback cycle, 346
grassland. *See also* annual grass invasion, perennial grass invasion
   and fire, 228
   California, 135–136
   early evidence, 237
   fire regimes, 189–194
   fuel characteristics, 48, 49
   fuel loads, 49
   Mediterranean Basin, 84, 108
   MTC regions, 11
   renosterveld, 186
grazing, 195, 208, 319
   and alien plants, 331, 335, 340
   and fuel structure, 353
   intense, 107, 120
Greece, 355, 361
   fire weather winds, 20
   fires and area burned, 92
   wildfires, 110
greenbelt, 377
Grevillea, 177
ground fires, 32, 33
growth bands, 208
gymnosperms, 253
   resprouting, 240
   serotiny, 75

habitats
  Australia, 204, 206–207
  core, 226
Hakea, 335, 342
Hakea sericea, 344
Halimium, 101
hard-seeded species, 69
Hawaiian Islands, 9
heat desiccation, 176
heat shock, 176, 251
heathlands, 10
  Australia, 201, 202, 315, 324
  fuel characteristics, 49
heat-stimulated germination, 72, 100, 214
  California, 127
  Mediterranean Basin, 93
heat-stimulated seeds, 69
Hedera, 96
Helianthemum, 101, 303
hemicryptophytes, 15, 60
herb species, 152, 160
herbaceous perennials
  Chile, 152
  fire history, 303
  fire–diversity interactions, 319
  origins, 295, 304
  resprouting, 60, 136
herbivores, 168, 216
  and fire, 30
  mass extinction, 351
herblands, 338
herophytes, 15
Hesperocyparis, 139
Heteromeles arbutifolia, 140
HighFire, 273
highly optimized tolerance (HOT) model, 47
Holocene, 43, 85, 118, 145
  human impact, 350, 351
home construction characteristics, 378
home ignition zone, 375
home protection guidelines, 368
homoplasy, 16
housing losses, 376, 378
human ignitions, 57, 349
  and fire frequency, 40–41
  California, 118
  Chile, 159
  Mediterranean Basin, 87
human impacts. *See also* land use, grazing
  and alien plants, 334
  early fire use, 349–353
  Mediterranean Basin landscape, 85–87
  on fire regimes, 46, 353–357, 363
  South Africa, 194–200
human population density, 110, 124, 354, 359
  and fire frequency, 357

and fire risk, 359
MTC regions, 355
human population growth, 353, 385, 396
human population, fire impact on, 357–360
hummock grasses, 228
humus development, 17
hydrological responses to fire, 382
hygrophilous forest, 154

I (probability of fire starts), 366
ignitions. *See also* lightning, human ignitions
  chaparral crown fires, 118–119
  conifer forests, 131
  fire, 30
  Mediterranean Basin, 91–92
  South Africa, 193
immaturity risk, 363, 365
in situ weathering, 206
inclusive fitness theory, 257, 389
independent crown fires, 32
inhibition of establishment, 214–216
insurance, 367, 379
insurance companies, 379
inter-fire interval. *See* fire return interval
International Biological Program (IBP), 15, 27
International Society of Mediterranean Ecologists (ISOMED), 29
International Union for the Conservation of Nature (IUCN), 29
invading species. *See* alien plants
isohydric behavior, 246
Israel
  climate, 7
  fire weather winds, 20
iteroparity, 265

jarrah forests, 220, 365, 377
Juglans californica, 138
Juniperus, 97, 107
Jurassic, 235
Jurassic fires, 234

karoo, 171, 187, 193
  taxa, 173–200
karrikinolide, 252
karsts, 95
katabatic (downslope) winds, 19, 21
kBP (thousands of years before present), 14, 145
Kho-Khoi pastoralists, 195
kwongan, 26, 189, 205
  biodiversity, 206, 210
  fire management, 365
  flammability, 224
  senescence risk, 216
kyr (a thousand years), 14

La Niña, 14, 157
ladder fuels, 130, 138
land abandonment, 86, 87, 95, 99, 396
land clearance, 208
land management practices, 352, 354
land planning, 377, 380
land use
    Australia, 204, 207–209
    California, 120, 121
    Mediterranean Basin, 86, 87–89, 94–107, 112
land zoning, 377
LANDIS model, 56
Landsat satellite imaging, 35, 37, 39, 120
landscape conversion. *See* type conversion
landscape models, 55–56
    Mediterranean Basin, 111
landscape scale events, 235
landscape variation
    and community diversity, 325–327
    California, 123
    MTC regions, 327, 328
landslides, 156
LANDSUM model, 55
Laurus, 96
Lavandula, 94
least squares regression analysis, 313
leave early or stay and defend, 368
leaves
    macrofossils, 283
    malacophyllous, 14
    sclerophyllous, 15, 17, 285
    size, 96
    small, 3, 84, 100–103
Lebanon, 20
life history and diversification, 268–273
life spans, 216
lightning, 57
    and fire frequency, 40–41
    California, 118, 119, 130
    Chile, 162, 284
    global distribution, 40
    Mediterranean Basin, 91
    Portugal and Spain, 91
    South Africa, 193–194
lightning flash density, 193
lightning strike density, 41
lignotubers, 14, 59, 60, 61
    Australia, 212
    chaparral, 127
    fynbos, 180
    matorral, 159
    origins of, 305
    taxa, 63
limestone, 95, 206
Lithocarpus densiflora, 139
litter fuels, 222–223

live fuels, 50, 51, 52
    live/dead ratio, 115
Loasa, 159
local extinction, 185
log barriers, 383
logging, 107, 147
longevity, soil-stored seed, 70
Lonicera, 96
Los Angeles, 4, 6, 117
    climate, 5, 7
LowFire, 273

Ma (million years ago), 14
MacArthur's broken stick model, 324
macrofossils, 283, 290, 295, 300
major fire events, 209, 353, 354, 357
    California, 4, 355, 359, 360, 361
    causes, 367
    climate impacts on, 360–361
    Mediterranean Basin, 360
    MTC regions, 355
    socio-political factors, 379–380
malacophyllous leaves, 14
malacophyllous species, 84, 100–103
Malibu Declaration, 29
mallee, 26, 203, 204, 205, 216, 229
    alien plants in, 347
    species richness, 315
mallets, 212
Malosma laurina, 140
map projections, 314
maquis, 48, 84, 95–97
    definition, 98
    fire–diversity interactions, 319
    species richness, 315
marl-limestone colluviums, 95
mast cones, 77
mast seeding, 254
mast year, 77, 78
masting cycles, 133
matorral, 49, 71, 152, 154, 156, 162
    alien plants, 341
    biodiversity, 321–322
    characteristic species, 153
    definition, 98
    fire in, 159–162
    resprouting, 240
    species richness, 315
McArthur Forest Fire Danger Index, 368
mechanical thinning, 147
Mediterranean Basin, 83–84, 84–106,
    see also northern hemisphere,
    individual countries
    alien plants, 331–335
    biogeography, 311
    boundaries, 83

Mediterranean Basin (cont.)
  climate, 7, 83, 94–107
  community diversity, 314
  fire history, 303, 304
  fire management, 109–112
  fire weather winds, 20
  fire–diversity interactions, 319
  fires and area burned, 92
  flora and postfire strategies, 93–94
  floristic comparisons, 24
  fuel characteristics, 48, 89–91
  human impact, 349, 350, 351, 352, 354
  ignition patterns and fire behavior, 91–92
  landscape, 10, 83, 84, 85–87
  lignotubers, 63
  major fire events, 355, 360
  Miocene climate, 282
  MTC region, 18
  MTV origins, 291, 294, 295
  non-flammable vegetation types, 107–109
  obligate resprouters, 59
  population density, 355
  postfire seeding, 69, 384
  postfire seedling recruitment, 101
  prescription burning, 354
  smoke-stimulated seed germination, 94
  socio-economic, land use and fire changes, 87–89
  species richness, 315
  summer precipitation, 5
  vegetation, 11
  vegetation patterns, 94–107
  vegetation types, 84
  winter precipitation, 5
  woody obligate resprouters, 97
Mediterranean Ecosystem conferences (MEDECOS), 27–29
mediterranean-type climate. *See* MTC
mediterranean-type ecosystems. *See* MTE
mediterranean-type vegetation. *See* MTV
meltemia winds, 18, 20
Mesembryanthemaceae, 176
meso-mediterranean zone, 94
Mesozoic, 235, 236, 239
  climate, 277
  fire history, 300
metapopulation dynamics, 44, 363
metapopulations, 328
metropolitan centers, 4
  expansion, 354
  fire-prone ecosystems, 3
  high density, 353
Mexico, 122, 140, 141, 142, 250
  Baja California, 120–124
  climate diagrams, 288
Mg ha^{-1} (megagrams per hectare), 14

microfossils, 283
Mimetes stokoei, 185
Miocene, 161, 166, 237, 270, 279–283, 293
  chaparral, 292
  fire history, 306
mistral winds, 18, 19, 20, 22
mixed (surface and crown) fires, 45, 131
models. *See* landscape models
Moderate Resolution Imaging Spectroradiometer (MODIS), 21, 44
molecular clock, 283, 298
molecular phylogeny studies, 304
monocots, 93
monocultures, 165
monsoonal storms, 131
montado, 107
montane conifer forests, 339–340, 356
montane matorral, 154
mosaic fuels, 371
mountain-mediterranean zone, 94
mountains
  and alien plants, 339
  fire suppression, 354
MTC
  Australia, 202
  California, 114
  Chile, 151
  Mediterranean Basin, 84
  origins, 276–283, 308, 394
  selective importance, 394
  South Africa, 168
MTC regions, 3, 4, 5–8, 14, 18–26, 81, *see also individual regions*
  biodiversity hotspot, 310
  biogeographic characteristics, 311
  boundaries, 8
  climate characteristics, 7
  crown fires, 394
  fire regimes, 353–361
  fire–alien plant interactions, 330
  flora evolution, 9
  floristic comparisons, 24
  forests, 13
  fuel loads, 386
  landscape variation, 327, 328
  major fire events, 355
  MTV outside, 140–144
  population densities, 355
  regional diversity, 310
  shrublands, 12
  species–area relationships, 312
  species richness, 315
  vegetation comparisons, 11
MTE, 11, 14
  climate, fire, geology convergence, 388–397
  ecosystem convergence, 10–18

ecosystem fire regimes, 56–57
postfire communities, 317
MTV, 11, 14, 81, *see also* sclerophyllous woody plants, vegetation types
  Australia, 201–230
  biodiversity, 209–225
  definition, 205
  fire and paleohistory, 300–308
  fire and the maintenance of, 225–230
  origins, 283–299, 308
  origins, evaluating, 289–291
  outside MTC regions, 140–144, 394
  selective environment for, 285–290
multistemmed growth, 262
multivariate modeling, 336
myrmecochory, 185, 256
Myrtaceae, 304
Myrtus, 96

Nassella pulchra, 135
Native Americans, 118, 135, 136, 140, 144–146
  and alien plants, 336
  early impact, 350
native grasslands, 135
native late successional increasers, 126
native postfire endemics, 126
native postfire opportunists, 126
native postfire specialists, 126
natural burning policy, 191
natural burning zone, 198
natural fire policy, 198, 364
natural fire zones, 198, 356
natural resource protection, 362, 364
natural resources, fire management, 362–366
Neolithic, 85
neophytes, 334
nested samples, 310, 313–314
net relatedness index, 273
new fire world, 228–230
New South Wales, fire impact, 360
niche conservatism, 259, 271, 272
niche construction, 187, 389, 395
niche space, 285, 313
Nicotiana, 71
nitrogen pollution, 10, 338, 385
non-random sorting, 271
non-resprouters, 100, 180
  evolutionary model, 265–267
  fire adaptations, 258–267
Normalized Differenced Vegetation Index (NDVI), 35
North Africa, fire weather winds, 20
northern hemisphere. *See also* California, Mediterranean Basin
  canopy seed beds, 75
  community diversity, 311, 316
  fire history, 300–304, 395
  fire-driven diversification, 269
  Mesozoic climate, 277
  MTV origins, 291–296, 308–309
  obligate resprouters, 243, 250
  serotiny, 253
  smoke-dependent seed germination, 73
Nothofagus, 156, 296
Nothofagus alpina, 155
Nothofagus antarctica, 156, 162, 164
Nothofagus dombeyi, 163
Nothofagus glauca, 163
Nothofagus obliqua, 25, 155
Nothofagus pumilio, 155, 164
nutrients
  and biodiversity, 210
  depletion, 365
  seeds, 255, 256

oak woodlands. *See also Quercus*
  California, 137–138
  crown fires, 33
  evergreen, 99–100
  Mediterranean Basin, 108
  non-flammable, 107
obligate resprouters, 15, 59, 60, 67, 76, 78, 242
  character syndromes, 247
  fire-independent recruitment, 243–250
  fruit dispersal, 244
  fynbos, 181
  global distribution, 307
  in MTC regions, 59
  Mediterranean Basin, 95, 96
  renosterveld, 187
  seedling/parent ratios, 248
obligate seeders, 15, 59, 68, 69, 241, 242
  Australia, 212, 213
  chaparral, 67
  character syndromes, 247
  enhanced flammability, 257, 258
  fire return intervals, 217–219
  fynbos, 180, 181, 184, 190
  life histories, 266
  life history hypotheses, 260–264
  non-resprouters, 258–267
  origins, 302–303
  seedling/parent ratios, 249
  semelparity, 265
old fire world, 227, 228–229
old-fields, 86, 95, 102, 384
Olea, 96
Olea europaea, 96
Oligocene, 278–279, 292
oro-mediterranean zone, 94
out-of-season burns, 364
out-of-season fires, 45

overstory species, fynbos, 182
oxygen levels, 236
ozone, 338

packing density, 50
packing ratio, 129
paleobotany, 237
Paleocene, 277
paleoclimates, 276–283
Paleofloras
   Creede, 292
   Piru Gorge, 279
paleohistory, MTV, 300–308
Paleolithic, 85
paleoprecipitation, 281
Paleozoic, fire history from, 233–238
Palmer Drought Severity Index (PDSI), 53
palynology, 283
parent trees, seedling recruitment from, 76–77
passive crown fires, 33
pasture burning, 92
patch mosaics, 371
peak diversity
   postfire, 320, 321
peatlands, 300
percolation models, 55
perennial grass invasion, 332, 347
persistence niche, 260
persistence, seed, 180–183, 212–213, 250
phanerophytes, 13, 15
Phillyrea, 96
Phoenix theophrasti, 97
phosphorus, 10
phrygana, 48, 98, 100–103
phylloclades, 97
phylogenetic distance, 273
phylogeny in ecological convergence, 16
Pinus
   as alien plants, 342
   bark thickness, 78
   California, 24, 128, 138
   Mediterranean Basin, 84, 95, 99, 101, 103–107
   origins, 295, 296, 303
   postfire seedling density, 104
   precipitation, 53
Pinus brutia, 103, 104
Pinus canariensis, 107
Pinus halepensis, 18, 22, 103, 104, 105, 315
Pinus nigra, 106
Pinus pinaster, 105
Pinus pinea, 105
Pinus ponderosa, 48, 129, 296
   California, 128, 129, 131, 133
   distribution, 143
   outside MTC regions, 142–144

Pinus radiata, 156, 165
Pinus sabiniana, 139, 391
Pinus sylvestris, 106
Pinus uncinata, 106
Pistacia, 96
Pistacia lentiscus, 96
plant diversity. *See* biodiversity
Pleistocene colonization, 351
Pliocene, 275
plot size, 313
plume-driven fires, 54, 117
Podocarpus drouynianus, 214
political factors in fire management, 378–380
pollarding, 99
pollen deposits, 42
pollen fossils, 283, 290, 291
poniente winds, 20
population. *See* human population
Portugal, 355
   alien plants, 335
   fire management, 111
   fires and area burned, 92
   lightning, 91
postfire aerial seeding, 382–383
postfire colonization, 78, 320
   annuals, 126
postfire management, 380–384
postfire recovery and seasonality, 45
postfire regeneration
   and fire severity, 37
   and species coexistence, 221
   chaparral, 125–128, 320, 321
   conifer forests, 132–135
   endogenous, 58–66
   four modes of, 58
   fynbos, 175–186
   Mediterranean Basin, 111
   sage scrub, 136
postfire resprouting, 241–258, 267–268
   character syndromes, 247
postfire restoration, 382–384
postfire seedling recruitment, 66–76, 241–258, 267–268
   enhanced flammability, 257–258
   factors driving, 242
   from parent trees, 76–77
   from resprouts, 76
   Mediterranean Basin, 94, 101
   MTC crown fires, 69
   origins, 302, 303
postfire strategies
   Mediterranean Basin, 93–94
   Native American, 145
power law relationship, 46, 47
power model, 323, 325

power tools, 367
powerlines, 367, 379
precipitation, 5–8, 9, *see also* summer precipitation, winter precipitation
  and fire cycle, 223
  and serotiny, 254, 255
  Australia, 207, 208
  biomodal rainfall regime, 285
  California, 6
  Chile, 151
  conifers and pines, 53
  Mediterranean Basin, 83
  paleo, 281
  South Africa, 191
predation, seeds, 216, 254, 255, 256
prescription burning, 120, 122, 147, 184, 369
  alien plants, 366
  and alien plant control, 343
  and alien plants, 339, 340
  and resource protection, 364
  California, 356
  in forests, 369–370
  Mediterranean Basin, 354
  pros and cons, 361
  South Africa, 196, 197, 356
  unexpected outcomes, 343
productivity, 46, 211, 221
  primary productivity, 47, 56, 286
property losses, 379
Protea, 177, 178, 183, 184–186
  fire management, 197
Proteaceae, 177–180
  stomatal crypt, 297
proteoid roots, 177
Prunus, 117
Pseudotsuga macrocarpa, 64, 119, 139
Pseudotsuga menziesii, 128, 139
Pteridium aquilinum, 239
Puya, 152, 161
pyro-endemics, 127, 134, 303, 304, 319
pyrogeography, 3

Quaternary, 226, 243, 250
  fire records, 351
  speciation in, 299
Quercus
  California, 117, 127
  epicormic sprouting, 64
  Mediterranean Basin, 84, 89, 96
  resprouting, 61, 65
Quercus agrifolia, 138
Quercus berberidifolia, 51
Quercus chrysolepis, 138
Quercus coccifera, 48, 97, 98
Quercus douglasii, 138

Quercus ilex, 48, 99
Quercus kelloggii, 138
Quercus suber, 99, 291

rainfall. *See* precipitation
rainforests, 188, 226, 227
  biodiversity, 219
  in Australia, 389
rangelands, 356
Raunkiaer life forms, 15
recolonization, 354
redwood forests, 139–140, 236
regeneration. *See* postfire regeneration
regional diversity
  and fire, 326–328
  MTC regions, 310
regional species density, 310
regulations
  Chile, 165
  fire-prone landscapes, 110
  home protection, 368
  South Africa, 198–199
relictual taxa, 243, 275
remote imaging technologies, 35, 39, 44, 120
  fire severity assessment, 381
  Mediterranean Basin, 111
renosterveld, 25, 49, 170, 186
  fire in, 186–187
  human impacts, 195, 196
  taxa in, 173
reproduction
  delayed, 66
  fire-dependent, 66
  fynbos, 175–180
residence time, 35
resource gradient, 263, 267
resource-based model, 261
resource-based theory, 388, 389, 390
resprouting, 58–66, 242, *see also* obligate resprouters
  Australia, 212
  capability, 391
  chaparral, 127
  genetic advantages, 269
  matorral, 159
  Mediterranean Basin, 112
  Mediterranean Basin woody species, 101
  origins of, 238–241
  postfire, 247, 267–268
  postfire resprouting and seeding, 241–258
resprouting capacity, 68, 99
resprouts
  postfire seedling recruitment from, 76
  small-leaved shrubs, 103
  stem ages, 65
Restionaceae, 181, 314

restoration methods
  postfire, 111
Rhamnus, 96, 117
rhizomes
  clonal spread, 181
  resprouting, 59
  survival of, 35
Rhus, 101, 117
riparian vegetation, 108, 137
road infrastructure, 367, 377
rock falls, 193
rock outcrops, 215, 224
root suckering, 181
Rosmarinus, 94, 101
Rubia, 96
running crown fires, 32
Rural Fire Service, 380
rural industries, 209
Ruscus, 96, 97

S (probability of fire reaching the urban environment), 366, 369
sage scrub, 24, 48, 136
  alien plants and type conversion, 335–339
  mosaic, 115
salt bushes, 108
sampling design, 313–314
San Francisco, climate, 7
Santa Ana winds, 19, 20, 21, 22, 23, 25, 119
saplings, vulnerability to fire, 133
Sarcopoterium spinosum, 91, 98, 101, 103
satellite species, 323
savannas
  and fire, 228
  Chile, 154
  fire regimes, 189
  oak, 137–138
sclerophyllous leaves, 15, 17, 285
sclerophyllous woody plants, 9, 10–12, 13, 26, 27, see also MTV (mediterranean-type vegetation)
  Australia, 227
  Chile, 153, 154, 156, 163
  Mediterranean Basin, 18, 96
  origins, 275, 394
scrublands
  Mediterranean Basin, 108
  sage scrub, 24, 48, 136, 335–339
seasonality. See fire seasonality, climate seasonality
seedbanks. See also soil seedbanks, canopy seedbanks
  Australia, 213–214
  canopy, 73–76
  storage, 254
  survival of, 35
  transient, 247

seed dispersal, 247, 256
seed persistence, 180–183, 212–213, 250
seed predation, 216, 254, 255, 256
seed retention, 214
seed supply constraints, 213–214
seedling density, 104
seedling recruitment, 66–76, see also fire-dependent recruitment, fire-independent recruitment, postfire seedling recruitment
  and gap size, 133
  Australia, 212, 214
  from parent trees, 76–77
  from resprouts, 76
  inhibition of, 214–216
  MTC crown fires, 69
  postfire recovery, 328
  riparian, 137
seedling/parent ratios, 248
selective environment for MTV, 285–290
self-organized behavior, 47
self-organized criticality, 47
self-pruning, 76
semelparity, 258, 265
senescence risk, 216–217, 363
Sequoia National Park, 357
Sequoia sempervirens, 139, 140, 236
Sequoiadendron giganteum, 132, 236, 369
serotiny, 73–76, 259
  angiosperms and gymnosperms, 74
  California, 139
  canopy seedbanks, 253–256
  inhibition of establishment, 215
  Mediterranean Basin, 93
  origins, 303
  Pinus, 104
  proteas, 177, 183, 184–185
  seedbanks, 213
shade intolerance, 182
shaded fuel break, 372
sharav winds, 19
shelter in place, 368
shepherding, 109
shrub encroachment, 338
shrublands, 10, 11, see also chaparral, sage scrub, garrigue, fynbos, matorral
  broadleaf evergreen, 95–97
  California, 24, 117
  canopy recover, 127–128
  crown fires, 32, 39, 46
  fire severity, 35
  fire size and distribution, 43
  fire–diversity interactions, 319
  fuel loads, 49, 50
  grass invasion, 332, 334
  Mediterranean Basin, 18, 84, 95, 100–103

MTC regions, 11, 12
  seedling recruitment, 66
Sierra Nevada, 279
silviculture, 165
site productivity, 45
slope orientation, 314
slope stabilization, 383
Smilax, 96
smoke-dependent germination, 71, 72, 73
smoke-stimulated seed germination, 71–72, 73, 100, 176, 214, 251–252
  fynbos, 176
  Mediterranean Basin, 93, 94
  origins, 303
snowpack, 124
social factors, fire management, 378–380
soil
  and biodiversity, 310
  Australia, 206–207, 210, 211, 296
  fertility, 243, 255, 322
  fire response, 325
  hydrophobicity, 381
  leaching, 206, 365
  nitrogen and phosphorus levels, 10
  South Africa, 168, 170, 177, 306
  southern hemisphere, 307, 395, 396
soil drought, 275, 285
soil duff consumption, 35
soil erosion, 382, 383
  and fire intensity, 37
  Mediterranean Basin, 111
soil hydrophobicity, 38
soil moisture, 210, 246, 286, 289
soil moisture stress, 287
soil seedbanks, 68–70, 101, 213, 214, 251–252
  seedling/parent ratios, 248, 249
  survival, 303
South Africa, 178–199, *see also* renosterveld, southern hemisphere, fynbos
  alien plants, 342–346
  biogeography, 311
  biome taxa, 173
  climate characteristics, 7
  fire history, 306–307
  fire impact, 359
  fire management, 196–199
  fire regimes, 189–194
  fire suppression, 356–357
  fire weather winds, 20
  fire–diversity interactions, 322
  floristic comparisons, 24
  fuel characteristics of vegetation, 49
  fynbos fire regimes, 174–186
  human impact, 194–200, 349, 350, 352
  landscape, 10
  lignotubers, 63

  major fire events, 355
  Miocene climate, 282
  MTC region, 25–26, 168–169
  MTV origins, 297–298
  obligate resprouters, 59
  Oligocene climate, 279
  population density, 355
  postfire seeding woody genera, 69
  scarcity of forests, 390
  species richness, 315, 328
  summer precipitation, 5
  vegetation, 11
  vegetation patterns, 168–174
  vegetation types, 172
  vegetation types, fire in, 186–189
Southern Australia, 202–222, *see also* Western Australia, *Eucalyptus*, southern hemisphere
  alien plants, 346–347
  alien plants from, 335
  biodiversity, 209–225
  biogeography, 311
  climate, 7
  fire and the maintenance of MTV, 225–230
  fire history, 304, 305–306, 307
  fire impact, 360
  fire management, 370
  fire regimes, 357
  fire regimes and land use, 207–209
  fire weather winds, 19, 21
  fire–diversity interactions, 322
  floristic comparisons, 24
  fuel characteristics of vegetation, 49
  habitats, 206–207
  home protection guidelines, 368
  housing losses, 378
  human impact, 350, 351, 352
  landscapes, 10, 11
  lignotubers, 63
  major fire events, 355, 360
  Miocene climate, 282
  MTC region, 26, 201
  MTV, 201–230
  MTV origins, 296–297, 298
  obligate resprouters, 59
  Oligocene climate, 279
  politics of fire management, 380
  population density, 355
  postfire events, 241
  postfire seeding woody genera, 69
  rainforests, 390
  summer precipitation, 5
  Tertiary transitions in vegetation, 227
  vegetation, 11
  vegetation groups, 204
southern hemisphere. *See also* Southern Australia, South Africa, Chile

southern hemisphere. (cont.)
  and serotiny, 254
  canopy seedbeds, 75, 76
  community diversity, 314
  fire history, 304–308, 395–396
  fire-driven diversification, 269
  MTV origins, 296–299, 309
  Proteaceae, 177
  seed dispersal, 256
  serotiny, 253, 255
  smoke-dependent seed germination, 73
Southern Oscillation, 14, 45
Spain
  climate, 7
  fire frequency records, 42
  fire weather winds, 20
  fires and area burned, 92
  lightning, 91
Spartium, 101
species–abundance curves, 323
species–area curves, 314, 322, 323, 324
species–area relationships, 310, 312, 322–325
species density, 310
species richness, 211, 217, 218, 310, 315, 319, 325, 328
specific leaf area, 16
spot fires, 54, 371
spotting distance, 371
sprouting, 59
stakeholders, 378
stand age, dead/live fuel ratio, 52
stand-replacing fires, 33
stand-thinning fires, 34
starch storage in roots, 262
stasis, 245
Station Fire, 4, 117
steep terrain, 54
stem ages, 65
stomatal crypt, 297
strandveld, 171, 188–189
stream flow, 196, 197
structural equation model, 327, 373
Styrax, 96
sub-mediterranean zone, 94
summer drought, 9
  and fire, 30
  and fire regimes, 56
  and fuel moisture, 52
  and major fire events, 360
  California, 280
  Chile, 157
  obligate seeders, 268
  plant adaptations, 246
summer precipitation, 5, 7
  and fire ignition, 92
  Miocene, 280

summer temperature, 7
Sundowner Winds, 19, 20
superposed epoch analysis, 53
supra-mediterranean zone, 94
surface fires, 32, 33, 34, 45
  conifer forests, 128–134
  delayed seedling recruitment, 76, 78
  fuel characteristics, 48
  Mediterranean Basin, 106, 109
  Pinus, 105, 106
  regeneration from, 58
surface fuels, conifer forests, 130
survival rates, 216

Teline (Genista) monspessulana, 342
temperate MTV, 201, 203
temperature
  California, 124
  Chile, 152
  Mediterranean Basin, 83
  MTC regions, 7
Terra satellite, 44
terraced slopes, 86
Tertiary, 209, 210, 226, 227, 228, 237, 243, 245, 275
  Australia, 297, 298
  climate, 283
  fire history, 300, 301, 305–306
  MTV origins, 290
  northern hemisphere, 291
Tetraclinis, 291
Tetraclinis articulata, 105
thermo-mediterranean zone, 94
thickets, 171, 187, 188–189
  taxa in, 173–200
thresholds, 386
thresholds of potential concern, 197, 363
thresholds of tolerance, 363, 365
Thymus, 94, 101
timber production, 354, 357
tissue hydration, 65
tomillar, 98, 100
top-down effects, 226, 241
topography, 55
Torreya californica, 139
tree felling, 344
tree ferns, 239
tree plantations, 165, 196
trees. *See also* forests
  Mediterranean Basin, 84
  mortality, 132, 392
  regeneration, 133–134
Triodia, 228, 229
Turkey
  fire management, 111
  fires and area burned, 92

twig diameter method, 35, 39
type conversion, 330, 337
   chaparral and sage scrub, 335–339

Ulex, 48, 101
Ulex europaeus, 341
Ulex parviflorus, 48, 89, 90, 103, 257
understory
   chaparral, 79
   conifer forests, 130
   evergreen oak woodlands, 99
   northern hemisphere, 311
understory fires, 32
understory recovery, 134
universal equable conditions, model of, 278
urban fuels, 375, 378
urban management, 378
urban planning, 367, 375, 377, 380
U.S. Forest Service (USFS), 371, 379
U.S. National Fire Danger Rating System, 192, 368

vegetation. *See also* flora
   ecological drift theory, 390
   regeneration modes, 58
vegetation patterns
   and fire regimes, 388
   evolution of, 289
   Mediterranean Basin, 94–107
vegetation structure
   and biodiversity, 310, 314
vegetation types. *See also* MTV
   Australia, 204
   California, 113, 135–136
   Chile, 152–156
   fuel characteristics, 48–49, 89–91
   Mediterranean Basin, 84
   non-flammable, 107–109
   South Africa, 168–174, 186–189
veld fires, 357
Very High to Extreme fire danger, 223
Viburnum, 96
Victoria, 202, 355, 368
   fire impact, 360
vine thicket, 188
Virgilia, 189
volcanic eruptions, 156, 161, 206

water availability, 207
water balance, 287
water deficit, 286
watershed stability, 38
Watsonia, 322
Watsonia pyramidata, 184
weather. *See* fire weather, climate
Western Australia. *See also* Southern Australia
   climate, 7
   fire–diversity interactions, 322
   fire frequency records, 42
   fire impact, 359
   flora evolution, 233
   human impact, 352
   major fire events, 355, 360
   MTC, 202
   population density, 355
   species richness, 315
wetland sites, fossil record, 290
Widdringtonia nodiflora, 240, 256
wilderness areas, fire management, 364
wildland fire management, 366–374
wildland–urban interface (WUI), 15, 87, 110, 118, 358
   California, 354
   management, 374–377
   South Africa, 199–200
winds. *See also* mistral winds, Santa Ana winds
   and fire danger, 54
   and major fire events, 360
   fire weather, 18–19, 20–21
winter precipitation, 5, 7, 9, 275
   and major fire events, 360
winter temperature, 7
Witch Fire, 23, 379
within-community diversity, 211
woodlands. *See also* forests, oak woodlands
   grass invasion, 332
   human impact, 352
   MTC regions, 11
wood-stimulated seed germination, 71–72
woody resprouters, 60
Working for Water programme, 197
Working on Fire, 197

Xanthorrhoea preissi, 76, 208

Zaca Fire, 365